Oceans of Kansas

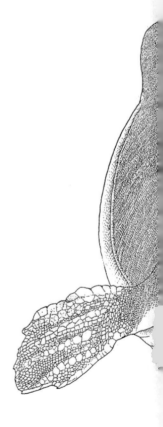

Life of the Past   James O. Farlow, editor

OCEANS OF KANSAS

A NATURAL HISTORY OF THE WESTERN INTERIOR SEA

SECOND EDITION

MICHAEL J. EVERHART

Indiana University Press

This book is a publication of

Indiana University Press
Office of Scholarly Publishing
Herman B Wells Library 350
1320 East 10th Street
Bloomington, Indiana 47405 USA

iupress.indiana.edu

© 2017 by Michael J. Everhart

This book is printed on acid-free paper.

Manufactured in China

Cataloging information is available from the
Library of Congress.

ISBN 978-0-253-02632-3 (cloth)
ISBN 978-0-253-02715-3 (ebook)

4 5 6 7 8 9 26 25 24 23 22

Dedicated to my parents, Jack McKay Everhart (1922–2000) and
Betty Lou Everhart (1923–1994)

"But the reader inquires, What is the nature of these creatures thus left stranded a thousand miles from either ocean? How came they in the limestones of Kansas, and were they denizens of land or sea?"

E. D. Cope, On the Geology and Paleontology of the Cretaceous Strata of Kansas (1872:319)

Contents

Preface to the Second Edition P

Twelve years have passed since I stopped researching and writing the original *Oceans of Kansas*. I was vaguely unsatisfied about not including all of the information that I wanted, but there had to be a stopping point in order to get the manuscript to the publisher. At the time, of course, I had no idea that I would have the opportunity to write a second edition and literally finish what I started. Since the original publication in 2005, many new discoveries have occurred in the Smoky Hill Chalk and other rocks deposited by the Western Interior Sea, and I've been fortunate to be able to collect and to describe my share of them. I've also benefited from the knowledge and experience of other paleontologists from outside Kansas and around the world in many co-authored papers. My research regarding the history of paleontology associated with the oceans of Kansas has also continued, and I have learned much more about the early collectors, as well as Cope and Marsh, and their discoveries in Kansas. The second edition presents a more complete view of the creatures that inhabited the oceans of Kansas during the Late Cretaceous, and also gives more credit to those less well known men who endured hardships on the Kansas prairie in the name of science. They are important because they represent the very beginnings of what we now consider to be modern paleontology. Amazingly, we are still finding new fossils in the Smoky Hill Chalk and experiencing some of the same thrills that I am sure were felt a hundred and fifty years ago.

Preface to the First Edition P

Although I am almost a native Kansan and proud of my state, I have to admit that the drive across Kansas on Interstate 70 is not a major scenic experience if you are expecting mountains or other dramatic landscapes. While there are low hills and river valleys to be crossed along the way, the most visible change from east to west is going from a moderate number of trees to almost no trees. That being said, Kansas has many charms that are well hidden from those who are traveling as fast as they can to get across the state, and even from those who have lived here their entire lives. To me, as a paleontologist, that means a wealth of rock exposures that faithfully reveal a fossil record compiled over millions of years when Kansas was covered by a succession of Paleozoic and Mesozoic oceans. Note that I use the word 'oceans' here in the broadest sense, since these bodies of water covered portions of a submerged continent (North America) and are more properly called 'seas.'

I have been interested in fossils for about as long as I can remember. In grade school I had the usual curiosity about fossil shells and crinoids that were found in the limestone rocks to the east of where I lived. Growing up south of Wichita, Kansas, I spent quite a bit of time exploring along the banks of the Arkansas River, other streams, and the spoil piles of local sandpits. Occasionally I would find the teeth or bones of Pleistocene mammals. I think that was when I realized that they represented the remains of extinct animals that lived long ago in a very different Kansas. One of my first paleontology books was *All About Dinosaurs* by Roy Chapman Andrews. Although his dinosaur-hunting adventures in the Gobi Desert were interesting, in two chapters in the book that discussed the marine reptiles and pterosaurs from the Kansas chalk, his description of life in the Cretaceous oceans started me wondering about what might be out 'there.'

Somewhere along the way, I saw my first shark teeth and fish bones from the chalk of western Kansas. A field trip during a vertebrate paleontology course in college provided me with my first experience collecting fossils in the Smoky Hill Chalk, and I have been hooked on that particular time span ever since.

Kansas has a wide variety of fossils, from very old Mississippian rocks (more than 340 million years old) in the extreme southeastern corner of the state to the Late Cretaceous rocks (about 75 million years old) of northwestern Kansas, which were deposited as bottom muds in the series of Paleozoic and Mesozoic oceans that covered Kansas. More recently deposited Tertiary sediments (nonmarine) containing the remains of extinct terrestrial animals are found along and in streams and rivers statewide. That means you can find fossils just about anywhere in the state.

If you are interesting in collecting fossils in Kansas, the first thing to decide is what kind of fossils you want to collect. The best place to start is a library or bookstore. I would recommend two books as "must have" references for amateur fossil hunters in Kansas: *Kansas Geology*, edited by Rex Buchanan (1984), and *Roadside Kansas*, by R. C. Buchanan and J. R. McCauley (1987, 2010). Once you have an idea of what kind of fossils you want to find and where to look for them, you're ready to get serious about it. There are many places where rocks are exposed and accessible to collectors. The most important thing to remember, however, is that most fossils in Kansas are on private property. You must have permission from the owner to enter the land and to collect. Always respect the property of others, take proper safety precautions, never leave your trash behind, and, most of all, have fun.

My intent in writing this book is to provide information about many of the animals that lived during the Late Cretaceous and, to some extent, the people who discovered their fossil remains and described them. Excellent sources of information are available in print and on the Internet regarding other kinds of fossils that can be found in Kansas, and I encourage you to spend some time learning about paleontology in general.

Acknowledgments

So many people, over the years, have helped me learn and understand the paleontology of Kansas during the Late Cretaceous that it is difficult to know where to begin in expressing my appreciation. You've all heard the adage that it takes a village to raise a child. I can certainly testify that it also takes one to write a book.

First, I thank my wife, Pamela Everhart, for her support and companionship in the field (not to mention her superior ability to find interesting things for me to dig up). She was also properly impressed when I brought a mammoth tooth to class back when we were in high school. This book and a lot of other projects in paleontology would not have been possible for me without her.

The rest of the list follows in no particular order. My fifth-grade teacher, Vivian Louthan, encouraged my interest in 'rocks' and took me to a gem and mineral show that left a lasting impression. John Ransom, Harry Rounds, and Don Distler, among many other teachers, guided my interest in the study of living things, and eventually the remains of things that lived millions of years ago. Paul Tasch introduced me to the Smoky Hill Chalk on a vertebrate paleontology field trip in 1968, and I immediately was hooked on it. David Parris, Barbara Grandstaff, and J. D. Stewart were all supportive as well as being excellent teachers and resources for otherwise unknown and mysterious information when we were getting started in our serious study of the Smoky Hill Chalk in the late 1980s. Although he may not remember it, David Parris sponsored my membership in the Society of Vertebrate Paleontology (SVP) many years ago. I owe a major debt of thanks to J. D. Stewart and Donald Hattin for our continuing discussion of the stratigraphy of the Smoky Hill Chalk. Another good friend, Pete Bussen (1927–2015), was a source of valuable information on history, paleontology, weather, and a variety of other useful subjects, which he gained in pursuit of his 'Doctor of Disagree-ology' degree. I found that Pete had a learned opinion on just about everything, whether I asked for it or not. A number of other landowners in western Kansas, including the Albins, Birds, Babcocks, Bentleys, Bodeckers, Bonners, Cheneys, Collinses, Millers, and Surratts, have generously allowed us access to their property over the years and obviously made many of our discoveries possible. Richard Zakrzewski, Larry Martin (1943–2013), and Ken Carpenter made the collections in their charge available for study and provided guidance and answers to even more questions. Their friendship has always been appreciated. Richard Zakrzewski assisted in getting my appointment as an Adjunct Curator of Paleontology at the Sternberg Museum. Dale Russell, Gorden Bell, Jim Martin, Bruce Schumacher, Kenshu Shimada, David Schwimmer, Glenn Storrs, Takehito Ikejiri, David Burnham, Mike Polcyn, and David Cicimurri indulged me in discussions of a variety of subjects, but mostly mosasaurs, plesiosaurs, and sharks. Earl Manning, in particular, has been a wonderful source for useful information on identification of specimens, reference material, historical issues in paleontology, and general all-around common sense. At one time or another, Bob Purdy, Michael Brett-Surman, Earle Spamer, Ted Daeschler, Charles Schaff, Larry Martin, Desui Maio, Michael Morales, George Corner, Greg Liggett, and many others have helped me track down fossils that originated in Kansas. Dan Varner (1949–2012), Gary Staab, Esben Horn, Doug Henderson, Russell Hawley, and other artists have provided me with their enlightened reconstructions of a lost world I can only dream about. Steve Johnson's questions regarding a few mosasaur vertebrae led us both to a once-in-a-lifetime discovery and a new species of mosasaur. The time spent in the field with Tom Caggiano, Steve Balliett, Andy Abdul, Fred Ackerman, and George Klein has been as educational for me as I hope it has been for them. I will always owe Tom Caggiano for his discovery of the missing dinosaur foot. Shawn Hamm and Keith Ewell dragged me 'kicking and screaming' into the study of fossil sharks from other formations in Kansas. A number of anonymous reviewers and some who aren't anonymous have made me aware (tactfully, of course) of my shortcomings as a writer and paleontologist. Their assistance has been invaluable to me, and I hope they see that I learned something in the process. Access to the Internet has allowed me to meet and converse with coworkers in distant places whom I will probably never meet in person. Nevertheless, I count them as good friends and thank them for their help over the years. A number of library people from all over the world, including Angie from Kansas and Steve from New Zealand, have helped me to accumulate necessary paleontology reference materials, some of which are extremely rare or difficult to find. And in closing, I certainly appreciate all those people from around the world who have communicated with me in English, because I know I would have never been able to return the favor in their languages. While this turned out to be a long list, I'm sure I've unintentionally left out several people, so please forgive the oversight. As I said earlier, it takes a village.

Second Edition Acknowledgments

Over the last twelve years I have made many new friends and acquaintances in paleontology. Unfortunately, I've also lost a few. I will certainly always remember, and owe a debt of gratitude to, Dan Varner (1949–2012), Larry Martin (1943–2013), and Jerome 'Pete' Bussen (1927–2015). I greatly valued their friendship and their contributions to the paleontology of the Late Cretaceous.

J. P. O'Neill, author of *The Great New England Sea Serpent*, has been my reader/reviewer on the text of the second edition. I certainly have benefited from her comments and corrections. I'm also indebted to my copyeditor, Carol Kennedy, for the huge amount of effort involved in cleaning up my writing. I've learned a lot from her in the final stages of getting this book ready for the printer.

The interest in Kansas paleontology shown by Keith Ewell, Ramo and Pam Decker, Shawn Hamm, Gale Pearson, Mike Urban, Fred Smith, Bob Levin, and others has provided me with unexpected opportunities to learn more about Kansas fossils. Ongoing discussions over the years with Ken Carpenter, Bruce Schumacher, Mikael Siverson, David Burnham, David Parris, Johan Lindgren, Kenshu Shimada, Ikejiri Takehito, and others have enriched me in many ways. Lastly, I would like to thank Mike Triebold, Anthony Maltese, and Jacob Jett of Triebold Paleontology for their timely and professional assistance in several important recent discoveries.

Abbreviations

AMNH	American Museum of Natural History, New York, New York
ANSP	Academy of Natural Sciences of Philadelphia, Philadelphia, Pennsylvania
CMC	Cincinnati Museum Center, Cincinnati, Ohio
CMNH	Carnegie Museum of Natural History, Pittsburgh, Pennsylvania
DMNH/DMNS	Denver Museum of Nature and Science, Denver, Colorado
ESU	Emporia State University Geology Museum, Emporia, Kansas
FFHM	Fick Fossil and History Museum, Oakley, Kansas
FHSM	Sternberg Museum of Natural History, Fort Hays State University, Hays, Kansas
FMNH	The Field Museum of Natural History, Chicago, Illinois
KU	University of Kansas Invertebrate Paleontology Collection, Lawrence, Kansas
KUVP	University of Kansas Vertebrate Paleontology Collection, Lawrence, Kansas
LACMNH	Los Angeles County Museum of Natural History, Los Angeles, California
MCZ	Museum of Comparative Zoology, Harvard, Cambridge, Massachusetts
NAMAL	North American Museum of Ancient Life, Lehi, Utah
NJSM	New Jersey State Museum, Trenton, New Jersey
RMDRC	Rocky Mountain Dinosaur Resource Center, Woodland Park, Colorado
ROM	Royal Ontario Museum, Toronto, Ontario, Canada
SDSMT	South Dakota School of Mines and Technology, Rapid City, South Dakota
UCM	University of Colorado Museum, Boulder, Colorado
UNO	University of New Orleans, New Orleans, Louisiana
UPI	Uppsala Paleontological Institute, Uppsala University, Sweden

UNSM	University of Nebraska State Museum, Lincoln, Nebraska
USNM	United States National Museum, Washington, D.C.
YPM	Yale Peabody Museum, New Haven, Connecticut

Oceans of Kansas

Introduction

An Ocean in Kansas?

The bright midday sun glinted off the calm waters of the Inland Sea and silhouetted the long, sinuous form of a huge mosasaur lying motionless amid the floating tangle of yellow-green seaweed. At 20 years old, more than 30 feet in length, and weighing over a ton, the adult mosasaur was almost full grown and was much larger than any of the fishes or sharks that lived in the shallow seaway. A swift and powerful swimmer over short distances, the mosasaur used surprise and the thrust of his muscular tail to overtake his prey with a short burst of speed. His jaws were more than four feet long and were lined with sharp, conical teeth that he used to seize and kill his prey. Several unusual adaptations in his lower jaws allowed them to flex in the middle and enabled him to easily swallow the large fish and other animals he caught. This adaptation to life in the ocean was essential to the mosasaur because he had to hold on to his prey with his teeth or risk losing it. If he let go of his prey in the middle of the ocean, there was a good chance a hungry shark would grab it or it would sink to the bottom and be lost.

The mosasaur was floating at the surface with his eyes and nostrils just above the water. His dark upper body absorbed the hot rays of the Late Cretaceous sun as dozens of tiny fishes emerged from hiding in the seaweed and darted cautiously around his submerged bulk. They were feeding on parasites and other small invertebrates that had attached themselves to his scaly hide. He breathed slowly and quietly through his nostrils as his ears and other senses remained on the alert for the telltale sounds made by approaching prey. A patient hunter, he preferred to let his victims come to him instead of wasting energy swimming around the vast seaway in search of food.

Overhead, winged reptiles of various sizes floated lazily through the cloudless sky, riding the thermals above the warm water while looking for schools of small fishes feeding near the surface. Occasionally, one would skim the surface of the water and grab an unwary fish with its narrow beak. The mosasaur had recently tried to eat the floating carcass of a dead pteranodon, but found the thin wings difficult to get into his mouth. After tearing off the small body, he had let the rest of the flyer sink to the bottom. The living ones overhead could see him clearly from above and avoided feeding near him.

Amid an ever-changing mixture of background noises made by a variety of creatures in the ocean, he noticed a faint buzz of clicking sounds that was getting louder, alerting him to a group of hard-shelled

One Day in the Life of a Mosasaur

ammonites feeding nearby. Though not his favorite prey, they were all that had approached him since he had taken a large, solitary fish early in the morning. Even with that recent meal, his appetite was still unsatisfied, and hunger was beginning to gnaw at him. The adaptations that made it possible for mosasaurs to return to the sea included an increased rate of metabolism; this kept him warm, but required large amounts of food to support a more active lifestyle.

Exhaling most of the air from his lungs, he slowly submerged his head, leaving behind only the faintest of ripples. His large eyes immediately located the brightly colored coiled shells of the ammonites as they approached, bobbing and darting below him. Propelled by water forced through their internal siphons, they moved generally backward through the water with their short tentacles trailing behind them. Instinctively, he knew that their large shells would hide him from their view until they had moved well past him. He would make his attack from above, long before they had a chance to sense the danger.

Using his four large paddles, the mosasaur carefully maneuvered his snakelike form into an attack position, watching intently for any indication that the ammonites had detected the danger from above. Singling out a slightly larger ammonite at the edge of the group, he dived downward with a powerful slash of his long, broad tail. The ammonites reacted quickly and instinctively to the disturbance, scattering in all directions below him, but not before his heavy jaws closed across the soft forebody of his victim. His sharp teeth shattered the front edge of the ammonite's shell, destroying its buoyancy and rendering the ammonite helpless.

Without his captive pocket of air, the ammonite would sink swiftly to the bottom of the seaway. With practiced ease, the mosasaur flexed his body upward and brought the ammonite toward the surface. Then he released it and grabbed the tentacles of the immobilized creature with his teeth as it began to sink. Far too late, the ammonite released a cloud of jet-black ink into the water. The mosasaur ignored the bitter taste of the ammonite's last defense as he gave a quick jerk of his head to pull the ammonite's soft body from its shell. The heavy shell and several fragments slipped sideways through the water and quickly disappeared into the murky depths. Opening and closing his jaws rapidly, the mosasaur swallowed the fleshy morsel in a single gulp.

Looking around for more prey, he saw another ammonite swimming in confused circles nearby. A swift lunge, and his sharp teeth crunched through the ammonite's hard shell. Moments later, the soft body of the second ammonite followed the first into the mosasaur's stomach. The rest of the ammonites had jetted away as fast as they could and were no longer in view. His hunger briefly satisfied, the mosasaur rose slowly to the surface to breathe and resume his ambush position rather than chase after the fleeing cephalopods.

He had hardly settled into waiting when he sensed noises made by the approach of another mosasaur. Female mosasaurs tended to band together in pods for the protection of their young, while males were solitary and territorial. The approaching mosasaur was probably a young male searching for his own place in the expanse of the Inland Sea. With one swift, fluid motion, the older mosasaur turned and began to swim toward the sounds made by the approaching intruder. With flippers held tightly against his body, he moved quickly through the water just beneath the surface. His tail broke through the water's surface repeatedly as he intentionally made as much noise as possible, wanting to sound threatening to the other mosasaur. Although he was prepared to fight for his territory, he would first try to frighten off this other male with his size and ferocity. Long-healed scars on his body showed that even the winners in such fights could be badly hurt. He had been lucky several times earlier in his life and had survived injuries that easily could have been fatal. As he had gotten older, he had learned to avoid such battles whenever he could.

The older mosasaur's course intercepted the other broadside in a patch of open water. Turning quickly to face the threat, the smaller animal displayed a mouth filled with sharp teeth. Despite being nearly 10 feet shorter and much less massive, the invader refused to turn and flee. The big mosasaur circled warily around his now stationary foe, watching intently as the other animal almost doubled back upon itself as it continued to show its open jaws. Trying to appear as threatening as possible, the younger animal still refused to turn and run. The larger mosasaur was in no mood for such tactics. Making a large splash with his tail to distract the intruder, he surged forward and seized the smaller animal across the throat and back of the head. For a moment, the smaller mosasaur struggled helplessly as the powerful grip of the larger animal threatened to crush his skull. Then the larger mosasaur moved his head quickly, snapping the other mosasaur's neck. The smaller mosasaur gave a brief shudder, then went limp. Angrily, the big mosasaur shook the slender body again, making certain that his foe was no longer a threat.

Realizing that his victim was too large for him to swallow, the mosasaur released his grip and moved away. The body of the dead mosasaur rose slowly toward the surface and floated there until most of the remaining air had escaped from its lungs. Then it began to sink headfirst toward the bottom. Still enraged by the invasion of his territory, the big mosasaur searched about for any other interlopers as he swam in a large circle back to his ambush site. The commotion caused by the brief battle had frightened any prey away and would certainly draw sharks to the area to feed on the remains of the dead mosasaur. Sharks also seemed to be attracted to the undulating movement of a mosasaur's tail. Although he was too large for them to be much of a threat to him, any shark bite could cause a wound that could become seriously infected. He already had several healed scars from past shark bites on his tail and flippers.

Later in the afternoon, he sensed the noisy approach of a group of swimming birds. Large and wingless, these birds migrated through the seaway every year during their journeys to and from their nesting grounds to the north. They were fast swimmers and fed on the abundance of small fishes and squid that lived in the sea, catching them in their toothy beaks.

He submerged quietly until he was well below the surface, then swam slowly toward the birds. From the sounds he heard, he could tell they were feeding. In the past, he had been able to ambush careless stragglers from below as they rested between dives for food. Nearing the flock, he could see the darker bodies of the birds silhouetted against the sunlit surface as they dived and fed on a school of small silvery fishes they had trapped. Slashing his powerful tail from side to side, he surged upward toward the body of the nearest bird. His mouth opened just before he reached the surface and quickly closed on the bird as his momentum carried his upper body several feet out of the water. Crushed by his powerful jaws, the bird struggled briefly and died.

When he was certain his prey would not escape, he moved the limp body around in his mouth until it was pointed headfirst into his throat. Then he lifted his head out of the water and allowed gravity to help him swallow the bird. The noise made by the rest of the retreating flock was already fading in the distance.

The hours passed by and the dark clouds of an approaching storm covered the sun as it sank toward the horizon. Driven by the changing weather, the waves became larger and larger. It became difficult for the mosasaur to maintain his stationary position and nearly impossible for him to sense the approach of possible prey against the increasing background noise caused by the wind and rain. Instinctively, he knew it was time to move to open water. Moving forward with rhythmic undulations of his tail, he headed toward the edge of the seaweed mat.

An Ocean in Kansas

Imagine, if you will, the middle of North America covered by a vast inland sea. Most of Texas, New Mexico, Oklahoma, Colorado, Kansas, Nebraska, South Dakota, North Dakota, Wyoming, and Montana; parts of Missouri, Iowa, and Minnesota; and the central regions of Canada were underneath a shallow ocean. Not just any ocean, but one that stretched for hundreds of miles from Utah to Minnesota, and from the Gulf of Mexico past the Arctic Circle (Fig. 1.1). At times, this ancient ocean was as large, though not as deep, as the present-day Mediterranean Sea and was the home of many kinds of strange creatures that have been extinct for more than 65 million years. This shallow, saltwater sea covered Kansas and the rest of the Midwest during most of the last 40 million years of the Age of Dinosaurs, and almost until the very end of the Cretaceous period, which lasted from about 144 Ma (million years ago) until 65 Ma. Drainage from the older North American continent to the east and the mountains rising from the new land to the west carried vast amounts of soil, sand, and gravel into this seaway, creating intermixed layers of sandstone, shale, and mudstones along the shorelines. In the clear waters at the center of the seaway, the calcium carbonate shells of billions and billions of microscopic, single-cell algae produced thick layers of chalk.

In Kansas, the geological record of the Cretaceous begins with marine and nearshore deposits of the Cheyenne Sandstone and Kiowa

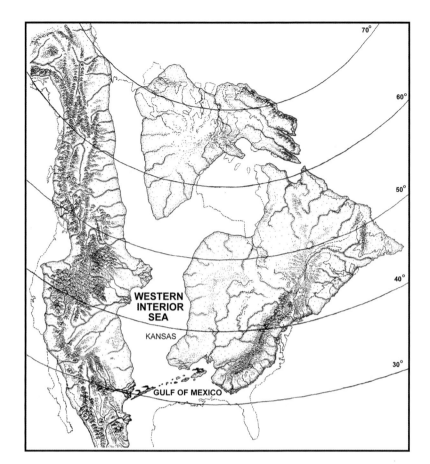

1.1. This map shows the approximate boundaries of the Western Interior Sea during the deposition of the Smoky Hill Chalk. Present-day exposures of the chalk are located just above the "K" in Kansas. Adapted from Schwimmer, 2002; base map by Ron Hirzel, used with permission.

Shale formations that lie on top of the Wellington Formation (Permian) in the central part of the state. The last Cretaceous rocks are the Sharon Springs, Weskan, and Lake Creek members of the Pierre Shale Formation, in the northwest corner of Kansas. Almost 30 million years of geologic history is preserved in between. One of the deposits near the top layer of the Cretaceous rocks in Kansas is referred to as the Niobrara Formation. It contains a unique upper member called the Smoky Hill Chalk. This chalk is composed mostly of calcium carbonate, very similar to the white cliffs near Dover, England. The Smoky Hill Chalk was deposited in Kansas during a 5- million-year time span, roughly between 87 and 82 Ma. During that time, the Western Interior Sea was gradually retreating from its greatest expansion. The deposition of these chalky marine sediments occurred during the last half of the Cretaceous period, and ended about 17 million years before the end of the Age of Dinosaurs.

In Kansas, the Smoky Hill Chalk is about 600 feet thick and lies above the Fort Hays Limestone and below the Pierre Shale (Fig. 1.2). For the most part, the chalk is composed of compacted shells (coccoliths) of microscopic, golden-brown algae (Chrysophyceae) that lived and died by the untold billions in the warm, shallow sea. Besides making up the chalk, these microscopic plants were the basis for a complex food web that

1.2. An exposure of the lower Smoky Hill Chalk in southeastern Gove County, Kansas.

supported vast numbers of small fishes and many large predators, including sharks, larger fishes, plesiosaurs, mosasaurs, pteranodons, and birds.

The Western Interior Sea, sometimes just called the Inland Sea, was formed by the flooding of low-lying areas of the North American continent during a period of the earth's history when there were no polar ice caps and sea levels were at their highest. Near the center of the sea, the water was probably less than 600 feet deep (Hattin, 1982) and the limey mud bottom was relatively flat and featureless. In the area where Kansas is now located, the sediments were deposited at a rate that would ultimately produce about an inch of solid chalk for every 700 years of time (ibid.). The chalk also contains more than 200 thin layers of bentonite clay, most of which are rusty red in color, that are the residuals of volcanic

ash deposited from periodic major eruptions in what is now Nevada, Utah, Idaho, and Montana. These ash deposits (Fig. 1.3) can be traced for miles across the chalk beds and are currently used as chronological markers when describing the stratigraphy of the formation. In addition, several species of vertebrate and invertebrate marine life that lived in the Western Interior Sea at different times during the deposition of the chalk are useful in determining the age and biostratigraphy of widely separated exposures (Chapter 13).

This shallow ocean was home to a variety of marine animals that are now extinct. These included giant clams, rudists, crinoids, squid, baculites, belemnites, ammonites, numerous sharks and bony fishes, turtles, plesiosaurs, mosasaurs, pteranodons, and even several species of primitive

1.3. The two thin bentonites separated by about 75 cm (30 in) of chalk, with a band of white chalk in between, make up Hattin's Marker unit 9. This is the Gove County exposure where the most complete specimen of *Bonnerichthys gladius* (FHSM VP-17428), a large filter-feeding fish, was collected in 2008. The age of the marker unit and the specimen is Middle Santonian, or about 85 Ma. Scale = 1 m.

marine birds with teeth. Although it seems unlikely that you would find dinosaur fossils in the middle of the Western Interior Sea, the partial remains of several of them (two hadrosaurs and a dozen or more nodosaurs) have been collected from the Smoky Hill Chalk since 1871. Relatively few in number, these specimens have been well documented (Marsh, 1872; Wieland, 1909; Eaton, 1960; Carpenter et al., 1995; Everhart, 2004; Everhart and Hamm, 2005; Liggett, 2005; Everhart and Ewell, 2006; Carpenter and Everhart, 2007). In order to get to where they have been discovered, the bodies of these dinosaurs must have somehow floated hundreds of miles out to sea before sinking to the bottom (Fig. 1.4). It is possible that they died during catastrophic floods and were carried out to sea in large, tangled mats of trees and other vegetation. However they arrived, there is no doubt that there are dinosaurs buried among the marine reptiles and fishes in the Smoky Hill Chalk.

Over a period of about 5 million years, the remains of many of these animals were preserved as fossils in the soft, chalky mud of the sea bottom. When this mud was compressed under the weight of hundreds of feet of overlying shale, it became a deposit of chalk that is about 600 feet thick in western Kansas. Much of the massive chalk formation that once covered Kansas has been eroded away over the last 65 million years, however, and is now exposed only in relatively small areas along the rivers in the northwest quarter of the state. The eastern edge of this part of Kansas is also known as the Smoky Hills, which provided the name for the Smoky Hill River that flows through it and, ultimately, for the geological formation known as the Smoky Hill Chalk.

During the last 140 years or so, the Smoky Hill Chalk has been the source of thousands of fossil specimens, many of which are on exhibit today in museums in Europe and around the world. A large number of these were collected by or for such famous paleontologists as E. D. Cope, O. C. Marsh, S. W. Williston, and members of the Sternberg family. These specimens include a large portion of the Yale Peabody Museum collection that resulted from the Yale College Scientific Expeditions of the 1870s. Much of the early work on the Cretaceous fossils from Kansas was published in volumes 2, 4, and 6 of the University Geological Survey of Kansas (1897, 1898, and 1900). Descriptions of these strange, 'prehistoric' animals from their often fragmentary remains were sometimes bizarre by today's standards and often resulted in inaccurate reconstructions drawn under the direction of the various paleontologists. E. D. Cope was one of the most imaginative in his descriptions of not only how the animals looked, but how they lived and interacted. Cope noted in regard to mosasaurs that

> their heads were large, flat, and conic, with eyes directed partly upward; that they were furnished with two pairs of paddles like the flippers of a whale, but with short or no portion representing the arm. With these flippers and the eel-like strokes of their flattened tail they swam, some with less, others with greater speed. They were furnished, like snakes, with four rows of formidable teeth on the roof of the mouth (Fig. 1.5). Though these were not designed for mastication, and, without paws for grasping, could have been little used for cutting, as weapons for seizing their prey they were very formidable. (1872:320)

Another good example of this sort of fanciful (and in this case highly inaccurate) prose was provided by the 'Father of American Paleontology,'

1.4. Nine vertebrae from near the end of the tail of a large (10 m) hadrosaur from the Smoky Hill Chalk in Gove County, Kansas. Bite marks (last vertebra at lower right) and partially digested bone indicate that this piece of the dinosaur's tail had been eaten by a large shark, most likely *Cretoxyrhina mantelli*, the ginsu shark.

1.5. Dorsal and ventral views of a mosasaur skull (*Clidastes* sp.) adapted from Williston (1893, Pl. III). This was one of the first drawings of a complete mosasaur skull. Note the two rows of pterygoid teeth on the roof of the mouth (Fig. 3).

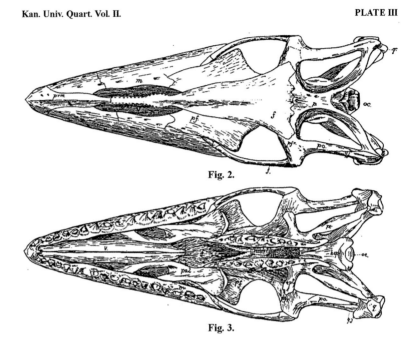

Fig. 2.

Fig. 3.

Joseph Leidy (1870:10), in his description of the first long-necked plesiosaur, *Elasmosaurus*: "We may imagine this extraordinary creature, with its turtle-like body, paddling about, at one moment darting its head a distance of upwards of twenty feet into the depths of the sea after its fish prey, at another into the air after some feathered or other winged reptile, or perhaps when near shore, even reaching so far as to seize by the throat some biped dinosaur." Wrong on all counts (see Chapter 7 for a discussion of these misconceptions).

Edward Drinker Cope and Othniel Charles Marsh are the famous paleontologists we usually associate with the fossils discovered in Kansas from the 1860s onward. Together with Joseph Leidy and Samuel W. Williston, they are the professional paleontologists who described and named dozens of extinct animals from Cretaceous rocks in the state. For the most part, however, these well-recognized figures are not the discoverers of those fossils. Many type specimens from the Pierre, Niobrara, Carlile, Greenhorn, and Dakota formations were discovered by amateurs, and many of these are significant additions to paleontology. Since the 1860s, this has included many important specimens from Kansas: the giant predatory fish *Xiphactinus audax* by Dr. George M. Sternberg; the first known elasmosaur, *Elasmosaurus platyurus*, by Dr. Theophilus Turner; the first mosasaur, *Tylosaurus proriger* (Kansas State Fossil), by Army Col. John B. Conyngham; the first known polycotylid plesiosaur, *Polycotylus latipinnis*, by railroad executive William E. Webb; the first specimen of the toothed bird Ichthyornis dispar and the first remains of the huge filter-feeding fish *Bonnerichthys gladius* by Professor Benjamin Mudge; the elasmosaur *Styxosaurus snowii* by the 70-year-old retired judge Elias P. West; the polycotylid *Dolichorhyn-chops osborni* by 17-year-old George F. Sternberg; the type specimen of the nodosaurian dinosaur *Niobrarasaurus coleii* by oil-field geologist Virgil Cole; the nodosaur *Silvasaurus condrayi* by landowner Warren Condray; the pliosaur *Megacephalosaurus eulerti* by teenagers Robert and Frank Jennrich; and most recently a rare and newly named mosasaur, *Selmasaurus johnsoni* (Polcyn and Everhart, 2008), by Steve Johnson. Sometimes having good eyes in the right place at the right time is more important than being the expert.

Both the Sternberg Museum of Natural History at Fort Hays State University in Hays, Kansas, and the Museum of Natural History at the University of Kansas in Lawrence, Kansas, have excellent collections and exhibits of fossils from the Smoky Hill Chalk. The Denver Museum of Nature and Science in Denver, Colorado; the Sam Noble Museum of Natural History in Norman, Oklahoma; the Field Museum in Chicago, Illinois; the Philadelphia Academy of Natural Sciences; and the American Museum of Natural History. A number of museums in Europe, including the Natural History Museum in London and the National Museum of Natural History in Paris, France, also have many Kansas fossils. One of the most complete exhibits of marine fossils from Kansas is the Rocky Mountain Dinosaur Resource Center in Woodland Park, Colorado. Many Kansas fossils were also sold to major museums in Europe and elsewhere around the world by the Sternberg family and others. Unfortunately, we still don't have a good record of where all the fossils collected by the Sternberg family have gone; even worse, we realize that some of them were likely destroyed during two world wars.

For the most part, this book will discuss discoveries regarding the natural history of the Western Interior Sea during the deposition of the Smoky Hill Chalk in a period roughly between 87 and 82 Ma. However, in order to better understand that time interval, it is useful to look at the Kansas oceans during that portion of the Cretaceous for which we have a geological record in the state.

The Mesozoic (Age of Reptiles) is divided into three major periods: the Triassic, the Jurassic, and the Cretaceous. Based on the 2014 geologic time scale (ver. 4.0) published by the Geological Society of America, the Mesozoic lasted roughly 186 million years. The Cretaceous period is the last of the three unequal divisions, roughly from 145 Ma to about 66 Ma, or a time interval just short of 80 million years. In Kansas, the geological record visible in the surface rocks is missing for all of the Triassic, Jurassic, and most of the Early Cretaceous. Rocks of the latter part of the Early Cretaceous lie nonconformably upon shales of the Permian period in the central part of the state. In this case, 'nonconformably' means there is a gap of about 140 million years in the geological record between the top of the Permian rocks and the bottom of the Cretaceous rocks (middle

Kansas during the Cretaceous: A Timeline

Albian). The gap is less under the western part of the state, where there are 'only' 40 million years of 'time' missing between the Morrison Formation (Late Jurassic) and the latter part of the Early Cretaceous. In other words, rocks that would have been formed during most of the Mesozoic (140 of 180 million years) are missing in most of Kansas. While we presume that Kansas was above sea level and eroding away for a least part of that time interval, whatever was going on in Kansas during the Triassic and Jurassic will never be known for certain because there is no geologic record.

We can, however, say quite a lot about the fossil record in the Cretaceous rocks that are preserved in Kansas. These layers of Cretaceous rocks, representing a fairly continuous progression of time from the oldest to the youngest, are stacked upon one another in an orderly fashion. The geology of Kansas (Plate 1) is relatively simple to visualize compared to that of places like Colorado or Arizona. There are no mountains to contend with, and everything is rather flat, at least in the large-scale view. The youngest Cretaceous rocks are in the northwest corner of the state, and the oldest rocks (Mississippian) occur in the southeast corner. Put another way, as you travel roughly 430 miles from St. Francis (Cheyenne County) in the northwest corner to Baxter Springs (Cherokee County) in the southeast, you are descending 250 million years through time (geologically speaking) at an average of about 580,000 years to the mile. It is almost like being in a time machine.

In this book I will describe the discovery of animals that lived during a much shorter period of time: the relatively brief geological period when the Smoky Hill Chalk was deposited near the middle of the Western Interior Sea. I will take occasional 'side trips' into other parts of the Cretaceous in Kansas, however, and even out of Kansas, where the rocks were deposited as a series of nearshore sandstones (including a river delta), offshore shales, and deeper water limestones and chalks from about 112 Ma to 75 Ma. I hope that when I am done, you will better understand how the geological record of that period was preserved at the bottom of the oceans of Kansas. Merriam's (1963) The Geologic History of Kansas is one of the better references in regard to the surface and subsurface rocks occurring in the state. Buchanan's Kansas Geology (2010), and Buchanan and McCauley's Roadside Kansas (2010), also provide introductory guides to the geology, fossils, and roadside rock exposures of Kansas. Additional information on the Cretaceous fossils and geology of Kansas is available on the Internet through the Oceans of Kansas Paleontology website (http://www.oceansofkansas.com).

The oldest Cretaceous formation in Kansas is the Cheyenne Sandstone. This is a relatively pure (beach?) sand and gravel layer that was deposited in south central Kansas (Clark and Kiowa counties) along the shore of the approaching sea as it advanced from the south. This flooding of the North American continent was due to a massive rise in sea levels that occurred about 110 Ma near the end of the Early Cretaceous during Aptian and Albian time. As the sea advanced, covering the much older and heavily eroded Permian shale, the sediments that were deposited

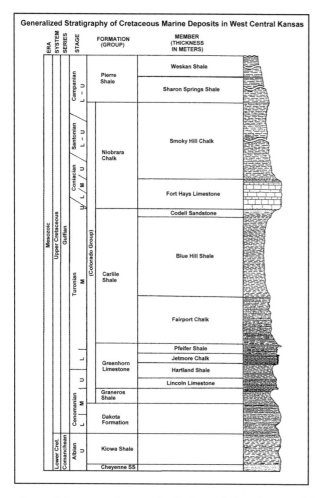

Generalized Stratigraphy of Cretaceous Marine Deposits in West Central Kansas

1.6. A generalized stratigraphic column of the Early and Late Cretaceous formations in west-central Kansas, roughly spanning a period from 110 to 75 Ma. Adapted from Shimada, 1996. (L = Lower; M = Middle; U = Upper).

changed from sand to sandy shale and then layers of dense gray shale (Scott, 1970). These shale layers represent the remnants of rocks that were being eroded from nearby land masses and being carried into the ocean by rivers. The gray shale that makes up the overlying Kiowa Shale preserves evidence of abundant life in the shallow sea and the nearby shoreline: many invertebrate fossils, teeth of sharks, and bones of fishes, turtles, plesiosaurs, and crocodiles. The sea continued to expand northward and eastward across Kansas throughout Albian time. As it did, it buried the Cheyenne Sandstone and Kiowa Shale under blankets of mud and other sediments deposited on the sea floor (Fig. 1.6).

About 99 Ma, the Albian stage (Early Cretaceous) ended and the Cenomanian stage (Late Cretaceous) began. In north-central Kansas, this is evidenced by sand and other sediments that were deposited in a huge delta by a major river or rivers flowing into the eastern edge of the sea from the northeast. Iron-rich rocks from as far away as Wisconsin and Michigan were being eroded away bit by bit and carried to the edge of the sea in central Kansas, where they were laid down layer after layer, forming the banded sandstones of the Dakota Formation. This formation is visible today as buttes and other multicolored (off-white to dark reddish-purple) erosional features to the northwest of McPherson and around Kanopolis Lake in the central part of the state (McPherson, Saline, Ellsworth, and Russell counties). Historically, major vertebrate fossils have been limited to those of a crocodile (*Dakotasuchus kingi*, Mehl, 1941; Vaughn, 1956), a nodosaurian dinosaur (*Silvasaurus condrayi*, Eaton, 1960) and a few general reports that mention the teeth of sharks and bony fishes. Recent collections of the upper Dakota Formation, however, indicate a rich marine fauna (Everhart et al., 2004) of sharks, rays, and bony fishes that existed along the shore during the transition from a nonmarine to a near-shore marine environment (Hattin and Siemers, 1978). Many thousands of leaf impressions were collected by G. M. Sternberg, B. F. Mudge, C. H. Sternberg, E. P. West (Everhart, 2015) and others from the Dakota (Lesquereux, 1868) beginning in the mid-1860s. Near the middle of the Cenomanian, sea levels rose again, and a dark gray shale called the Graneros covered the sand, effectively burying the river delta under some ten meters of mud (Hattin and Siemers, 1978). The sea continued to deepen and the Graneros Shale was replaced by the Lincoln Limestone Member of the Greenhorn Formation during the Upper Cenomanian. The deposition of alternating limestones and shales continued thorough the Lower Turonian. At the high-water mark of this expansion (transgression) of the sea, an 8–10 inch layer of resistant limestone was laid down, forming the Fencepost Limestone bed at the top of the Greenhorn Formation. Then, at the beginning of the middle Turonian, the sea began to recede (regress) from the middle of the continent. A chalky limestone called the Fairport Chalk Member of the Carlile Formation was deposited for a time and then was replaced by the dark-gray Blue Hill Shale Member. Near the end of the middle Turonian, the coastline was approaching rapidly as the seaway narrowed, and the shale was replaced by the nearshore Codell Sandstone.

By the beginning of the Upper Turonian, the ocean was gone again from parts of Kansas, or at least no record of deposition remains from that time. Major erosion occurred in the central part of the state, where almost all the Codell Sandstone has been removed. During the following transgression, when the seas returned near the beginning of the Coniacian, they deepened rapidly, and the clear-water Fort Hays Limestone was deposited on top of the remaining Codell Sandstone (Merriam, 1963). The Fort Hays Limestone is the lower member of the Niobrara Formation, formed during a period when the sea was at its widest and the water near the center (Kansas) was at its deepest. By the middle of the Coniacian, the sea was again slowly regressing, and the limestone was replaced by the Smoky Hill Chalk.

As mentioned earlier, the Smoky Hill Chalk was deposited over a period of about 5 million years, between 87 and 82 Ma. This period of time includes the late Coniacian, all of the Santonian, and the beginning of the early Campanian stage. By the early Campanian, sea levels were still dropping, and the chalk was eventually replaced by gray shales of the Sharon Springs Member of the Pierre Shale. In Kansas, there are

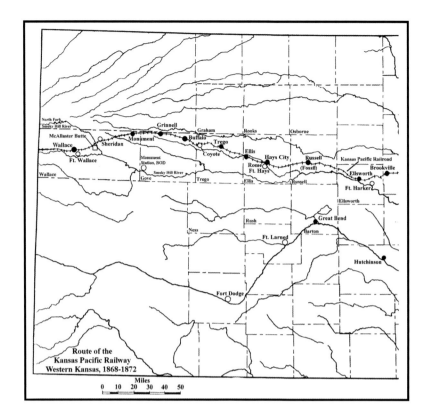

1.7. A map of western Kansas, circa 1868–1869, just prior to the completion of the Kansas (Union) Pacific Railway. Many of the discoveries described by Cope and Marsh from Kansas occurred south of the railroad, along the Smoky Hill River in Wallace, Gove, and Trego counties. Note that present-day Logan County is the eastern half of what was Wallace County in 1870. Open symbols indicate habitations that no longer exist.

indications that the western portion of the chalk may have been raised to or slightly above sea level for a brief time. When the sea returned, a dark-gray shale was deposited throughout most of the rest of Campanian age across Kansas. The geological forces forming the Rocky Mountains, however, were also lifting much of the Great Plains to the west. At some point before the end of the Cretaceous, Kansas rose above the sea for the last time, and the sea bottom was exposed to the forces of nature. Over the last 65 million years or so, surface erosion has removed much of the upper Pierre Shale in Kansas, and the geological record of the Western Interior Sea in the state ends well before the end of the Cretaceous. What remains, however, is one of the best records that we know of concerning the marine life that flourished in the oceans of Kansas.

In the following chapters, I will discuss the fossil discoveries that have occurred during the last 150 years (1867–2017) in the rocks covering the western third of Kansas (Fig. 1.7). Kansas was a very different place during the 1870s, and paleontology was as much an adventure as it was a scientific pursuit. The fossils that were being discovered there were new to science and generated an intense interest in paleontology that continues today. To me, the early and continued discovery of these fossils provides a fascinating look at the strange creatures that lived and died in the seas of the Late Cretaceous. The stories of their discovery and identification also provide an interesting view of the early years and the growth of paleontology in the United States.

The long, slender fish swam rapidly toward the large school of minnow-sized prey that had been corralled into a shimmering bait ball by larger predators. His appearance and undulating movement were similar to those of an eel, except for the sharp, swordlike spike projecting from his lower jaw. Although located on his lower jaw, the 'sword' functioned not unlike that of more normal-appearing swordfishes. When he reached the edge of the bait ball, he slashed his head rapidly back and forth, trying to strike and stun as many of the quickly moving smaller fishes as he could. Then he turned around and swallowed as many of the injured fishes as he could find. He had competition from the other large fishes feeding on the school, but his tactic was generally successful. When he could find no more injured fishes to eat, he reentered the school, repeating the process again and again. It was a good opportunity, and he soon filled his stomach. He never saw the massive shark that came out of the free-for-all and grabbed him. In one quick bite the shark severed the fish's head from its body. The detached head with its heavy spikelike lower jaw sank rapidly to the bottom of the sea, where it was partially buried in the soft, chalky mud.

Discovery of the Western Interior Sea

This kind of predator-and-prey encounter, in which a smaller fish gets eaten by a larger predator, happened continuously as part of the food web in the Western Interior Sea, but this incident was special. A portion of the skull of this odd fish from the Late Cretaceous would become the first fossil collected and named from the Smoky Hill Chalk Member of the Niobrara Formation.

When Lewis and Clark set out in 1804 on their westward trek to explore the Louisiana Purchase, they had no idea they would also be crossing the expanse of an ancient ocean that once covered the middle of North America. Early in the expedition they discovered the only fossil that survives today (Chapter 5). Along the Missouri River, near the northwest corner of what is now Iowa, they came across a fossil that Meriwether Lewis described, in a note that is curated along with the specimen, as "the petrified jaw bone of a fish" (Fig. 2.1; see Spamer et al., 2000, for a more detailed account). The exposures along the river in this area are not far from the Late Cretaceous Niobrara Formation located in southeastern South Dakota. The 'fish jaw' of Meriwether Lewis was eventually presented to the American Philosophical Society, where it was studied and then misidentified some years later by Dr. Richard Harlan (1824) as the jaw of a new species of marine reptile, *Saurocephalus lanciformis*.

He believed it to be most closely related to the marine reptiles called ichthyosaurs. The jaw ended up in the collection of the Academy of Natural Sciences of Philadelphia (ANSP) where Joseph Leidy (1856:302) noted that the specimen (ANSP 5516) was a fragment of a "maxillary bone with teeth, of a peculiar genus of sphyrænoid fishes, from the cretaceous formation of the Upper Missouri."

The journals of the Lewis and Clark expedition also tell of another mysterious fossil. In 1818, Dr. Samuel Mitchell (406) wrote, "What shall we think of the genus and species of that petrified skeleton of a very large fish, seen in the Sioux county, up the Missouri by Patrick Gass? In his *Journal to the Pacific ocean with Messrs. Lewis and Clark in 1804–06*, he relates that it was forty-five feet long and lay on top of a high cliff." As noted on a copy of Clark's original map made years later for the Maximilian-Bodmer western journey (Moulton, 1983–1997: vol. 1, Clark-Maximilian Sheet 9, 1983), the remains were discovered along a stretch of the Missouri River in what is now northwest Gregory County, south-central South Dakota, and probably came from the Late Cretaceous Pierre Shale Formation. At least four members of the Lewis and Clark expedition noted the discovery of the large skeleton in their journals on Monday, September 10, 1804 (see Moulton, 1983–1997). Clark's description (ibid., 3:61, 1987) is perhaps the most complete: "Below the Island on the top of a ridge we found a back bone with the most of the entire [length] laying Connected for 45 feet. Those bones are petrified, some teeth & ribs also connected." John Ordway (ibid., 9:57, 1996a) described the remains simply as "the rack of Bones

2.1. The type specimen of a fish (*Saurocephalus lanciformis*) collected from the Niobrara Formation by the 1804 Lewis and Clark expedition. The specimen is a partial left upper jaw (maxilla) with teeth. It was described and misidentified as the jaw of a marine reptile by Harlan (1824).

2.2. An illustration of the rostrum (premaxilla) of a mosasaur from the Pierre Shale of South Dakota misidentified by Harlan (1834) as the type specimen of *Ichthyosaurus missouriensis*. The specimen was later donated by Harlan to the Muséum National d'Histoire Naturelle of Paris, France, where it was rediscovered in 2004 just prior to the First Mosasaur Meeting (Chapter 9).

F.3.   *F.4.*   *F.5.*

F.6.   *F.8.*   *F.7.*

Trans. Amer. Philos. Soc. Plate XX vol. 4.

of a verry [*sic*] large fish" while Joseph Whitehouse (ibid., 11:72, 1997) wrote that they "saw lying on the banks on the South side of the River, the Bones of a monstrous large Fish, the back bone of which measured forty-five feet long." Gass (ibid., 10:38, 1996) also noted that "part of these bones were sent to the City of Washington." While the bones they collected and sent back to Washington were apparently lost, the description appears most likely to be that of a large mosasaur. Moulton (ibid., 3:63, 1987) also speculated it may have been a large plesiosaur (elasmosaur?), but provided no further evidence in that regard.

Ten years after naming of *Saurocephalus*, Dr. Harlan misidentified fragments of another fossil from the Western Interior Sea. In this instance, Harlan (1834:405) noted that the remains had been discovered by "a trader from the Rocky mountains . . . [who] observed, in a rock, the skeleton of an alligator-animal, about seventy feet in length; he broke off the point of the jaw as it projected, and gave it to me. He said that the head part appeared to be about three or four feet long." Ignoring the field observations of the fur trader, just as he had those of Meriwether Lewis, Harlan (ibid.) decided that the remains were those of an ichthyosaur and gave it the name *Ichthyosaurus missouriensis*. An examination of the accurately drawn figure published with his paper clearly shows the fragment to be the anterior end of the premaxilla of a mosasaur skull (Fig. 2.2; F3–F5). The mistake was noted relatively quickly by his contemporaries, but that is only the beginning of this rather fantastic fossil story.

From this point, the tale becomes more complicated. Several years later, the articulated skull, lower jaws, and vertebrae of a strange beast (a marine lizard or mosasaur) were recovered from the same Big Bend of the Missouri area in South Dakota. In this case, 'articulated' means that the bones of the skull of the mosasaur were still arranged in their original or natural positions. The remains came into the possession of a retired U.S. government Indian agent named Major Benjamin O'Fallon (1793–1842) and were displayed in the formal garden of his home in St. Louis (Goldfuss, 1845:3). The specimen eventually attracted the attention of Prince Maximilian zu Wied (1782–1867) during his travels through

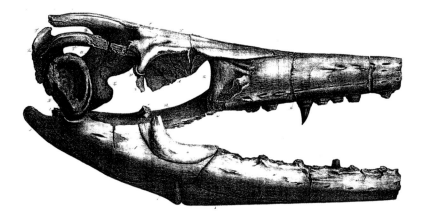

2.3. The skull of *Mosasaurus maximiliana* as published by Goldfuss (1845). Discovered in a concretion, this skull was the first articulated mosasaur skull ever collected. Note that the anterior ends of the premaxilla and both dentaries are missing from this specimen and were described earlier by Dr. Richard Harlan (1834) as the remains of his mistaken *Ichthyosaurus missouriensis* (Chapter 9).

the American West from 1832 to 1834, and he acquired the specimen. The prince shipped the specimen back to Germany, where a well-known naturalist, Dr. August Goldfuss, spent several years preparing and describing it. Although it was encased in a hard limestone concretion, the skull was preserved fully articulated and uncrushed. It was the best example of a mosasaur skull collected until that time in terms of understanding the construction of the mosasaur skull, much more useful than the disarticulated specimen of *Mosasaurus hoffmanni* from the Netherlands that was still enclosed in its limestone matrix. It was, however, missing the tips of the lower jaws and the anterior end of the premaxilla (Fig. 2.3). In what was an excellent paper that was subsequently ignored by many other early workers on mosasaurs, Goldfuss (1845; see also Goldfuss, 2013, for a recent translation) described the specimen completely and gave it the name *Mosasaurus maximiliana* in honor of his benefactor.

Russell (1967) noted that soon after the Goldfuss paper was in printed in 1845, a letter from Hermann von Meyer to "Professor Bronn," published in the German journal *Neues Jahrbuch für Mineralogie, Geognosie, Geologie und Petrefaktenkunde*, provided the first indication that Harlan's 'ichthyosaur' fragment was probably the missing premaxilla of the Goldfuss mosasaur. Although the Goldfuss skull is still in the collection of the Institut für Geologie und Paläeontologie in Bonn, Germany, Harlan's fragments were thought to be lost (Russell, 1967). In 2004, however, the missing premaxilla was rediscovered quite unexpectedly by Gordon Bell and Mike Caldwell in the National Museum of Natural History, Paris, France, where it had been safely stored for more than 150 years (see Chapter 9).

Harlan's legacy remains, however, because the name *Mosasaurus maximiliana* Goldfuss 1845 became the junior synonym of *Mosasaurus missouriensis* (Harlan, 1824). While it is unlikely that Goldfuss will ever receive the credit he deserves for his meticulous work on the first articulated skull of a mosasaur ever collected, Baur did note that "if this important paper had been studied more carefully by subsequent writers [e.g., Cope, Marsh, and others], much confusion could have been spared" (1892:2). Williston elaborated further on the subject when he said, "As

Baur has said, had later authorities studied this paper more attentively they would not have claimed as new a number of discoveries made and published long before, among which may be mentioned the position of the quadrate bone, the presence of the quadratoparietal and malar arches, and the sclerotic plates" (1895:165).

Following the American Civil War, the pace of westward expansion in the United States increased significantly. Gold had been discovered in Colorado, and Denver was growing rapidly. Communication between Kansas City and Denver was largely by a stagecoach and wagon freight line along the Butterfield Trail that ran westward across the prairie through western Kansas and eastern Colorado. At about the same time, the Union Pacific portion of the transcontinental railroad was being completed across Nebraska, and a southern route, the Kansas Pacific Railway, was being built from Kansas City to Denver. Along with the survey crews and construction workers for the railroads in this westward expansion came the first settlers. The resulting encroachment on Indian lands in the West resulted in conflicts and made it necessary for the government to establish a military presence between Kansas City and Denver. Several forts were built along the Butterfield Trail and elsewhere in the western half of Kansas. Along with the troops and guns came military doctors, who were arguably among the best-educated men in Kansas at the time. During the period between 1866 and 1872, two of these doctors were also among the first fossil collectors and paleontologists in Kansas.

Fort Harker was established originally in 1866 as Fort Ellsworth in central Kansas, near present-day Kanopolis, on the Butterfield Trail to Denver and the eventual route of the Kansas Pacific railroad. Dr. George M. Sternberg (1838–1915), an older brother of the Charles H. Sternberg who would later become famous as a fossil hunter, was the military surgeon assigned to the fort. Dr. Sternberg, who would later be better known for his work in bacteriology and as the surgeon general of the army during the Spanish-American War, began routinely collecting fossils in western Kansas before anyone else. According to Rogers (1991:130), Dr. Sternberg's younger brother Charles credited him with alerting "O. C. Marsh, Joseph Leidy and other paleontologists to the existence of Kansas's vast fossil beds, worthy of exploration. It was his brother who made possible the first placement of Sternberg fossils in the halls of the Smithsonian." Dr. Sternberg began by collecting fossil leaf impressions from the sandstone of the Dakota Formation (early Late Cretaceous) near Fort Harker, and then made significant collections of vertebrate fossils from the Smoky Hill Chalk and the Pierre Shale of western Kansas while serving with General Sheridan's military campaign against the Indians from 1868 to 1870.

Almost all of Dr. Sternberg's specimens were donated to the U. S. Army Medical Museum in Washington, D.C. From there they were transferred to the Smithsonian (United States National Museum—USNM). By my rough count during a visit in 2001, Dr. Sternberg is attributed as

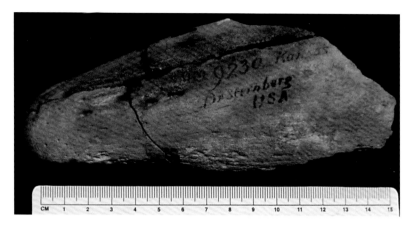

2.4. Dr. George M. Sternberg collected many fossils from western Kansas and sent them back to the U.S. Army Medical Museum (AMM) in Washington, D.C. His specimens were subsequently transferred to the United States National Museum (Smithsonian). The photo shows the premaxilla of a *Tylosaurus proriger* (UNSM 3884). The inscription reads: "AMM 9230 Kansas, Dr. Sternberg, USA."

the collector of more than 30 mosasaur specimens in the USNM collection and actually signed his name to each bone in most cases (Fig. 2.4). He also discovered the large fin ray (Fig. 2.5) that Leidy (1870) described as the type specimen for the giant Late Cretaceous fish, *Xiphactinus audax*. Cope (1872b) later more fully described and named the same fish from more complete specimens, but his *Portheus molossus* Cope 1872 name will always be the junior synonym of Dr. Sternberg's discovery of *Xiphactinus audax* Leidy 1870 (Chapter 5).

Dr. Sternberg was also one of the first collectors of fossils from the Pierre Shale in far western Kansas in 1868 and 1869. However, the specimens attributed to him in the USNM collection are poorly preserved and not useful. The first major vertebrate fossil collected from the Pierre Shale and the first major Cretaceous vertebrate to be described from Kansas was the type specimen of *Elasmosaurus platyurus* Cope 1868, which was discovered by another army surgeon in the spring of 1867. Dr. Theophilus H. Turner (1841–1869), the assistant surgeon at Fort Wallace in western Kansas, discovered the remains of a very large marine reptile eroding from a ravine in the Pierre Shale about 12 miles northeast of the fort (Almy, 1987). Later that summer, he gave three of the vertebrae to John LeConte, a member of a party that was in the process of surveying the route for the Union Pacific Railroad (LeConte, 1868). After the survey was completed, LeConte delivered the vertebrae to E. D. Cope at the Academy of Natural Sciences of Philadelphia (ANSP) in November 1867. Cope immediately recognized the bones as belonging to a large plesiosaur and wrote to Turner, asking him to procure the remainder of the specimen and send it to Philadelphia at the expense of the ANSP (Almy, 1987).

With the help of other soldiers and civilian employees at Fort Wallace, Turner returned to the site in late December 1867 and secured some 900 pounds of bones and concretions. Near the end of February 1868, at the urging of Cope, Turner arranged to transport the specimen by military wagon train some 90 miles east to where the approaching Kansas Pacific Railway was being built. From there, the remains were shipped by rail to Philadelphia. Cope received the crates containing the specimen

2.5. Leidy's (1873:pl. 17) figure of the large fin ray (USNM 52) discovered by Dr. George M. Sternberg in western Kansas. It was from this specimen that Leidy (1870) named the giant teleost fish *Xiphactinus audax*. The fin ray is shown in dorsal and ventral view and is approximately 40 cm (16 in) in length.

Xiphactinus audax - **Pectoral fin ray**
Type specimen - USNM V-52

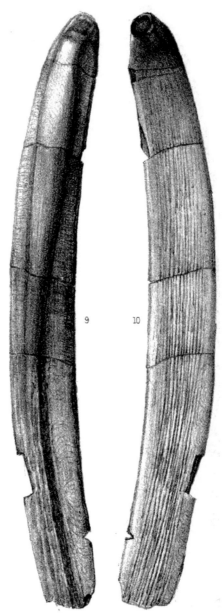

Adapted from Leidy 1873 (Plate XVII)

in March, examined the remains, and, as would soon become the custom in his rivalry with O. C. Marsh, hurriedly described and named the specimen. At the March 24, 1868, meeting of the ANSP, Cope (1868a:92) reported the discovery "of an animal related to the *Plesiosaurus*" which he called *Elasmosaurus platyurus*. At about the same time, a short note from Cope (1868b), also including the new name, was published in LeConte's (1868) railroad survey report.

2.6. A recent photograph of a dorsal vertebra of the giant plesiosaur *Elasmosaurus platyurus* (ANSP 10081), discovered and collected in western Kansas by Dr. Theophilus Turner in 1867, and subsequently figured by Cope (1869b:pl. II). The centrum of this vertebra is about 12 cm (5 in) across.

The controversy that followed regarding the restoration of *Elasmosaurus* with the head on the wrong end (Cope, 1869b; Chapter 7) completely overshadowed the fact that Dr. Turner had discovered and successfully collected one of the largest vertebrate fossils known at the time, under primitive conditions, and with no prior experience (Fig. 2.6). Although he was formally thanked by Cope (1869) at the December 15, 1868, meeting of the ANSP, Turner certainly deserves more recognition for this feat than he has so far received.

The next major fossil to be reported from Kansas was a partial mosasaur skull discovered in the Smoky Hill Chalk in western Gove County. Williston reported that the partial skull of the type specimen of *"Tylosaurus proriger* (Cope 1869) was collected by Colonel Cunningham [*sic*]

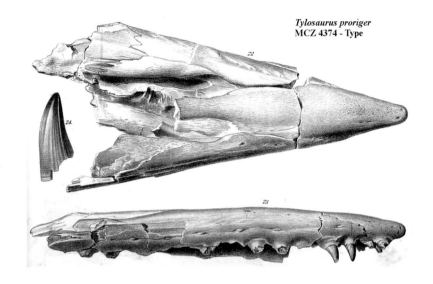

2.7. Detail of Plate XII from Cope (1870), showing the anterior portion of the skull of the type specimen of *Tylosaurus proriger* (MCZ 4374) in dorsal (fig. 22) and right lateral view (fig. 23), and a single tooth (fig. 24) recovered with the remains collected by "Col. Connyngham [*sic*] and Mr. Minor" (Cope, 1875:166). According to Cope, the skull fragment is about 78 cm (31 in) in length (Everhart, 2016).

Tylosaurus proriger
MCZ 4374 - Type

and Mr. Minor in the vicinity of Monument station [Gove County], and sent by them to Prof. Louis Agassiz (Fig. 2.7). The locality is probably Monument Station of the overland route, in the vicinity of Monument Rocks, in the valley of the Smoky Hill River" (1898a:28). This story is only partially true, and there is much more to it, which I will tell in Chapter 9. It was, however, William E. Webb, the land manager for the National Land Company, working for the Kansas Pacific Railway, who actually acquired the specimen and sold it to Agassiz at the Harvard Museum of Comparative Zoology (Everhart, 2016).

After becoming the 34th state in 1861, Kansas established the Kansas State Agricultural College (now Kansas State University) at Manhattan in 1863. It was there that Professor Benjamin F. Mudge (1817–1879) began the first systematic collection of fossils from the Western Interior Sea. Professor Mudge (1866a) reported on fossil footprints he had collected in 1865 from the Dakota Sandstone (early Late Cretaceous) 50 miles north of Junction City, Kansas. Mudge was also the first to note and publish (1866b) the presence of fossil leaves in the Dakota Sandstone, although at the time he was unsure of their age or geological provenance (Fig. 2.8). In a footnote, LeConte (1868:7) reported the presence of the leaf imprints and indicated that "Professor Mudge, of the Kansas Geological Survey, has procured specimens from the same locality." Lesquereux (1868:91) credited B. F. Mudge for a collection of leaf fossils from the Dakota Formation that he examined in the Smithsonian, but Mudge actually collected the fossil leaves from just north and east of Fort Harker in Kansas, not from near "Fort Ellsworth in Nebraska" as Lesquereux wrote.

According to S. W. Williston, who was a student of Mudge's in the late 1860s and early 1870s, the "first expedition, as I remember, was up the Republican and Solomon rivers into the wholly uninhabited region, the home then of the bison and roving bands of marauding Indians. It was made shortly after the close of the college year in 1870" (1898a:28). Mudge sent many of his fossils to E. D. Cope in Philadelphia, giving Cope the

2.8. Two partial leaf imprints preserved on a piece of sandstone from the Dakota Formation, Ellsworth County, Kansas. In life, the leaves would have been long and narrow, similar to, but not related to, the modern willow tree (*Salix* sp.). Scale bar = 1 cm.

opportunity to name most of the mosasaur species from the Smoky Hill Chalk even though it was O. C. Marsh who actually collected specimens in Kansas first. However, after being fired from his teaching job at the agricultural college in 1873 over a dispute with the administration, Mudge began collecting fossils for O. C. Marsh and Yale College. His discoveries include a toothed bird (*Ichthyornis dispar*), several species of mosasaurs, and other important specimens that are now in major museums in the United States, including at least 303 specimens credited to him in the Yale Peabody Museum collection.

Peterson (1987:228) characterized B. F. Mudge as the "most active" of the early fossil collectors in Kansas. In 1866, Mudge went even further west to the area around Ellsworth (near Fort Harker) and north to near the fork of the Solomon River to collect invertebrates and plant fossils. Peterson said that

> only Mudge is known to have done much fossil collecting in 1869, and he did not venture very far into western Kansas. During the summer he went up the Republican River as far as the northern state line where he found many fossil plants and other specimens to add to the KSAC [Kansas State Agricultural College] collection. In October, he accompanied Kansas Senator Edmund Ross and two others on an expedition, with a military escort, up the Solomon River to visit troops stationed in the area and to determine the valley's potential for agricultural and railroad purposes. Mudge concentrated on soils and geology, including fossils, but also noted evidence of ancient Indian occupation. In what is now Phillips County, Mudge found a number of fossils in Cretaceous formations

including the vertebrae and other portions of an eight-foot saurian. Returning down the South Solomon River, Mudge noted much exposed magnesium limestone but found fewer fossils. Although the results of this trip were not spectacular, it was important for introducing Mudge to the large, well preserved fossils of new species awaiting discovery in western Kansas. (Ibid.:228–229)

In 1870, following completion of the Kansas Pacific Railway to Denver, Mudge and his partner, Joseph Savage of Lawrence, collected fossils from the Cretaceous formations around Fort Wallace; the vertebrates he forwarded to Cope in Philadelphia and the invertebrates he sent to F. B. Meek at the Smithsonian (Peterson, 1987:229–230). Savage reported by letters to newspapers in Lawrence on their collecting in western Kansas. One of his notes, published in the *Lawrence Daily Journal* (December 4, 1870, 1) tells of being close to Marsh and the first Yale College Scientific Expedition: "Prof. Marsh, of Yale College, was camped three miles below us on the Smoky [Hill River], with a scout from Fort Wallace. He has been out all summer on the Union Pacific road, and started in the spring with twelve students. He has only three left with him at present. We found his tracks on our entire route to-day down the North branch [of the Smoky Hill River]." Both Mudge and Savage would later donate fossils to Marsh at Yale.

E. D. Cope's relationship with Mudge was further solidified in late 1871 when Cope visited him at the Kansas State Agricultural College and examined Mudge's collection of mosasaur and fish specimens (Cope, 1872a). Cope (1871:405; later figured in Cope, 1875:pl. 26, fig. 3) also named a new species of mosasaur (*Liodon mudgei*) in Mudge's honor "in recognition of the valuable results of his investigations as State geologist of Kansas." In his review of American mosasaurs, however, Russell (1967:181–182) considered *L. mudgei* to be "*Platecarpus nomen vanum*, either *P. ictericus* or *P. coryphaeus*."

Mudge and Savage continued to collect together in 1871. In his letter published in the *Lawrence Daily Journal* (August 1, 1871), Savage reported on the discovery of two associated specimens that they collected from the Pierre Shale near the railroad town site of Sheridan, Kansas:

> July 22, 1871. We had a very successful day hunting fossils around this place yesterday. . . . We also found the remains of two saurians, in such position as to show without a doubt that they died in the act of one swallowing the other. Their bones were all in place, except that the jaws and teeth were gone, . . . and were nearly in a straight line with each other, the tails pointing in opposite directions. The neck vertebrae overlapped each other a little, and one passed about two inches under the other. They were of the species of mososaurians [*sic*], and were probably about 15 feet long when living, but very nearly the same size. I secured for our university the one whose eyes were too big for his throat, and gave Professor Mudge the one which came so near being food for his fellow. (Ibid.:2)

Cope (1872e) borrowed the specimens from the University of Kansas collection and described a new species of plesiosaur (*Plesiosaurus gulo*) from the remains of the larger individual (KUVP 1329). Originally the specimen was fairly complete, comprising 34 cervical and dorsal vertebrae, parts of the pectoral and pelvic girdles, and ribs. However, based on Cope's measurements, the string of vertebrae would only have been about 2 m (6.5 ft) long, indicating a small individual. Cope noted that the smaller specimen was a mosasaur similar to *Clidastes*. Unfortunately, the specimen was not returned to the University of Kansas. A handwritten note in the storage drawer indicates: "Ichthyosaurus examined by Cope and this is what's left."

According to Peterson:

> [Mudge] went to western Kansas where he found many more vertebrates. Although he kept specimens for the KSAC cabinet [collection], he sent the best items to experts in the East for identification and publication. The molluscs [*sic*] were sent to Meek, most of the plant material to Lesquereux, and most of the vertebrates to Cope. Louis Agassiz at Harvard and James D. Dana at Yale also received items. Many of Mudge's finds were published by Cope and Lesquereux in the yearly report of the U. S. Geological Survey of the Territories. His fossil plants of 1871 included seven new species, including one species of oak that Lesquereux named *Quercus mudgeii*. [See also Preliminary Report of the United States Geological Survey of Montana and Portions of Adjacent Territories Washington, 1872:301–304.]
>
> The collectors had another very productive year in 1872. Professor Mudge again ventured north of the Smoky Hill River. After stopping near Hays, where he found fossil shells and fish, he went north to Smith County, where he met the rest of his party: Prof. G. C. Merrill of Washburn [University in Topeka, Kansas]; Prof. P. H. Felker of Michigan Agricultural College; R. Warder of the Indiana Geological Survey; and seven KSAC students. They explored the geology of the valleys of Prairie Dog Creek and several branches of the Solomon River and found many vertebrate fossils. Later in the year, Mudge spent two weeks examining the geology of the Arkansas River valley that became a subject for a paper he presented before the Kansas Academy of Science. In the Fall, he had visits from both Marsh and Cope. It appears that he gave most of his saurian fossils to Cope, who found fourteen new species among them. The rarer birdlike fossils he passed to Marsh. In July a third visitor was Leo Lesquereux, who spotted a number of new species in Mudge's plant material. Lesquereux spent the summer examining plant fossils and sites in the West. On his way to the Rocky Mountains, he stopped to collect fossil leaves at Fort Harker where he met Charles Sternberg who thereafter sent all his plant material to Lesquereux. (1987:230–231)

Leidy (1868) made a brief report to the Academy of Natural Sciences of Philadelphia regarding the discovery and photographs of a huge Kansas mosasaur. The recorded remarks (ibid.:316) are brief but enigmatic: "Dr. Leidy exhibited some photographs of fossil bones, received from Mr. W. E. Webb, Sec. [Secretary] of the National Land Co., at Topeka, Kansas. They represent vertebrae, and fragments of the jaws with teeth of a skeleton of *Mosasaurus*, reported by Mr. Webb to be about 70 feet in length, recently discovered on the great plains of Kansas, near Fort

Wallace." Stories regarding a huge but unverified mosasaur from western Kansas are recorded in various other publications, including many newspapers in Kansas and Webb's (1872) book *Buffalo Land*, a semifictional account of early explorations of the plains, and by Cope (1872c:333, 1872d:279).

The specimen is actually the type specimen of *Tylosaurus proriger* first described by Cope (1869a) (and figured in Cope, 1870) from a specimen (MCZ 4374) in the collection of the Museum of Comparative Zoology. Supposedly the remains were obtained by Prof. Agassiz during a congressional fact-finding trip to western Kansas in 1868 (see Almy, 1987:193; Everhart, 2016). Cope (1869a:123) stated that the type specimen "belonged to Prof. Agassiz" and then noted (1875:161) that the original description "was based on material in the Museum of Comparative Zoology, Cambridge, Mass., brought by Prof. Agassiz from the Cretaceous beds in the neighborhood of Monument, Kans. and near the line of the Kansas Pacific Railroad." The present-day location of Monument, Kansas, is about 35 miles north-northwest of the original Butterfield Despatch stage line station near Monument Rocks, which is near the Smoky Hill River. (Again, I'll explain this complicated story in greater detail in Chapter 9.)

When the stage line company went out of business in the late 1860s, the civilian employees at the station literally picked up and moved some of the small buildings to the original townsite of Antelope (now Monument, Kansas) on the railroad. The small U.S. Army garrison at Monument Station, commanded by Captain Conyngham (Wetzel, 1960), also marched to the town of Monument. There are no Late Cretaceous marine fossils to be seen in or near present-day Monument, Kansas, which is surrounded by miles of relatively flat prairie used today for raising wheat and corn.

An August 13, 1872, newspaper description of the town of Monument (Topeka Weekly Leader:2) does not paint a pretty picture of the living conditions in western Kansas at the time: "There are twenty cloth houses at Monument, besides the quarters tents [U.S. Army] . . . four saloons, two fruit stands, one boarding house and news depot and post office. At Monument is a level plain as far as the eye can reach on either side. Twelve miles westward is Sheridan, and yet seventeen further and you reach Fort Wallace, where Harrison Nichols keeps hotel. We'd soon be in mid ocean in a wash-tub as in the centre of these plains on foot."

While Kansas has often been suggested as the place where the "Bone Wars" between E. D. Cope and O. C. Marsh began, it is more likely that their rivalry started over access to Cretaceous fossils being recovered from marl pits in New Jersey in the late 1860s. The competition and other confrontations between the two men certainly intensified in Kansas during the early 1870s, and then reached their peak farther west and north in Colorado and Nebraska with the discovery of dinosaurs and giant Pleistocene mammals. Marsh was the first to actually collect fossils in Kansas, although by the time he got there, Cope had already been the recipient of many specimens, including the giant plesiosaur *Elasmosaurus platyurus* (Chapter 7) and the type specimen of *Tylosaurus proriger* (Chapter 9), from various amateur collectors in Kansas.

Marsh arrived in November 1870 with his first Yale College Scientific Expedition on the way back from San Francisco, and collected for two weeks near Fort Wallace. He returned to Kansas with another expedition composed of Yale students in the summer of 1871 and again collected many fossils from the chalk. Cope's only visit to Kansas to collect fossils was in the fall of 1871. (Historical trivia: According to the *Lawrence Daily Journal* (November 18, 1886:3; see also Osborn, 1931:318 for details) Cope was also in Kansas on the previous day, November 17, as the keynote speaker at the dedication of the Snow Hall science building at the University of Kansas. The *Lawrence Daily Journal* (August 17, 1893:4) reported that Cope had been the guest of Chancellor Snow at the University of Kansas. Then in late August, on his return trip from fossil collecting in Texas and Oklahoma, Cope (1894:67–68) wrote that he had stopped briefly in Wellington, Kansas, in 1893 to examine Pleistocene elephant remains collected from a nearby sand pit.) For more details on Cope's last trip to Kansas, see Davidson (2016).

Although Marsh returned again in 1872, both men would subsequently hire others (Fig. 2.9) to collect for them on more or less a full-time basis. Marsh hired Professor B. F. Mudge to collect for him in Kansas in 1873. Mudge then recruited several of his students, including 22-year-old Samuel W. Williston, to help him in the field. After the death of Mudge in 1879, Williston went to work for Marsh at Yale for several years and then returned to the University of Kansas in 1890. Once he was back in Kansas, Williston began to collect extensively for the university and established a paleontology program that produced many well-recognized paleontologists, including Barnum Brown and Elmer Riggs.

Cope, on the other hand, hired a young, relatively inexperienced Charles H. Sternberg in 1876 as his man in Kansas. While initially at a disadvantage due to the superior knowledge and experience of Marsh's collectors, Sternberg would become a well-recognized supplier of fossils for museums all over the world. His sons, George F., Levi, and Charles M. Sternberg would later go on to careers in paleontology of their own, with George eventually returning to Kansas to continue the exploration of the Cretaceous and Tertiary deposits in the state.

After setting up his fossil business initially in a vacant school building in Oakley, Kansas, George Sternberg (1883–1969) was hired by the Kansas State Teachers College (now Fort Hays State University) in 1927. He worked out an arrangement with the college to serve as museum director for the school term (nine months) and then collect for himself during the summer months. He would work for the college and be a part-time commercial collector for the rest of his life. The museum would grow and eventually become the Sternberg Museum of Natural History (Rogers, 1991; Liggett, 2001).

2.9. Many people collected fossils in western Kansas, starting about 1866 with Dr. George M. Sternberg. Here are some of the more notable ones (left to right, top row): Edward D. Cope (1840–1897); Othniel C. Marsh (1831–1899); Benjamin F. Mudge (1817–1879); (left to right, bottom row): Charles H. Sternberg (1850–1943); Samuel W. Williston (1851–1918), and George F. Sternberg (1883–1969).

There were many other 'collectors' who made significant contributions to our knowledge of the paleontology of western Kansas, but whose names are relatively unknown. I will credit people like Judge E. P. West, Joseph Savage, Handel T. Martin, William E. Webb, Marion Bonner, and others for their discoveries in the following chapters of this book.

Stratigraphy

One of the important issues that will surface repeatedly in this book is the lack of stratigraphic information on the occurrence of many of the fossils that have been collected from the Smoky Hill Chalk in the past. In this usage, 'stratigraphic occurrence' refers to what approximate chronological level the fossil was discovered within the 200 m (650 ft) thick chalk unit (Fig. 2.10). As noted earlier, this chalk was deposited in the Western Interior Sea over a period of about 5 million years. Fossils located near the bottom of the chalk are as much as 5 million years older than those near the top. Knowing the ages of specimens is useful in understanding when species appeared and when they became extinct or were no longer present, and what ecological relationships may have existed.

Relatively few of the fossil remains collected since 1868, including important type specimens, have even good locality data, let alone stratigraphic information. This means that we cannot establish when the animal lived in relation to other remains that we find in the chalk, which limits the usefulness of any fossil in the study of the ecosystem of the Western Interior Sea. Even though an accurate frame of reference for locating fossils within the 5-million-year depositional period of the chalk

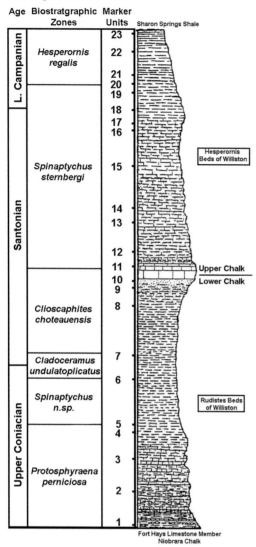

Generalized Stratigraphy of the Smoky Hill Chalk in Western Kansas

Age	Biostratigraphic Zones	Marker Units	
			Sharon Springs Shale

L. Campanian — *Hesperornis regalis* — 23, 22, 21, 20, 19, 18, 17, 16

Santonian — *Spinaptychus sternbergi* — 15, 14, 13, 12, 11, 10, 9, 8

Hesperornis Beds of Williston

Upper Chalk
Lower Chalk

Clioscaphites choteauensis

Cladoceramus undulatoplicatus — 7, 6

Upper Coniacian — *Spinaptychus n.sp.* — 5, 4

Rudistes Beds of Williston

Protosphyraena perniciosa — 3, 2, 1

Fort Hays Limestone Member
Niobrara Chalk

2.10. A generalized stratigraphic column for the Smoky Hill Chalk Member of the Niobrara Formation in western Kansas, including established geological ages, biostratigraphic zones from Stewart (1990) and marker units described by Hattin (1982). The Smoky Hill Chalk is roughly 200 m (650 ft) thick and was deposited during a 5 million year period in the Late Cretaceous (87–82 Ma).

has been in place since Hattin (1982), few people were aware of it and even fewer used it in the field.

The geology of western Kansas was not well understood when most of the early collecting was done. According to Zakrzewski (1996), geological studies of the western interior of the United States had begun as early as the 1850s, but had proceeded slowly and sporadically. Meek and Hayden (1861) first referred to the chalk and limestone strata as the Niobrara Division in their description of exposures along the Missouri River near the mouth of the Niobrara River in Nebraska. In Kansas, these Upper Cretaceous strata were referred to simply as the "Niobrara" by the geologists and paleontologists of the day (Hattin, 1982). E. D. Cope (1872a) wrote the earliest substantial account of vertebrate fossils from the

"Niobrara Beds." Though he described the geology of the chalk and each of the various species in some detail, he made no attempt to delineate their stratigraphic occurrence within the formation.

In 1889, the University Geological Survey of Kansas, which eventually became the Kansas Geological Survey (Buchanan, 1989), was established by the state legislature. Samuel W. Williston was appointed to the faculty of the University of Kansas the following year. Williston, who had been a student of B. F. Mudge and worked for O. C. Marsh for at least a decade, was well qualified to oversee the continuing effort to understand the geology and paleontology of Kansas. These two events provided the basis for much of the early progress in these sciences in Kansas.

At the time, even reaching an agreement on what to name the formations that cropped out in the western part of the state was no easy matter. Most of the early terminology used to describe the Smoky Hill Chalk was based on the occurrence of the predominant fossils and added little stratigraphic information to individual specimens. The 'Niobrara Division' of Meek and Hayden (1861) was made up of two distinct units, a lower limestone member and an upper chalk member. Logan (1897) was the first to call the upper chalk member the 'Pteranodon Beds,' in apparent recognition of the abundance of well-preserved *Pteranodon* remains that had already been discovered there (Chapter 10). That same year, demonstrating his support for Logan's descriptive terminology, Williston (1897) further divided the Pteranodon Beds into the lower Rudistes Beds and the upper Hesperornis Beds, providing essentially the first biostratigraphic subdivisions of what was to become the Smoky Hill Chalk.

Williston (1897) briefly discussed the stratigraphic occurrence of mosasaurs in the Pteranodon Beds for the first time. He also made the observation that *Clidastes* does not occur in the lower Rudistes Beds, indicating that other genera (*Platecarpus* and *Tylosaurus*) probably occurred within 100 feet of the contact of the chalk with the underlying Fort Hays Limestone. Williston (ibid.:245) was certainly aware of the lack of good stratigraphic data for Niobrara vertebrate fossils when he wrote, "I need not call the attention of future collectors to the importance of locating the horizon of specimens more accurately than has been done heretofore."

Williston published the first comprehensive description of the systematics and comparative anatomy of mosasaurs from the Smoky Hill Chalk, and also discussed their range and distribution in comparison with specimens discovered earlier in New Zealand and Europe. He commented that *Tylosaurus*, "so far as was known, begins near the lower part of the Niobrara [Smoky Hill Chalk] and terminates at its close or in the beginning of the Fort Pierre [Pierre Shale]" (1898b:88). Of *Platecarpus*, he stated that the species on which the genus is based are "known nowhere outside of Kansas and Colorado, and are here restricted exclusively to the Niobrara" (ibid.) He again concluded that the lowest horizon of *Clidastes* "is the upper part of the Niobrara in Kansas" (ibid.:89). It was not until after the turn of the twentieth century that the exploration for oil and gas in western Kansas enabled rapid advances in understanding the geology of the entire Niobrara Formation. According to Hattin (1982), Moore and Haynes (1917) were the first to regard the Kansas Niobrara as a formation and the first to give member status to the currently recognized divisions, the Fort Hays Limestone and the Smoky Hill Chalk.

Russell (1967) reviewed mosasaur specimens in the Yale Peabody collection and suggested that the Smoky Hill Chalk could be divided into a lower, *Clidastes liodontus—Platecarpus coryphaeus—Tylosaurus nepaeolicus* zone and an upper, *Clidastes propython—Platecarpus ictericus—Tylosaurus proriger* zone. He also suggested that the increased abundance of *Clidastes* specimens in the upper portion of the chalk was an indication of a gradual change from a mid-ocean to a nearshore environment. Russell (1970) noted significant differences between the distribution of mosasaur species in the Smoky Hill Chalk and that of the Gulf Coast species occurring in the Selma Formation of Alabama. In his initial paper concerning the biostratigraphy of the Smoky Hill Chalk, Stewart (1988:81) stated that he was aware of several exceptions to Russell's stratigraphic distribution of mosasaurs in the Smoky Hill Chalk that caused him to regard it with "a degree of skepticism."

It was only after Hattin (1982) published his composite measured section of the Smoky Hill Chalk that significant progress could be made in understanding the vertebrate biostratigraphy of this formation. Hattin used bentonites and other geological features to delineate his 23 lithologic marker units, and he divided the chalk into five biostratigraphic zones based on the occurrence of invertebrate species. In doing so, he provided field workers with the first dependable method of determining their stratigraphic location in the section.

Stewart (1988) initially used the distribution of the primitive swordfish genus *Protosphyraena* to further delineate biostratigraphic zones in the Smoky Hill Chalk. Stewart (1990) then incorporated Hattin's marker units as upper and lower boundaries for his six proposed biostratigraphic zones (Table 2.1). This report provided the first comprehensive description of the distribution of known invertebrate and vertebrate species in the Niobrara Formation and was the first attempt to assign specific stratigraphic ranges for mosasaur species within the Smoky Hill Chalk. Even with the substantial improvements over previous attempts, Stewart believed that his biostratigraphy was flawed by the lack of reliable stratigraphic data for thousands of specimens collected in the 1870s, and even those specimens collected during the previous 20 years. He stated that his framework was "submitted in the hopes that other researchers will test it and improve upon it." More recently, publications on the stratigraphic occurrence of mosasaurs in the Smoky Hill Chalk (Schumacher, 1993; Sheldon, 1996; Everhart 2001) have benefited from the framework provided by Hattin (1982) and Stewart (1990). Most of the other papers published since 1990 regarding newly discovered specimens from the chalk have generally included accurate stratigraphic information.

Age	Marker Units	Ma	Biostratigraphic zone of:
Early Campanian	MU 16 to MU 23	83–82	*Hesperornis*
Late Santonian	MU 11 to MU 16	84–83	*Spinaptychus sternbergi*
Middle Santonian	MU 8 to MU 11	85–84	*Clioscaphites vermiformis* and *C. choteauensis*
Early Santonian	MU 6 to MU 8	86–85	*Cladoceramus undulatoplicatus*
Late Coniacian	Base to MU 6	87–86	*Protosphyraena perniciosa* / *Spinaptychus* n. sp.

Even though fossils have been collected from the Smoky Hill Chalk for more than 150 years, there is still additional collecting and describing to be done. A number of new species have been discovered in the past 25 years, and probably others will be described in the future. In addition, many species that were discovered and/or named by early workers need better specimens and stratigraphic information if we are to fully understand their occurrence and significance in the Western Interior Sea.

Recommended Reading about Early Paleontologists and Kansas Fossils

Edward D. Cope (1840–1897)

Davidson, J. P. 1997. The Bone Sharp. The Academy of Natural Sciences of Philadelphia, Philadelphia, 237 pp.

Davidson, J. P. 2002. Bonehead mistakes: The background in scientific literature and illustrations from Edward Drinker Cope's first restoration of *Elasmosaurus platyurus*. Proceedings of the Academy of Natural Sciences of Philadelphia 152:215–240.

Osborn, H. F. 1931. Cope: Master Naturalist. Princeton University Press, Princeton, New Jersey, 740 pp. (Reprinted by Arno Press, New York, 1978.)

Joseph Leidy (1823–1891)

Warren, L. 1998. The Last Man Who Knew Everything. Yale University Press, New Haven, Connecticut, 320 pp.

Othniel C. Marsh (1831–1899)

Schuchert, C., and C. M. LeVene. 1940. O. C. Marsh: Pioneer in Paleontology. Yale University Press, New Haven, Connecticut, 541 pp.

Benjamin F. Mudge (1817–1879)

Page, L. E. 1994. Benjamin F. Mudge, the state geological surveys, and fossil collecting in Kansas, 1864–1870. Earth Sciences History 13(2):121–132.

Peterson, J. M. 1987. Science in Kansas: the early years, 1804–1875. Kansas History Magazine 10(3):201–240.

Williston, S. W. 1899. Prof. Benjamin F. Mudge. American Geologist 23(6):339–345.

Charles H. Sternberg (1850–1943) and George F. Sternberg (1883–1969)

Liggett, G. A. 2001. Dinosaurs to Dung Beetles: Expeditions through Time. Guide to the Sternberg Museum of Natural History. Sternberg Museum of Natural History, Hays, Kansas, 127 pp.

Rogers, K. 1991. A Dinosaur Dynasty: The Sternberg Fossil Hunters. Mountain Press Publishing Company, Missoula, Montana, 288 pp.

Sternberg, C. H. 1909. The Life of a Fossil Hunter. New York: Henry Holt and Company. (Reprint, Bloomington: Indiana University Press, 1990.)

Sternberg, C. H. 1917. Hunting Dinosaurs in the Badlands of the Red Deer River, Alberta, Canada. World Company Press, Lawrence, Kansas, 261 pp.

George Miller Sternberg (1838–1915)

Sternberg, M. L. 1920. George Miller Sternberg: A Biography. American Medical Association, Chicago, 331 pp.

Theophilus Turner (1841–1869)

Almy, K. J. 1987. Thof's dragon and the letters of Capt. Theophilus Turner, M.D., U.S. Army. Kansas History Magazine 10(3):170–200.

William E. Webb (1838–1906)

Everhart, M. J. 2016. William Edward Webb (1838–1906)—war correspondent, railroad land baron, town founder, Kansas legislator, adventurer, fossil collector, author, mining engineer. Transactions of the Kansas Academy of Science 119:179–192.

Elias P. West (1820–1892)

Everhart, M. J. 2015. Elias Putnam West (1820–1892)–lawyer, attorney general, militia commander, judge, postmaster, archaeologist, and paleontologist. Kansas Academy of Science, Transactions 118(3–4):285–294.

Samuel W. Williston (1851–1918)

Martin, L. D. 1994. S.W. Williston and the exploration of the Niobrara Chalk. Earth Sciences History 13(2):138–142.

Shor, E. N. 1971. Fossils and Flies: The Life of a Compleat Scientist—Samuel Wendell Williston, 1851–1918. University of Oklahoma Press, Norman, Oklahoma, 285 pp.

Invertebrates, Plants, and Trace Fossils

The school of minnow-sized fishes darted about, feeding on the abundant plankton clouding the moonlit water. At this time of year many species spawned, and the water swarmed with tiny larval forms. The little fishes had been fortunate not to attract any predators that night in the great expanse of the Western Interior Sea (WIS) and were sharing their feast only with two giant filter-feeding fishes that ignored them. As the sun began to rise, the school bunched closer together, moving in unison, and then began their descent into the depths and their daytime shelter.

Total darkness enveloped them, but their large eyes adapted and allowed them to see small sources of phosphorescence, including members of their school, glowing dimly around them. The little fishes gradually descended nearly 400 feet and reached the sea floor, where they could see the large, faintly phosphorescent mounds that they used for shelter. As they reached the muddy bottom they slipped through a narrow horizontal opening and entered the darkened refuge.

Sometimes when they entered a shelter, they encountered other species, including long, sinuous eel-like forms, but this time the dark cave was empty. Thin sheets of tissue moved back and forth inside the space in response to gentle pulses of current emanating from deeper in the darkness. The school of fishes settled in for the day, moving closer together in the limited space. Their eyes could make out the vague shapes of their companions and the faint phosphorescent glow from other similar shelters outside their own. This was how these little fishes in the middle of a big ocean spent their lives, feeding in the upper reaches when it was dark and relatively safe and then sheltering on the bottom during the daylight hours.

This day, however, would be fatally different. Later in the day, the gentle pulsing current slowed and then stopped. Without warning, the upper portion of the shelter closed suddenly, shutting off the dim glow from outside. Startled, the little fishes panicked and searched for a way out of their shelter. There was none. Their rapid movements quickly used up the available oxygen in the surrounding water. Soon their frantic searching slowed and then stopped. One by one they died in the darkness, entombed inside the giant clam . . . for the next 85 million years (Fig. 3.3).

Invertebrates and Other Fossils

Little of the bright sunlight above ever reached the bottom of the Western Interior Sea near its middle over the land that is now western Kansas. The depth to which enough light can penetrate the water to support photosynthesis (the photic zone) was probably variable in the WIS, but most likely not beyond the first 100–150 feet because of the abundant growth of single-celled algae nearer the surface. However, the steady rain of organic detritus from the algae, plankton, and other organisms in the water column supported an unusual abundance of life on the soft, muddy bottom of the sea. Huge, flat clams called inoceramids sometimes literally covered the bottom, nearly edge to edge in many places, stretching outward in all directions for mile after monotonous mile.

Life usually had been good for these strange shelled creatures for many thousands of years, because the sluggish circulation pattern of the shallow sea overhead brought just enough food and oxygen to the bottom to supply their meager needs. At times the oxygen levels decreased, killing most of the shelled creatures living in the nearly total darkness that enveloped the limey mud surface of the sea floor. Layer after layer of the flattened shells of previous generations were already hidden below the surface. The surrounding mud itself was made up mostly of calcitic shells from untold billions of dead microorganisms that formed the base of the food chain prospering more than a hundred feet overhead in the sunlit waters near the surface. Living in a precarious balance, invertebrate life on the bottom came and went in irregular pulses measured in hundreds of thousands of years.

The animals living here on the bottom weren't picky about their food. Most were filter feeders, continuously moving large quantities of seawater through their various feeding mechanisms to remove small bits of whatever happened to be suspended near them. Over many millions of years, inoceramid clams in the middle of the seaway had evolved into larger and larger forms, eventually reaching a maximum diameter of nearly 1.5 m (5 ft; Fig. 3.1) in Kansas. Along the western edge, in what is now Colorado, they were even larger and more elongated, up to 3 m (9 ft) in total length (Kauffman et al., 2007). Whether this increase in size provided them with more surface area for their huge gills in a low-oxygen environment or a more efficient system for filtering bits of food out of the water is unknown. What is known is that their large, flat shells at times covered the bottom; these provided the only available hard substrate for many other animals to colonize, and, especially during the Santonian when the middle portion of the chalk was deposited, they even provided shelter for schools of small fishes.

These inoceramids were often covered with a layer of smaller oysters packed edge to edge. Frequently several generations of oysters lived together, stacked on the empty shells of the previous occupants. There was little diversity among the major invertebrates on the sea bottom; the giant clams and their attached oysters made up nearly all of the biomass. Smaller epizoans, such as cirripeds, lived wherever they could find an attachment point. A few, mostly solitary, rudists lived in heavy, funnel-shaped shells set deep in the mud. Occasionally, rudists lived together in colonial masses (Fig. 3.2); they were, however, at the extreme north end of their preferred range, and their shells never occur in the Smoky Hill

3.1. The lower valve of a giant inoceramid, *Platyceramus platinus* (FHSM IP-532), in the Sternberg Museum of Natural History, Hays, Kansas. The specimen was collected in the 1960s by G. F. Sternberg. The hinge of this shell, representing both where the two valves are joined and where growth started, is at lower left. The shell measures about 1.2 m (4 ft) across its widest dimension.

Chalk in large numbers. In warmer waters closer to the equator, they were colonial reef builders, growing together edge to edge.

An observer standing on the sea floor would have initially been able to see little or nothing in the surrounding darkness. As the watcher's eyes adjusted to the darkness, tiny motes of colored light may have become visible, and appeared to float or dart around unattached to a recognizable life form. As in modern oceans, it is likely that many forms had evolved bioluminescence as a means of attracting prey or a mate in the permanent darkness. The squid and their cousins, the ammonites, baculites, and belemnites, that lived in these dark waters may have had faint, glowing markings.

During those times when oxygen levels were sufficient, schools of small fishes sheltered in and around the open shells of the larger bivalves covering the sea floor (Fig. 3.3). Rarely, there were also small, deep-bodied fishes called pycnodonts (Chapter 5) that nibbled at the variety of smaller invertebrates (epibionts) growing on the oyster-encrusted larger shells. The presence of coprolites composed of nearly 100% ground-up oyster shells suggests an as yet unknown predator that fed exclusively on these small oysters (see *Martinichthys*, Chapter 5). Occasionally, a mud-grubbing ptychodontid shark (Chapter 4) would emerge from the

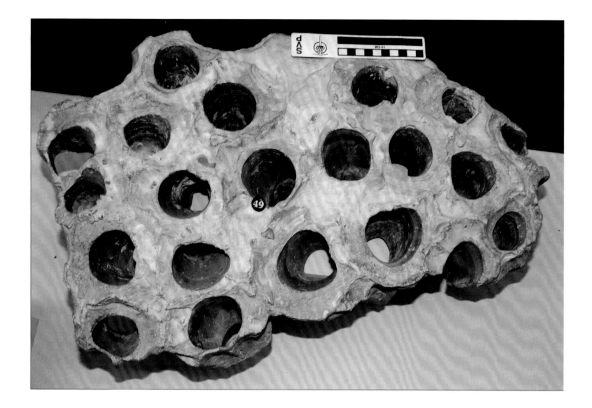

3.2. Exhibit in the Sternberg Museum showing an example of a rudist clam (*Durania maxima*) in the rare colonial form. Most rudist remains from Smoky Hill Chalk are those of individuals. Scale bar = 10 cm (4 in).

gloom to feed on the young, thinner-shelled inoceramid clams scattered among the larger ones.

The disk-shaped calcite 'scales' called coccoliths that surrounded each of the single-celled golden-brown algae (phytoplankton) forming the base of the food chain and the coccolith-containing fecal pellets (compacted waste products of animals that fed on the phytoplankton) compose the bulk of the chalk. The next most common biological remains preserved in the chalk are the shells of the larger invertebrates. In some areas, especially in the lower chalk, it is impossible to walk across an exposure without stepping on numerous fragments of giant inoceramids and oysters. Loose bits and pieces of their shells occur in layers that accumulate and blanket the surface in some areas because they are more resistant to erosion than the chalk that surrounds them. In that regard, they can easily confuse both the novice and the expert alike, appearing in odd shapes and textures that mimic the bones, and especially the jaws, of vertebrates. In the years I have collected in the chalk, I am sure I've picked up and examined thousands of shell fragments just to be sure I wasn't missing something more important.

The first Cretaceous fossils reported from western Kansas were actually not those of invertebrates or vertebrates. They were instead the imprints of the leaves of deciduous trees that commonly occur in the Dakota Sandstone north and west of Salina (Fig. 3.4). While the Dakota is now recognized as being deposited in the early part of the Late Cretaceous, its age was confusing to the early geologists who were exploring in

3.3. The partial remains of two small fishes (*Kansius sternbergi*) are preserved on the lower valve of a *Platyceramus platinus* shell from the middle Santonian chalk of southeastern Gove County, Kansas.

the wilds of Kansas and Nebraska. Hawn (1858) had initially indicated that the darkly stained Dakota sandstones were of Triassic age, but Meek and Hayden (1859) countered that suggestion with the report of the discovery of fossil leaves that came from trees unknown earlier than the Cretaceous. Their source, Dr. J. S. Newberry, an "authority on fossil botany," noted that "they include so many highly organized plants, that were there not among them several genera exclusively Cretaceous, I should be disposed to refer them to a more recent era. . . . A single glance is sufficient to satisfy any one they are not Triassic" (ibid.:33).

B. F. Mudge (1866) also noted the presence of fossil leaves in the Dakota Sandstone (Late Cretaceous) that is exposed in north-central Kansas. At the time, however, he was unsure of their age. John LeConte described the geography/geology of Kansas in his survey report for the Union Pacific Railroad and noted a bed of rocks (the Dakota Sandstone) that "continue to Fort Harker, where the sandstone becomes less ferruginous, of a reddish and pale yellow color, and contains leaves of trees of exogenous [trees that grow annual rings] growth" (1868:7). He also indicated that a collection had been made there in November of 1866 and sent to Leo Lesquereux (1806–1899) for examination.

It is quite likely that some of these leaf imprints were also collected and sent to the U.S. Army Medical Museum in Washington, D.C., by Dr. George M. Sternberg, the surgeon at Fort Harker. Sternberg collected many fossils while serving with the U.S. Army in Kansas, but kept few field notes. However, he did write that "the bluffs north and east of the fort (Fort Harker) are composed of a recent red sandstone [the Dakota Sandstone] which contains the impressions of the leaves of trees of existing species (oak, ash, willow, etc.)" (Sternberg, 1920:12).

3.4. This imprint of a fossil leaf in the dark red sandstone of the Dakota Formation was identified by Lesquereux (1868) as a variety of *Sassafras*. Recent studies, however, conclude that these trees were not closely related to the modern genus, if at all.

In a footnote, LeConte indicated that "Professor Mudge, of the Kansas Geological Survey, has procured specimens from the same locality" (1868:7). That same year, Lesquereux credited B. F. Mudge for a collection of leaf fossils from the Dakota Formation that he examined in the Smithsonian for his 1868 publication, even though he had the locality confused. The title of the paper, "On Some Cretaceous Fossil Plants from Nebraska," was misleading since the fossil leaves were actually collected by Mudge from a few miles north and east of Fort Harker (present-day Kanopolis, in Ellsworth County) in Kansas, and not from near "Fort Ellsworth in Nebraska." The collection is mentioned again by Hayden: "Species of sweet-gum, poplar, willow, birch, beech, oak, sassafras, tulip-tree, magnolia, maple and others have been described from the fossils" (1873:266). It should be noted here that current research by Wang (2002) and others indicates that these tree species are not actually related to those modern species as originally believed by Lesquereux and others of the day.

Literally thousands of these fossil leaf imprints were collected by various individuals, including retired Judge E. P. West, who collected 8,000 leaf impressions for the University of Kansas (Everhart, 2015). Large collections were made by the young Charles H. Sternberg (1909) in the early 1870s and sent to museums around the world. Sternberg's collecting localities near Fort Harker, however, are still something of a mystery, and in any case, all are certainly located on private land. Fossil leaf imprints occur occasionally in road cuts and other exposures. Years ago I visited

3.5. A ground-level view of the lower valve of *Volviceramus grandis*, a large, bowl shaped inoceramid from the Late Coniacian chalk in Trego County, Kansas. The shells of this species are very thick compared to those of the much larger *Platyceramus platinus*. Scale bar = 10 cm (4 in).

one locality west of Salina where in a rocky pasture just about every piece of sandstone I picked up on the surface had an impression of at least one leaf.

LeConte (1868:7–8) was certainly one of the first to mention invertebrates from western Kansas, listing fourteen 'new' species of shells from the Kiowa Shale (?) near Salina as he started his westward survey for the Union Pacific Railroad. Then from the "clay and limestone," probably the Greenhorn Formation, near present-day Bunker Hill in Russell County, he discovered "a Belemnite not described and *Inoceramus problematicus*," along with shark teeth and vertebrae. Further west, he noted a thin bed of oysters (*Pseudoperna congesta*) and "many teeth of fish and sharks." Several days later, still traveling west, he collected fragments of "gigantic *Inocerami*" in the lower Smoky Hill Chalk.

In his *Notes on the Tertiary and Cretaceous Periods of Kansas*, Mudge mentions that the fossils of the Fort Hays Limestone include "*Inocerami*, fragments of *Haploscapha* [*Volviceramus grandis*, Fig. 3.5], *Ostrea*, with occasional remains of fish and Saurians" (1876:214). Of mollusks in the "Niobrara proper," he noted that the "most common are *Ostrea congesta* and *Inoceramus problematicus*. Less common but still seen in many strata, are the fragments of the large *Haploscapha* [*Platyceramus platinus*], with occasionally a perfect specimen from 30 to 33 inches in length. It is thin,

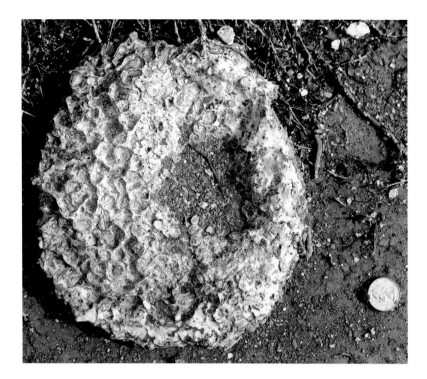

3.6. Upper valve of *Volviceramus platinus* from the Late Coniacian chalk in Trego County, Kansas, covered edge to edge with *Pseudoperna congesta* oysters. At least two layers of oysters are present. It is easy to see where the species name, *congesta*, was derived. Scale = coin = 2.5 cm (1 in).

with a transverse fiber like the *Inocerami*, and always lies crushed flat in numerous fragments, but lying in their normal position. A few *Gryphea*; also fragments, frequently weighing ten pounds or more of a large *Hippurites* near *H. Toncasianus* [the rudist, *Durania maxima*]. Near Sheridan, we discovered a bed of *Baculites ovatus*" (ibid.:216).

According to Williston (1897), the *Baculites* seen by Mudge would have been from the Pierre Shale Formation, a geological division that had not been identified in western Kansas during Mudge's time. Elias (1931) and Gill, Cobban, and Schultz (1972) provide the most recent information regarding the occurrence of invertebrates in the Pierre Shale of western Kansas. More recently, limestone concretions containing large numbers of *Baculites maclearni* have been collected from an exposure of the Sharon Springs Member of the Pierre Shale near McAllaster Butte in northwestern Logan County (pers. obs., 1997; Turner et al., 2001). They also occur fairly commonly in the layer of septarian concretions near the top of the Sharon Springs Member.

Williston (1897) noted that the small oyster, *Ostrea congesta*, was much more abundant in the 'Rudistes Beds' (lower chalk) than in the overlying 'Hesperornis Beds' (upper chalk). It is worth noting here that *Ostrea congesta* (Fig. 3.6), the species of oysters mentioned in several early accounts, was described and named by T. A. Conrad (1803–1877). At the time, Conrad was recognized as an expert on modern and fossil shells and had published numerous articles on modern freshwater species from the northeastern United States. The Cretaceous oysters and other specimens were collected during a survey of the Missouri River by J. N. Nicollet, a French mapmaker working for the U.S. government.

The name was published as a footnote: "*Conrad's description of the *ostrea congesta*: Elongated; upper valve flat; lower valve venticose, irregular; the umbo truncated by a mark of adhesion; resembles a little *gryphea vomer* of Morton" (Nicollet, 1843:169). *Ostrea congesta* was placed in the genus *Pseudoperna* by Stenzel (1971). While the description and publication of the name are not up to current standards, they are accepted. The correct citation should be listed as *Pseudoperna congesta* (Conrad in Nicollet, 1843).

Williston wrote, as one of the first notes on biostratigraphic changes within the Smoky Hill Chalk, that while "several species of *Inoceramus* are found in all horizons, the Haploscaphas (*Volviceramus grandis*) are abundant only in the lower horizons" (1897:241). J. D. Stewart (1990a; pers. comm., 1992) indicated that *V. grandis* becomes extinct in Kansas just below Hattin's (1982) marker unit 6. This observation has been verified by the author several times at various localities over the years, and it is useful as a stratigraphic marker. If you cannot find pieces of the thick, heavy shells of *V. grandis*, you are almost certainly above marker unit 6. As you move above Hattin's (1982) marker unit 3, however, the shells are not as large and are less likely to be preserved intact (pers. obs.), suggesting that some kind of change in the environment was occurring. Williston goes on to say that on the Smoky Hill River in Trego County, near the mouth of Hackberry Creek,

> there are places where these shells can be gathered by the wagon load, often distorted, but not rarely in extraordinary perfection. A very thin shelled Inoceramid [*Platyceramus platinus*] measuring in the largest specimens forty-four by forty-six or eight inches is not rare over a large part of the exposures. Invariably where exposed, as they sometimes are in their entirety on low flat mounds of shale, they are broken into innumerable pieces (Fig. 3.7). For that reason, I have never known of one being collected complete or even partially complete. Not withstanding their great size, the shell substance is not more than an eighth of an inch in thickness. Fragments of Rudistes [*Durania maxima*] are not rare in some places in the lower- most horizons and I have seen specimens near the Saline River northwest of Fort Hays into which one could thrust his arm to the elbow. They are totally wanting in the [upper] Hesperornis beds. (1897:241)

Williston's biostratigraphic observations regarding the occurrence of invertebrates are particularly interesting because they are quite accurate even after a hundred years and have served as the basis for more recent work by Stewart (1990a), Everhart (2001, 2003), and others. Thin-shelled *Platyceramus platinus* bivalves occur throughout the Smoky Hill Chalk, reaching the 1.2 m (48 in) size noted by Williston and other early collectors at about the midpoint of the formation (middle Santonian). The smaller-diameter (60 cm [24 in] maximum) but much more robust *Volviceramus grandis* became extinct in the lower chalk (late Coniacian). Another large but unusual inoceramid with a thin, rippled shell that is often referred to as the 'snowshoe clam' for its odd wavy shape, *Cladoceramus*

3.7. Remains of a large *Platyceramus platinus* shell covered with *Pseudoperna congesta* oysters. Scale = 10 cm (4 in).

undulatoplicatus, is limited to a narrow zone (early Santonian) in the chalk. It occurs for a brief (geologically speaking, that is—probably no more than a couple hundred thousand years) time after the extinction of V. *grandis*. Its occurrence, however, is used as a stratigraphic marker for the beginning of Santonian-age deposits (see Blaire and Watkins, 2009) around the world. Inoceramids are much more abundant in the lower half of the chalk, suggesting that bottom conditions were more favorable for them prior to the late Santonian and early Campanian. While the presence of these invertebrates indicates that minimal conditions for their survival were present, the very rare occurrence of small fishes preserved inside these large clams suggests that oxygen levels were generally not favorable for vertebrates (nor other invertebrates) to survive on the sea bottom (Stewart, 1990b, 1990c). Rudists are represented in the chalk by a single species, *Durania maxima*. They apparently were at the northern edge of their distribution during the Late Cretaceous and were not very successful in colonizing the mud bottom of the Western Interior Sea in Kansas. From what we are able to discover, it appears they lived with most of their shell buried in the mud, with only the expanded surface (collar) of the lower shell and a small upper shell visible on the surface (Hattin, 1988). In the chalk they mostly occur as cone-shaped masses that represent a single bivalve, but occasionally they lived cemented together in groups that look much like a giant honeycomb (Fig. 3.8). Although they are an unusually modified clam, in the warmer seas to the south and around the world they were colonial reef builders, similar in that regard to modern corals. Stewart (1990a) noted that they occur through the lower one-half of the Smoky Hill (late Coniacian through middle Santonian), then disappear for most of the upper chalk, then reappear in small numbers at the top of the formation. My personal observation has been that they become very rare well before the middle of the chalk (around marker

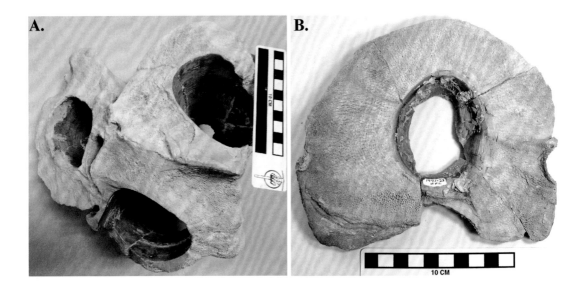

3.8. Examples of the rudist *Durania maxima* from the lower Smoky Hill Chalk. A, a colony of three individual clams; B, the collar of the lower valve of a solitary *D. maxima.*

unit 6). I have also collected a fragment of a *Durania* lower valve (FHSM IP-1518) just below the contact with the Sharon Springs Member of the Pierre Shale. The invertebrate collection at the Sternberg Museum of Natural History includes several large *Durania* specimens.

Hattin (1988) noted what he believed to be evidence of predation on *Durania*, possibly by something like the shell-crushing shark *Ptychodus* (Chapter 4). I have never observed "bite marks" or anything similar in the field, but I have noted that the thick-walled, heavy shells of *Durania* are seldom collected intact. Most of the specimens observed in the field are fragments of larger shells that were broken up before being preserved in the chalk. It is difficult for me to imagine something living in the Western Interior Sea that was powerful enough (and hungry enough) to crush a bivalve that has a shell 5 cm (2 in) thick. It may well be that the larger ptychodontid sharks, such as a 7 m (23 ft) *Ptychodus mortoni* (Shimada et al., 2009; Chapter 4) could do just that to get at the body of the clam inside. That is one of the mysteries that I hope to resolve in the future.

Concerning cephalopods, Williston (1897) noted that the shells of ammonites occur only rarely in the chalk and that usually only their impressions remain. He recalled seeing impressions of a shell about 0.3 m (1 ft) in diameter. This is consistent with my observations over the years. Outside of the small collection in the Sternberg Museum, I've seen only two fragmentary impressions of ammonite shells (*Texanites?*) that had been donated by an amateur collector (Stewart, pers. comm., 1993). Surprisingly, however, one of the fragile aptychi (jaw apparatus; see Lehman, 1979; Morton, 1981) of the ammonite was preserved on the surface of one of the molds. This occurred because the ammonite shell is made of aragonite, a calcium carbonate mineral that was chemically dissolved (diagenesis) before it could be preserved as a fossil, but after it had lain on the mud long enough to form the mold. The thin, delicate-appearing aptychus is composed of much more stable calcite and was more likely

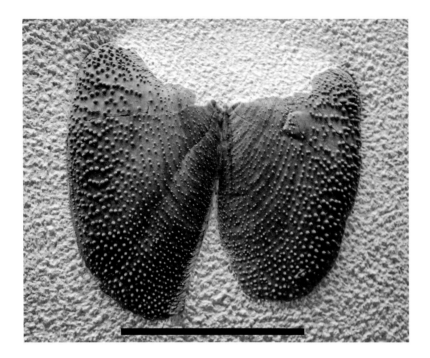

3.9. The paired aptychi (FHSM IP-528; *Spinaptychus* n. sp.) of an as yet unidentified ammonite collected from the Smoky Hill Chalk. The delicate aptychi are composed of calcite and are preserved even though the much larger and thicker shell of the ammonite (aragonite) was dissolved. Scale bar = 10 cm (4 in).

to be preserved (Fig. 3.9). These aptychi are given an informal name (*Spinaptychus* n. sp. in this case), but have not been formally associated with a described species of ammonite.

Stewart (1990a; Fig. 2.10) used the occurrence of ammonite aptychi to define two of his biostratigraphic zones in the chalk. The zone of *Spinaptychus* n. sp. (n. sp. = new species, for a yet-undescribed type of ammonite aptychi) occurs in the low chalk and is late Coniacian–early Santonian in age (ibid.:22). We have collected a slightly different variety of aptychi that occurs just below this zone (designated *Spinaptychus* n. species "B," Stewart, pers. comm., 1992), which indicates the probable presence of another, different species of ammonite.

The presence or absence of certain species of cephalopods, particularly ammonites, scaphites, belemnites, and baculites has been used by many workers to define the age of the rocks where they occur. Their use as stratigraphic markers is limited in the Smoky Hill Chalk because their aragonite shells were dissolved away and their molds or impressions were seldom preserved. Still, even the limited evidence of their presence that was preserved can be useful. Hattin (1982) described the biostratigraphy of the Smoky Hill Chalk in regard to the occurrence of invertebrates, including a small ammonite called *Clioscaphites*, and compared it with equivalent strata in other localities.

Stewart's (1990a; Fig. 2.10) zone of *Clioscaphites vermiformis* and *C. choteauensis* is early Santonian and occurs below the middle of the chalk. In this case, the shells of *Clioscaphites* are represented only by highly detailed molds left in the chalk after the shells were dissolved. Hattin (1982:28, also pl. 8, figs. 4, 5, 6) noted that "*Clioscaphites choteauensis* is not only common through several meters of strata that represents its zone but is also preserved excellently" (Fig. 3.10). Stewart and Carpenter (1990)

3.10. The mold of the shell of *Clioscaphites choteauensis*, a small ammonite from the lower Santonian Smoky Hill Chalk, Gove County, Kansas. Scale = 5 cm (2 in).

reported the remains of a small ammonite, most likely *Clioscaphites*, in a coprolite (KUVP 25870) from this zone, suggesting that at least the juvenile forms were preyed upon by larger fishes and marine reptiles. The zone of *Spinaptychus sternbergi* (Stewart, 1990a:23) is Santonian in age and located immediately above the middle of the chalk, or Hattin's (1982) marker unit 10.

Stewart (1990a:30) also noted that the aptychi (*Rugaptychus* sp.) of a smaller ammonite occurs in the uppermost chalk (zone of *Hesperornis*). More recently, Everhart and Maltese (2010) reported the first specimen of a heteromorph ammonite (cf. *Glyptoxoceras*) preserved under a mosasaur skull in the middle Smoky Hill Chalk. Although only the impression of the ammonite shell is preserved, an internal, tubular structure called the siphuncle is still visible as a darkened trace on the chalk (Fig. 3.11). Unlike the tight spiral coil of most ammonites, the heteromorph ammonites grow in a variety of open forms.

Occasionally the bullet-shaped back portion (guard) of belemnites occurs in the chalk. These guards are rare discoveries (I have collected only one small fragment in 30 years), but they are often preserved well enough that new species have been named. Based on a single damaged specimen in the University of Kansas collection, Jeletzky (1955) reported on the first belemnite from the Smoky Hill Chalk, which he identified as *Belemnitella praecursor*. Two years later, Miller (1957a) also described *Belemnitella praecursor* from 12 specimens collected by G. F. Sternberg and M. V. Walker (Fig. 3.12). Those specimens apparently had not been available to Jeletzky. Following Miller's publication, Jeletzky (1961) was able to examine the material in the collection of the Sternberg

3.11. Flattened internal mold of a heteromorph ammonite (cf. *Glyptoxoceras*; FHSM IP-1484). Probable anterior end of the shell is at the bottom of the photo. A possible trace left by internal siphuncle appears as a darkened trace at the top. Scale in mm.

Museum, and subsequently redescribed them as belonging to the genus *Actinocamax*, including two new species, *A. sternbergi* and *A. walkeri*. Hattin (1982) mentions that belemnites are rare in the chalk, but did not figure them.

Williston was puzzled by fragments he saw occasionally in the chalk that were of a "glistening, fibrous nature" (1897:242). A specimen subsequently collected by H. T. Martin solved the mystery when it became apparent that the fragments represented the gladius (internal support) of a "large cuttlefish, apparently different from any described species." The remains discovered by Martin were those of a fragmentary specimen that was about 15 cm (6 in) wide and 30 cm (12 in) long, with a small "sepia bag" (ink sac) preserved below it.

Logan (1898:497) briefly described a new genus and species of squid (*Tusoteuthis longus*) from the specimen and noted that the specimen had been collected from near the top of the chalk ("Hesperornis beds"). While the root of the genus name, *teuthis*, is from the Greek, meaning 'cuttlefish,' it is also feminine in gender. Miller (1968a) modified the masculine species name, *longus*, of Logan (1898) to the feminine *longa* to correct the oversight.

The internal support structure (pen) of *Tusoteuthis* (Fig. 3.13) was shaped somewhat like a tennis racquet, with a round expanded area (gladius) at one end attached to a long central shaft (rachis). Fragmentary

3.12. Examples of belemnite guards collected from the Smoky Hill Chalk by G. F. Sternberg. Scale = mm.

remains are fairly common in the chalk but are extremely fragile and seldom collected intact. The presence of large numbers of squid in the Western Interior Sea poses an interesting question regarding their adaptations to the warm marine environment, because many modern squid seem to prefer much deeper, colder waters than would have covered Kansas during the Late Cretaceous.

Little additional work was done on the study of squid remains from the Smoky Hill Chalk for nearly 60 years following the initial description by Logan (1898). Miller (1957a:809) described a new species (*Niobrarateuthis bonneri*) from a nearly complete specimen collected in Logan County by M. C. Bonner and noted that "the shape of its gladius differs materially from that of any previous described genus." The type specimen (Fig. 3.13) is on exhibit in the Sternberg Museum of Natural History, Hays, Kansas. Two additional new species, *Enchoteuthis melanae* and *Kansasteuthis lindneri*, were named from single specimens discovered in Logan and Rooks counties, respectively, and are also in the collection of the Sternberg Museum (Miller, 1968b). Stewart (1976:74) noted that the original description of *Tusoteuthis longa*, as well as those of *N. bonneri* and *E. melanae*, was "based on ventral views mistaken for dorsal views" and questioned the validity of the additional species. Another new species,

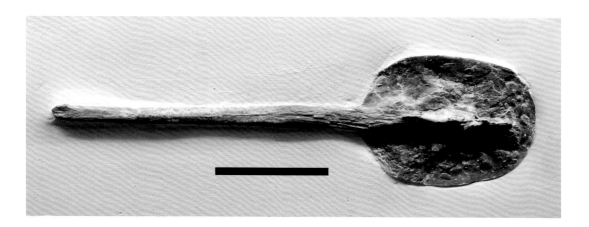

3.13. A fossilized squid pen of *Tusoteuthis longa* (FHSM IP-710) in the exhibit at the Sternberg Museum of Natural History. This specimen was collected by Marion Bonner in Logan County and was originally described as the type specimen of *Niobrarateuthis bonneri* by Miller (1957a). The pen of a squid (rachis) is made up of chitin, not bone or cartilage. Scale bar = 10 cm (4 in).

Niobrarateuthis walkeri, was subsequently collected and described by Green (1977) from the lower chalk of Ellis County. Nicholls and Isaak (1987) reported on several specimens of *T. longa* from the Pembina Shale Member of the Pierre Shale (Canada) and noted that their examination of the Kansas material indicated that only one species, not five, was present in the Western Interior Sea during the Late Cretaceous. In their paper regarding predation on cephalopods, Stewart and Carpenter (1990:205) suggested that all teuthid species previously "described from the Niobrara Formation . . . are assignable to the taxon *Tusoteuthis longa*."

Occasionally the remains of these squid show bite marks from some unknown predator, most likely a large fish or a mosasaur. In one specimen (Stewart and Carpenter, 1990; Carpenter, 1996), the gladius of a squid was collected inside the remains of a 1.5 m (5 ft) fish called *Cimolichthys* (Chapter 5). The prey had been swallowed tail first, and the open jaws of the predator suggest that the fish died with the much too large to swallow body of the squid lodged in its mouth. Stewart and Carpenter (1990) also reported on several coprolites containing fragments of a squid rachis (KUVP 65095; 65728; 65729). My experience indicates that fragmentary (partially eaten) squid remains, as previously noted by Logan (1898) and others, are relatively common occurrences in the chalk (Fig. 3.14).

The only echinoderms known from the chalk are an unusual form of crinoid called *Uintacrinus*. Related to starfish and sea urchins, these animals were apparently colonial and free-living (nonstalked; not attached to the sea bottom). The first *Uintacrinus socialis* specimens were discovered in the Uinta Mountains of Utah by the Yale College Scientific Expedition of 1870 (Marsh 1871a), but it was mentioned only as a "new and very interesting crinoid, allied apparently to the *Marsupites* of the English Chalk" (Marsh, 1871b:195). Marsh's material from Utah was apparently so fragmentary that the species was only described several years later by one of Marsh's students, George Bird Grinnell (1876), based largely on complete, articulated specimens collected from the Smoky Hill Chalk of Kansas. Springer (1901) published one of the most complete descriptions of these unusual crinoids.

The description of *Uintacrinus* provided by Charles H. Sternberg is of some interest here because he and his sons collected many of the

3.14. Fragment of a rachis of *Tusoteuthis longa* preserving evidence of the damage caused by a predator's bite (circle). This specimen was collected from the Late Coniacian Smoky Hill Chalk of Gove County, Kansas. Scale bar = 10 cm (4 in).

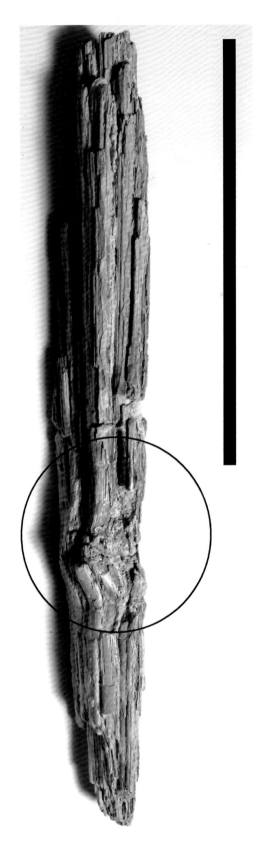

original specimens: "Their bodies were about the shape of half an egg, with an opening in the center, and ten arms radiating from the margin. These arms were three feet long, with feathered edges. Over the mouth, too, were smaller arms used to comb off into the mouth the tiny animal life of the sea, that was strained through, and caught in the meshes of the feathered arms" (1917:156). In the chalk, *Uintacrinus* appears to have been colonial; most of the remains indicate many individuals died at the same time, and settled to the sea bottom, where they were preserved as thin layers of limestone. Miller et al. (1957:163) reported that the slabs are usually between "a quarter inch and one inch in thickness." Williston (1897) noted that he had discovered a number of specimens in 1875 while collecting in the chalk with B. F. Mudge, and he indicated that some of these served as the types for the genus and species described by Grinnell (1876).

According to Williston (1897), the large slab of *Uintacrinus* now on display at the Museum of Natural History at the University of Kansas was collected in 1891 near Elkader in Logan County. This specimen has more than a hundred individuals preserved in a slab measuring about 1.8 m x 1.2 m (6 ft x 4 ft). The Sternberg Museum also has a nice specimen (Fig. 3.15) on exhibit that was collected 4 miles west of Elkader (Miller, Sternberg, and Walker, 1957:fig. 2). Williston (1897) noted that all the specimens he was familiar with came from the vicinity of Elkader (eastern Logan County). Miller, Sternberg, and Walker (1957) noted that the most eastern known occurrence of *Uintacrinus* was at Castle Rock in eastern Gove County. While they suggested that the specimens from Gove County were from lower in the chalk (older) than those from Logan County, more recent information (Hattin, 1982; Stewart, 1990a) indicates that all of the known occurrences were from approximately the same stratigraphic level. In Kansas *Uintacrinus* is known only from the early to middle Santonian, or just below the middle of the Smoky Hill Chalk (Hattin's marker unit 10). *Uintacrinus* is also useful as a stratigraphic marker in marine rocks from the Late Cretaceous around the world.

Although the view that these crinoids floated together near the surface with their feeding arms dangling downward is currently the most widely accepted, there are other interpretations of their life and preservation. Hess believes that "*Unitacrinus socialis* lived gregariously on the soft bottom and was buried in life position; its colonies may have resembled dense patches of tall eel grass" (1999:231). I would argue that although this is certainly a possibility, the lack of other fossil materials (teeth, bones, etc.) that should have accumulated over time in association with these crinoid slabs is hard to explain if indeed they were sitting on the sea bottom.

Apart from Williston's (1897) brief comments regarding invertebrates from the Smoky Hill Chalk, a number of other authors have also described their occurrence in some detail. Logan (1897; 1898; 1899) provided a lengthy discussion of the geology and the invertebrates of the Kansas Cretaceous. In addition to his summary of the geology, depositional environment, and invertebrate fauna, Miller (1968a; 1969) provided

3.15. A close-up of the Late Cretaceous crinoid *Uintacrinus socialis*. This specimen, collected near Elkader in Logan County, is part of a slab containing dozens of individual crinoids, on exhibit in the Sternberg Museum of Natural History, Hays, Kansas. The individual crinoid was preserved on the underside of a large mass of similar crinoids that died and were buried together. Such occurrences are not uncommon in some exposures near the middle of the chalk.

a review of previous work on invertebrates of the Niobrara Formation in Kansas. Most recently, Hattin (1982) described the invertebrate fauna of the Smoky Hill Chalk in relation to his stratigraphic references and provided excellent photographs of many species. Stewart (1990a) incorporated Hattin's stratigraphic markers into his biostratigraphy of the chalk.

Other Invertebrate Remains and Biostratigraphy

Other, less common invertebrates are known from the Smoky Hill Chalk but are seldom seen or collected. In addition to those mentioned above, Logan (1898:481) included several smaller species of inoceramids, sponges (coelenterates), serpulid worms, and arthropods (cirripeds and barnacles). In addition to listing the genera of Foraminifera that occur in the chalk, Miller (1968a) noted the occurrence of several species of oysters, *Baculites*, a *Pecten* specimen (a genus of scallops), scaphites (*Clioscaphites*), four different kinds of ammonite aptychi, and several belemnites, including *Actinocamax walkeri* and *Belemnitella praecursor*. A single occurrence of a small (23 mm) decapod crustacean (*Linuparis*?; KU 7295) is also documented (ibid.:61–62). This may be the same specimen that Mudge (1876:217) had reported as a "rare crustacean" in a coprolite. I am not aware of any other crustacean specimens.

Hattin (1982:71) provides a listing of invertebrates in the chalk by their method of preservation, with notes on their relative abundance. The bivalves, crinoids, belemnites, and serpulids are examples of the preservation of calcareous skeletal material. Molds of scaphites, baculites, and ammonites make up the second category. Borings by sponges are the third form of fossil evidence. Squid pens represent the preservation of a combination of calcareous and organic matter. The crustacean noted by Mudge (1876) is an example of the rare fifth group, preserved chitin.

Hattin (1982:26–29) was also the first to propose a biostratigraphy of the Smoky Hill Chalk based on the occurrence of the invertebrates. In Stewart's (1990a) paper on the biostratigraphy of the Smoky Hill Chalk, he notes the occurrence of invertebrate species in each of his six biostratigraphic zones. My field experience validates Stewart's conclusions and indicates that *Volviceramus grandis* occurs only in the lower two zones, while *Platyceramus platinus* dominates the upper four zones. *Pseudoperna congesta* occurs abundantly in the lower five zones and probably occurs throughout the chalk. The rudist *Durania maxima* is present in the lower four zones, disappears, then reappears in the uppermost zone, just below the contact with the Pierre Shale. The squid *Tusoteuthis longa* occurs throughout the chalk and, based on a specimen I collected in 1990 (Stewart, 1990a; pers. comm., 1990), also occurs just below the contact with the Fort Hays Limestone. Based on the occurrence of their aptychi, ammonites were probably present throughout the deposition of the chalk. The fluctuations noted in the presence or absence of various invertebrate species may be due to changes in the bottom environment (i.e., decreased oxygen), a changing climate (i.e., cooling, Stewart, 1990a:26), other factors that we do not understand, or just the lack of serious collecting.

Pearls

Brown (1940) reported on the presence of fossil pearls in the Niobrara (Smoky Hill Chalk) and Benton (Greenhorn Limestone/Carlile Shale), noting that many of the specimens had been collected by G. F. Sternberg west of Hays. These pearls are tan to light gray in color, have lost their shiny outer layer (nacre) and occur in round, unattached forms or hemispherical shapes attached to a piece of inoceramid shell (ibid.; Fig. 3.16). In some cases, the attachment occurred postmortem when the pearl was pressed against the surface of the shell by the pressure of the overlying rocks. The largest of these pearls examined by Brown (ibid.:367) had a diameter of 2 cm. A clipping from the Hays newspaper in 1940 indicated that George Sternberg had donated 50 fossil pearls to the Smithsonian. These are probably the same ones reported by Brown (ibid.), and include many that were given to him by other collectors. While not rare in the chalk, they are difficult to find because they normally occur in strata where inoceramids are most common and where they are obscured by thousands of other shell fragments. Kauffman (1990:66) described several giant pearls from the Smoky Hill Chalk, including a pendant pearl in the collection of the Museum of Comparative Zoology at Harvard that was 11 cm long and 6.5 cm in maximum width. The largest pearls apparently come from the deeper, bowl-shaped shells such as *Volviceramus grandis* (ibid.:68).

I have collected several pearls from the Smoky Hill Chalk over the years, but I have to admit that I wasn't specifically looking for them at any point. The largest one I have ever collected was located in the low chalk of Gove County in 2003. The specimen (FHSM IP-1451) represents

3.16. Pearls I have collected from the Smoky Hill Chalk in Trego and Gove counties, Kansas. The pearls at left have been crushed into the shells of the inoceramid clams where they were formed. The three pearls at right were found unattached. Note the layering visible in the pearl at upper right. Scale bar =10 cm (4 in).

an apparently huge, hemispherical pearl, 4–5 cm in diameter and 2.5 cm high, or roughly the size of half a golf ball (Fig. 3.17). With the exception of a large, very irregular specimen, all the rest of the Smoky Hill Chalk pearls in my personal collection are less than 1.5 cm in diameter. In late 2003 I collected several smaller pearls from the base of the Lincoln Limestone Member of the Greenhorn Limestone (Upper Cenomanian) in Russell County in association with numerous shark teeth. These pearls were much smaller than those collected in the Smoky Hill Chalk, averaging 4–5 mm in diameter. A polished cross-section of one of them shows concentric layers of calcite formed around a central nucleus, nearly identical to the one published by Brown (1940:fig. 16).

Coprolites

'Coprolite' is a scientific term for the fossilized excrement, feces, or droppings of ancient animals. It was coined by Dr. William Buckland (1829). Coprolites are trace fossils, rather than body fossils (bones, teeth, etc.). Hawkins (1834:pls. 27–28) illustrated several coprolites in his book on ichthyosaurs and plesiosaurs from the Jurassic of England. In the Smoky Hill Chalk, the coprolites that we find are probably from sharks, bony fishes, and marine reptiles (mosasaurs). This fossilized waste can often tell us much about what kind of prey these animals ate and even how their digestive systems worked. B. F. Mudge (1876) and S. W. Williston (1898) were among the first to discuss the presence and origin of coprolites in the Smoky Hill Chalk.

Coprolites are fairly common in the chalk and may contain partially digested bones (usually fish bones), teeth (Shimada, 1997; Everhart,

2007), and even fragments of squid pens (Stewart and Carpenter, 1990). Sharks are thought to be the source of the so-called spiral coprolites (Stewart, 1978), but most of the other larger ones (Fig. 3.18) are likely to be from marine reptiles (mosasaurs). Coprolites tend to be more resistant to erosion than the chalk because they are composed largely of less-soluble calcium phosphate instead of calcium carbonate. Fossilized gut contents (Miller, 1957b; Stewart, 1978) and regurgitated prey material (see Hattin, 1996; Everhart, 1999) are two other categories of unusual biological materials preserved in the chalk.

One of the first notes of coprolites in Kansas comes from Prof. Benjamin Mudge, the first state geologist of Kansas: "Coprolites of fish and Saurians are frequently found, containing the remains of the food of the animal. Small fish appear to be the most common food; but in one instance a rare crustacean was found preserved this way. The coprolites are not so hard as those of Europe, being a little firmer than chalk, and finer grained. . . . In some cases the undigested organic material (bones) was one-fourth the whole weight" (1876:217). Williston also mentioned coprolites in his paper on mosasaurs: "Coprolites which I have always had reason to believe were from these animals are in some places very abundant, weighing from an ounce or two up to a half a pound or more. They are ovoidal in shape with sphincter or intestinal impressions upon them, and contain very comminuted parts of fish bones, fish scales, etc." (1898:214). Williston added that "coprolites, evidently of mosasaurs, are frequently found, with large, undigested fish bones, and fragments of fish bones" (1902:261). Although I accept Williston's account, I have never seen a coprolite from that chalk that contains large fish bones.

Stewart noted that "spiral coprolites, argued to be enterospirae by Stewart (1978), occur in this horizon [zone of *Spinaptychus* n. sp.]. McAllister (1985) has conclusively demonstrated that such structures occur in the colon of a dissected *Scyliorhinus canicula* [a small modern shark]. Presumably, they can also be expelled intact from the rectum. Their presence in this horizon and absence in the other horizons implies the presence of some heretofore undetected shark taxon in this zone. Other types of coprolites occur in all horizons of the Smoky Hill Chalk Member" (1990a:22). Spiral coprolites are not often collected from the chalk, but they are there.

3.17. Top, side, and bottom views of a giant inoceramid pearl (FHSM IP-1451) from the late Coniacian chalk of Gove County, Kansas. This may be the second-largest pearl known from the chalk after the pendant pearl specimen (11 cm x 6.5 cm) cited by Kauffman (1990) in the collection of the Museum of Comparative Zoology at Harvard. Scale bar = 5 cm (2 in).

3.18. This large coprolite was discovered in the lower chalk of Gove County and probably is from a mosasaur, although there is no way to tell for certain. Coprolites occur frequently in the lower chalk and sometimes contain the bones of fishes or other prey. Scale = 10 cm (4 in).

Other things we find in the low chalk (late Coniacian) are partially digested bone fragments of fishes and marine reptiles, and even limb bones of small dinosaurs (Everhart, 1999; Everhart, 2003; Everhart, 2004a). The surface layer of the bone is usually heavily pitted, and any teeth in fragments of jaws appear to have been dissolved by acid back into their sockets. Because some of the fragments include embedded shark teeth or identifiable bite marks, I interpret them as regurgitated stomach contents (Fig. 3.19). In most cases, the specimens represent pieces of the prey that were big enough to require a large shark to swallow them. Why do I think the specimens are the regurgitate from sharks and not some other large predator? With notable exceptions (Sternberg, 1922; Martin and Bjork, 1987; Everhart, 2004a), mosasaurs ate fishes. However, as reptiles they had digestive systems that were capable of digesting heavy bone. They were also more likely to swallow their prey whole, and their teeth were not capable of shearing through large bones. Sharks, on the other hand, are known to carve up their prey, and big sharks like *Cretoxyrhina* had teeth and jaws strong enough to shear bone. We know this from pieces of mosasaur vertebrae (FHSM VP-13283) that have been severed and retain fragments of embedded shark teeth (Everhart, 1999). The digestive system of sharks, however, may not be able to completely dissolve dense reptile and dinosaur bone, and modern tiger sharks are known to be able to regurgitate materials from their stomach when stressed (Gudger, 1949).

Sometimes we collect ovoidal structures that look like flattened coprolites but are composed almost entirely of ground-up fragments of oyster shell (Fig. 3.20). They are oval and are always flattened, with average

measurements of 4 cm x 2.5 cm x 1 cm (1.5 x 1 x 0.4 in) (Everhart and Everhart, 1992). The smooth underside sometimes bears impressions that appear to be from the anal sphincter of some predator. Superficially, they resemble burrow structures but do not include any other debris that might be expected from winnowing of sediments by burrowing crustaceans. Interestingly, they occur contemporaneously with an unusual plethodid fish called *Martinichthys* (see Chapter 5) in the lower chalk from just below Hattin's marker unit 4 to just above marker unit 5. I suspect but have been unable to prove that *Martinichthys* fed on the abundant oysters that were attached to inoceramids and that these oyster shell–filled structures are the coprolites of that genus. Other, less defined masses of tiny inoceramid prisms appear to be either coprolites or regurgitates from another predator, possibly *Ptychodus*, feeding on young inoceramids.

3.19. A, a partially digested mosasaur dorsal vertebra (FHSM VP-16356) with the remains of a shark tooth embedded at right, scale bar = 5 cm (2 in); B, close-up showing the damaged tip of a shark tooth (FHSM VP-16357), most likely *Cretoxyrhina*, embedded in the vertebra, no scale.

Mudge was among the first to note the presence of

Remains of Logs and Tree Limbs

> an occasional fragment of fossilized wood. . . . This wood was, in a few instances, bored before fossilization by some small animal. This might have been done by the larva of an insect (a 'borer') when the tree was living, or later by a teredo [naval shipworm; the clam *Teredo navalis*]—when the trunk floated in water. In either case it shows that the Cretaceous vegetation was subject to the same enemies as that of the present period. Some of this wood was in charred condition [carbonized] and would burn freely. Other specimens were changed into almost pure silica, the cavities studded with quartz. In one case a log, weighing about 500 pounds, had all conditions of the transformation; a portion had the appearance of soft decayed wood, which crumbled in handling, and other parts ringing like flint under a hammer. Occasionally specimens were converted into chalcedony, but the annual growth of the wood distinctly remained. In a single instance we detected the fibrous structure of the palm. (1877:283–284)

Williston (1897:243) noted that "fossil wood is occasionally found in the formation. A tree about thirty feet long was discovered near Elkader a

3.20. These masses of ground-up oyster shell (*Pseudoperna congesta*) appear to be the coprolites of an unknown predator that apparently fed on the abundant colonies of oysters attached to inoceramids. They occur only in a limited stratigraphic zone in the late Coniacian chalk. Scale = 10 cm (4 in).

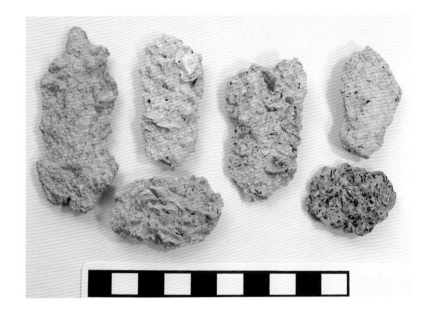

year or two ago [about 1895]" and that fragments of amber occur in association with the bark of these trees. I have occasionally come across places where small fragments of a shiny, coal-black material are eroding out of the chalk. When the source is uncovered, it turns out to be the remains of a log or tree limb that has been greatly compressed by the weight of the overlying chalk. Unlike 'petrified wood' that has been replaced with agate or other minerals, this wood appears to have been 'carbonized' during preservation to the point that little is left but the carbon that made up the original cellulose in the wood (Fig. 3.21). Carbonized wood from the chalk is fragile and breaks up readily into blocky chunks. It will burn if placed in a flame. The inside of a carbonized log when first opened smells faintly like burnt wood, but the wood was never in a fire. These are simply the remains of waterlogged trees that sank to the bottom of the sea and were buried in the chalky mud. Sometimes the logs may be 6–7 m (20+ ft) long but are crushed to a thickness of less than 5 cm (2 in). Inside the coal-like, carbonized portions, small pockets of delicate fibrous wood remain. The outside of a log may be covered with patches of black calcite crystals that incorporated some of the underlying carbon as they grew. Occasionally, attached shells, mostly oysters, are visible. *Teredolite* (shipworm) borings, if any, are rare in the Smoky Hill Chalk, although they are quite common in driftwood that occurs in the underlying Fairport Chalk Member of the Carlile Shale (pers. obs.).

Collecting fossils in the Smoky Hill Chalk is much like walking along a beach. If you look long enough, you are likely to come across the remains of just about anything that lived in the sea during the time that the chalk was being deposited. We try to collect everything that looks unusual, and if we cannot immediately identify it, we save it until we can. As our friend J. D. Stewart once told us (pers. comm., 1990), "Go ahead

and pick up everything you see. You can always throw it away later. If you don't pick it up, you may never see it again." A number of our discoveries over the years, including the first known coelacanth remains (Chapter 5) from the Smoky Hill Chalk, were collected in large part because we didn't know what they were at the time.

3.21. Examples of fossil wood from the Western Interior Sea: A, exterior of a log showing chalk-filled openings made by borings of *Teredo* clams; B, interior of same log showing the crushed living spaces originally occupied by the *Teredo* clams; C, a piece of silicified wood; D, a portion of a carbonized log. Scale bars = 5 cm (2 in).

The young adult male mosasaur struggled wearily as it swam slowly just below the calm surface of the warm, shallow sea. It was the peak of the mosasaur breeding season. A chance encounter earlier that day with an enraged and much larger bull mosasaur had left his left front flipper crippled and useless. The 6 m (19.7 ft) marine reptile kept rolling over on its left side because the damaged flipper no longer provided the necessary steerage to keep him upright. Surfacing to breathe was difficult, and his efforts to bring his head above the surface of the water were noisy. Trying to swim without the use of the flipper had tired the creature to the point of exhaustion; worse, the struggle had attracted the attention of other predators in the area.

Even though bleeding from the superficial wounds had stopped hours ago, a swarm of small *Squalicorax* sharks still cruised warily in circles around the struggling mosasaur, occasionally darting in closer as if sizing up the best opportunity to attack. The largest of these sharks was less than a third the length of the mosasaur and normally would not have represented a threat. In fact, the mosasaur had killed and eaten many such sharks over the years.

Now the tables had turned and the wounded mosasaur was potential prey instead of the predator. There was no place to hide in the middle of the Western Interior Sea and no protection to be had from others of his own kind.

From deep below and far outside the circle of scavenger sharks, another much larger shark moved steadily toward the mosasaur. The shark's acute senses had detected the wounded reptile's thrashing noises from more than a mile away and alerted the shark to the potential feeding opportunity. Almost as long as the snakelike mosasaur and twice as heavy, the giant shark scattered the smaller scavengers in its path as it approached the mosasaur from below at high speed. The open jaws of the shark hit the mosasaur near the center of its body, on the right side, just behind the rib cage, and the impact lifted the wounded animal almost completely out of the water. As the massive jaws closed, rows of razor-sharp, 5 cm (2 in) teeth sliced out a dinner plate–sized chunk of skin, muscle, bones, and intestines from the mosasaur, opening a fatal wound.

Then, before the reptile could even react to the first bite, the jaws closed again, this time across the 5 cm (2 in) wide bones of the mosasaur's spinal column. Razor-sharp teeth sliced through muscle and bones, severing the spinal cord and major blood vessels. As the shark's powerful jaws drove its teeth through the mosasaur's vertebrae, the tips of several teeth were broken off in the bone. The water quickly filled with blood as the mosasaur thrashed briefly in its death throes, its body cut almost completely in two by the shark's massive bite.

Then, even as the big shark opened and closed its mouth rapidly to swallow the large chunks of flesh and bone, the other scavengers flashed into the blood-filled waters and attacked the carcass of the dead mosasaur. The big shark started to move back into the feeding frenzy, then abruptly changed directions, swimming slowly away from the swarming scavengers. Its immediate hunger satisfied, the shark turned its attention away from the battle over the remains of the mosasaur and resumed its westward journey through the warm waters. Days later, the shark opened its mouth and regurgitated a foot-long chunk of indigestible mosasaur vertebrae.

The five vertebrae, severely eroded in places by the shark's stomach acid, but still held together by tough ligaments, sank rapidly through 400 feet of water and landed on the soft chalky mud of the sea bottom. The impact partially buried the vertebrae and protected them from further attack by scavengers. Before long, they had completely disappeared beneath the surface of the mud, where they would remain for almost 85 million years. Thousands of feet of sediment eventually covered the remains before the land rose and the ocean receded hundreds of miles south to the present Gulf of Mexico.

Over the span of millions of years, the chalky sediments surrounding the vertebrae were gradually compressed into layers of soft rock and then lifted nearly a mile above sea level. As this was happening, erosion by wind and water wore away more than a thousand feet of shale and chalk before the vertebrae were exposed to the light of day.

Then in 1995 a sharp-eyed bipedal mammal wandering across the dry, eroded surface of the exposed sea bottom happened to chance upon the lumps of reddish-brown bone weathering from the gray chalk. Quickly recognizing them as mosasaur vertebrae, the human took a pick and a brush out of his pack and carefully exposed the rest of the 12 inch long specimen. Disappointment clouded the human's face as he searched in vain along the chalk exposure for the rest of the mosasaur. Five vertebrae are not much of a consolation when you are hoping to find a 20 foot long mosasaur with more than 130 vertebrae. Giving up, he marked and bagged each vertebra, put them in his pack, and resumed his search. It was more than a week later, when he cleaned up the specimen, that he saw the bite marks and embedded sharks' teeth and realized the significance of what he had collected.

The preceding story is fiction, of course, except for the ending where I discovered the mosasaur vertebrae (Shimada, 1997c; Everhart, 1999). We will never know the exact circumstances regarding what happened to this particular mosasaur some 85 Ma. Dan Varner's painting (Plate 6) shows

Sharks

4.1. Five mosasaur dorsal vertebrae (FHSM VP-13283) with ribs in ventral view that have been eaten and partially digested by a large *Cretoxyrhina mantelli* shark. Note that the anterior (left) and posterior (right) vertebrae have been sheared through by teeth of the shark. Blue triangles indicate bite marks on the vertebrae, and the red rings denote the location of the broken tips of embedded shark teeth. The specimen is an exhibit at the Sternberg Museum of Natural History, Hays, Kansas. Scale = 10 cm (4 in).

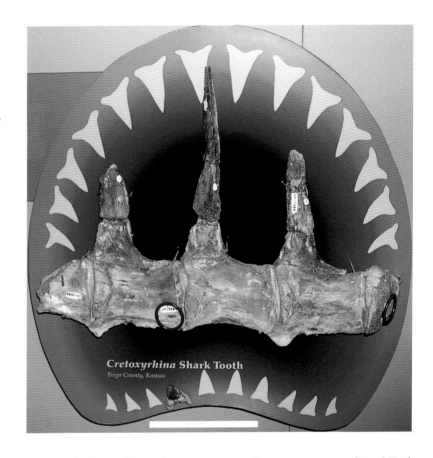

a ginsu shark attacking a living mosasaur. In contrast, as my friend Earl Manning likes to remind me, the actual scenario may have been much less dramatic. Manning (pers. comm., 2002) suggested that it was far more likely that the mosasaur in question was a long-dead and bloated carcass floating at the surface and slowly coming apart. In such cases, sharks and other scavengers would have plenty of time to remove the more readily detachable parts like the limbs and tail, generally leaving only the skull and dorsal vertebrae to be preserved as fossils. 'Bloating and floating' is certainly the case in many instances and is the only reasonable explanation for how the remains of large dinosaurs, such as *Claosaurus agilis* and *Niobrarasaurus coleii* (see Carpenter, Dilkes, and Weishampel, 1995; Chapter 12) could have been transported across hundreds of miles of ocean to be buried in the middle of the Western Interior Sea. However, it is also reasonable to assume that these Cretaceous sharks would have behaved much like their modern cousins. Sharks are an essential part of Mother Nature's 'clean-up crew' and play a vital role in recycling organic resources back into the marine ecosystem. Sick, wounded, and otherwise vulnerable animals likely were suitable prey for ancient sharks just as they are for modern sharks.

The limited remains of this mosasaur (FHSM VP-13283; Fig. 4.1) provide a number of clues that lead me to believe that my fictional story in this case is not far from the truth. The specimen consists of five vertebrae from the lower back of what was probably a young adult mosasaur named *Platecarpus planifrons*. The vertebrae are 5 cm (2 in) in diameter and were still articulated when I collected them. The three center vertebrae have a partial rib remaining on the left side. The front and back vertebrae of the series have been severed by the shark's bite—the anterior one is cut across at an angle of about 45°, the posterior one at right angles to the mosasaur's spine. Most, but not all, of the surface of the bones is corroded in a way that is consistent with the attack of stomach acids (Varricchio, 2001; Everhart, 2004b). This is especially evident on the cut ends of the two vertebrae at the front and back of the specimen that would not have had a protective covering of skin, muscle, and connective tissue and so would have been exposed to the shark's digestive process initially and for the longest time.

The 'smoking gun' that identifies the attacker in this case is the broken pieces of three teeth that were left behind. Fragments of two *Cretoxyrhina mantelli* teeth are still embedded in the mosasaur vertebrae (FHSM VP-13284). The tip of one tooth is in the cut and eroded surface of the last vertebra, and a second, much larger piece is broken off in the right side of the second vertebra. A third tooth fragment, which was apparently preserved but not recovered, left its impression pressed into the right side of vertebra number four (Shimada, 1997c:fig. 4). Several scratches on the bones are interpreted as the marks left by other teeth as the shark opened and closed its jaws in the process of swallowing the large piece of the mosasaur. The dorsal processes are missing on all the vertebrae. It is impossible to tell if the bite or bites that severed these five vertebrae from the spine of the mosasaur occurred while the mosasaur was living or as the result of scavenging on a rotting, floating carcass. In either case, however, the shark's ability to bite through large solid bones is evident in this and similar remains preserved from the Smoky Hill Chalk (Everhart 2004c, 2005). Further evidence of this feeding behavior was described by Shimada and Hooks (2004) in regard to the bones of protostegid turtles occurring in the Mooreville Chalk of Alabama that had been bitten by *Cretoxyrhina mantelli*.

Of course, these large sharks didn't feed only on mosasaurs; they ate anything and everything available. And they weren't always successful at biting through bone. A nearly complete elasmosaur (FFHM 1974.823; Chapter 7) paddle in the Fick Fossil and History Museum, Oakley, Kansas, exhibits multiple slashing bite marks across the shaft of the propodial (humerus or femur). It appears that a big *Cretoxyrhina* literally swam up the paddle and tried to bite through the upper limb bone by head shaking. The shark's teeth cut deeply into both sides of the bone, but did not cut through it (Fig. 4.2). However, since the paddle is fairly complete and still articulated, it is likely that the attacking shark or other sharks were able to detach it at the shoulder/hip. From there, it dropped free of the elasmosaur carcass and went to the sea bottom, where it was preserved intact, to be discovered and collected by George F. Sternberg in 1925 (Everhart, 2005).

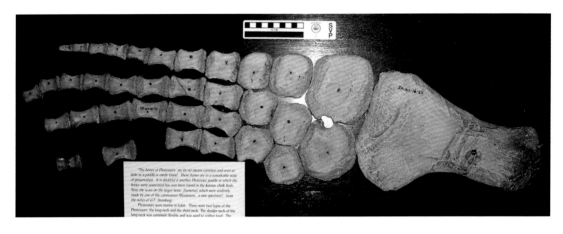

4.2. Left fore-paddle of a large elasmosaur (FFHM 2–25) that was scavenged by large sharks and detached from the body of the plesiosaur. Bite marks on the shaft of the humerus (right) indicate that the shark had the entire paddle in its mouth when it tried to sever the bone by head shaking. The bites left deep scars across the distal end and mid-shaft of the humerus but did not cut through the large, thick bone. Scale = 10 cm (4 in).

These shark-bitten specimens are similar to many other partial remains that occur in the chalk (Everhart, 1999). FHSM VP-13744 is a series of five cervical (neck) vertebrae from a medium-sized (species unknown) mosasaur. Again, the first and last vertebrae in the series have been bitten through, but there are no embedded teeth. All the vertebrae show the effects of a lengthy exposure to stomach acids (Varricchio, 2001). In another specimen, a series of about 20 much smaller vertebrae from the end of a mosasaur's tail were discovered lying on the surface of the chalk (FHSM VP-13750). In this case, not only is the final dismemberment of the mosasaur's tail recorded by the presence of numerous severed vertebrae, but there is also preserved evidence of an earlier, healed bite near the end of the tail that includes the tip of an embedded shark tooth (Rothschild and Everhart, 2015). The first wound became infected and fused at least five vertebrae together into a club-like mass on the end of the tail. Martin and Bell (1995) reported on the occurrence of similar 'club-tailed' mosasaur vertebrae in the Cretaceous of South Dakota where the tip of the tail had been bitten off and the resulting infection fused some of the remaining vertebrae. In the Kansas specimen, the mosasaur apparently survived the first shark attack and lived long enough for the wound to heal. Martin and Rothschild (1989) reported a similar fused series of mosasaur vertebrae (KUVP 1094) as a likely result of an infection following a shark bite (see also Shimada, 1997c:fig. 3). In this case, the broken tip of at least one shark tooth is also embedded in those vertebrae.

Several years ago, my friend Tom Caggiano (see Chapter 12) discovered the front one-quarter or so of the skull of a small mosasaur (FHSM VP-13748), probably a young *Platecarpus tympaniticus* (Fig. 4.3). The muzzle had been sheared off right behind the third tooth in both upper jaws. The anterior ends of both the right and left maxilla were still sutured tightly on the premaxilla, but all the teeth had been dissolved back into their sockets and the surface of the bone was corroded by stomach acids. In this case, the front of the mosasaur's skull had been bitten off by a shark, swallowed, and partially digested before being regurgitated. There were no other bite marks or imbedded teeth, so the identity of the attacker/scavenger is not known for certain. However, the remains are

more consistent with the bone-shearing bite of a large *Cretoxyrhina* than with the flesh-slicing bite of a smaller *Squalicorax*.

Most of the other partially digested specimens consist of a jumble of smaller bones or smaller remains. One or two vertebrae are fairly common, and so are small sections of mosasaur tail vertebrae. Fragments of jaws have been collected with the teeth always dissolved down into the sockets. Limb bones and girdle elements (shoulder and pelvis) are also parts that are readily detachable from a mosasaur carcass. Hamm and Everhart (2001) reported on the isolated discovery of the radius and ulna of a small dinosaur, a juvenile *Niobrarasaurus coleii*, from the Smoky Hill Chalk. Two long, unserrated bite marks suggest that the lower limb bones had probably been severed from a floating carcass by a large *Cretoxyrhina*, and the appearance of the surface of the bones indicated that they had been partially digested (Fig. 4.4). Beyond that, there is no other explanation for two juvenile dinosaur limb bones occurring in the chalk hundreds of miles from the nearest shore.

In 2005, I was collecting in Gove County with a friend, Keith Ewell, when he discovered and picked up a string of nine articulated vertebrae. When he showed them to me, he was thinking that they came from a big fish called *Xiphactinus* (Chapter 5), but I initially thought they were more likely from the neck of a plesiosaur. After they were cleaned up and examined, it became apparent that he had discovered a shark-bitten fragment of the tail of a dinosaur (Chapter 12). Bite marks and corroded bone indicated that this part of a dinosaur's tail had been eaten by a shark. Even better, it was part of the tail of a hadrosaur, only the second set of

4.3. Dorsal (top) and ventral (bottom) views of the remains of the anterior end (premaxilla and both maxillae) of a juvenile mosasaur skull (FHSM VP-13748) that had been bitten off and partially digested, most likely by a large shark. Note that the teeth have been digested completely and are dissolved deeply into their sockets. Scale = 5 cm (2 in).

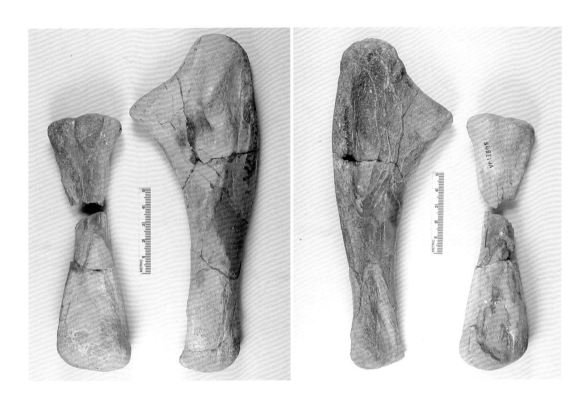

4.4. Front and back views of the ulna and radius of a juvenile nodosaur (*Niobrarasaurus coleii*; FHSM VP-13895, collected from the Smoky Hill Chalk near Hattin's Marker unit 8 in Gove County, Kansas. The eroded surface of the bones and bite marks on the ulna suggest that at least the forelimb of the dinosaur was eaten by a large shark. Scale bars = 5 cm (2 in).

hadrosaur remains ever collected from the chalk (O. C. Marsh [1872] discovered the first in 1871; see Everhart and Ewell, 2006).

Another specimen intrigues me because it indicates an attack on a mosasaur but leaves us without knowing the immediate outcome. In 1990, I discovered the 1 m (39 in) skull of a large *Tylosaurus kansasensis*. (FHSM VP-13742) in Gove County. A skull of this size would have come from a tylosaur that was about 7 m (23 ft) long (Everhart, 2002). The skull had weathered out of the lower chalk but appeared to be relatively complete. There were, however, no postcranial elements in the remains except for a few paddle bones (phalanges). Either the skull had fallen off a drifting carcass or scavengers had separated it from the rest of the skeleton before it was buried. The interesting part is that the tips of three broken *Cretoxyrhina* teeth were embedded in the skull—one in the top of the premaxilla, and one each in the outside (lateral) surface of both dentarys. In addition, there were bite marks on the top of the premaxilla, and the tip of the premaxilla had been sheared off at an angle (Fig. 4.5). None of the bites appeared to be fatal, although the teeth embedded in the lower jaws may indicate a bite to the throat of the mosasaur. No healing of the bites was indicated. Damage from weathering precluded much further analysis of this skull, but it was evident that the large mosasaur had been bitten several times by a fairly large shark.

Two years later, my wife, Pam, discovered three dorsal vertebrae from a medium-sized (5–6 m, 16.4–19.7 ft) mosasaur eroding from the edge of a gully. Our usual division of labor in the field is fairly simple: she finds them and I get to dig them. A quick check showed that at least

4.5. Left oblique view of the premaxilla of a large *Tylosaurus kansasensis* skull (FHSM VP-13742) that had been bitten by a ginsu shark (*Cretoxyrhina mantelli*). Note that the end of the premaxilla (far left) was sheared off by the bite of the shark and that another bite mark is visible on middle of the bone. The broken tip of a *Cretoxyrhina* tooth is still embedded in the top of the premaxilla (T). Scale = 10 cm (5 in).

two more vertebrae continued into the chalk. They were coming out "round end" (condyle) first, which indicated that they were pointing me toward the head of the animal, not the tail. Finding articulated vertebrae is good news; it usually means that there are more bones still buried in the chalk. I had to remove about 0.5 m (20 in) of chalk that was sitting on top of the layer containing the vertebrae. Several hours of hard digging passed before I was able to see what was hidden underneath. Twenty-two vertebrae were eventually uncovered, along with ribs lying on either side of the vertebral column. It was looking really good until I got to the 21st vertebra and ran into empty chalk. I dug some more and located one more vertebra, the cervical at the base of the neck, lying off to the side. Mosasaurs have seven cervical vertebrae. In this case, I was six vertebrae away from where the skull should have been. I dug deeper into the chalk, looking for the skull, but nothing remained. Tired and disgusted, I gave up looking for the skull and got busy recovering the vertebrae and ribs. After all that work, what I had was a headless, tailless, limbless string of vertebrae with a few ribs attached (FHSM VP-13746), possibly from an *Ectenosaurus* (G. Bell, pers. comm., 2007).

Much later, when I prepared the bones, I began to notice some obvious signs of 'foul play.' There was good evidence of at least three bites by a big shark along the 1.5 m (5 ft) string of vertebrae (Everhart, 2004c). That the tooth marks were spaced about 3 cm (1.25 in) apart strongly indicates the bites had come from a very large shark. As the result of one bite, the broken tip of a *Cretoxyrhina* tooth was still embedded in a vertebra. The ribs on one side of the carcass had been bitten off cleanly, but they were reasonably intact on the other side. The most anterior vertebra of the specimen, the last one I located, had been twisted to the side and showed

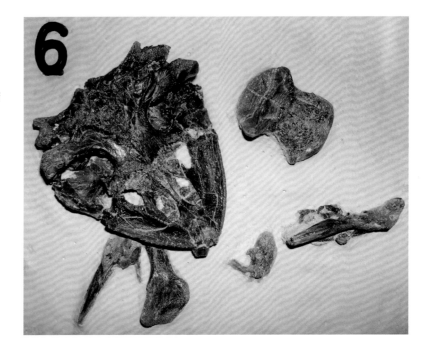

two deep, long gouges from a ripping shark bite. And, as if to add insult to the injury, nearly all the ribs showed the serrated bite marks of another shark, *Squalicorax falcatus*. In some areas it appeared as if the smaller shark(s) had grabbed hold of the ribs with serrated teeth and then slid downward, peeling off the flesh and leaving long, parallel scratches on the bone. I'm sure the mosasaur was long dead at the time, but the bite marks looked particularly painful. Although it's probably not possible to establish the sequence of who was dining when, I think that one or more large *Cretoxyrhina* tore the carcass apart first and then the smaller *Squalicorax* sharks finished cleaning the bones. Or the reverse order is possible: the smaller sharks may have fed first on a dead mosasaur and then been driven off by the larger predator. In any case, the remains preserve the interaction of three different species, unusual in the chalk.

From my observations over the years, I know that *Squalicorax* probably scavenged any and all kinds of food sources (fish, turtle, mosasaur, plesiosaur, pteranodon, and even other sharks) that were available. Druckenmiller et al. (1993) reported the discovery of a Kansas *Squalicorax falcatus* specimen that was closely associated with the remains of a fish (*Ichthyodectes ctenodon*), a small turtle (Fig. 4.6), and mosasaur paddle elements, which they interpreted as stomach contents of the shark. Bourdon and Everhart (2011) suggested that the presence of several *Squalicorax* teeth within the remains of a large *Cretoxyrhina* indicated scavenging of the carcass by a smaller shark or sharks.

The serrated teeth of *Squalicorax* left distinctive parallel scratch marks on the bones of many kinds of marine animals (Fig. 4.7). In fact, at least in the low chalk, it is unusual *not* to find *Squalicorax* bite marks on larger specimens. Even the large (9 m; 29 ft) *Tylosaurus proriger* (FHSM

METRIC

1 2 3

VP-3) specimen on exhibit at the Sternberg Museum has bite marks from serrated teeth on the right ilium, and a broken *Squalicorax* tooth was collected with the remains. Schwimmer, Stewart, and Williams, (1997) documented the evidence of *Squalicorax* scavenging on various species of vertebrates from specimens located in museum collections. As indirect evidence, their list of attributable bite marks (ibid.:77–79) includes 28 museum specimens, from a shark (*Ptychodus mortoni*, KUVP 59041) through a dinosaur (Nodosauridae, indet., in the San Diego Natural History Museum, SDNHM 33909). The list also includes many bony fishes, turtles, mosasaurs, and plesiosaurs from the Smoky Hill Chalk of Kansas.

More directly, they also note the tip of a *Squalicorax* tooth embedded in a mosasaur vertebra and in a juvenile hadrosaur metatarsal (Schwimmer, Stewart, and Williams, 1997:76). Both specimens are from Alabama. During the 1997 Society of Vertebrate Paleontology meeting in Chicago, I was with David Schwimmer when we discovered serrated bite marks on a juvenile hadrosaur limb bone (FMNH PR 27383) from the Selma group of the Mooreville Formation (Late Cretaceous), Dallas County, Alabama, in the Field Museum collection. Since then, I have noted many more instances of *Squalicorax* bite marks and embedded teeth in association with vertebrate remains in the Smoky Hill Chalk.

Although it is about as far from an ocean as you can get, Kansas is a great, if generally unrecognized, place to collect shark teeth from Pennsylvanian through Cretaceous time. See Shimada and Fielitz (2006) for an updated, comprehensive listing of the fishes from the Smoky Hill Chalk, including sharks. However, as happened with the first edition of this book, their list is slowly becoming outdated as we discover the remains

4.7. This is a close up of the proximal end of a mosasaur rib that has been scavenged by *Squalicorax* sp. It preserves the typical bite marks made by serrated teeth of this genus of sharks. Scale in cm.

Sharks in the Late Cretaceous Western Interior Sea

of more species in the chalk. I'll mention additions as we go further into this chapter and others in the book.

The first vertebrate remains actually reported from the state were two shark teeth and a dorsal fin spine from three different species of paleo-shark from the 'Carboniferous of Kansas' collected near Manhattan (Leidy, 1859). As settlements and the Kansas Pacific Railway moved westward, the first remains of Late Cretaceous sharks were recovered along the route of the railroad west of Salina. LeConte (1868:8) reported the "teeth and vertebrae of a species of shark belonging to the genus *Lamna* [*Cretalamna*] from rocks near Bunker Hill station (now Bunker Hill, Kansas, in Russell County)." Further west, on the bank of a small stream near Walker Creek, LeConte also noted that "the strata of slaty [*sic*] clay contains many teeth of fishes and sharks" (ibid.). Rock exposures in this part of Kansas are mostly from the Greenhorn and Carlile formations (Upper Cenomanian to Middle Turonian) that underlie the Niobrara Formation.

Leidy (1868) was the first to report on a tooth of *Ptychodus* (a shell-crushing shark) from the "Cretaceous series" of western Kansas when he described a single, damaged tooth from "a few miles east of Fort Hays, Kansas." The tooth was collected by Joseph LeConte during the survey for the Kansas Pacific Railway in 1867 and donated to the Academy of Natural Sciences of Philadelphia (ANSP) upon his return to Philadelphia. Leidy considered it to be a new species and named it *Ptychodus occidentalis*. Although the type specimen was reportedly lost from the collection of the ANSP (Spamer, Daeschler, and Vostreys-Shapiro, 1995), I was able to examine the collection at the ANSP in 2009 and photographed a large tooth from the same locality that was definitely *P. occidentalis* (ANSP 1465; Fig. 4.8). This tooth is figured in Leidy (1873:pl. 17, figs. 7, 8). A handwritten note with the tooth indicates that it was "collected 3 miles east of Ft. Hays by Dr. LeConte." Although the exact locality has not been identified, exposures along Big Creek in that area of eastern Ellis County are most likely Middle Turonian Blue Hill Shale. The dark color of the tooth is typical for fossils from the Blue Hill Shale.

Leidy (1873:pl. 18, figs. 1–10 [*Ptychodus*], figs. 21–23 [*Cretoxyrhina*] and figs. 29–31 [*Squalicorax*]) was also the first to publish figures of shark teeth collected from the Smoky Hill Chalk, reporting on specimens collected and donated by Dr. George M. Sternberg during his tour of duty with the U.S. Army in Kansas in the late 1860s. Dr. Sternberg, the older brother of Charles H. Sternberg, was one of the first fossil collectors in western Kansas, discovering the type specimen of *Xiphactinus audax* (Chapter 5), but he is better known for his work in bacteriology and as the surgeon general of the army during the Spanish-American War (Sternberg, 1920).

While O. C. Marsh appears to have purposefully ignored all fishes and sharks in Kansas, E. D. Cope did just the opposite, and as a consequence he named a number of new species, mostly bony fishes. Possibly without realizing it, Marsh recovered two *Squalicorax kaupi* teeth (YPM

56409) in association with *Pteranodon* remains (*Pteranodon longiceps?* YPM 1169; Shimada, pers. comm., 2003) that he personally collected in July 1871 (Marsh, 1871).

Cope (1872a:324) noted that in the "Cretaceous ocean of the West, . . . sharks do not seem to have been so common as in the old Atlantic." Later that same year, Cope stated that "the remains of sharks and rays are far less abundant in the Cretaceous of Western Kansas than in New Jersey," but then added that "in the region near Fort Hays and Salina [the Greenhorn and Carlile formations], shark's teeth are more frequently found" (1872b:355–357). Following brief descriptions, Cope then named two new species of sharks from the Niobrara "near Fort Wallace" (ibid.:357). *Galeocerdo crassidens* Cope (*Squalicorax kaupi*; AMNH FF 1908) was named from two teeth, and *G. hartwelli* Cope (*Squalicorax falcatus*; AMNH FF 1909) from a single tooth located beneath the bones of Cope's giant *Protostega gigas* specimen (Chapter 6). Appropriately, the *hartwelli* species was named for one of Cope's U.S. Army assistants on that dig, Martin V. Hartwell. Photographs of the teeth (Fig. 4.9) were published in Hussakof (1908:fig. 6). Siverson (pers. comm., 1998) suggested that *Squalicorax hartwelli* (Cope 1872b) should probably be used for the teeth of this shark genus collected in North America instead of *S. falcatus* (Agassiz, 1843), which was originally authored to describe similar teeth from Europe. The revision, although it makes good sense to me, is not without some disagreement by other shark experts and has not yet been published.

In Cope's (1875) *Vertebrata*, he again mentions the two *Galeocerdo* species along with naming a new species of shell-crushing shark, *Ptychodus janewayi* (named for Dr. John H. Janeway, an army surgeon at

4.9. The type specimens of *Galeocerdo crassidens* Cope and *G. hartwelli* Cope in the collections of the American Museum of Natural History. They are considered to be junior synonyms of *Squalicorax kaupi* Agassiz and *S. falcatus* Agassiz. Adapted from Hussakof (1908, fig. 6).

Fort Hays who assisted Cope when he visited Kansas). He further notes that the *P. janewayi* teeth had been collected near Stockton, in Rooks County, "in a bed containing many teeth of 'Oxyrhina' (*Cretoxyrhina*) and 'Lamna' (*Cretalamna*)" (1875:243). For some reason, instead of illustrating those specimens, Cope described and figured (ibid.:pl. 42, figs. 9–12) several teeth of *Lamna* (*Leptostyrax*) *macrorhiza* Cope from Ellis County and *L. mudgei* Cope from the "Niobrara Epoch of Kansas" (ibid.:297). It seems fairly apparent from his cursory remarks that Cope was not interested in shark teeth while dealing with the wealth of new species of fishes and marine reptiles coming from Kansas.

Expressing a different point of view on the abundance of shark teeth, B. F. Mudge reported that the "teeth of Selachians are quite common. At one locality, over 400 were collected in an area of 30 inches, and apparently from the jaws of one individual—a *Ptychodus*—and all in excellent preservation" (1876:217). In a report to the Kansas Academy of Science at the annual meeting, Mudge noted the discovery of "the teeth, cartilaginous jaw and vertebrae of a shark, *Galeocerdo* [*Squalicorax*] *falcatus*" (1877a:5). In another version of that paper, Mudge reported the evidence of scavenging on marine reptiles by sharks, noting that "frequently . . . the bones of Saurians were found with the marks of the serrate teeth of *Galeocerdo*, which could not have been made unless the bones were still fresh and unhardened" (1877b:287). A fragmentary *Xiphactinus audax* skull (KUVP 155) collected by B. F. Mudge in the early 1870s from the Carlile Shale (middle Turonian) of Russell County shows numerous *Cretoxyrhina* bite marks on the lower jaw.

Earlier, Cope had also mentioned scavenging by sharks in his description of mosasaur remains he had discovered: "While lying on the bottom of the Cretaceous sea, the carcass had been dragged hither and thither by the sharks and other rapacious animals, and the parts of the skeleton were displaced and gathered into a small area" (1872a:322). Williston noted that the bones of mosasaurs "frequently bear the impression of teeth of post-mortem origin, and in many cases I have found the teeth of small sharks imbedded in them" (1898:214). Unfortunately, none of these early specimens were properly identified and curated into a collection when they were described.

Williston (1900) was probably the next author to discuss in detail the occurrence of sharks in the Cretaceous of Kansas, listing eight species of *Ptychodus*, three species of Scyllidae (cat sharks), and twelve species of lamnids (*Cretoxyrhina*, *Cretalamna*, etc.) that had previously been identified from the state. Three of the *Ptychodus* species (*P. anonymous*, *martini*, and *Ptychodus* sp.) were newly described by Williston himself. Only three of the *Ptychodus* species were reported from the chalk (*P. mortoni*, *martini*, and *P. polygyrus*); the rest were from the 'Benton Cretaceous,' the older rocks that underlie the Niobrara Formation. The 'Benton Cretaceous,' or 'Fort Benton,' is an obsolete stratigraphic term (Hattin, 1982; see also Schumacher and Everhart, 2005) that includes the Graneros, the Greenhorn, and the Carlile formations (middle Cenomanian through middle Turonian time).

All of the '*Scyliorhinus*' specimens identified by Williston were collected from the Kiowa Shale (Early Cretaceous) in Kiowa County and are currently housed in the USNM collection. When I asked Robert Purdy (pers. comm., 2003) to reexamine them, he indicated that they were more likely to be *Leptostyrax*. More than half of the lamnid species reported by Williston (1900) were also from the Kiowa Shale (Early Cretaceous) of Clark County, Kansas. Williston (ibid.) was unable to substantiate Leidy's *Lamna sulcata* or Cope's *L. mudgei* from the specimens he examined. Four other species—*Lamna* (*Odontaspis?*) sp., *Lamna* sp., *L. quinquelateralis*, and *Leptostyrax bicuspidatus*—are indicated to have been collected from the Kiowa Shale of Clark, Kiowa, and McPherson counties. *Corax* (*Squalicorax*) *curvatus* and *Lamna* (*Leptostyrax*) *macrorhiza* were reported from the 'Benton Cretaceous' of Ellsworth and Ellis Counties, respectively. The only shark species named from the Niobrara (here meaning the Smoky Hill Chalk) were the three ptychodontids, *Isurus* (*Cretoxyrhina*) *mantelli*, *Lamna* (*Cretalamna*) *appendiculata*, and *Corax* (*Squalicorax*) *falcatus*. From its description, Williston (ibid.:251–252) did not believe that the single specimen of *Scapanorhynchus raphiodon* collected by Hayden and figured by Leidy (1873:pl. 18, fig. 49) was even from Kansas. However, Hattin (1962) reported *Scapanorhynchus raphiodon* from the Carlile Shale, and more recently, *Scapanorhynchus* has been reported from the lower Smoky Hill Chalk (Hamm and Shimada, 2002) and the Blue Hill Member of the Carlile Formation (Everhart et al., 2003).

One of the most interesting aspects of Williston's shark paper was the publication of photographs of the shark teeth in the University of Kansas collection that he was describing. So far as I am aware, this was

4.10. The teeth and replacement teeth of *Squalicorax kaupi* (FHSM VP-2213) in medial view as preserved in place on the calcified cartilage of the right upper jaw.

the first time such photos had been made available in the study of shark teeth. I also appreciate being able to visit the Museum of Natural History at the University of Kansas and examine many of the same teeth that were cited and photographed well over a hundred years ago. Lane (1944) summarized Williston's information on sharks and reprinted several of his figures without significant modifications or additions. Schultze et al. (1982) inventoried the type and figured shark specimens in the collection of the University of Kansas Museum of Natural History, but other than modernizing the names and listing many more specimens, his list of species was essentially the same as in Williston's time.

One curious omission from the report by Schultze et al. (1982) was *Squalicorax kaupi*, a species first collected and described more than a hundred years earlier from the chalk by Cope (1872b:355–357) as *Galeocerdo crassidens*. This is not a reflection on their work since the teeth would have to have been figured in a publication to be included in the review. *Squalicorax kaupi* is currently well recognized in the Santonian and Early Campanian chalk, and a nicely preserved set of teeth is still attached to the calcified cartilage of a jaw (FHSM VP-2213) in the Sternberg Museum collection (Fig. 4.10).

Stewart (1990) was the first to define the biostratigraphic occurrence of sharks in the Smoky Hill Chalk (Fig. 4.11). From his review of museum collections and experience in the field, he (ibid.:29) noted that *Cretoxyrhina mantelli, Cretalamna appendiculata, Scapanorhynchus raphiodon, Pseudocorax laevis,* and *Squalicorax falcatus* occur in the lower one-third of the chalk, along with the ptychodontids *P. mortoni, anonymous,* and *P. martini.* Stewart (ibid.:27) also noted that he had recovered the teeth of *Rhinobatos* sp. (a guitarfish) from acidized

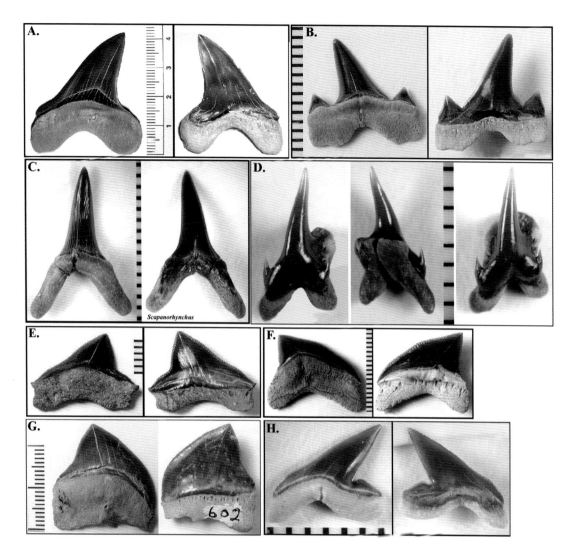

samples of chalk, but he did not assign them to a biostratigraphic zone. In an updated but unpublished list, Stewart (pers. comm., 1993) indicated that *Rhinobatos incertus* was present in the uppermost chalk (zone of *Hesperornis*). This made me curious, and I started looking for *Rhinobatos* in the Cretaceous rocks of Kansas. Although very small, the teeth of this extinct guitarfish happen to be present in nearly every rock strata that I examined (Everhart, 2007).

While all the ptychodontids were extinct in the chalk by the middle Santonian, the teeth of *Cretoxyrhina*, *Cretalamna*, and *Squalicorax* occur through the rest of the deposition of the chalk. *Squalicorax falcatus*, however, appears to have been largely replaced by S. *kaupi* by the early Campanian, which appears as early in the chalk as the middle Santonian (pers. obs.). S. *pristodontus*, a larger species, first appears right at the very top of the chalk. Two teeth of S. *pristodontus* (FHSM VP-15009 and 15010) were collected by Pete Bussen just below the contact of the Smoky Hill Chalk with the Pierre Shale near Twin Buttes in Logan County.

4.11. The teeth of several species of sharks occur in the Smoky Hill Smoky Hill Chalk with the greatest variety occurring in the lower half of the formation. *Cretoxyrhina* and *Squalicorax* are the most common, while *Scapanorhynchus* and *Johnlongia* are generally the smallest. A, *Cretoxyrhina mantelli*; B, *Cretalamna appendiculata*; C, *Scapanorhynchus* cf. *raphiodon*; D, *Johnlongia* sp.; E, *Squalicorax falcatus*; F, *Squalicorax kaupi*; G, *Squalicorax pristodontus*; H, *Pseudocorax laevis*. Scale in mm.

Stewart (1990:23) also noted that the last occurrence of *Cretoxyrhina* in the middle of North America is near the top of the chalk (early Campanian). In addition, my observation is that the largest shed teeth of *Cretoxyrhina* are located in the lower half of the chalk and that *Cretoxyrhina* teeth become generally smaller and fewer as you go higher in the formation. It is interesting that this geologically sudden decline of a large and successful shark coincides with the dramatic increase in the number and size of mosasaurs in the Western Interior Sea, but proving that one was the cause of the other is probably not possible.

Shimada (1996) reported on the occurrence of selachians in the Fort Hays Limestone Member of the Niobrara. As might be expected, the "usual suspects" or species that occur in the Smoky Hill Chalk also occur in the underlying limestone, including *Ptychodus mortoni*. Shimada (ibid.) added a single tooth of a new and apparently rare species, *Paranomotodon* sp. (FHSM VP-12463) to the Niobrara Chalk shark species list. The following year, Shimada discussed the stratigraphic occurrence (1997a), the dentition (1997b), and the paleoecology (1997c) of *Cretoxyrhina mantelli* in three papers that resulted from extensive studies conducted as part of his graduate research. Shimada (2008) also described a new, smaller species of *Squalicorax* (*S. microserratodon*; KUVP 141876) from the Smoky Hill Chalk. It was originally collected by the Sternberg family, and the stratigraphic occurrence of this species is unclear.

In addition, Shimada (1997d) reported on the very large vertebral centra (KUVP 16343) of an unknown species of shark from the Kiowa Shale (Upper Albian) of Clark County. The centra, if complete, would have been 14.5–18 cm (6–7 in) in diameter. As a comparison, the largest centra from a recently discovered 5.5–6 m (18–19.7 ft) *Cretoxyrhina mantelli* (FHSM VP-14010) were 10 cm (3.9 in) in diameter. The Kiowa specimen is the largest centra of a Mesozoic shark known to date and came from a shark estimated by Shimada (ibid.) to be 8.3–9.8 m (27–32 ft) long. Unfortunately, no additional specimens of this huge shark have been collected to date.

Two other publications from outside Kansas regarding Cretaceous sharks should be mentioned here. Meyer (1974) wrote his Ph.D. dissertation on the sharks of the Gulf Coast (Mississippi and East Texas Embayments), and Welton and Farish (1993) published a well-illustrated book on the species of sharks and rays from the Cretaceous of Texas. Both are excellent resources for identifying shark teeth from the Cretaceous of Kansas and will be even more useful as shark faunas from other formations in the state are investigated.

In addition to shed teeth, the "soft parts" of some larger sharks from the Smoky Hill Chalk are well represented. The largest and most complete specimens ('shark mummies') of *Cretoxyrhina* are nearly all from the low chalk of Gove and Trego counties. These specimens usually consist of a significant portion of the shark, including calcified cartilages of the cranium and jaws, teeth, dermal denticles, vertebral centra, and even fin elements. Mudge (1877a:5) was among the first to mention this unusual preservation. In 1891, Charles H. Sternberg discovered the remains of a very large *Cretoxyrhina* in the lower chalk. The specimen was acquired by the Ludwig-Maximilian University of Munich, where it was examined by Charles R. Eastman, an American student studying in Germany. Eastman (1895) wrote his doctoral dissertation on the specimen and later returned to the United States, where he studied fossil fishes.

Sternberg later commented on the discovery of this "unusual specimen of an ancient shark, *Oxyrhina mantelli*, discovered on Hackberry creek, Gove County, Kansas, in 1890 [*sic*]. The remarkable thing about this specimen is that the vertebral column, though of cartilaginous material, was almost complete and that the large number of 250 teeth were in position. When Chas. R. Eastman, of Harvard, described this specimen, it proved so complete as to destroy nearly thirty synonyms used to name the animal, and derived from many teeth found at various former times" (1907:124). Two years later, Sternberg again described the discovery of this 6 m (20 ft) shark, probably from Gove County, south of Park, Kansas, and noted that the specimen was nearly complete, with "over 250 teeth." "This is the first time and, I believe, the only time that so complete a specimen of this ancient shark has been discovered. The column and other solid parts were composed of cartilaginous matter which usually decays so easily that it is rarely petrified. I suppose my specimen was old at the time of its death, and bony matter had been deposited in the cartilage. It is not very likely that such a specimen will ever be duplicated. Dr. Eastman's study of this skeleton enabled him to make synonyms of many species which had been named from teeth alone" (1909:113–114).

Sternberg mentions the discovery again and then describes a new find by his son George F.:

> The crowning discovery of our work here was the discovery by George Sternberg of a nearly complete skeleton of a great shark, Lamna. . . . In 1891, while employed by the Munich Museum, I discovered the first and most complete skeleton known of the shark *Oxyrhina* [*Cretoxyrhina*] *mantelli* in the same vicinity. This was made the subject of Dr. C. R. Eastman's inaugural address delivered before the Ludwig-Maximilian University of Munich for his Ph.D. degree. The specimen we collected on the south side of Hackberry creek, South of Banner post office, in Trego County, includes the plates of the mouth holding the teeth, of which some 150 are in sight, and the entire column of flattened disk-like vertebrae to within five feet of the end of the tail. The total length is about twenty feet of the preserved head and column. I think this will prove one of the greatest scientific discoveries of the year, because the sharks are cartilaginous, and consequently their skeletons are not preserved. But in the two cases mentioned enough bone was deposited to preserve the greater part of the skeleton. (1911:71)

Unfortunately, the *Cretoxyrhina* specimen in the Munich museum was destroyed in World War II. The second specimen (KUVP 247) mentioned by Sternberg is currently on exhibit in the collection at the University of Kansas. The shark was mentioned again by Sternberg when he noted that "I have preserved in the Museum of the University of Kansas a shark twenty-five feet long, and mingled with his remains were the bones of a *Portheus* [*Xiphactinus*]" (1917:162). Shimada noted that the specimen was a "nearly complete skeleton of *Cretoxyrhina mantelli*" and "is associated with many bones of *Xiphactinus audax* (KUVP 245), including the cranium, jaws, ribs and other elements throughout the matrix. They most likely represent stomach contents" (1997c:927, figs. 1–2). Contrary to Sternberg's description, shark cartilage does not become 'ossified' or filled with bone. The cartilage does in rare cases, however, accumulate enough calcium to be preserved as a fossil.

The remains of another George Sternberg specimen, a 5 m (16 ft) *Cretoxyrhina mantelli* (FHSM VP-2187; see Shimada, 1997a, 1997b, 1997c) are on exhibit in the Sternberg Museum of Natural History. The 'head' of this specimen consists of calcified cartilage (cranium and jaws) and several hundred teeth, and is covered with scales (shagreen or dermal denticles). The rest of the specimen is made up of a nearly continuous series of calcified vertebrae. The fins and tail cartilages are missing. The largest teeth in the front of the jaw are nearly 5 cm (2 in) in height.

In April 2002, I discovered an even larger specimen (FHSM VP-14010) in the lower chalk (late Coniacian) of southern Gove County. The remains, including skull, pectoral fins, and vertebrae, were spread out over a distance of about 5 m (16 ft) and did not include the tail. Based on comparisons of tooth and vertebrae sizes with FHSM VP-2187, Corrado et al. (2003) conservatively estimated the length of FHSM VP-14010 to be 5–5.5 m (16.5–18 ft). However, because the posterior vertebrae were still more than 7 cm (3 in) in diameter where the tail had eroded out of the hill, I believe that about 2 m (6.5 ft) of the shark's tail had eroded away years earlier, and that it would have been between 6 and 7 m (19.7–23 ft) long. In either case, it was a very big shark.

Reasonably complete remains of smaller species of Late Cretaceous sharks are also known from the chalk, including a very nice Kansas specimen of *Squalicorax falcatus* (USNM 423665) on exhibit in the Smithsonian (United States National Museum) and the remains of the *S. kaupi* specimen (FHSM VP-2213) mentioned above. In 1992, I discovered another specimen in Gove County that preserved the anterior half of a 2 m (7 ft) long *S. falcatus*. The shark remains were upside down and had possibly been complete before eroding out of the chalk at the edge of a gulley. As it was, I noticed a number of half-dollar-sized (3 cm) vertebrae lying on the surface of the chalk. They had come from the midsection of the shark. If present originally, the back half of the remains had probably been washed down the gully over the previous several years. Fortunately, the anterior half, including a complete skull and most of the teeth, was still safely buried in chalk. After picking up the loose pieces, I was able to remove the head and anterior vertebrae in one piece on a slab of chalk and carry it out of the field. The specimen (CMC VP5722) is now in the collection of the Cincinnati Museum Center (Fig. 4.12).

These 'shark mummies' provide much more information about these extinct sharks, but they are relatively rare. The teeth of these Late

4.12. The well preserved "head" of a *Squalicorax falcatus* shark (CMC-VP5722) in ventral view from southeast Gove County, Kansas. The specimen preserves teeth from the upper and lower jaws at left and center, and six vertebrae at right. Much of the surface of the specimen is covered with patches of denticles from the sharks skin. Scale bar = 10 cm (4 in).

Cretaceous sharks, however, are the most common vertebrate fossils collected from the chalk. Collecting shark teeth is a hobby for many of the local residents of western Kansas, and a number of extensive collections have resulted. The collection in the Fick Fossil and History Museum in Oakley, Kansas, exhibits more than 10,000 shark teeth, most of which were picked up one at a time during the lifetimes of Ernest and Vi Fick from chalk exposures near Hell Creek on their ranch in southwestern Gove County. A donation of fossils collected from the chalk in the 1930s by C. Y. Stout that is housed in the Sternberg Museum contains more than 2,000 shark teeth. Where do all these teeth come from?

Consider that a ginsu shark (*Cretoxyrhina mantelli*) has 42 rows of teeth in its upper jaws (21 on each side) and 38 in the lower jaws (Shimada, 1997b). The working teeth in the front of each row are shed and replaced continuously throughout the life of the shark, at intervals of a few weeks to several months. This means that an older shark would shed or otherwise lose thousands of teeth during its lifetime. These teeth are replaced by what amounts to a shark tooth assembly-line process. Behind each "working" tooth in the front row is a series (family) of replacement teeth in various stages of completion. The crowns of the teeth develop first and the roots are completed last as the tooth moves downward or upward toward the edge of the jaw. Each succeeding tooth in the family is also slightly larger than the tooth it replaces, reflecting the continued growth in the size of the shark and its jaw.

When a tooth is lost, the next tooth in the series is moved over the edge of the jaw, rotated about 180°, and put to work. In *Cretoxyrhina*, there may be up to 6 replacement teeth in each row, so a single shark could have up to 560 (a total of 80 teeth in the upper and lower jaws x 7 teeth in each row) teeth, including developing bud teeth, in its jaws at any one time. The actual number is probably less because the tiny symphysial

4.13. Three views of the second tooth of *Johnlongia* sp. (FHSM VP-17602) collected from the Smoky Hill Chalk, Trego County, Kansas. Scale in mm.

teeth (8 upper, 2 lower) at the center of the jaws are more or less vestigial and probably aren't replaced in the same manner. Other sharks have differing numbers of teeth rows and differing rates of replacement, but all shed their teeth continuously throughout their lives.

If the collections I have seen are an accurate indication, the large, razor-edged teeth of the ginsu shark (*Cretoxyrhina mantelli*) and smaller, serrated teeth of the crow sharks (*Squalicorax falcatus* and *S. kaupi*) are picked up more frequently in the Smoky Hill Chalk than all the other species combined. Teeth of other species, including *Cretalamna appendiculata*, *Scapanorhynchus raphiodon*, and *Pseudocorax laevis*, are collected much less often. Generally this is because the smaller size of these teeth makes them harder to see, but it is also because these sharks occurred in smaller numbers in the deeper waters present during the deposition of the chalk near the center of the Western Interior Sea. A number of teeth of a new species of *Galeorhinus* (cow shark) were recovered (Mike Triebold, pers. comm., 2002; J. D. Stewart, pers. comm., 2002) several years ago during the preparation of the remains of a very large (5.1 m; 17 ft) *Xiphactinus audax* that I discovered in 1996. As of 2004, however, the description has yet to be published. In 2003, a single tooth of a recently named genus of shark (*Johnlongia* sp.; FHSM VP-15721) first reported from Canada and Australia was collected from the lower chalk of Trego County by Keith Ewell (Cappetta, 1973; Siverson, 1996; Cicimurri, 2004; Shimada, Ewell, and Everhart, 2004; Shimada and Fielitz, 2006). A second *Johnlongia* tooth was discovered in 2010 (FHSM VP-17602; Fig. 4.13), which provides an idea of just how rare these tiny teeth really are in the Kansas chalk.

Since then, additional collecting has continued to produce more species from the chalk as well as from the underlying Dakota, Greenhorn, and Carlile formations and the overlying Pierre Shale. Siverson et al. (2015) described two new species of *Cretalamna* from the Smoky Hill Chalk: *C. ewelli* sp. nov. (holotype = FHSM VP-18510) was described from individual teeth collected in the lower chalk (late Coniacian) of Trego County by my friend Keith Ewell, and *C. hattini* sp. nov. was

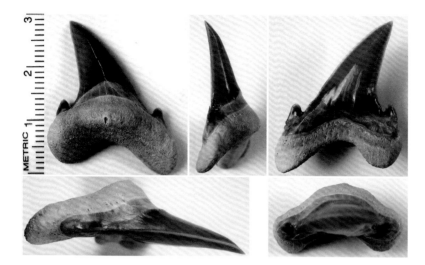

4.14. A nearly perfect tooth of *Cardabiodon* sp. that I collected from the Pfeifer Shale Member of the Greenhorn Formation in Russell County, Kansas. Scale in mm.

named from the upper chalk of Logan County based on a partial shark skeleton and about 120 teeth (LACM 128126; see also Shimada, 2007). The new species is named for Donald E. Hattin, who published (1982) the most current stratigraphic reference to the Smoky Hill Chalk.

In 1999, Siverson described a new genus and species of a large lamniform shark (*Cardabiodon ricki*) from Australia. Based on the size of the vertebrae in the type specimen, it is probably larger than *Cretoxyrhina*, but with relatively smaller teeth for its size. A second new species (*C. venator*) was subsequently described by Siverson and Lindgren (2005) from the Middle Turonian Western Interior Sea deposits of Montana. Almost immediately we began to recognize that there were *Cardabiodon* specimens from our Carlile and Greenhorn formations here in Kansas. Keith Ewell collected several *Cardabiodon* teeth from the Greenhorn Formation in 2005 that Siverson was able to identify from pictures on my Oceans of Kansas web site. I collected my first specimen (FHSM VP-17141) from the Pfeifer Shale Member, just below the Fencepost Limestone in Russell County in July 2007 (Fig. 4.14). Dickerson et al. (2012) described an associated set of *Cardabiodon* teeth from a specimen (FHSM VP-425) in the Sternberg Museum collection that had been collected in 1982, and then misidentified as *Cretoxyrhina mantelli*. Although *Cardabiodon* was a successful genus around for millions of years, with apparently worldwide distribution, both it and another large shark called *Cretodus* become extinct during Turonian time in the Western Interior Sea and are replaced by *Cretoxyrhina* at the top of the food chain. We have no idea why these large sharks went extinct.

Ptychodus

The domed, shell-crushing teeth of ptychodontid sharks such as *Ptychodus mortoni* and *P. anonymous* are occasionally collected one at a time like the teeth of other sharks. Sometimes, however, they occur in large numbers where they have eroded out on the surface of the chalk, or even as articulated jaw plates from a single individual (Fig. 4.15). These

4.15. Assorted teeth of *Ptychodus rugosus* (FHSM VP- 2223), a shell-crushing shark from the late Coniacian chalk of Kansas. Scale in cm.

unusual sharks became extinct by the middle Santonian in Kansas, and their teeth occur only in the lower one-half of the chalk. Due to a lack of complete specimens, we were uncertain what they looked like. In the past, they have been described as looking like modern rays, or like "normal" sharks.

Woodward noted that the dentition of *Ptychodus* "is that of a true Ray, and does not bear the slightest resemblance to that of the Cestraciont [hybodont] Sharks" (1887:128). Stewart briefly described a partial specimen of *Ptychodus mortoni* (KUVP 59041) that included calcified vertebral centra and suggested that "all living sharks and rays (including *Heterodontus*) are members of the monophyletic Neoselachii, united by synapomorphies including the presence of calcified centra. Since *Ptychodus* shares this derived state, it must be regarded as a neoselachian and not as a hybodont" (1980:154). One *P. anonymous* (now *P. rugosus*) specimen (AMNH 19553) collected recently in Gove County included more than 200 teeth, 5 vertebral centra, and hundreds of oral scales (denticles) from inside the mouth of the shark (Everhart and Caggiano, 2004). An examination of the vertebral centra and oral denticles from this specimen indicated that they were similar to those of *Squalicorax*.

Shimada et al. (2009) included *Ptychodus* among the rays (Hybodontiformes). The following year, Shimada et al. (2010; I was a coauthor) carefully avoided classifying *Ptychodus* as a hybodont, or neoselachian shark. Instead we opted for a possible compromise body form (the nurse shark, *Ginglymostoma*) in preparing a figure showing the estimated (gigantic) size of the specimen (FHSM VP-17415) we were describing. Shimada then cautioned against assigning *Ptychodus* to a specific group, noting that "the claim that ptychodontids are neoselachians has not been satisfactorily substantiated yet, whereas the alternative hypothesis that ptychodontids may be nested within Hybodontiformes remains plausible" (2012:1250–1251). Up to that point, there was no conclusive evidence that would assign them to one order or the other.

The breakthrough came at the at the Annual Meeting of the Kansas Academy of Science in 2015, when Brian Hoffman (Hoffman, Claycomb, and Hageman, 2015) presented information about their discovery that *Ptychodus* (*P. rhombodus*) teeth had a triple layer of enameloid, appearing identical at the microscopic level to the teeth of neoselachian sharks (e.g., *Cretoxyrhina* or *Squalicorax*). This is a very different structure from the single crystallite enameloid that occurs in the teeth of batoids and hybodontids. Their follow-up paper (Hoffman, Hageman, and Claycomb, 2016) was published in the Journal of Paleontology and apparently closes the issue; *Ptychodus* was a neoselachian shark.

Because the 'pavement' toothed upper and lower jaws of *Ptychodus* are superbly adapted for crushing hard-shelled prey such as inoceramid clams and other bivalves, it is generally assumed that they must have fed on them. However, there are no specimens of this shark that preserve stomach contents. Wear facets on some teeth do indicate that some hard-shelled items were routinely consumed. Kauffman (1972) interpreted a series of depressions along the edge of a single, 12.5 cm (4.9 in) long shell of *Inoceramus tenuis* from the English Chalk (Lower Cenomanian) as bite marks of *Ptychodus decurrens*. Kauffman suggested that the marks were probably made by the lateral teeth along the edge of the shark's jaw. If so, the size of the depressions, about 1.5 cm (0.6 in) across, would indicate that the shark had been quite large. It is difficult to explain why the shell was not completely crushed by the bite of the shark, but Kauffman (ibid.:441) suggests that the bivalve may have already been dead and was rejected by the shark. While this scenario is certainly possible, the lack of similar remains from other localities, including the Western Interior Sea, raises questions as to the cause of these presumed bite marks.

Although I agree that ptychodontid sharks most likely fed on clams and other hard-shelled prey, I suspect that they avoided the larger ones in favor of the more easily crushed and readily available smaller individuals. I think these sharks cruised near the bottom and scooped up mouthfuls of mud containing thin-shelled younger clams instead of trying to attack the thicker-shelled, full-grown inoceramids. That said, the large size of these sharks suggests that as opportunists, they could have eaten anything they wanted, including hard-shelled ammonites and baculites in the water column, or other smaller fishes.

In the Late Cretaceous of Kansas, several lines of evidence support the hypothesis that ptychodontid sharks were primarily bottom-feeders. First, from the Cenomanian through the early Coniacian, when ptychodontid sharks probably had their greatest diversity, inoceramid clams were generally fairly small. By the middle of the Coniacian in the ocean over Kansas, however, one species, *Volviceramus grandis* (see Fig. 3.5), became larger and dominated the sea bottom. Growing to a maximum of about 50 cm (20 in) across, their shells were quite thick (2–3 cm; 0.8–1.2 in), and thus were probably more massive than all but the largest of the shell-crushers would want to routinely tackle. However, these

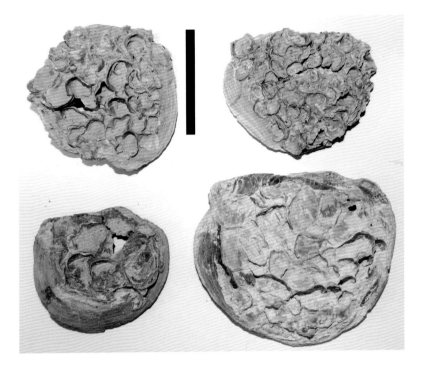

4.16. Examples of small *Volviceramus grandis* inoceramid shells from the late Coniacian chalk of Gove County, Kansas. Note that all are heavily colonized by *Pseudoperna congesta* oysters. Scale bar = 5 cm (2 in).

large inoceramids must have taken many years to grow to that size. From the time they hatched from tiny eggs as mobile (free-swimming) larvae until they reached a certain size, they were vulnerable to being eaten by a variety of predators, including the crusher sharks. We find small, fragile masses of jumbled calcite prisms from tiny inoceramid shells occasionally in the low chalk. I interpret these as either the coprolites or regurgitate of a ptychodontid shark (see Hattin, 1996, for further discussion). It is readily apparent that something had been feeding on the abundance of young inoceramids and left the indigestible remains of their shells behind. Incidentally, the smallest complete shells of *Volviceramus grandis* that we see (rarely) in the lower chalk are only about 6–8 cm (2.5–3.3 in) in diameter (Figure 4.16).

Coincidentally with the increased growth in size and thickness of the shells of *V. grandis*, most of the ptychodontids became extinct in the Western Interior Sea by the end of the Coniacian. I do not believe that one was necessarily the result of the other, since the remaining species of *Ptychodus* would still have been able to feed on the younger, more fragile juvenile bivalves. However, following the extinction of *V. grandis* in the early Santonian, another species of giant inoceramid, *Platyceramus platinus*, became the dominant bivalve on the sea bottom. It is this species that is often mentioned as growing to 1.2 m (4 ft) or more in diameter (Chapter 3). However, this species was very thin-shelled; obviously it was able to put less energy and resources into building its large shells than did *V. grandis*, possibly because of the absence of predators such as the ptychodontid sharks.

Although ptychodontid sharks reach their greatest diversity in Kansas in the lower one-third of the Smoky Hill Chalk (late Coniacian), their

4.17. A small lateral tooth of *Ptychodus occidentalis* that I collected from the upper Dakota Formation (Middle Cenomanian) of Russell County, Kansas. This specimen is the earliest known example of a ptychodontid shark in Kansas. Scale in mm.

first occurrence in Kansas occurred much earlier (middle Cenomanian), where *Ptychodus occidentalis* first occurs at the transition between the Dakota Sandstone and Graneros Shale (Everhart, pers. obs., 2004; Fig. 4.17). By the beginning of the upper Cenomanian *P. whipplei* and *P. anonymous* are also present, in addition to *P. decurrens*, in the basal Lincoln Limestone Member of the Greenhorn Limestone that overlies the Graneros Shale (Everhart, pers. obs.; Liggett et al., 2005; Shimada and Martin, 2008). Williston (1900) described *P. anonymous* as a new species of ptychodontid shark from the 'Benton Formation' (= obsolete term for the Graneros Shale, Greenhorn Limestone, and Carlile Shale formations) but Herman (1977) later reidentified the teeth as *P. mammillaris*. For a number of reasons, including the loss of the type specimens, Herman's suggestion has been ignored. Until recently *P. anonymous* was also regarded as a valid species in the Smoky Hill Chalk, including the description of a specimen by Everhart and Caggiano (2004) that included associated teeth, vertebrae, and dermal denticles.

That said, our 2004 identification was overturned by Hamm (2010a), who determined that what Williston (1900) had identified as *Ptychodus anonymous* as a new species in the 'Benton Formation' was not the same species as what we were seeing in the Smoky Hill Chalk. Hamm concluded that the name should be changed to *P. rugosus*, originally described by Dixon (1850) from the English Chalk. However, Hamm did suggest that *P. anonymous* is still a valid species for specimens from the Carlile and Greenhorn formations. Viewed another way, it may be that *P. anonymous* evolved into *P. rugosus* before going extinct in the Late Coniacian–Early Santonian.

4.18. A tooth of *Ptychodus martini* from the lower chalk of Gove County, Kansas. *P. martini* is one of a few species of ptychodontid sharks with low, nearly flat crowns on their teeth. This suggests that they were feeding on different prey than the species with the higher, more robust crowned teeth. Scale in mm.

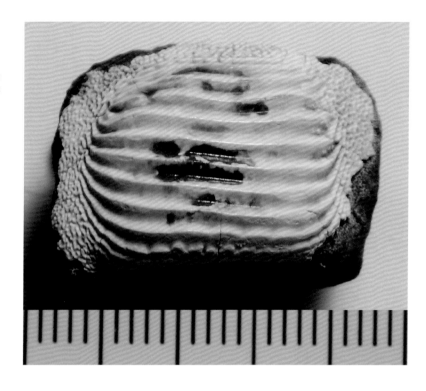

Cope (1874:48) was the first to report *Ptychodus polygyrus* from Kansas, based on a single tooth "found by Professors Mudge and Merrill" in the Niobrara of Ellis County. Although the fate of that specimen is unknown, another large tooth (KUVP 55237) was provisionally reported by Williston (1900:240) as *P. polygyrus* from "the lower beds of the Niobrara Cretaceous of the Smoky Hill River" and is presently in the collection of the University of Kansas. Three other specimens from the Smoky Hill Chalk (FHSM VP-76, VP-2123, and VP-15008) are in the collection of the Sternberg Museum. In a second paper, Hamm (2010b) reidentified these sometimes very large teeth as *P. marginalis* (originally described and named by Agassiz in 1839). Schumacher (pers. comm., 2003) showed me pictures of an associated set of 264 *P. marginalis* teeth from Russell County in a private collection. If the specimen is from the Fairport Chalk (middle Turonian) as we suspect, it would be the best and earliest known example of the species in Kansas.

Another rare species with relatively flat tooth crowns, *Ptychodus martini*, also occurs rarely in the lower third of the chalk (Late Coniacian). *P. martini* is represented by two relatively complete jaw plates: the type specimen (KUVP 55271) described and figured by Williston (1900) consisting of 110 associated teeth and another specimen of a smaller individual reported by Hamm and Everhart (1999) with 170 associated teeth. The second specimen occurred below Hattin's marker unit 4 in southeastern Gove County. Over the years, Pam and I have collected several individual teeth of *P. martini* (Fig. 4.18), and other individual teeth are curated in the collection of the Sternberg Museum, including FHSM VP-2121.

4.19. The first known associated set of *Ptychodus mammillaris* teeth from Kansas (FHSM VP-17989). Scale bar in mm.

What we know about the stratigraphic occurrence of *Ptychodus* species in Kansas has improved significantly in recent years. The earliest occurrences of the genus *Ptychodus* in Kansas are specimens of *P. occidentalis* Leidy 1868 from the top of the Dakota Formation (Middle Cenomanian) in Russell County collected by Keith Ewell and I in 2004. Shimada and Everhart (2003, unpublished) reported what was then the youngest-occurring *P. mammillaris* tooth from the base of the Fort Hays Limestone (Lower Coniacian), and Everhart and Darnell (2004) reported an even earlier specimen from the Fairport Chalk (Middle Turonian). However, the earliest and first associated set of *P. mammillaris* teeth (FHSM VP-17989; Fig. 4.19) was discovered by a private collector in Russell County, just below the Fencepost Limestone Member of the Greenhorn Formation (lower Middle Turonian; Hamm, 2013).

While *P. anonymous* has been limited to pre-Niobrara deposits (Hamm, 2010a), the earliest occurrence of *P. mortoni* has been extended to the base of the Fort Hays Limestone by the discovery of a large associated specimen by Ramo and Pam Decker in Jewell County, Kansas Shimada et al. (2010). The very large size of this specimen (FHSM VP-17415) strongly suggests that *P. mortoni* had evolved elsewhere before moving into the Western Interior Sea. *P. mortoni* is the last species of ptychodontid shark present in the Western Interior Sea (Stewart, 1990; Shimada, 1996; Fig. 4.19). In 2007 I had the opportunity to view the upper and lower jaw plates of a *Ptychodus mortoni* shark that had been collected from the middle (Santonian) chalk near Hell Creek in Gove County. Almost all of the teeth were in place, just as they were when the shark died about 85 Ma (Fig. 4.20).

In 2010, Steve Mense contacted me in regard to a *Ptychodus* specimen that he had discovered near the Smoky Hill River in eastern Logan County. I met him on the site, and the two of us dug up the remains of a fairly large *Ptychodus mortoni* shark. We collected over 550 teeth, but the jaw plates were scattered over an area about a meter square. There were large pieces of calcified cartilage, including parts of the palatoquadrate

4.20. Nearly complete upper (left) and lower (right) jaw plates of a *Ptychodus mortoni* shark from the Hell Creek area of southwestern Gove County. Scale bar = 10 cm (4 in).

(upper jaw), a couple of vertebrae, and thousands of tiny dermal denticles that once were part of the shark's skin. Of some interest to me were the large crowns of bud teeth that were being formed when the shark died. The upper portions of the teeth were perfect, but they did not have roots (Fig. 4.21). Steve generously donated the specimen (FHSM VP-17606) to the Sternberg Museum of Natural History.

Of some interest here is the fact that *Ptychodus mortoni* was the first crusher shark to be described from North America. The specimen (a single tooth) was collected in Alabama and figured by Morton (1834). Morton then sent the tooth to England, where Gideon Mantell (1836), named it *Ptychodus mortoni* in honor of Morton. The tooth subsequently was examined by Agassiz in 1834–1835, and then figured in a chapter of his *Poissons Fossiles* in 1839, but not described in print until 1843. For years researchers have given Agassiz credit for naming *P. mortoni*, but the credit really belongs to Mantell. The specimen is presently in the Mantell collection at the British Museum of Natural History (Natural History Museum of the United Kingdom) as NHMUK PV OR 28394. The whole story behind the naming of *P. mortoni* is more complex than I can explain here, but if you are interested, I refer you to Everhart (2013) and to Ikejiri and Everhart (2015).

Chimaeroids

Chimaeroids (including modern-day ratfishes) are cartilaginous fishes, but they are not sharks. There are many modern species, and they have a very complete fossil record going back to at least the Triassic. That said, they were not part of the fauna of the Smoky Hill Chalk until 2006, when Gary Olson sent me three unusual 'bones' that had been collected the middle Santonian chalk of southwestern Gove County. I didn't have any idea what they were and contacted Ken Carpenter at the Denver Museum of Nature and Science. Ken immediately recognized the specimens as jaw plates of a chimaeroid and noted that they were not bones, but

4.21. Occlusal view of four crowns of teeth in the process of being formed from a large specimen of *Ptychodus mortoni* (FHSM VP-17606) collected from the middle Santonian chalk of Logan County in 2010. Scale in mm.

METRIC 1 2 3 4 5

rather calcified cartilage. I contacted David Cicimurri and David Parris, both of whom had previously worked with chimaeroid fossils. We were able to compare the specimen with other material from the East and Gulf coasts, and we determined that the jaw plates most likely resembled a chimaeroid specimen described and figured by Leidy (1873), *Eumylodus laqueatus*. In our paper (Cicimurri, Parris, and Everhart, 2008), we described the partial dentition of the Kansas specimen (FHSM VP-15685; Fig. 4.22) and noted that a chimaeroid fin spine had been collected over a hundred years ago by H. T. Martin and sent to the British Museum of Natural History. That specimen (BMNH P10343) had been previously identified as *Edaphodon* (Stahl, 1999). Although rare in the Smoky Hill Chalk, chimaeroid remains appear to be fairly common in the nearshore deposits of the Western Interior Sea in the Fox Hills Formation (Late Campanian and Early Maastrichtian) in North Dakota and Colorado (Hoganson and Erickson, 2005; Hoganson, Erickson, and Everhart, 2015).

This book is generally focused on a narrow (geologically speaking), 5-million-year window of time during the Late Cretaceous when the Smoky Hill Chalk was deposited on the bottom of the Western Interior Sea. The fauna from that interval (87–82 Ma) is the most widely collected and the most thoroughly studied of any period from the Cretaceous of Kansas. The chalk is accessible in many localities, preservation is excellent, and the chalk matrix is relatively easy to remove. It is also my favorite because of the occurrence of mosasaurs and the large number of their remains that have been collected from the chalk since the late 1860s. But many other productive Cretaceous rocks are relatively unstudied, and

Other Times, Other Sharks

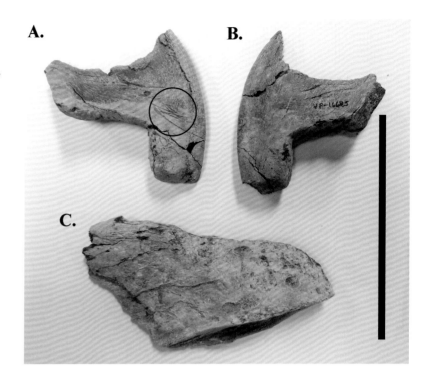

4.22. Three elements, left and right vomerines and right palatine plate of a chimaeroid fish identified as *Eumylodus* sp. (FHSM VP-16685) from the Santonian chalk of western Gove County, Kansas. A, lateral view of right vomerine, note *Squalicorax* shark bite marks (circled); B, lateral view of left vomerine; C, aboral (lateral) view of right palatine. Scale bar = 10 cm (4 in). See Cicimurri et al. (2008) for a detailed description.

until recently Kansas has lagged behind other places in the Midwest, such as Texas and South Dakota, in the number of Cretaceous shark species that have been documented from the state.

In the past 15 years, I have had a number of opportunities for further study of Kansas sharks. My discovery of a huge *Cretoxyrhina mantelli* specimen (FHSM VP-14010) in 2002 was the beginning of a series of discoveries and studies about Kansas sharks. The opportunities even extended outside the Cretaceous with the collection of the first documented ctenacanth shark (FHSM VP-15012) remains, including fin spines, dermal denticles, teeth, and preserved cartilage, from the Grant Member of the Winfield Formation (lower Permian) of Morris County, Kansas, near Herington, Kansas, in August 2002 (Everhart and Everhart, 2003). The specimen is now on loan for study to John Maisey at the American Museum of Natural History (Fig. 4.23).

In October 2002, my wife and I began working with a 'fish tooth conglomerate' from the Blue Hill Member (Turonian) of the Carlile Shale in Jewell County, Kansas, that had been discovered in 1958 by Donald Hattin (1962). I was able to collect quite a few pieces of the conglomerate from the south shore of Lovewell Reservoir, but unlike our usual field trips, all of the collecting of specimens was done in our fossil lab. The conglomerate was dissolved very slowly in dilute acetic acid, and then the teeth were collected on a daily basis and sorted while we peered through a microscope. Although the matrix was mostly composed of thousands of tiny teleost teeth (mostly *Enchodus*), over a period of several months it produced a number of selachian (shark) species that had never been

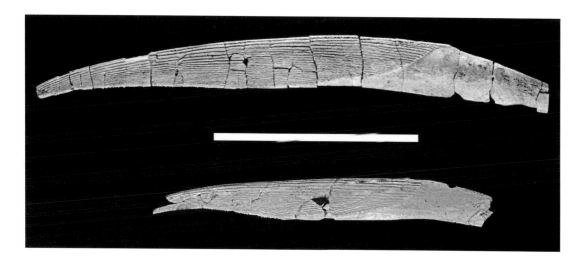

reported from Kansas (among them, *Lonchidion* sp., *Chiloscyllium greeni*, *Ischyrhiza mira*, and *Ptychotrygon triangularis*). Many of the shark teeth were so small (1 mm or less) that they had to be picked and sorted using forceps and a binocular microscope. Imagine 500 or more tiny *Rhinobatos incertus* (a guitarfish) teeth in a volume about the size of an aspirin tablet! This extinct little ray occurs throughout the Cretaceous strata in Kansas (Everhart, 2007), and its modern relatives are common in the ocean, and even in some freshwater environments. After *Rhinobatos*, *Scapanorhynchus raphiodon* and *Squalicorax falcatus* were the most common species collected. The study generated an abstract/poster for the 2003 Society of Vertebrate Paleontology meeting (Everhart et al., 2003) and I hope will result in a future paper.

In the late spring of 2003, I began looking at the Kiowa Shale in McPherson County with Shawn Hamm, then a student at Wichita State University. Except for a thin layer of sand called the Cheyenne Sandstone in southern Kansas, the Kiowa is the oldest Cretaceous formation (Albian) exposed in Kansas. It was deposited at a time when the Western Interior Sea was advancing across Kansas from the south and had not yet covered the entire state. Unlike the deeper, blue-water ocean in which the Smoky Hill Chalk was deposited, the Kiowa Shale was laid down as nearshore mud and fine sand in a shallow, relatively high-energy environment.

Working some 10 meters below the level of the surrounding wheat fields in an active shale mining quarry near Marquette, Shawn and I collected the much older shark fauna along with the scattered fragmentary remains of fishes, turtles, crocodiles, and plesiosaurs. Again, many of the teeth were quite small, though not as small as those from the Lovewell fauna. Most of the teeth and other remains were literally handpicked from the surface of a rapidly weathering layer of pyritized sandstone. Expanding on the work of Beamon (1999), several relatively "new to Kansas" species were collected, including *Polyacrodus* sp. (Everhart, 2004a), *Onchopristis dunklei*, and *Pseudohypolophus mcnultyi*, with *Leptostyrax macrorhiza* and *Carcharias amonensis* being the most common sharks.

4.23. The two dorsal fin spines from a ctenacanth shark (FHSM VP-15012) from the lower Permian of Morris County, Kansas. (Photo credit: American Museum of Natural History, New York, used with permission). Scale bar = 10 cm (4 in).

In August 2003, Keith Ewell visited the Sternberg Museum and asked for assistance in identifying Permian-age shark teeth he had collected near Manhattan, Kansas. After making contact with Keith, I visited the site with him and saw his specimens. We then assisted him in collections made at three sites in Geary County that produced more than 200 teeth of *Cladodus* sp., *Petalodus* sp., *Acrodus* sp., and *Chomodus* sp.; dorsal spines of *Ctenacanthus*, *Physonemus*, and *Hybodus*; and calcified cartilage (Ewell and Everhart, 2004). The most productive site was in the Neva Limestone (Council Grove Group, Lower Permian). Although the species of sharks identified were not necessarily new to Kansas, the number of teeth, their large size, and their excellent condition was certainly unexpected. The teeth of *Cladodus* sp. are the most common and, in size, rival the largest of the *Cretoxyrhina* teeth we have collected from the Smoky Hill Chalk.

As a result of our meeting and subsequent conversations, Keith became interested in Cretaceous fossils further west. In October, as noted earlier in this chapter, Keith discovered the tooth of a new species of shark (*Johnlongia* sp.) in the Smoky Hill Chalk (Shimada, Ewell, and Everhart, 2004). as well as numerous shark teeth in the Greenhorn Formation. Later in the year Keith discovered a rich accumulation of shark teeth, fish remains, reptile bones (coniasaur), and even some very early bird bones (cf. *Ichthyornis*) in the basal Lincoln Limestone Member (Upper Cenomanian) of the Greenhorn Limestone in Russell County. Together we collected samples for further examination. Again, the teeth were small (more picking with a microscope) but the numbers were large (literally hundreds of teeth), with *Carcharias amonensis* and *Squalicorax falcatus* being the most common species, along with *Cretoxyrhina mantelli*, *Cretalamna appendiculata*, at least two species of *Ptychodus*, and numerous sand-grain-sized teeth of the ray *Rhinobatos incertus*. One of the rarest finds was Keith's discovery of a single 1.5 mm (0.06 in) rostral tooth of the primitive sawfish *Onchopristis dunklei*.

About 8 miles away, in eastern Russell County, we also collected samples from a layer of poorly cemented sand in the upper Dakota Sandstone (middle Cenomanian). The sand produced a number of small *Carcharias amonensis* teeth, along with a large *Cretodus semiplicatus* tooth, several *Squalicorax* sp. teeth, many *Hybodus* sp. and *Rhinobatos* teeth, and more rostral denticles of the sawfish *Onchopristis dunklei* (Everhart, Everhart and Ewell 2004). The preservation is occasionally poor due to diagenesis and the intrusion of selenite crystals, and contrasts markedly with the pristine condition of the teeth in the nearby basal Greenhorn.

These specimens are significant because few vertebrate remains (outside of the type specimens of the crocodiles *Hyposaurus vebbii* and *Dakotasuchus kingi*, and a dinosaur, *Silvasaurus condrayi*) have ever been reported from the Dakota Sandstone in Kansas. Although Scott (1970) suggested that the type specimen of *Dakotasuchus* came from the underlying Kiowa Formation, in 2015 I was able to locate the concretion containing the bones at Kansas Wesleyan University in Salina, Kansas, and confirm that it did come from the Dakota Formation.

In 2008, I led a field trip to a new locality in the Blue Hill Shale of Mitchell County with Gail Pearson to hunt ammonites. While it wasn't the most successful of my field trips, I did collect a lot of bone fragments from a weathering concretion. When I assembled them, they represented the first remains of a plesiosaur ever reported from the Blue Hill (Everhart, 2009). I wasn't able to return to the site until March of 2010, but this time we brought along Fred Smith, a local resident and the editor of the local newspaper (Tipton Times). He had never collected fossils in the shale and just wanted to go along. The ground was pretty saturated from melting snow, but we had a six-wheeler and drove through the pasture with no problem. Once there, Gail and I got busy collecting a few more bone fragments, but Fred gave up and wandered up the hill. Pretty soon he came back down holding a long, narrow concretion and said it was a piece of petrified wood. Sure enough, there was a round shape on both ends, but it wasn't a fossilized tree branch. I took one look and realized that the round shapes were shark vertebrae and that the concretion had been formed around the vertebral column of an ancient shark.

I was, of course, excited at the discovery, but Fred's look seemed to say, "Are you nuts or something?" However, it was too muddy to do any digging that day, and a blizzard came through after we left. We were rained out in April (it was a wet spring), but we finally were able to get back to the site in June. Once we got the dig started, we quickly discovered more large shark vertebrae, and then the first of many teeth (Fig. 4.24). By the time the dig was completed, we had collected (Shimada et al., 2011) more than 160 teeth and 60 vertebrae from a single *Cretodus* sp. (FHSM VP-17575), the first associated shark remains ever recovered from the Blue Hill Shale. The specimen is currently being described and will probably be named as a new species of *Cretodus* from the Western Interior Sea.

In 2011, I gave an Oceans of Kansas talk for a group in Concordia, Kansas. My host brought a specimen from her rock garden to show me. The chunk of limestone (two pieces) from the Greenhorn Formation had been collected from a road cut by their father years earlier and was shared between her and her brother. When I saw it, I was amazed to discover that it contained a section of vertebrae from a large shark. More than that, I could see that there were patches of articulated dermal denticles (shagreen, or shark skin) preserved on the surface. After my initial amazement, she asked me if I wanted it for the museum (of course!). Closer examination showed me that not only were patches of dermal denticles preserved, but some places even showed that layers of the underlying fibers of the shark's skin were also fossilized. A month or so later, the piece that her brother had kept was also donated. Altogether the 0.8 m (2.6 ft) piece of limestone (FHSM VP-17978) includes 27 articulated vertebrae from a shark that would have been 4–5 m (13–16 ft) long in life, and a significant amount of the skin of the shark (Everhart 2012). Study of this specimen is still in progress.

Bice and Shimada (2016) reported on the fauna from the middle Turonian Codell Sandstone Member of the Carlile Shale. As noted

4.24. Field photo of the initial discovery of shark teeth (circles) and vertebrae on the dig for the FHSM VP-17575 *Cretodus* sp. specimen. The shark specimen was discovered by Fred Smith in the Blue Hill Shale (Middle Turonian) of Mitchell County, Kansas. Scale bar = 10 cm (4 in).

in Chapter 2, the Codell Sandstone was largely eroded away as the Western Interior Sea retreated during at the end of the Turonian. In some places, this erosion actually concentrated fossils into a relatively thin lag deposit between the underlying Blue Hill Shale and the overlying Fort Hays Limestone Member. Ramo and Pam Decker, who discovered the early *Ptychodus mortoni* specimen mentioned earlier in this chapter, had also extensively collected sharks teeth and other fossils from the Codell in Jewell County. When Kenshu Shimada indicated an interest in their collection, they generously donated the specimens to the Sternberg Museum. Graduate student Kelly Bice wrote her masters thesis on the Codell fauna, and then she and Shimada wrote a comprehensive paper on the subject. Published in 2016, the paper described 38 taxa, including 22 species of sharks and rays from this nearshore deposit. In the process, they also named a new species of shark, *Squalicorax deckeri* (holotype = FHSM VP-18579) in honor of the Decker family.

Since the publication of the first edition of this book, a lot of additional collecting of shark teeth has resulted in many new papers describing the occurrence of Late Cretaceous sharks in the rocks of Kansas. With the discovery of these new and productive localities, the Cretaceous (and Permian) shark faunas of Kansas appear to be quite similar to those in Texas, New Mexico, Nebraska, South Dakota, and elsewhere in the Midwest. We look forward to additional collecting and study of Kansas sharks in these areas.

Fishes, Large and Small

5

The big *Xiphactinus* swam effortlessly through the clear, warm waters of the Western Interior Sea in a solitary never-ending search for its next meal. Ten years old and nearly 4 meters long, it had not yet reached full size, but it was already larger than any of the other species of fishes in this ocean except the giant ginsu sharks. Although still wary of the occasional prowling shark, the X-fish had only one other major competitor for the larger fishes that it preyed upon: the large marine lizards that shared the same waters. The X-fish had lived long enough and grown too large for any of the big marine lizards to be interested in it as a food item. Capable of taking very large prey as they were, the lizards were unlikely to attack anything larger than they could swallow, and the X-fish's massive body was well beyond what they could eat. The X-fish had on occasion taken a newly birthed mosasaur, but it generally avoided their family groups. Its favorite prey was other fishes, especially some of the larger varieties that were closely related to his own kind. A large meal satisfied the X-fish's hunger and provided the energy necessary to support its constant swimming.

The X-fish's senses alerted it to a group of *Gillicus* feeding nearby and close to the surface. As it swam closer, it could see that the 2 meter long predatory fish had surrounded a school of much smaller fishes, compressing them into a shimmering globe of millions of tiny, darting forms. The *Gillicus* swam in circles around the trapped school, keeping it from spreading out or escaping. Occasionally some of the *Gillicus* would dart through the mass of little fishes with their jaws wide open, gorging themselves on the sudden abundance of prey. Their attention diverted, they did not notice the X-fish's approach from the dark waters beneath them. Gauging their movements, the X-fish selected its victim and accelerated swiftly upward, meeting the other fish almost head-on. Opening its large mouth at the last moment, it seized the smaller fish's head from below. As its massive lower jaw closed, the large, conical teeth at the front punctured the thin bones covering the head of the *Gillicus* and kept it from getting away. The smaller fish struggled briefly, but the shock of impact had stunned it into submission. The X-fish held on to its prey for several moments, then repositioned it so that it was pointed headfirst into the larger fish's mouth. Then the X-fish began to rapidly open and close its jaws, drawing the smaller fish deeper and deeper into its gullet. A shower of silvery scales, dislodged from the *Gillicus* by the teeth of the X-fish, glittered in the sunlight as they drifted away.

5.1. A photograph of the famous fish-within-a-fish specimen of a large *Xiphactinus audax* (FHSM VP-333) and its last meal (*Gillicus arcuatus*; FHSM VP-334) in the Sternberg Museum of Natural History, Hays, Kansas. The specimens were collected from the Smoky Hill Chalk of Gove County by George F. Sternberg in 1952. The exhibit as shown is about 4.3 m (14 ft) in width.

At first, swallowing the big *Gillicus* was quite easy, but then the X-fish began to have problems. At nearly 2 meters in length, the *Gillicus* weighed more than 80 pounds. It was probably the largest prey the X-fish had ever eaten. Once the head and pectoral fins of the *Gillicus* had passed into the larger fish's throat, however, there was no turning back. The bony pectoral fin rays of the *Gillicus* were folded back tightly against its body. Any attempt to reverse direction would cause them to unfold and catch in the X-fish's muscular esophagus. The large body of the *Gillicus* now filled the mouth and throat of the X-fish and made it difficult for water to reach its gills. The last two feet of the *Gillicus*, including its bony tail, still protruded beyond the X-fish's jaws (Plate 5). Barely alive but tightly confined within the bony skull and forebody of the larger fish, the *Gillicus* thrashed about occasionally as it was swallowed. A final spasm caused one of its sharp fin rays to pierce the esophagus of the X-fish.

Twice as long as the *Gillicus* and much more massive, the X-fish struggled to swallow its prey. Unable to adequately replenish the oxygen in its body with the smaller fish lodged in its throat, however, it was quickly weakening. Slowly the smaller fish was moved deeper and deeper into the gullet of the larger one. Finally the bony tail moved entirely into the mouth, and the gills of the X-fish began to function normally again. Now the prey began to move more easily into the stomach of the X-fish,

but something was wrong. The large fish swam in ever-slower circles as it finished swallowing the *Gillicus*. During the prey fish's futile struggle to break free, one of its fin spines had punctured something vital inside the X-fish. Within a few minutes, the X-fish stopped swimming, rolled over on its back as it died, and then sank headfirst toward the muddy bottom.

While this story is fiction, the famous 'fish-within-a-fish' specimen re-covered by George Sternberg is a fact (Walker, 1982, 2006; Rogers, 1991; Liggett, 2001). While two visitors from the American Museum of Natural History (AMNH) were on a brief field trip with George Sternberg in the spring of 1952, Walter Sorenson discovered the tip of the caudal (tail) fin of a large *Xiphactinus audax* (FHSM VP-333) eroding out from the chalk in eastern Gove County. Recovering the fossil would have taken far more time than the visitors had, so they graciously gave the specimen to Sternberg. After they left, Sternberg would return to the site and con-tinue removing the chalk that covered the bones. He would spend the entire month of June camped near the fossil as he worked on it to ensure that it wasn't damaged by livestock or curious visitors. Several times, as he realized how complete and spectacular the fish was, he called the AMNH and tried to give it back to them. In fact, the museum already

Fishes

had a larger specimen that had been provided by Charles Sternberg and said they didn't need another one. They thanked Sternberg for the offer, then reiterated that he had done all the work on it and that it was his fish.

Hundreds of hours of meticulous field work under a hot summer sun by Sternberg and others from the university eventually uncovered what was certainly one of the most complete and well-preserved specimens of *Xiphactinus audax* Leidy 1870 known up to that point (Bardack, 1965:40). Sternberg's patient work also revealed the last meal of the larger fish, a well-preserved *Gillicus arcuatus* (FHSM VP-334). The smaller, 2 m (6.6 ft) long fish had been swallowed headfirst and rested entirely within the ribs of the 4 m (13 ft) *Xiphactinus* (Fig. 5.1). We know that the smaller fish had not been there long before the larger fish died because it had not yet been digested. What caused the death of this *Xiphactinus*, and a surprising number of other specimens of the same species, cannot be determined from their fossils, but it most likely involved an injury that occurred when the large prey was swallowed. In the case of the Sternberg specimen, a fin from the struggling *Gillicus* could have pierced the heart or a major blood vessel of the *Xiphactinus* and killed it quickly, or there may have been rapidly fatal damage to the gills of the *Xiphactinus*. In any case, the fossil remains have preserved a very interesting and puzzling moment in time. A black and white photograph of the big fish and its last meal was even published as a two-page spread in the July 19, 1954, issue of *Life* magazine, and for many years the specimen was regarded as the most photographed fossil of all time.

While *Xiphactinus audax* remains as complete and well preserved as the fish-within-a-fish specimen are relatively rare, many of this species have been collected with a *Gillicus* as the last meal. Bardack (1965) surveyed 18 relatively complete *Xiphactinus* specimens in museum collections and indicated that at least three (including the Sternberg specimen) had identifiable *Gillicus* remains inside. A similar-size specimen (DMNH-1667) in the Denver Museum of Nature and Science had lived long enough to partially digest its last meal, also a large *Gillicus*. In fact, seven of the 18 specimens he surveyed included fish as stomach contents. Preservation of stomach contents is rare in the fossil record (Cicimurri and Everhart, 2001), and the relatively frequent discovery of *Xiphactinus* specimens containing the remains of their last meals raises the question of why so many of them died so quickly after eating.

The giant (5.3 m; 17 ft) *Xiphactinus* (NAMAL 2000–0925–009) that I discovered in 1996 also included the partially digested remains of another *Gillicus*. The X-fish was collected, prepared, and reconstructed in a dramatic 3-D mount by Triebold Paleontology (Fig. 5.2). For a brief time, this was the largest specimen of *Xiphactinus audax* that had ever been collected, but records are meant to be broken. At least two 5.5 m (18 ft) specimens are now known.

Like in modern oceans, there were many different species of fishes living in the Western Interior Sea. Some, such as *Xiphactinus*, were large and are well represented in the fossil record. Others are smaller or known

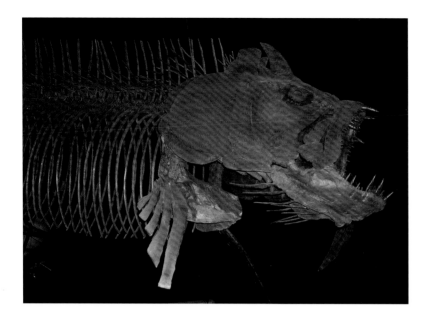

5.2. The business end of Triebold Paleontology's 3-D reconstruction of a very large (5.3 m; 17 ft) *Xiphactinus audax* on exhibit in the Rocky Mountain Dinosaur Resource Center, Woodland Park, Colorado. Unlike the Sternberg fish-within-a-fish specimen, the bones were scattered over a large area of the chalk. The specimen was discovered by the author in Gove County, Kansas in 1996.

from single specimens, and still others remain unknown because they were not preserved as fossils or have yet to be discovered. Shimada and Fielitz (2006) provided a comprehensive listing of all of the known fishes from the Smoky Hill Chalk. The list is extensive, comprising at the time at least 16 chondrichthyans (sharks and their relatives) and 54 osteichthyans (bony fishes). As with all other faunal lists of fossils, however, the numbers keep changing over time with the addition of new discoveries, including the addition of a giant filter-feeding fish, *Bonnerichthys gladius*. I will not repeat their entire list in this chapter, but instead will comment on some of the more commonly collected or unusual species.

After shark teeth (Chapter 4), fish remains are the most common vertebrate fossils occurring in the Smoky Hill Chalk. Russell (1988) suggested that about 60 percent of the Late Cretaceous specimens in museum collections were fishes. While this number probably includes some cartilaginous fish specimens (sharks and ptychodontids) in those collections, my experience is that the percentage of fish specimens observed in the field probably approaches 80 or 90 percent of the total number. The difference can be explained by the fact that museum specimens are generally more complete, while in the field the smaller bits and pieces of fishes we see are likely not to be collected, even though they may be identifiable to species. If your field time or storage space is limited, you tend to become a bit more choosy about what you collect.

After reviewing my field notes for the years 1988–1995 when my wife and I were vacuuming up ("Hoovering") almost everything we saw, I discovered that 78 percent of 430 vertebrate remains we had collected were fishes (not including isolated shark and fish teeth) and 16 percent were mosasaur. The other 6 percent were more or less equally divided between pteranodons (2 percent), turtles (2 percent), and plesiosaurs (1 percent), with a single bird bone making up a small fraction of a percent. I admit that I picked up every mosasaur specimen that I saw, but not every fish. It is important to note, however, that these specimens ranged from complete fishes to a few vertebrae or just an identifiable piece of a fin or jaw. In size, the fish specimens ranged from tiny 10 cm (4 in) fishes inside clam shells to giant *Xiphactinus* specimens more than 4 m (13 ft) long.

Far and away the most common remains we observed during that time, however, were fish tails. These generally consisted of a complete, mostly undamaged caudal fin with a few attached vertebrae. Mostly these were from medium-size fishes, with an occasional fairly large *Ichthyodectes* thrown in. An unknown fish-eating predator was apparently biting or breaking off the bony, relatively nutritionless tails while swallowing the rest of the fish. Carpenter (1996:44) reported severed tails as evidence of predation on fishes in the Pierre Shale. Our unofficial count over the years has been that we locate about ten tails for every one skull. It was pretty frustrating to discover a nice caudal fin eroding out of the chalk, start to uncover the bones, and see them end with just a couple of inches of vertebrae. Needless to say, we gave up collecting fish tails a long time ago.

Relatively few remains of fishes (or anything else) are complete when located in the Smoky Hill Chalk. The multitudes of smaller fishes that must have been present as prey for larger species were most likely entirely consumed, while the larger ones were sometimes torn apart by sharks or other predators. Sometimes the fish bones we discover were partially digested (see Chapter 4 on sharks) before being buried or occur inside coprolites.

In the Western Interior Sea once the remains of the animal reached the bottom, they generally were not disturbed by large predators. The sea floor was simply too low in oxygen most of the time to support the variety of scavengers, like crabs, that you might expect today to be feeding on a dead fish. In most cases, due to their size and density, the bones of larger fishes had a better chance of being preserved as fossils. The major exception to that rule occurred when schools of small fishes were trapped inside giant inoceramids when they died (Chapter 3 introduction), and were preserved intact (Stewart, 1990b). In some cases, notably the fish-within-a-fish *Xiphactinus* mentioned above, the remains of a last meal are identifiable inside the larger fish. Evidence of feeding or scavenging is also preserved as bite marks, severed bone, and associated shed or embedded teeth (usually shark).

The ecosystem of the Late Cretaceous was probably not unlike that of the Gulf of Mexico or similar warm-water areas around the earth today. Immense populations of sardine-size fishes were necessary to support larger fishes and the predators, including marine reptiles, that fed on them. A Late Cretaceous marine food web for the Western Interior Sea might look something like this: Microscopic single-celled algae, as the primary producers, converted sunlight, carbon dioxide, and nutrients into biological materials for growth and reproduction. The shells that surrounded these algae, called coccolithophores, were made up of many tiny

disk-shaped coccoliths composed of calcite (calcium carbonate). It was the accumulation of vast numbers of these coccolithophores that formed the limey mud on the sea floor and eventually became the Smoky Hill Chalk (Hattin, 1982)

Many species of microscopic single-cell or small multicellular animals fed on the algae. Their fecal (waste) pellets containing the indigestible coccolithophores and coccoliths of the algae (Hattin, 1982) also added to the limey mud below.

Tiny fishes, small crustaceans, and the larvae of other marine animals, including young squid and ammonites, fed on the microscopic protozoa and other primary consumers. Until recently, we believed that there were no large filter-feeding vertebrates (similar to modern baleen whales and basking sharks) during the Late Cretaceous. An 'ah, hah!' moment by Matthew Friedman, a graduate student studying fossil fishes in museum collections, resulted in the identification and description of a new genus of filter-feeding Late Cretaceous fish from North America called *Bonnerichthys gladius* (Friedman et al., 2010). Previously, the massive pectoral fins of these fishes had been misidentified as being from a giant swordfish (originally *Portheus gladius*, then *Protosphyraena*; Fig. 5.3) by Cope (1873a). (There is more of that story later in this chapter.)

Small fishes, young fishes of many species, and many invertebrates (ammonites and squid) fed on the tiny fishes and smaller larvae. Medium-size fishes fed on the smaller fishes. Pteranodons, marine birds, young marine reptiles, ammonites, and squid also fed on smaller prey at this level.

Larger fishes, sharks, short- and long-necked plesiosaurs, and small mosasaurs like *Platecarpus* and *Clidastes* fed on the medium-size fish and on invertebrates such as ammonites, belemnites, and squid. Sharks, cephalopods, and possibly mosasaurs also served as scavengers to recycle the biological materials from the remains of dead animals. Large mosasaurs, such as *Tylosaurus*, and the ginsu sharks fed on everything as the top (apex) predators.

Because we discover the remains of the large apex predators (mosasaurs) quite often, we are fairly certain that there must have been a very productive ecosystem running at full tilt to support them. These large marine animals simply could not survive without huge amounts

5.3. Isolated pectoral fin of *Bonnerichthys gladius* (FHSM VP-212), collected by George F. Sternberg from the Santonian chalk of Gove County in 1949. Originally identified as *Protosphyraena gladius* Cope. Scale bar = 20 cm (8 in).

of smaller animals as prey. The coastlines, estuaries, and swamps that bordered both coasts of the Western Interior Sea may have served as semiprotected hatcheries for untold numbers of smaller fishes and invertebrates. Unfortunately, those places were not well preserved in the fossil record, but they had to be there to support the base of the food web. The remains we see preserved as fossils in the Smoky Hill Chalk are only a small fraction of the abundant life that must have existed in the Western Interior Sea during the Late Cretaceous.

The specimens that are preserved are important indicators of what was going on in the ocean over Kansas. Large predators, such as mosasaurs and *Xiphactinus*, are probably overrepresented in the fossil record because their more robust, heavier bones greatly improved their chances for preservation. Because they are there, however, we can view them as indirect evidence of the vastly larger number of smaller fishes and invertebrates that lived but were consumed as prey, and as a result left only a scant record.

A Brief History of Fossil Fish Collecting in Kansas

Though not from Kansas, the first known fossil fish from the Niobrara Chalk was collected by the Lewis and Clark expedition in August of 1804 from the bank of the Missouri River in what is now Harrison County, Iowa. The only fossil specimen surviving from that expedition is a fragment of a fish jaw (ANSP 5516; see Fig. 2.1) that is currently in the collection of the Academy of Natural Sciences of Philadelphia (Spamer et al., 2000; Fig. 2.1). The history of this fossil is somewhat confusing, however, because it was originally misidentified by Dr. Richard Harlan as the jaw of an 'Enalio Saurian' (a marine lizard thought then to be similar to ichthyosaurs and plesiosaurs). Harlan (1824) described the jaw and gave it the name *Saurocephalus lanciformis*. It wasn't until six years later when another medical doctor, Isaac Hays (1830), described a similar species of Cretaceous fishes (*Saurodon leanus*) from a more complete New Jersey specimen, allowing the mistake to be officially noticed and corrected. Leidy subsequently noted that the *Saurocephalus* type specimen was a fragment of a "maxillary bone with teeth of a peculiar genus of sphyraenoid fishes from the Cretaceous formation of the Upper Missouri" (1856:302).

This group of fossil fishes would continue to give other paleontologists classification problems over the years as more discoveries were made. The latest misidentification occurred when a jaw fragment discovered in a concretion from a beach in western Canada was described as a new species of toothed pterosaur (Arbour and Currie, 2011). The specimen was reviewed and reidentified as a saurodontid fish related to *Saurocephalus* by Vullo, Buffetaut, and Everhart (2012).

Although the specimens were not from the Cretaceous, Joseph Leidy (1859) described a dorsal spine (*Xystracanthus*) and two teeth of two different species of Paleozoic sharks (*Cladodus* and *Petalodus*) from eastern Kansas two years before Kansas became a state. The teeth had been collected in 1858 by the U.S. Geological Survey expedition (Meek and Hayden) while traveling through Kansas during their exploration of what would become several Midwestern states. This paper is important because it is the first description of vertebrate fossils from Kansas and is one of the first descriptions of North American vertebrates from the Permian.

Professor Benjamin Mudge (1817–1879), the first state geologist of Kansas, could rightly be considered the first paleontologist in Kansas. He began collecting fossils in the central and western parts of the state several years before Marsh and Cope became interested. As was customary for the times, Mudge sent many of his specimens to the scientists 'back East' for identification. For the most part, he was communicating with E. D. Cope during the late 1860s and early 1870s. In an 1870 letter (Williston, 1898:29–30; see also Lawrence, Kansas, Republican Journal, November 16, 1870, 3), Cope complimented Mudge, his "Esteemed Friend," on the scientific value of the material he had sent.

In fact, many of the new species of fishes and mosasaurs described by Cope were from the specimens sent to him by Mudge (Everhart, 2002). During his only trip to Kansas in late 1871, Cope (1872a) visited Mudge in Manhattan, Kansas, and examined his collection. By then, however, Mudge had begun sending some specimens, including the remains of the first known toothed bird (*Ichthyornis*) to Cope's rival, O. C. Marsh (see Chapter 11). In 1874 Mudge began collecting fossils exclusively for O. C. Marsh and Yale College. Unfortunately for Kansas, and for paleontology, Professor Mudge died in 1879.

In late 1867, Dr. Theophilus Turner, the assistant surgeon assigned to Fort Wallace in western Kansas, also began communicating with Cope about his discovery of a huge plesiosaur from what we now call the Pierre Shale. After receiving the specimen, Cope (1868) reported seeing the remains (scales, vertebrae, and teeth) of six species of fishes, including *Enchodus*, that he believed to be stomach contents from the new plesiosaur (*Elasmosaurus platyurus*—see Chapter 7). In 2002 when I examined the remains that had been collected by Turner (ANSP 10081), there were still fish scales, teeth, and bone fragments visible in some of the concretionary matrix that was associated with the plesiosaur bones. However, I consider them more likely to be the normal detritus that might be expected to accumulate on the sea bottom next to the plesiosaur's bones, rather than evidence of the plesiosaur's last meal. In any case, the fragments did provide evidence of several species of fishes that were living at the same time (Middle Campanian) as the elasmosaur.

In addition to performing his military duties during the late 1860s, Dr. George M. Sternberg (older brother of Charles H. Sternberg) collected fossils from western Kansas and sent them to the Army Medical Museum in Washington, D.C. The fossils were eventually forwarded to the Smithsonian Institution (United States National Museum—USNM), where they were more readily available for study by the paleontologists of the day. When I examined the Sternberg collection at the USNM in

2001, I was amused to see that Dr. Sternberg had written his name in ink on each of the bones that he donated.

Leidy (1870) described a new genus and species of fishes (*Xiphactinus audax*) from a large (40 cm; 16 in) fragment of a pectoral fin ray (USNM 52) that Dr. Sternberg had collected from the chalk. After examining the same specimen, however, Cope (1871) disagreed with Leidy (his mentor at the ANSP) and renamed it *Saurocephalus audax* Leidy. Cope's suggestion of a name change was apparently not recognized by anyone else. Cope (1872b) subsequently described another "new" genus and species from the remains of several relatively complete specimens of fishes that had been discovered by one of his U.S. Army assistants (M. V. Hartwell) near Fort Wallace during his 1871 visit to western Kansas. He called his 'new' fish *Portheus molossus* without recognizing that it was the same species that Leidy had named from a pectoral fin almost two years earlier. It is not known if Leidy ever recognized that it was the same species or not.

Almost 30 years later, O. P. Hay (1898) published a short note in *Science* suggesting that *Xiphactinus* Leidy was the correct genus, but by then it was too late. *Portheus molossus* Cope was too well established in the literature and in museum collections to be banished that easily. Even the prominent paleontologist H. F. Osborn (1904) was unaware of, or even worse, ignored, Leidy's name in favor of Cope's when he wrote about a *Portheus molossus* specimen (AMNH 322199) collected by Charles Sternberg and recently placed on exhibit at the American Museum of Natural History. Since that time, even though it is more widely known to be a junior synonym of *Xiphactinus audax* Leidy, the name *Portheus molossus* Cope is much better recognized, and some (although fewer) museum collections still have their specimens mislabeled.

There is another, unusual twist to this story. *Xiphactinus (Polygonodon) vetus* is a sister species of *X. audax* that was named by Leidy (1856) on the basis of large, oddly fluted teeth occurring in the Cretaceous marls of New Jersey. In naming them, Leidy believed that the teeth were those of a reptile (Schwimmer, Stewart, and Williams, 1997) and apparently didn't consider that they might be from a large fish. To confuse the issue, the scientific name of the big fish that occurs in the Kansas chalk really should be *Polygonodon audax* Leidy 1870. That is unlikely to happen, however. As a side note, the Turonian-age ichthyodectid teeth described by Mkhitaryan and Averianov (2011:fig. 7) from Uzbekistan appear very similar to those of *X. vetus*.

Following the discovery of *Elasmosaurus platyurus* in 1867 and the first Kansas mosasaur, *Tylosaurus proriger*, in 1868, Kansas quickly became a popular place to search for fossils. In 1870, O. C. Marsh mounted the first of three organized fossil-collecting trips that he led to the western states, including Kansas, with his students. Although they spent less than two weeks in Kansas, the Yale College Scientific Expedition of 1870 was an immediate success, recovering dozens of specimens, including several mosasaurs, part of the wing of a previously unknown giant pterosaur from

Kansas (Marsh, 1871), and even a fragment of an equally unknown bird (*Hesperornis*).

Strangely, Marsh was not at all interested in fishes and instructed his field workers to ignore them (Shor, 1971:78). This allowed his rival, E. D. Cope, to describe and name many of the species from the Smoky Hill Chalk. Marsh's instructions may have been the source of a statement by B. F. Mudge, who wrote, "the least interesting are the fish, which have, however, given many new species and some new genera. The small ones are near entire, but the larger are represented by well-preserved portions of the skeletons" (1876:216). Years earlier, Mudge had collected a number of fish specimens for Cope (see Cope, 1872a) and Cope, in turn, had credited Mudge with the discovery of a number of new species. After being hired by Marsh, however, Mudge's collecting was focused more on toothed birds, pteranodons, and marine reptiles. Chris Bennett (pers. comm., 2004) indicated that when he examined the Yale fish collection in the 1980s, "there were still unopened packages from Mudge."

Cope (1872b:385) published a list of the families of fishes he saw in the "Cretaceous Rocks of Kansas" that included most of the common varieties that we are familiar with today, even though only limited collections had been made from the chalk at that time. Many of Cope's 'species' were described from fragmentary remains or even single teeth, and the family names would be changed again and again as more complete specimens were collected. Cope (ibid.:357) also noted that 24 species had been described from the Kansas chalk and that the same genera had been located in the chalk of Europe (see Table 5.1 for an updated listing of fishes from the Smoky Hill Chalk). Cope's 1872 list included:

Saurodontidae (Ichthyodectidae): *Portheus* [*Xiphactinus*], *Ichthyodectes*, *Gillicus*, and *Saurocephalus*
Pachyrhizodontidae: *Pachyrhizodus* and *Empo* [*Cimolichthys*]
Stratodontidae: *Stratodus*, *Enchodus*, and *Apsopelix*

The history of collecting, identifying, describing, and naming the many species of fishes that have been located in the chalk is confusing at best. In most cases, the early paleontologists involved were working with scraps and seldom had the luxury of seeing even a large portion of the complete fish. Rather than try to relate the history of the discovery of each genus, I will go into the details of those I believe are the more interesting ones.

The family Ichthyodectidae is represented by three species in the Western Interior Sea. Most of the history of the discovery of *Xiphactinus* (*Portheus* of Cope) has already been discussed in the preceding paragraphs. The remains of *Xiphactinus audax* Leidy, *Ichthyodectes ctenodon* Cope, and *Gillicus arcuatus* Cope are still collected commonly as fossils in the

Xiphactinus, Ichthyodectes, **and** *Gillicus*

Table 5.1 Classification of bony fishes of the Western Interior Sea: (Adapted from Romer, 1966; Everhart, 2005; Shimada and Fielitz, 2006; and others)

Class Osteichthyes		
Incertae sedis	*Aethocephalichthy shyainarhinos*	Fielitz, et al., 1999
Infraclass Holostei		
Order Semionotiformes		
Family Lepisosteidae		
	Lepisosteus	(Wiley and Stewart, 1977)
Order Pycnodontiformes		
Family Pycnodontidae		
	Micropycnodon kansasensis	Hibbard & Graffham, 1941
	Hadrodus marshi	Gregory, 1950
	Anomoeodus cf. *A. barberi*	Hussakof, 1947
Order Pachycormiformes		
Family Pachycormidae		
	Genus *Protosphyraena*	Leidy, 1857
	Protosphyraena perniciosa	(Cope, 1874)
	Protosphyraena nitida	(Cope, 1872)
	Protosphyraena tenuis	Loomis, 1900
	Bonnerichthys gladius	Friedman et al., 2010
Order Amiiformes		
	Paraliodesmus guadagnii	Dunkle, 1969
Order Aspidorhynchiformes		
Family Aspidorhynchidae		Woodward, 1896
	Asarotus arcanus	Schaeffer, 1968
Infraclass Teleostei		
Superorder Elopomorpha		
Order Crossognathiformes		
Family incertae sedis		
	Laminospondylus transversus	Springer, 1957
Family Crossognathidae		
	Apsopelix anglicus	Dixon 1850
Family Pachyrhizodontidae	*Pachyrhizodus caninus*	Cope 1872
	Pachyrhizodus kingi	Cope 1872
	Pachyrhizodus leptopsis	Cope 1875
	Pachyrhizodus minimus	Stewart 1899
Family Abulidae		Genus / species undetermined
Order Anguilliformes		
Family Urenchelidae		
	Urenchelys abditus	(Wiley and Stewart, 1981)
Order Ichthyodectiformes		
Family Ichthyodectidae		Crook, 1892
	Ichthyodectes ctenodon	Cope, 1870
	Xiphactinus audax	Leidy, 1870

Table 5.1 Continued...

	Gillicus arcuatus	(Cope, 1875)
Family Saurodontidae		
	Saurocephalus lanciformis	Harlan, 1824
	Saurodon leanus	Hays, 1830
	Prosaurodon pygmaeus	(Loomis, 1900)
Order Tselfatiformes		Nelson 1994
Family Plethodidae		Loomis, 1900
	Bananogmius aratus	(Cope, 1877)
	Bananogmius ornatus	Woodward, 1923
	Bananogmius ellisensis	Fielitz and Shimada, 1999
	Pentanogmius evolutus	(Cope, 1878)
	= B. polymicrodus	Stewart, 1898
	Martinichthys ziphioides	(Cope, 1877)
	Martinichthys brevis	McClung, 1926
	Syntegmodus altus	Loomis, 1900
	Thryptodus zitteli	Loomis, 1900
	Luxilites striolatus	Jordan, 1924
	Niobrara encarsia	Jordan, 1924
	Zanclites xenurus	Jordan, 1924
	Pseudanogmius maiseyi	Taverne, 2002
Order Aulopiformes		
Family Cimolichthyidae		
	Cimolichthys nepaholica	(Cope, 1872)
Family Enchodontoidae		
	Enchodus gladiolus	(Cope, 1872)
	Enchodus petrosus	Cope, 1874
	Enchodus dirus	Leidy, 1857
	Enchodus shumardi	Leidy, 1856
Family Dercetidae		
	Anguillavus hackberryensis	Martin, 1922
	Stratodus apicalus	Cope, 1872
	Leptecodon rectus	Williston, 1899
Family Ichthyotringidae		
	Apateodus busseni	Fielitz and Shimada, 2009
Order Polymixiiformes		
Family Polymixiidae		
	Omosoma garretti	Bardack, 1976
Order Beryciformes		
Family Holocentridae		
	Kansius sternbergi	Hussakof, 1929
	Caproberyx sp. Dixon?	(See Stewart, 1990)
	Unnamed holocentrid	(See Stewart, 1990)
Order Crossopterygii		
Family Coelacanthidae		
	Unnamed coelacanth	(Stewart et al., 1991)
	Megalocoelacanthus dobiei	Schwimmer et al., 1994

Smoky Hill Chalk. Bardack (1965) recognized four species of *Xiphactinus* and two species each of *Ichthyodectes* and *Gillicus* worldwide.

Xiphactinus is the largest of these three fishes in the Smoky Hill Chalk, reaching lengths of more than 5.5 m (18 ft). Besides its extremely large size, it is readily distinguished from the other genera by the teeth of unequal size in its jaws (Fig. 5.2). The anterior teeth, particularly those in the premaxillae, are the largest in the jaws. *Ichthyodectes* grew to about half the size of *Xiphactinus*, reaching lengths of around 3 m (10 ft). Its teeth are medium in size and are generally about the same size across the jaws. We have some evidence that *Ichthyodectes* fed on smaller fishes such as *Enchodus* because of a partial fish-within-a-fish specimen collected by Charles H. Sternberg (Everhart, Hageman, and Hoffman, 2010).

Gillicus grew to about one-third the length of *Xiphactinus*, or about 2.5 m (8 ft) long. The teeth in *Gillicus* are so small that at first the jaws appear to be nearly toothless. This has led to some speculation that *Gillicus* was a filter feeder, but I contend that a 2 m long fish with a big mouth does not need big teeth to feed on the large number of small fishes that were readily available. A number of modern marine and freshwater species of predatory fishes do quite well without large teeth. Eventually we may collect a *Gillicus* preserved with the remains of its last meal.

The postcranial skeletons of *Xiphactinus*, *Ichthyodectes*, and *Gillicus* are nearly indistinguishable except by size. Vertebrae (and tails) are probably the most common remains collected. The remains of *Xiphactinus* appear to be the most common of these three fishes in the chalk. This is probably due to a preservational bias that favors larger animals. *Gillicus* remains are relatively uncommon, in part because they were more likely completely consumed by larger predators such as *Xiphactinus* and mosasaurs. There is little fossil evidence regarding what *Ichthyodectes* preyed on or what preyed on it other than the specimen reported by Druckenmiller et al. (1993) that included the remains of an *Ichthyodectes ctenodon* as stomach contents of a shark (*Squalicorax falcatus*).

Shimada (1997; see also Sternberg, 1917) reported on the remains of a *Xiphactinus* as the stomach contents of a large *Cretoxyrhina mantelli* (KUVP 247) that was collected by Charles Sternberg and is currently on exhibit in the University of Kansas Museum of Natural History (Fig. 5.4). The shark apparently ate most of the big fish, including the skull, but then died soon afterward, before the bones could be digested. Many *Xiphactinus* bones are commingled with the vertebrae and teeth of the shark in this large wall-mounted exhibit.

The broken tip of a *Cretoxyrhina* tooth lodged between two vertebrae in association with the skull of another *Xiphactinus* specimen (ESU 1047) was described by Shimada and Everhart (2004). Many *Cretoxyrhina* teeth were collected in association with another fragmentary *Xiphactinus* skull (KUVP 12011) that also exhibited unserrated tooth marks on the lower jaw. A relatively early *Xiphactinus* specimen (KUVP 155) collected by B. F. Mudge from the Carlile Shale of Russell County and described

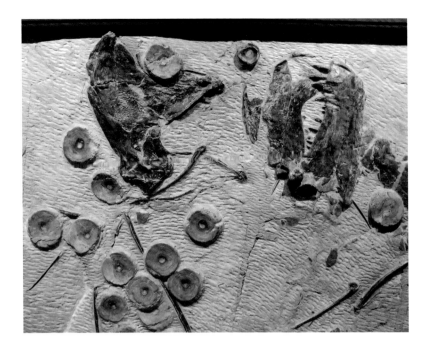

5.4. The tooth bearing dentary, skull fragments, and ribs of a large *Xiphactinus audax* (KUVP 245) inside the remains of an even larger *Cretoxyrhina mantelli* (KUVP 247) shark, discovered by George F. Sternberg in Trego County.

by Stewart (1900:293, pl. 45) also shows numerous unserrated shark bite marks on the lower jaw.

Bardack (1965) notes that the genus *Xiphactinus* occurs worldwide, including Australia, and may occur as early as Albian time in Europe. In 2001, fragments of a *Xiphactinus* skull were collected in the Czech Republic (central Europe) by a 15-year-old student, Michal Matějka. Those remains are curated in the National Museum in Prague. The southernmost specimen in the Americas that I am aware of was discovered in the mountains of Venezuela (Carrillo-Briceño, Alvarado-Ortega, and Torres, 2012) and the northernmost specimen in North America was reported from Alberta, Canada, by Vavrek, Murray, and Bell, (2016).

Perhaps the most unusual occurrences that I am aware of are the *Xiphactinus* vertebrae and jaw fragments collected from gravels deposited by retreating glaciers in northwest Missouri (Everhart, 2014, unpublished data). Associated with mosasaur teeth, plesiosaur teeth bones, and *Cretoxyrhina* shark teeth, the *Xiphactinus* remains had been scraped from where they had been originally deposited in Cretaceous-age rocks in Minnesota and Iowa and carried southward by the advancing ice sheet.

Illustrations in Mantell (1822:241, plate XLII), *The Fossils of South Downs*, and Mantell (1833) *Geology of the South-East of England*, undoubtedly represent a small specimen of *Xiphactinus*. The total length of the jaw is given as 5.5 in (14 cm). In 1822, Mantell thought the jaw came from an unknown fish, but then he changed his mind and regarded the jaw as belonging to an unknown reptile (1833). My contact with the British Museum of Natural History indicates that the specimen is still there and is curated as BMNH (OR)4066. Note that this specimen was fully reported by O. P. Hay:

> The earliest reference which we have to any remains of the genus of fishes usually called *Portheus* is that found in Mantell's *Geology of Sussex*:241, Pl. XLII, 1822. No systematic name is there assigned to this fish. Later, Louis Agassiz, in his *Poissons Fossiles*, vol. v:99, referred to Mantell's description, and refigured the materials (*op. cit.*, Pl. XXV b, figs. I a, I 6), presenting at the same time additional figures of remains from the same locality (Pl. XXV a, fig. 3; Pl. XXV b, figs. 2, 3). All these he included, with other remains, under the name *Hypsodon lewesiensis*. (1898:25)

Bardack (1965:37) later noted that the first *Hypsodon* species named by Agassiz was a pachyrhizodid fish, not an ichthyodectid, so Leidy's (1870) genus name is valid and has priority.

In North America, remains of the genus range from the Middle Cenomanian (Greenhorn Limestone) in Kansas into the middle Campanian Pierre Shale (Bardack, 1965:10; Carpenter, 1996:33). In 2009, I collected three *Xiphactinus* vertebrae (FHSM VP-17462) from the Pfeifer Shale Member of the Greenhorn Formation, just below the Fencepost Limestone, in Russell County. The specimen is much more complete than that but is inaccessible because the remaining portion goes straight into the side of the roadcut, and would be unrecoverable without heavy machinery. Shimada and Martin (2008) reported *Xiphactinus* teeth and a vertebra from the basal Lincoln Limestone Member (Upper Cenomanian) of the Greenhorn Limestone. *Xiphactinus* teeth were also reported from the deltaic Dakota Formation (Middle Cenomanian) by Everhart, Everhart, and Ewell, (2004). Williston (1894) described a 4.2 cm (1.6 in) fish vertebrae from the Kiowa Formation (Albian, Early Cretaceous) which he referred to *Portheus*, but that specimen has not been relocated. I have collected similar vertebrae from the Kiowa Formation associated with a caudal fin, but would only refer them to an unknown ichthyodectid. *Ichthyodectes* and *Gillicus* remains also occur about the same time span (Bardack, 1965), including teeth fragments of an ichthyodectid similar to *Xiphactinus* or *Ichthyodectes* in the Early Cretaceous (Albian) Kiowa Shale reported by Everhart et al. (2003).

One issue that is notable for these fishes, and for several other species, is that there are very few, if any, remains of juvenile or young individuals represented in collections from the Smoky Hill Chalk deposited in the middle of the Western Interior Sea. It appears that all of the remains of *Xiphactinus, Ichthyodectes,* and *Gillicus* occurring in the chalk are those of at least subadult fishes. This strongly suggests that the early stages of their lives were spent elsewhere. Obviously, it could also indicate a preservational bias against smaller individuals, but by now some evidence of their presence should have been recovered in the fossil record. The only specimens of a young *Xiphactinus* that I know of are what appear to be the fragmentary remains of two tiny skulls, one of which I collected (Fig. 5.5). It is worth noting here that freshwater ichthyodectid fish

5.5. The left dentary and premaxilla of a small (less than 0.05 m) *Xiphactinus audax* in left lateral view from the lower Santonian chalk of Gove County, Kansas. Young *Xiphactinus* would have been preyed upon by many other species. Scale bar in mm.

remains have been discovered half a world away in the Central Kyzyl-kum Desert of Uzbekistan (H. Sues, pers. comm., 2016). Mkhitaryan and Averianov (2011) reported on abundant specimens of the ichthyodectid *Aidachar paludalis* from the brackish to freshwater Bissekty Formation (Turonian). This suggests that *Xiphactinus* may have spawned in freshwater rivers, like modern salmon, and that their young may have spent some time there before maturing and entering the ocean.

The family Saurodontidae includes three species from the Western Interior Sea that are closely related to the Ichthyodectidae described above (Stewart, 1999). All were small to medium-size (1–2 m; 3–6 ft) predatory fishes and are uncommon as fossils in the chalk. As noted in the introduction to this chapter, the type specimen of *Saurocephalus lanciformis* Harlan 1824 was the first fossil fish known to be collected from the Niobrara Formation. The most unusual character of these fishes is the predentary bone that projected forward from the lower jaw. In life, it is likely that this bone was covered by something similar to the beak on a modern swordfish (Fig. 5.6). The function of this underslung sword is unknown; it was possibly used for slashing at smaller prey, like the feeding tactics employed by modern billfish, but I've never seen any apparent wear or damage on the predentary bone.

Another characteristic of these fishes is their flat, bladelike teeth, set in a single row in the jaws. In *Saurocephalus* and *Saurodon*, the teeth are closely set, nearly vertical, and have a keyhole-like notch at the base of each

Saurocephalus, Saurodon, and *Prosaurodon*

5.6. Skull of *Saurodon leanus* (KUVP 161) from the Smoky Hill Chalk in right lateral view. The specimen was originally described and figured by Stewart (1898:pl. XIV) as the type specimen of a new species (*S. 'xiphirostris'*). Scale = 10 cm (4 in).

tooth on the inside of the jaws. In *Prosaurodon*, the teeth are more rounded and do not have the notch that is a characteristic in the other two species. In the lower jaw of *Prosaurodon*, the teeth are inclined slightly forward except for the anterior-most three or four teeth, which are inclined posteriorly (Loomis, 1900:pl. 23, fig. 10). Stewart's (1999) phylogenetic analysis of *Saurocephalus*, *Saurodon*, and *Prosaurodon* placed *P. pygmaeus* as the closest relative of *Gillicus arcuatus* among the saurodontids.

The three species apparently occurred at different times in the Western Interior Sea, with little overlap. According to Stewart (1999), *Saurocephalus lanciformis* appears only in the uppermost chalk (early Campanian); *Saurodon leanus* Hay 1830 is known from the lower through the middle chalk (late Coniacian through the Santonian). *Prosaurodon pygmaeus* appears first about the middle of the chalk and continues upward into the Sharon Springs Member of the Pierre Shale (ibid.; Carpenter, 1996). Pre-Niobrara records of the Saurodontidae are unknown in Kansas. However, a much older specimen from the Cenomanian Eagle Ford Group of Texas (Stewart and Friedman, 2001) suggests that family existed for some time prior to the deposition of the Niobrara. The extended lower jaw of these saurodontids might suggest that it could have been used in digging for prey in the bottom muds, but that seems pretty unlikely in the Western Interior Sea over Kansas, especially considering that the teeth were not suitable for hard-shelled prey. Like many other fish species of the time, and unlike *Xiphactinus*, we have never collected them with preserved stomach contents.

In 2013, my son Matthew was with us on a field trip. We had been pretty successful on the previous two days, with a couple of mosasaur specimens and other fossil remains. As luck would have it, a couple hours before we were headed home, Matt spotted a *Saurodon* predentary exposed on the edge of deep gully. Then, while he was waiting for me to come up and see it, he discovered a big *Xiphactinus* skull and another fish specimen within 6–9 m (20–30 ft) of where the *Saurodon* was eroding out. The rest of the group started working on the other fish while Matt and I examined the *Saurodon* site. The predentary bone was the only

thing visible at first, but we quickly located the end of the dentary where it had been attached. That was a good sign; we hoped it meant the rest of the fish or at least the skull was still safely contained in the chalk. After digging places to stand on the edge of the gully, we started removing the chalk from where we thought the skull should be. The chalk came loose easily in layers, and within fifteen minutes or so, we could see the outline of the skull (Fig. 5.7). The specimen came to an end about five vertebrae behind the skull. The skull, including the predentary, was about 0.4 m (16 in) in length, and represented a good-size fish. I was initially disappointed to see that the whole fish wasn't there, but on the other hand, we didn't have time for a major excavation. We cleared a channel around the skull, then used plaster bandages to make a jacket. While it was drying we went over to see how the other discoveries were panning out. The *Xiphactinus* skull was disarticulated, but huge and well preserved; the other fish, not so much.

After the plaster had hardened, the jacket containing the *Saurodon* skull was rolled without any problems. I spent a few hours prepping on it, and then Matt donated it to the Sternberg Museum (FHSM VP-18690; See Fig. 13.6). It's one of the best *Saurodon leanus* skulls in the Sternberg collection, comparable to the exhibit specimen (FHSM VP-168), collected by G. F. Sternberg from Logan County.

Xiphactinus has already been mentioned as the largest of the bony fishes in the Western Interior Sea. However, the skull of a large (3 m; 9.8 ft) specimen of *Pachyrhizodus caninus* (FHSM VP-2189) at the Sternberg

Pachyrhizodus

Pachyrhizodus with Preserved Intestine
Trego County, Kansas

Intestines and other soft body parts rarely are
preserved as fossils. Perhaps this fish was a
deposit feeder, swallowing sediment to gather
worms or other soft-bodied, burrowing creatures.
Mud packed into the gut might have retained the
shape of the intestines while the flesh decayed.

5.8. A exceptionally well preserved specimen of *Pachyrhizodus minimus* (FHSM VP-326) in ventral view collected by George F. Sternberg from the low chalk of Trego County in 1949. Lighter colored area in the center of the fish represents preserved stomach and intestinal contents. Scale bar = 10 cm (4 in).

Museum is certainly as large as any *Xiphactinus* skull that I have seen. It is 56 cm (22 in) long and has teeth 4 cm (1.6 in) long. The entire fish, however, would have been proportionally shorter than a *Xiphactinus* with the same size head. Still, at over 2.7 m (9 ft) in length, it would have been a large and impressive fish. As a comparison to modern fishes, *Xiphactinus* would be a giant tarpon, and *Pachyrhizodus caninus* would be more like a big grouper or sea bass.

Pachyrhizodus caninus Cope 1872 and *P. leptopsis* Cope 1874 were in the 2–3 m (6.6–9.9 ft) size range and could be considered in the middle echelon of predatory fishes. They were equipped with heavy jaws that were filled with sharp, curved teeth. The teeth were noted to be "short and stout" by Cope (1872b:343), who later said that "they bear a superficial resemblance to those of a mosasauroid genus" (1875:220). Stewart and Bell (1994) reported that some specimens from Texas initially identified as the remains of the earliest mosasaurs (Cenomanian) by Stenzel in 1945 were, in fact, jaw fragments of *Pachyrhizodus leptopsis*. In addition, the material cited by Thurmond (1969) as the jaw of an early (Turonian) mosasaur from Texas was also reidentified by Stewart and Bell (1994) as the remains of *P. leptopsis*. The source of the confusion is that the teeth have large bases and, as Cope noted, look very much like those of a mosasaur. Closer examination, however, shows that they are attached directly to the bony surface of the jaw, and not implanted in sockets like mosasaur teeth. Stewart and Bell (1994:8) noted that other specimens of *Pachyrhizodus* in museum collections have been misidentified as mosasaurs, and vice versa.

Pachyrhizodus minimus (Stewart 1899) is a much smaller species (less than 1 m; 3.3 ft). The holotype (KUVP 327) is maintained in the collection of the University of Kansas. *P. minimus* has the distinction of being the most commonly collected 'complete' fish fossil in the Smoky Hill Chalk, including scales and preserved gut contents. Miller (1957) reported on a 0.8 m (34 in) specimen (FHSM VP-326; Fig 5.8) from Trego County in the Sternberg Museum collection that includes a natural cast of portions of the stomach and the intestine and an expelled coprolite. In 1990, my wife discovered a complete (0.6 m; 24 in) specimen (CMC-7552) in the early Santonian Chalk of Lane County that we donated to the Cincinnati Museum Center.

Fish remains attributable to the genus *Enchodus* Agassiz 1835 occur worldwide, and the genus apparently survived for a time past the Cretaceous/Tertiary extinction event, based on specimens from Lebanon. *Enchodus* has been erroneously referred to as the 'Saber-Toothed Herring' because of the oversized teeth located on the palatines of the skull and the ends of the lower jaw. The genus is not related to modern herrings but instead is placed in the same order (Salmoniformes) as modern salmon. *Enchodus* is represented in the Smoky Hill Chalk by at least four species, ranging from small (*E. shumardi*) to medium-size (*E. petrosus*) fishes. However, the exact number of species is unclear even today because, although *Enchodus* remains are far from being rare, they are seldom complete. This led early workers such as Cope to name several new species from fragmentary specimens. However, the isolated tiny jaws of *Enchodus*, with their oversized teeth, are common occurrences when layers of chalk are split during a dig.

Stewart (1990a) reported the occurrence of *Enchodus* in all but the highest level of the Smoky Hill Chalk, but their absence there is more of a collecting issue than anything else, because they occur as relatively common fossils in the overlying Pierre Shale. Goody (1976) reported on *Enchodus gladiolus* and *E. petrosus* remains in the Pierre Shale and discussed the occurrence of the genus in North America in general. In another study, *Enchodus* species represented about 25 percent of the remains observed in the Pierre Shale of Wyoming (Carpenter, 1996:34). The most recent phylogenetic analysis of the family Enchodontidae was done by Fielitz (1999).

The large palatine teeth of *Enchodus* indicate that it was certainly a predator, but as yet we are uncertain what it ate. I am not aware of any *Enchodus* remains preserved with stomach contents. Although such oversized teeth might be useful in impaling small, soft-bodied prey such as squid, this raises the question of how the fish got the prey off the teeth and into the mouth to be swallowed. Most likely the long, slender teeth were used as a 'fish basket' to more effectively trap smaller fishes, similar to the interlocking teeth of some plesiosaurs. While the largest species, *Enchodus petrosus*, had fangs that were nearly 6 cm (2.5 in) long on a body that was no more than about 1.5 m (5 ft) in length, even the smallest species had oversized fangs. They would have looked much like some of the modern deep-sea fishes, which also have gaping mouths filled with long, sharp teeth. *Enchodus* teeth are usually well represented in collections of micro-vertebrate fossils, and in one instance they composed nearly all (literally millions of tiny teeth) of a thin layer of 'fish tooth conglomerate' discovered in the Blue Hill Member (middle Turonian) of the Carlile Shale of Jewell County, Kansas (Hattin, 1962:102, pl. 27B; Everhart et al., 2003).

Enchodus and Cimolichthys

5.9. The fangs of *Enchodus petrosus* collected from the Upper Coniacian chalk of Gove County. The specimen includes the two palatines with large teeth and a large anterior tooth on the dentary. Scale bar in mm.

Whatever the function of these unusual fangs (Fig. 5.9), they were apparently not a deterrent to other predators. *Enchodus* remains occur as stomach contents of *Cimolichthys* (below), *Ichthyodectes* (Everhart, Hageman, and Hoffman, 2010), mosasaurs, and plesiosaurs (Cicimurri and Everhart, 2001), and their bones occasionally occur in coprolites that are usually attributed to sharks. The skull of a large *Enchodus petrosus* at the Sternberg Museum (FHSM VP-2939) collected in Trego County preserves evidence of scavenging by *Squalicorax falcatus*, with serrated bite marks on the bones of the skull and a shed *Squalicorax* tooth lying against the underside of the cranium. It seems apparent that the smaller *Enchodus* species existed in large numbers and were an important part of the ecology of the Western Interior Sea. In that regard, they were probably very much the ecological equivalent of the modern herrings or sardines.

One of the more common species of fishes that is preserved in the chalk is a medium-size predator called *Cimolichthys nepaholica* (Cope 1872b). The genus name *Cimolichthys* (from Greek 'cimoli,' a white chalky clay; and 'ichthys,' fish) had been coined by Leidy in 1857 to describe fish remains (*C. levesiensis*) from the English chalk. 'Cimoli' is the Latinized version of the Greek word 'kimolia,' a white clay from the Aegean island of Kimolos in the archipelago of the Kyklades. Leidy apparently named the English species *C. levesiensis* for the town of Lewes in Sussex, England, where the remains were first recognized.

The genus name has been confused over the years in large part because E. D. Cope named a new genus and species, *Empo nepaholica* (1872b:347), and four new species of *Cimolichthys* (ibid.:351–353: *C. sulcatus, C. semianceps, C. anceps,* and *C. gladiolus*) from Kansas in the same paper. Cope later described *E. nepaholica* as "a fish as large as pike of forty pounds" (1872c:345). He also noted that *Cimolichthys* "was applied by Dr. Leidy to a fish erroneously referred by Agassiz and Dixon to *Saurodon,* Hays. He [Leidy] did not characterize it; and until the barbed palatine teeth, characteristic of it, are discovered in our species, their reference to it will not be fully established" (ibid.:347). Two years later, Cope (1874:46) revised the spelling of *nepaholica* to *nepaeolica* [*nepæolica*], changed the genus name of *C. sulcata* to *Empo sulcatus,* changed *C. semianceps* to *Empo semianceps,* and added two new species of *Empo: E. merrillii* and *E. contracta.* While *C. gladiolus* eventually became *Enchodus gladiolus, C. anceps* disappeared from the literature along with the type specimen. Goody (1976:102) noted that the type specimen of *C. anceps* could not be relocated. Goody also determined that, based on Cope's description of the type, the specimen was more likely the ectopterygoid of *Enchodus petrosus.*

As noted above, Leidy's *Cimolichthys lavesiensis* is older (Turonian) than the specimens of *Empo nepaholica* examined by Cope from the Smoky Hill Chalk (late Coniacian–early Campanian). It doesn't appear that Cope ever accepted that *Cimolichthys* Leidy and *Empo* Cope were the same genus. Cope noted that he had "formally referred some of the species of *Empo* to the genus which embraces the fish called by Leidy *Cimolichthys levesiensis*; but I find that they do not possess the same type of teeth. . . . The genus therefore takes this name [*Empo*]" (1875:230). Much of the confusion comes from having to work with fragmentary material. *Cimolichthys* specimens are usually poorly preserved, and the skulls are quite fragile. However, Cope (1875:pls. 52 and 53) was the first to illustrate the American *Cimolichthys* and one of a few to date to figure the dermal scutes (ibid.:pl. 53, fig. 9).

Empo nepaholica and the other new *Cimolichthys/Empo* species of Cope were later determined to be fragments of the same species of fishes and were recognized as such by Loomis (1900) and by Hay (1903). By virtue of being the first of the species names published in Cope's (1872b) paper, the species is rightfully called *nepaholica.* Since the genus name *Cimolichthys* Leidy 1857 takes precedence over that of *Empo* Cope 1872, the correct name for Cope's many 'species' became *Cimolichthys nepaholica.* Like the similar problem with *Xiphactinus* Leidy and *Portheus* Cope, the name *Empo* has had a long, if ill-deserved, life of its own and is still seen on labels in museum exhibits and collections. As for the other species names, Goody said that "there is no reason for retaining Cope's various species that are based mainly on isolated teeth and fragments of jaw bones" (1970:2).

Cimolichthys can be visualized as a Cretaceous barracuda (or a fresh-water pike as suggested by Cope), even though it is not related to either

5.10. The skull and jaws of *Cimolichthys nepaholica* in ventral view, crushed dorsoventrally, and still embedded in chalk. Anterior is to the right; the occipital condyle of the skull is indicated by the oval at left. Scale in mm.

(like *Enchodus*, *Cimolichthys* is in the same order as modern salmon). This fish grew to almost 2 m (6 ft) in length, and the narrow, triangular skull, with three rows of teeth set in the lower jaws (Hay, 1903), is readily recognizable. Cope noted in his description of the genus *Empo* that "the dentaries support several series of teeth; one of the large ones on the inner side and several smaller on the outer" (1875:228). His illustration (ibid.:pl. 53, fig. 6) accurately shows three rows of teeth on the dentary.

Portions of the skull of *Cimolichthys nepaholica* have been figured by Cope (1875), Loomis (1900), Hay (1903), and Goody (1970). While their remains are often seen in the field, good specimens are rare because the skull was lightly constructed and tended to either come apart prior to preservation or as a result of weathering (Fig. 5.10).

We can assume that *Cimolichthys* had a voracious appetite for fairly large prey because a number of specimens have been discovered with the remains of an undigested last meal inside. One of the strangest 'death by gluttony' occurrences in the fossil record was reported by both Kauffman (1990) and Stewart and Carpenter (1990; see also Carpenter, 1996) regarding a specimen of *Cimolichthys* (UCM 29556) discovered in the Pierre Shale of Wyoming. This *Cimolichthys* apparently died with a large squid (*Tusoteuthis longa*—UCM 20556) lodged in its mouth. The squid had been swallowed tail first, as evidenced by the rachis (squid pen) being located inside the body of the *Cimolichthys*. However, the wide-open jaws of the fish appear to indicate that part of the head and/or tentacles of the squid were still outside the mouth. This probably meant that the fish's gills were blocked from getting oxygen from the water, causing death by suffocation.

Another unusual *Cimolichthys* specimen (FHSM VP-15065) in the Sternberg Museum was collected by Greg Winkler and Pete Bussen in

the early 1990s from the lower chalk (Upper Coniacian) of western Gove County. In this case, the skull of a large (1.8 m; 5.9 ft) *Cimolichthys* was discovered eroding out of the chalk. Much of the skull was already lost, but the rest of the fish was complete back to the tip of the tail. When the specimen was initially prepared, it contained not only a large *Enchodus* (FHSM VP-15066) but also the remains of another unidentified, smaller fish (FHSM VP-15067) as a last meal. It is uncertain whether both fish inside the *Cimolichthys* were consumed by the larger fish or this is a classic case of a 'fish within a fish within a fish.' In either case, the partially digested condition of the *Enchodus* leads me to believe that the *Cimolichthys* died within a few hours after eating it.

In a similar situation, from the lower chalk of southeastern Gove County in 1994 I collected a 1.3 m (4 ft) *Cimolichthys* (FHSM VP-14024) with a 0.7 m (2 ft) long, partially digested *Enchodus petrosus* (FHSM VP-14025) inside. Both palatine bones of the *Enchodus*, minus the large fangs, were located near the anus of the larger fish. Besides being an indication that the prey had been swallowed headfirst, it also explains our field observation of collecting an unusual number of partially digested *Enchodus* palatine bones (Fig. 5.11), and no other associated remains. Being the heaviest bones in the skull of the *Enchodus*, mostly indigestible and located at the anterior end of the skull, they were probably expelled separately while the rest of the prey was still being digested. The large teeth of the FHSM VP-14025, however, appeared to have been broken off the palatines and were not located in the rest of the remains.

It is probable that *Cimolichthys* was preyed upon by larger fishes and mosasaurs. Although they were not noted by Bardack (1965) as being

5.11. At left are a pair of *Enchodus petrosus* palatine bones that have lost their fangs and been partially digested by a predator, possibly *Cimolichthys*. These damaged bones were collected as a pair, most likely originally from a coprolite, but not associated with other material from the skull. This is a fairly common occurrence in the Smoky Hill Chalk. At right is a normal *E. petrosus* palatine bone with a fang from a skull. Scale bar in mm.

5.12. Six vertebrae from a *Cimolichthys nepaholica* specimen. The two vertebrae at upper left have a calcite 'cone on cone' (bicone) artifact that filled the space between two adjacent vertebrae. Scale bar in mm.

stomach contents in any of the *Xiphactinus audax* specimens he surveyed, I did locate several partially digested *Cimolichthys* vertebrae in the abdominal region of a *Tylosaurus proriger* (FFHM 1997–10) I collected in 1996–97. While their remains have not been reported as stomach contents of other predators, many of the severed tails which we commonly see in the chalk are from *Cimolichthys*. One interesting feature often observed in *Cimolichthys* remains in the field is the "cone-on-cone" calcite crystals (Fig. 5.12). that often fill the conical hollows between their deeply cupped vertebrae.

Protosphyraena
(Pachycormidae)

One fairly common genus of fishes in the Smoky Hill Chalk that was noticeably missing from Cope's 1872 list was *Protosphyraena*, a primitive Late Cretaceous 'swordfish' that is well represented in the fossil record from as far away as Europe, Australia, and Japan. A fragment of the pectoral fin was first figured and described from the English chalk by Mantell (1822), and Leidy (1857) authored the name *Protosphyraena ferox* for the English specimens. Leidy (1865:pl. XX, figs. 7–9) also illustrated a *Protosphyraena* tooth from the Navesink Formation in New Jersey, but he erred in stating it was from a dinosaur. After examining specimens at the Smithsonian that were collected by Dr. G. M. Sternberg, Leidy (1870:12) clearly described fragments of the distinctive fin of *Protosphyraena perniciosa*, but attributed them to the crusher shark *Ptychodus*. At the time, he noted (ibid.) that he was following the same association as reported by Agassiz in 1837.

Many of the first Kansas specimens of *Protosphyraena* were collected by B. F. Mudge in the early 1870s from Rooks County. Some of them were sent to E. D. Cope, possibly as early as 1872. As the collector, however, it is apparent that Mudge had a better idea of what the fish looked like than did Cope. Mudge wrote of the skull:

5.13. A, fragment of the anterior portion of a *Protosphyraena* jaw showing two of the three kinds of teeth described by Mudge (1874); note that the large, bladelike tooth at the end of the jaw is broken; scale bar = 2.5 cm (1 in); B, these bladelike teeth are unique to *Protosphyraena* and are a common occurrence in the chalk; it is likely that they were broken off while the fish was feeding on prey; scale bar = 5 cm (2 in).

The most remarkable species of fish which we have found, the present season, are of a genus new to me, and I think to science. They are armed with a long, strong weapon at the extremity of the upper jaw, something like that of a swordfish, but round and pointed and composed of strong fibres [*sic*]. The jaws are provided with three kinds of teeth. On the outer edge is a row of large, flat, cutting teeth, somewhat resembling those of a shark [Fig. 5.13]. Inside, and placed irregularly, are small, blunt teeth; while in the back portion of the palate is the third set—small, sharp and needle-like in shape, forming a pavement. The jaws are also fibrous, like the snout. There are three species of this genus. Prof. Marsh has them for critical scientific examination. (1874:122)

At least fourteen specimens (YPM 42137, 42138, 42152, 42200, 42285, etc.) were collected by Mudge in 1874 from Ellis and Rooks counties and sent to Marsh. More specimens would be collected and sent to Yale in 1875 and 1876. However, the three species mentioned by Mudge had already been described by the end of the 1874 field season by Cope, and Marsh was not interested in fishes.

The first species of *Protosphyraena* from Kansas was described by Cope (1874) from teeth and skull fragments and named *Erisichthe nitida*. The type specimen (AMNH 2121) was "discovered by Prof. B. F. Mudge near the Solomon River" in Phillips County (Cope, 1875:217–218) and consisted of portions of the dentary and a pectoral fin. This specimen and other specimens of *Protosphyraena* in the Cope collection were figured by Hay (1903).

Portheus gladius Cope was named from a fragment of a large (1 m; 3.3 ft) pectoral fin that Cope (1873) believed to be similar to that of *Portheus* (*Xiphactinus*). The fin ray (AMNH 1849) was later figured by Cope (1875:pl. 52, fig. 3) and the genus name was changed to *Pelecopterus*. According to Cope (1873:338), this specimen was also collected by B. F. Mudge "near the Solomon River." Loomis (1900) eventually changed the name to *Protosphyraena gladius*. That said, 110 years later, it turned out not to be *Protosphyraena* at all, but rather a new genus of filter-feeding fishes called *Bonnerichthys gladius* (Friedman et al., 2010). More on that in a following section.

5.14. A fragment of the sawtoothed edge of the pectoral fin of *Protosphyraena perniciosa*. These fins reach 0.9 m (36 in) or more in length. Each 'tooth' marks the beginning of one of the long elements (rays) that make up the fin. Scale in cm.

The next year, Cope (1874:41) named *Ichthyodectes perniciosus* from fragments of sawtooth-edged pectoral fins that were discovered by B. F. Mudge. In Latin, the species name *perniciosus* means 'destructive, harmful, or dangerous,' and is certainly descriptive of the jagged leading edge of the pectoral fins (Fig. 5.14). It is apparent from Cope's descriptions at the time, however, that he did not associate the unusual pectoral fins with the swordfish-like skulls mentioned by Mudge (1874). Cope admitted, however, that he "had already been in receipt of fragments of these beaks, associated with loose teeth of the genus *Erisichthe*, but it was Prof. B. F. Mudge who first pointed out that both belong to one and the same genus" (1877b:821).

By then, however, Cope (1875:244) had changed the name of this species to *Pelecopterus perniciosus*. In his description of another American species, Cope noted that another "species has been found in England, and figured by Dixon in the '*Geology of Sussex.*' The portions represented in this work are the mandibles, which resemble those of the *E. nitida*, and which were supposed at that time to belong to a species of *Saurocephalus*. A muzzle, perhaps of the same species, was regarded as a Sword-fish, which was called *Xiphias dixonii* by Agassiz. It should be now termed *Erisichthe dixoni*" (1877b:823).

In the case of Cope's three Kansas species in three different genera, and *Xiphias dixonii*, however, they were soon all determined to be junior synonyms of the genus *Protosphyraena* Leidy 1857. Newton (1878:788) was one of the first to comment: "Dr. Leidy in 1856 [published 1857] proposed, as already mentioned, the generic name of *Protosphyraena*, for those British specimens; and this name therefore must be adopted, and not that of *Erisichthe*, which was given by Prof. Cope in 1872 to the American specimens." Cope's barely pronounceable *Erisichthe* genus eventually faded into oblivion, although he was still referring to it as late as 1886.

5.15. The proximal portion of the left pectoral fin of *Protosphyraena nitida*. Note the smooth leading edge of this fin compared to that of *P. perniciosa*. Scale bar = 10 cm (4 in).

Loomis (1900) was the next to work on describing the various species of *Protosphyraena*. His profusely illustrated paper in *Palaeontographica* is still one of the best references available on these fossil fishes from the Kansas chalk. As noted above, Hay (1903) also produced a valuable reference on these Cretaceous fishes, especially with illustrations of the specimens in the E. D. Cope collection at the American Museum of Natural History. Hay (ibid.:9) also noted that it was A. S. Woodward (1895) who placed Cope's *Pelecopterus perniciosus* into the genus *Protosphyraena* and first published the name *Protosphyraena perniciosa* in his Catalog of Fossil Fishes in the British Museum.

With the deletion of *P. gladius* (Friedman et al., 2010), there are three species of *Protosphyraena* that are currently recognized from the Smoky Hill Chalk: *P. nitida* (Cope, 1873), *P. perniciosa* (Cope, 1874), and *P. tenuis* (Loomis, 1900). *P. perniciosa* occurs in the late Coniacian lower chalk, where its 90 cm (3 ft) long, sawtoothed fins and heavy pectoral girdle are fairly common discoveries. The skull, with its long, swordlike snout and flat, blade-shaped teeth, is also discovered frequently, although is it difficult to identify for certain to species because is it so similar to that of *P. nitida*. The pectoral fins of *P. nitida* are readily recognizable (Fig. 5.15) because they lack the sawtoothed edges and are much smaller than those of *P. perniciosa*. Upon close examination of the fin of *P. nitida*, numerous fine ridges can be seen running at right angles to the edge of the fin. *P. tenuis* occurs in the early Santonian and appears to continue through the deposition of the chalk. It is a smaller species than *P. perniciosa* but has fins with a similar sawtoothed appearance.

Another, earlier species of *Protosphyraena* was described by Albin Stewart (1898:27) from a specimen (KUVP 415) that had been collected by S. W. Williston in 1897 in southern Mitchell County. The remains were discovered in the Lincoln Member (Upper Cenomanian) of the Greenhorn Limestone (part of the obsolete term 'Fort Benton Cretaceous'), and Stewart named the new species *Protosphyraena bentonianum*. Note that the species name was first published as *bentonia* in 1898 by Stewart, but then was noted to be a typographical error by Stewart (1900). The specimen consists of a poorly preserved rostrum and fragmentary bones from the skull and does not appear to differ greatly from *Protosphyraena perniciosa*. In 2003, Keith Ewell and I collected the fragments of numerous *Protosphyraena* teeth (unreported data) from the upper part of the Dakota Sandstone in Russell County. The Dakota Sandstone is middle Cenomanian in age.

Stewart (1979, 1988, 1990a) was the most recent North American author to report on the occurrence of *Protosphyraena*. In doing so, however, he unintentionally introduced a change in the species name for *Protosphyraena perniciosa*, dropping the second "i" and spelling it "*pernicosa*" (Stewart, pers. comm., 2004). As he has been the only worker to study this species in the last century or so, his version of the name came to be the accepted usage for a short time. However, the rules of the International Commission on Zoological Nomenclature (ICZN) normally require that such changes be made for good reason and be approved. The spelling of the species name authored by Woodward (1895:414) is used here and has been used in other publications since 2005.

Stewart (1979) was also the first to associate the occurrence of the various species of *Protosphyraena* with the stratigraphy of the chalk. He reported that *P. perniciosa* was restricted to the lower chalk (late Coniacian); the other three [two] species generally occurred during the Santonian and Campanian. Noting that the limited occurrence of *P. perniciosa* was useful as a stratigraphic marker, Stewart (1990a) designated the late Coniacian lower chalk as the biostratigraphic zone of *Protosphyraena perniciosa*. However, Stewart also acknowledged that "consultation and field observations with Mike and Pam Everhart convince me that *Protosphyraena nitida* is a rare member of the lowest Smoky Hill Chalk fauna" (ibid.:21). The change was due in part to Pam's 1988 discovery of a complete *P. nitida* skull and fins (LACMNH 129752) in the lower chalk just above the contact with the Fort Hays Limestone in Ellis County (Fig. 5.16). Two other partial *Protosphyraena* sp. skulls (NJSM 15021, 15839) we collected from the same locality were donated to the New Jersey State Museum. Since then we have discovered a number of *P. nitida* remains in the lower chalk, and in 2003 I collected the associated partial skull and fins (FHSM VP-17562) of a small (subadult) individual in Gove County below Hattin's (1982) marker unit 3. It also appears that *Protosphyraena perniciosa* reappears in the upper chalk (Pete Bussen, pers. comm., 1995; Everhart, pers. obs.; FHSM VP-17424) in the early Campanian, although it is much more common in the late Coniacian.

More than a hundred years after its discovery, it is still unusual to collect *Protosphyraena* skull and fin material together, and until recently a complete specimen had never been collected. In 2003, a nearly complete *Protosphyraena perniciosa* specimen was discovered in southwestern Gove County by Triebold Paleontology (M. Triebold, pers. comm., 2003). Unlike some shark specimens (Chapter 4) where the cartilage became

more calcified with age, the postpectoral skeleton of *Protosphyraena* was generally poorly ossified, and their carcasses apparently tended to fall apart as they decomposed, or were torn apart by predators and scavengers. Only the skull and fins and a small bone (hypural) at the base of the tail were ossified enough to be preserved in most cases. Unfortunately, the skulls and pectoral fins are usually collected separately. McClung (1908) described and figured a caudal fin of *Protosphyraena* from a specimen (KUVP 55500) discovered by C. H. Sternberg. A second, more complete specimen (KUVP 41419; Fig. 5.17) is also in the collection of the University of Kansas, but is as yet unreported. It has been tentatively identified as both *P. nitida* and *P. tenuis,* but without the pectoral fins, no one really knows what species it is.

Most likely, *Protosphyraena* was a fast-swimming predator on smaller fishes, similar to the behavior observed in modern billfish. It seems unlikely that we will ever recover a complete specimen of *Protosphyraena,* with stomach contents, but it would be a very nice discovery.

As noted earlier in this chapter, *Protosphyraena gladius* had a rather confusing journey lasting more than 135 years through the naming process. The skull of *P. gladius* was unknown, but the large, heavy fins, up to a meter in length and quite thick, are common occurrences in the Smoky Hill Chalk and around North America. Cope noted that the fin of *P. gladius* was "a formidable weapon, and could readily be used to split wood in its fossilized condition" (1875:244). While this is somewhat of an

Bonnerichthys gladius—
New genus—
(Pachycormidae)

5.17. A caudal fin of *Protosphyraena* sp. (KUVP 41419) from the Upper Coniacian chalk of Trego County, Kansas, in the collection of the University of Kansas. Scale bar = 10 cm (4 in).

exaggeration because of their relatively fragile condition as fossils, the sharp, axe-blade shape of the heavy fin is quite evident.

In 1971, Chuck Bonner collected a relatively complete specimen of this big fish, including the pectoral fins and some unusual but not readily identifiable bones from the skull, and donated the partially prepared material to the University of Kansas. Anthony Maltese saw the specimen as a student at KU and wanted to get a closer look at it. Years later, at Anthony's request, Dr. Larry Martin agreed to the loan of the specimen to the Rocky Mountain Dinosaur Resource Center (Triebold Paleontology) for further preparation and casting. Once the remains had been cleaned, it was apparent that this fish was not a swordfish like *Protosphyraena*, and he noted that "there was no sword-like rostrum or teeth present in the fossil, two traits found in *Protosphyraena*." (Anthony Maltese, pers. comm. 2007). Coincidentally, a visiting graduate student, Matt Friedman, who had been studying pachycormid fishes from the Jurassic of Europe, recognized the unusual bones of the skull as similar to other ancient filter-feeding fishes (like the giant *Leedsichthys* fish from the Jurassic). *Protosphyraena gladius* did not need a spikelike rostrum or large, sharp teeth because it used its big basket-shaped mouth to filter small prey from the water as it swam. It quickly became apparent to Friedman that a long-empty niche (filter feeding) in the Late Cretaceous was now filled by this previously misidentified fossil. The discovery of a large filter-feeding fish in the Western Interior Seaway was then reported at a paleontology meeting in Scotland (Friedman, Shimada, and Maltese, 2007). Contrary to past times, there was simply not then enough information available to actually describe and name a new genus of filter-feeding fishes from the single specimen. Friedman set out to examine other specimens in museum collections and gather more data about this enigmatic fish.

On a very hot day in July 2008, I was on a field trip in the Smoky Hill Chalk with Kenshu Shimada from DePaul University, Chicago. I was busy with a mosasaur specimen when Shimada came over and told me about a large fin that he had discovered protruding from the side of a chalk gully. When we went over to examine it, it was quickly apparent that it was the broken end of a *Protosphyraena gladius* fin, much larger than any *P. perniciosa* fin that I had ever seen. There was about 18 inches of chalk over the top of it, but it was too good a specimen to pass up, even though the temperature was nearly 36°C (100°F) and there was no shade. Grabbing our tools, Shimada and I started removing chalk from over the specimen. The chalk was tough, but after a couple of hours digging, we were able remove the block containing the specimen. In the process, we discovered that there were even more bones going back into the chalk. Not unlike the story behind George Sternberg's acquisition of the fish-within-a-fish specimen in 1952, the rest of the dig became my responsibility.

The remaining six months of the year was somehow consumed with other projects. I didn't get back to the locality until June of the following year. This time I brought along my friends from the East Coast (annual field trip) and put them to work clearing the overburden off a large area over the fish remains (thank you, Tom, Fred, and George). Once that was done, I went to work uncovering the bones. I immediately ran into the second pectoral fin and many other unrecognizable bones. Work progressed fairly rapidly, but I kept hitting new bones and didn't seem to be coming to the edge of the fossil. This was obviously a big fish. Then, totally unexpectedly, I uncovered a third large fin. Something was really wrong. Fishes don't have three pectoral fins. I continued working and eventually uncovered a large field of bones covering an area of about 1.8 x 2.4 m (6 ft x 8 ft), but without reaching a clearly defined edge. Surveying the remains, I realized that we were not prepared to remove any of it. So, after taking pictures and measurements, I covered the remains and made plans for another dig.

There was no doubt that I was going to need help getting the fossil out of the ground in one piece. I contacted Mike Triebold (Rocky Mountain Dinosaur Resource Center—RMDRC) and made arrangements to meet them at the site the following month. Mike brought along Anthony Maltese and Jacob Jett, plus some heavy-duty equipment (jack hammer and chainsaw). Once on-site, Mike went to work clearing more overburden with the jack hammer while we decided how to isolate a big jacket containing the two fins and what appeared to be cranial bones. By noon, we were ready to start on the jacket. Mike first marked the block with fluorescent paint (better to see through the dust) and then cut a deep channel around the bones with the chainsaw. Once that was done, Anthony and Jacob applied plaster and burlap to the block while Mike and I mixed the plaster as quickly as we could. It was drying fast and we needed to get it completed in hurry. Hollow pipes were added during the process to reinforce the jacket and to provide handles for carrying it once we turned it. Then we waited an hour or so to let the plaster harden. Finally we started driving long chisels under the jacketed block to free it from the underlying chalk. This was the part that we were worried about because the entire layer of chalk containing the bones was badly fractured. Finally, after tying a rope under part of the jacket (belt and suspenders) that contained a fin, we were ready to flip it over—our moment of truth.

Mike and I went to the gully on the south side of the jacket to lift it from below while Anthony and Jacob lifted from above. On the count of three, we started turning the jacket. As it was being raised, Mike and I both saw a block of chalk drop out of the center portion, but we could not stop the lift. As soon as the jacket was rotated upward and out of our reach, Mike and I hurried around to the other side to help lay the jacket down gently. The first thing we checked was the spot where the block of chalk had dropped out; we breathed a big sigh of relief when we saw that the heavy fin that had been supported by the fallen block had stayed in the jacket and was undamaged.

After that, cleanup was uneventful. The jacket went off to the RMDRC for preparation, and I returned to the site three more times that summer to finish collecting the remains of the big fish, including the caudal fin. As it turned out we had collected the most complete specimen of *Protosphyraena gladius* ever discovered (Figure 5.18). Probably more than 4.5 m (14.8 ft) in length, it included nearly everything from the bones at the front of the skull (premaxillae) to the caudal fin. The mysterious third 'pectoral fin' discovered by Dr. Shimada is now recognized as the dorsal fin of the fish, something that had not been previously described or even considered. Now we need to sort out those other single fins in various museum collections that may have been mislabeled as pectoral fins.

The specimen was generously donated by the landowner to the Sternberg Musuem and is curated as FHSM VP-17428. The remains were then sufficiently prepped out to be included in our paper in *Science* describing *Bonnerichthys gladius*, the new genus of filter-feeding fishes from the Western Interior Sea (Friedman et al., 2010). Two years later we did a follow-up report on the distribution of *B. gladius* remains across North America (Friedman et al., 2013). There's still more work to be done on this previously unknown fish from the Late Cretaceous, including a description of the pectoral fin, the teeth (yes, tiny ones) and the identification of the stomach contents that were discovered by Anthony Maltese during preparation.

5.18. The pectoral fins (left), pectoral girdle, and skull elements (right) of *Bonnerichthys gladius* (FHSM VP-17428) from the Lower Santonian chalk of Gove County, Kansas. Scale bar (lower left) = 50 cm (20 in).

As if finding an unexpected filter-feeding fish in the Western Interior Sea wasn't enough, another friend and coauthor of mine discovered a second, earlier-occurring genus in the Greenhorn Formation in southeastern Colorado (Schumacher et al., 2016). *Rhinconichthys purgatoirensis* sp. nov. (DMNH 63794) is the first North American representative of what is apparently a group of filter-feeding fishes that occurred worldwide during the Late Cretaceous. Two other species, *R. taylori* sp. nov., reported by Friedman et al. (2010), from England, and *R. uyenoi* from Japan are included in this genus.

Another species of extinct fishes named by Cope (1877b), *Erisichthe ziphioides* (AMNH 2131), was eventually determined to be a completely new and unrelated genus, *Martinichthys* by McClung in 1926. Cope's type specimen consisted of a single bone (or possibly a pair of fused bones) forming the blunt rostrum or snout of the skull that Cope, assuming it was a swordfish, erroneously described as the "muzzle of an old individual, which has lost a good deal of its apex by attrition" (1877b:822). Hay believed it was simply "a species having a short and blunt snout" (1903:22), and renamed it *Protosphyraena ziphioides*. However, the specimen had fooled both men because it was described upside down by Cope (1877b:822–823) and figured in a similar fashion by Hay (1903:figs. 13–14).

Based on differences that he observed in the shape of rostra, McClung (1926) identified six new species from a number of similar specimens as belonging to a new genus of plethodid fisheses, which he called *Martinichthys* in honor of H. T. Martin (1862–1931), chief preparator at the University of Kansas. The best of these specimens, a nearly complete skull (KUVP 497) and associated vertebrae, was designated the holotype of *M. brevis*. At the time, this specimen and another specimen (KUVP

Martinichthys **(Plethodidae)**

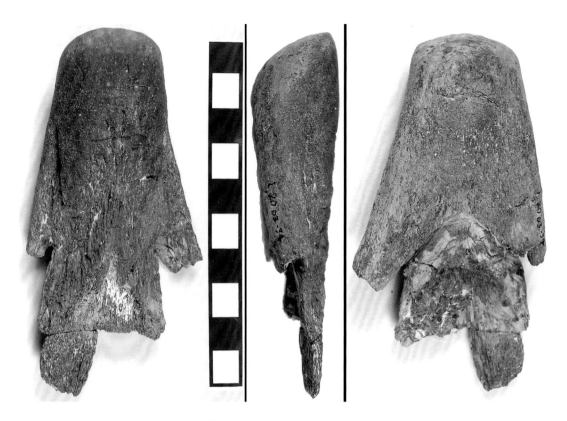

5.19. The rostrum of *Martinichthys brevis* (FHSM VP-15567) in dorsal, left lateral, and ventral views. This specimen was collected in 2003 from the lower chalk of Gove County. Note the typical worn appearance of the anterior end of this rostrum. Until recently, only two reasonably complete skulls were known of *Martinichthys*. Both are in the collection of the University of Kansas Museum of Natural History. Scale in cm.

498) called *M. ziphioides* were the only remains known with skull elements and vertebrae. Since then, several dozen isolated rostra have been collected, and there are now at least two fairly complete skulls collected by Anthony Maltese yet to be described.

Martinichthys is a rare genus of plethodid fishes that has been collected only from the lower Smoky Hill Chalk (late Coniacian) in Kansas. It is currently known only from a few skulls and a fairly large number of preserved, bony rostra (snouts) in the collections of the University of Kansas Museum of Natural History and the Fort Hays State University Sternberg Museum of Natural History. It now appears that, except for the heavy rostrum, the skeleton of the fish was poorly ossified (even less than *Protosphyraena*) and therefore unlikely to be preserved. The fact that most of the rostra examined to date, including Cope's type specimen, are heavily worn on the anterior end (Fig. 5.19) also raises questions about the feeding habits of the fishes.

Martinichthys is also unique in that it is known only from the chalk in western Kansas, and there only from a very limited stratigraphic interval. Almost all of the known rostra occur within a time span of less than 150,000 years near the end of the Coniacian. Besides the 10 specimens reported by McClung (1926), at least 12 other rostra had been added to the University of Kansas collection by 1993. In addition, there were a dozen or so in the Sternberg Museum collection. Almost all were from the low chalk in Gove and Trego counties. My wife and I recently donated 19 specimens (FHSM VP-15549 to 15568) that we had collected

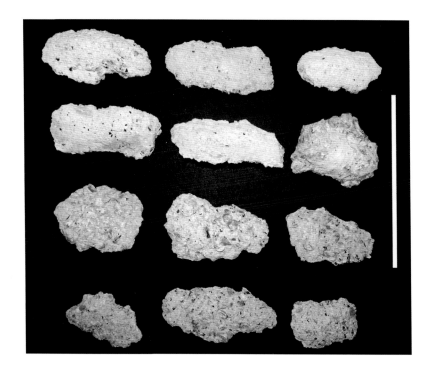

in Gove County between 1990 and 2003. While I cannot be certain of the occurrence of the other specimens, I know nearly all of ours were collected from just below Hattin's marker unit 4 to just above marker unit 5 (Everhart and Everhart, 1993). This zone is also the only place that unusual coprolites composed of oyster (*Pseudoperna congesta*) fragments are known to occur (Everhart and Everhart, 1992; Fig. 5.20). We are still looking for the link between these coprolites and the unusual wear seen on the rostra of *Martinichthys* (and a related species called *Thryptodus zitteli*). Strange as it may seem, I am fairly certain that these fishes were using their bony 'noses' or rostra to batter open the small oyster shells that were living on inoceramid clams so that they could feed on the oysters inside. Nothing else that I can think of can account for the obvious wear on the rostra that is readily observed on most of these specimens.

After his review of McClung's 10 specimens and two new specimens at the University of Kansas, Taverne (2000a) reduced the number of species to two, *Martinichthys ziphioides* (Cope) and *M. brevis* (McClung), and chose to validate only the two species with associated skulls (KUVP 497 and 498). Taverne notes that "*Martinichthys* is a valid genus of Tselfatiiformes, characterized by at least five unique characters and which is comprised of two species: *M. brevis* with a short and thick rostrum (Fig. 5.21) and *M. ziphioides* with a long and narrow rostrum. The other species described in this genus [McClung, 1926], *M. acutus*, *M. alternatus*, *M. gracilis*, *M. intermedius* and *M. latus* are junior synonyms of *M. ziphioides*" (2000a:10).

While I agree with Taverne (2000a) that McClung's number of species (seven) is almost certainly too high, more work needs to be done to explain the wide variety of shapes that can be observed in the other 50 or

more specimens now in museum collections. At least two 'Martinichthys' specimens in the Sternberg Museum collection, one that I collected in 1992 (FHSM VP-15568) in Gove County and another collected by J. R. Green in Trego County in 1973 (FHSM VP-3248), appear to represent an as yet undescribed taxon of plethodid fishes between *Martinichthys* and *Thryptodus* (Shimada, pers. comm., 2003).

Other Plethodids

Martinichthys is only one of a diverse family of primitive bony fishes called Plethodidae, which was established by Loomis (1900). Many of the species are known only from single specimens in the collection of the University of Kansas Museum of Natural History, and some of those are incomplete. As a group, they were generally medium-size fishes with many small comblike teeth in the jaws and on the bones of the palate. The skull is relatively flat and broad. No plethodid specimens have be discovered with stomach contents, so it is unknown what they ate. It is possible that they fed on invertebrates on the sea floor, but more likely that they preyed on small ammonites and other cephalopods nearer the surface.

The most common plethodid genus is *Bananogmius* Whitley 1940, which includes four species from the chalk. Note that the name *Anogmius* had originally been given as a genus of Saurodontidae (Cope, 1872a:170) and is therefore a junior synonym of that genus. The genus *Anogmius* was thus 'pre-occupied' and could not be used again, even by Cope, as he attempted to do in 1877.

While *Bananogmius* was the most common genus of plethodid in the Smoky Hill Chalk, its most common species has now been placed in another genus. According to Taverne (2000b; 2004), *Bananogmius evolutus* Cope 1877a is specifically excluded from the original genus and placed into a new genus, *Pentanogmius*. Taverne (2001a) noted that the three species remaining in the genus *Bananogmius* are *B. aratus* Cope 1877, *B. favirostris* Cope 1877, and *B. ornatus* Woodward 1923. About the

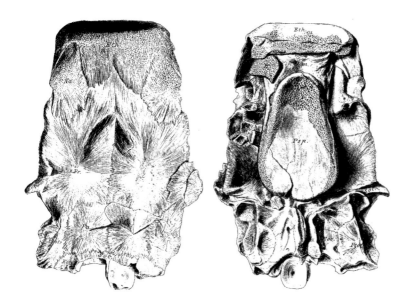

5.22. The skull of *Thryptodus zitteli* as figured by Loomis (1900:pl. 21) in dorsal (left) and ventral views. This specimen was the most complete skull ever collected of this species. No scale, but total length probably less than 10 cm (4 in).

same time as *B. evolutus* was placed into the new genus by Taverne, a new species (*B. ellisensis*) was described by Fielitz and Shimada (1999) from an unusual three-dimensional specimen (FHSM VP-2118) in the Sternberg Museum from the Blue Hill Shale of Ellis County. This brings the total number of species within the genus *Bananogmius* back to four.

Loomis (1900) named two new plethodids: *Syntegmodus altus* and *Thryptodus zitteli*. While these are regarded as synonymous with *Bananogmius* by some workers, Taverne (2001c:251) noted that the specimen of *Syntegmodus altus* (AMNH 2112) that he examined did represent a valid genus that is intermediate between *Bananogmius* and another genus called *Niobrara*. The type specimen of *Thryptodus* described and figured in great detail by Loomis (1900:pls. 21–22) from the Kansas chalk (Fig. 5.22) was apparently destroyed in Germany during World War II, and so far as I am aware there has not been as complete a specimen collected since. Shimada and Schumacher (2003) described what is probably the earliest occurrence (Lincoln Limestone; Upper Cenomanian) of *Thryptodus* from a specimen (FHSM VP 13996) in the Sternberg Museum.

Taverne (2003) also redescribed *Thryptodus zitteli* from fragmentary specimens at the University of Kansas Museum of Natural History (KUVP 456, 457, and 459) and a specimen in the American Museum of Natural History (AMNH 19557), and characterized the species as being more closely related to *Plethodus* than to *Martinichthys*. While it is rare, the remains of *Thryptodus* are collected occasionally from the lower chalk. Over the years, I have collected four fragmentary specimens in southeastern Gove County. In most cases, the remains consist only of part of the heavy, rounded end of the snout (fused ethmoid bones), but one specimen (FHSM VP-15571) included most of the skull and several vertebrae. Based on what I have seen in collections from the Late Cretaceous of Texas, *Thryptodus* is much more common further south in the Western Interior Sea than in Kansas.

Jordan (1924) added three more plethodid species to the fauna of the Smoky Hill Chalk. All are based on fairly complete specimens of small fishes in the collection of the University of Kansas Museum of Natural History. His reconstruction (ibid.:pl. 14) of *Niobrara encarsia* is shaped somewhat like a modern tuna. The type specimen (KUVP 179) was collected from the chalk of Trego County by H. T. Martin and is 69 cm (27 in) long. Taverne (2001b) noted that *Niobrara* was the most primitive genus within the order Tselfatiiformes (Bananogmiiformes) and that it was closely related to the genus *Bananogmius*. A similar species, *Zanclites xenurus* (KUVP 52) described by Jordan (1924) is slightly smaller (55 cm; 21.5 in) and was collected in Gove County, also by H. T. Martin. Taverne (1999) confirmed its placement in Tselfatiiformes but did not comment on the family to which it belonged. The type specimen of *Luxilites striolatus* (KUVP 295) was described from a poorly preserved skull (Jordan, 1924). Taverne (2002a) noted that *Luxilites* was a valid genus and that it belonged to a subgroup of the Plethodidae, which includes *Bananogmius*, *Syntegmodus*, and *Niobrara*. Taverne (2002b) also described the newest plethodid species from the Kansas chalk (*Pseudanogmius maiseyi*—AMNH 8129) from a specimen originally identified by Hay (1903) as *Anogmius* sp.

Holosteans

The holosteans are an intermediate group between primitive bony fishes and the more advanced teleosts. Modern holosteans include freshwater forms such as the bowfin (*Amia*) and the gar (*Lepisosteus*) and are notable for their heavy scales. Another group of holosteans called pycnodonts lived in the oceans covering Kansas during the Cretaceous. Pycnodonts have tall, narrow bodies, heavy, diamond-shaped scales, and jaws filled with small nipping and crushing teeth. They apparently fed on small invertebrates (epifauna) that were attached to the inoceramid bivalves on the sea floor. Most pycnodont specimens in the chalk consist of isolated jaw elements with their rows of round, smooth teeth.

A single species, *Micropycnodon kansasensis*, was described by Hibbard and Graffham (1941) from the lower chalk (late Coniacian). The type specimen (KUVP 1019; Fig. 5.23) was described from Rooks County by Claude Hibbard in 1939, and a more complete specimen (KUVP 7030) was collected in northeast Trego County about 1935 by George Sternberg. Dunkle and Hibbard (1946) described the second specimen and indicated that it was probably from the Fort Hays Limestone. The only *Micropycnodon* specimen that we have collected (a jaw fragment; FHSM VP-16484) was discovered by my wife in 1991 in the low chalk of Gove County. Stewart noted that "all the Smoky Hill Chalk Member specimens of *Micropycnodon* known to me are from the zone of *Protosphyraena perniciosa* [Late Coniacian]" (1990a:24).

I described a specimen of *Micropycnodon* (FHSM VP-16583) that I collected as a coprolite (Everhart, 2007). The remains had passed through the gut of the unknown predator and were discovered as a single black tooth exposed on the surface of a small coprolite (Fig. 5.24). Inside were

5.23. An occlusal view of the small, crushing teeth on the vomer of the type specimen of *Micropycnodon kansasensis* (KUVP 1019) collected in 1939 from the low chalk in Rooks County, Kansas. Scale bar = 2 cm (0.8 in).

more fish bones, including a *Micropycnodon* premaxilla, and fragments (prisms) of small inoceramid shells. The coprolite was located in the lower part of the Santonian-age chalk, just above Hattin's marker unit 6, and as such is the uppermost record (youngest) of *Micropycnodon kansasensis* in chalk. Exactly who (predator or prey) had been feeding on small inoceramids is unclear, but I suspect it was the *Micropycnodon*.

Hadrodus is another, much larger pycnodont that has been collected only once from the chalk. Stewart also mentioned that the "single published record of *Hadrodus marshi* [Gregory, 1950] was probably from the zone of *Hesperornis* [early Campanian]. Within the zone of *Clioscaphites vermiformis* and *C. choteauensis* is an undetermined pycnodont that is not *Micropycnodon* and is probably not *Hadrodus*" (1990a:24). While I am not aware of any additional specimens of *Hadrodus* from the chalk, Shimada (2006) reported on a single premaxillary tooth (FHSM VP-14006) from the underlying Blue Hill Shale. A fragment of a large pycnodont jaw plate (FHSM VP-16865), possibly *Hadrodus*, was collected from the Greenhorn Formation, and donated by Ramo Decker.

In addition, Shimada and Everhart (2009) described a partial prearticular tooth plate (FHSM VP-17319) of *Anomoeodus* cf. *A. barberi* Hussakof, a pycnodont species not previously described from the chalk. This may be the "undetermined pycnodont" mentioned by Stewart (1990a:24, above). The specimen had been collected in the upper Smoky Hill Chalk of western Logan County in 1974 by Jerome 'Pete' Bussen (1927–2015), who was a long-time friend of J. D. Stewart and probably showed him the tooth plate.

While the remains of pycnodonts are rare in the Smoky Hill Chalk, their remains, especially teeth, have been collected almost continuously from Albian through late Coniacian strata in Kansas. Jaw fragments and teeth of *Coelodus brownii* and *Coelodus stantoni* were reported by Williston (1900) from the Early Cretaceous Kiowa Shale of Kiowa and Clark counties. Isolated teeth of *Coelodus* sp. have been collected from the Kiowa Shale of McPherson County (Beamon, 1999; Everhart, 2004).

5.24. Single pycnodont tooth discovered on the surface of a fish coprolite in the Lower Santonian chalk of Lane County, Kansas. The coprolite also contained fish bones and calcite prisms from small inoceramid clams. Scale in mm.

An examination of my 2003 collection from the Kiowa Shale of McPherson County indicates that pycnodont teeth make up about 21 percent of more than 500 teeth collected (unreported data). I have also recently collected pycnodont teeth (cf. *Coelodus* sp.) from the Upper Dakota Sandstone (middle Cenomanian), and was present when a pycnodont jaw plate (FHSM VP-15548; cf. *Micropycnodon*) was collected in 2003 from the base of the Lincoln Limestone Member (Upper Cenomanian) of the Greenhorn Limestone. Hibbard (1939) described a jaw fragment of *Coelodus streckeri* (KUVP 946) from the Fairport Chalk Member (Carlile Shale) of Russell County, and Zielinski (1994) reported on the discovery of the splenial tooth plate (FHSM VP-6728) of a pycnodont (cf. *Anomoeodus* sp.) from the Blue Hill Shale Member of the Carlile Shale. Everhart et al. (2003) noted numerous teeth and jaw fragments of a pycnodont-like fish, *Hadrodus priscus*, in addition to teeth of an unidentified pycnodont from a fish tooth conglomerate in the Blue Hill Shale. Not surprisingly, pycnodonts (cf. *Coelodus*) were also present in the earliest marine formation in Kansas, the Champion Shell Bed in the basal Kiowa Shale of Kiowa County, Kansas (Everhart, 2009).

Wiley and Stewart (1977) reported the discovery of remains of a gar (*Lepisosteus* sp.; KUVP 36243) from the lower chalk in western Trego County. The specimen included skull and fin elements, teeth, and scales. Gars normally occur only in fresh or brackish water, and this was the first report of a gar in the marine environment of the Western Interior Sea (ibid.:761). The authors were unable to explain the occurrence of the remains of a freshwater fish in a marine deposit, and no additional specimens have been collected since that time. Everhart et al. (2003) reported gar scales from the upper Blue Hill Member of the Carlile Shale (middle Turonian), a nearshore coastal environment. In the summer of 2003, I also collected gar teeth and a single ganoid scale (cf. *Lepidotes* sp.) from a similar coastal locality in the Kiowa Shale of McPherson County. Williston (1900:pl. 30, fig. 4) included two teeth from the Kiowa Shale that he tentatively identified as *Mesodon* but noted that they probably should be referred to as *Lepidotes* sp. (ibid.:256). The Sternberg Museum has a section of articulated scales of a large *Lepidotes* (FHSM VP-5114) on exhibit from the Kiowa Shale of Clark County, Kansas.

Dunkle (1969) described a new species of amioid fishes (*Paraliodesmus guadagnii*—USNM 21083) from a small (17 cm; 6.7 in) specimen located inside a *Volviceramus grandis* shell from the chalk of Gove County in 1954. Stewart (1990a:24) noted that *Paraliodesmus* was also from his biostratigraphic zone of *Clioscaphites vermiformis* and *C. choteauensis* (early Santonian). The only other amioid remains I am aware of from the Cretaceous of Kansas are seven cf. *Pachyamia* sp. teeth that I collected from the Kiowa Shale of McPherson County in 2003. A similar tooth (FHSM VP-13540) was collected by Beamon (1999) and identified as "lepisosteid" (30).

Stewart noted that "*Kansius sternbergi, Caproberyx* sp., *Trachichthyoides* sp. and one or more undescribed holocentrid genera occur within inoceramid bivalves (*Platyceramus platinus*)" (1996:390). These fishes are all small (10 cm or less; 3.9 in or less) and are generally known only from remains preserved inside the clam shells. Most occur in the Lower Santonian when the inoceramids were large and plentiful. A list of the other species of fishes documented from the Smoky Hill Chalk includes *Apsopelix anglicus* Cope 1871, *Stratodus apicalis* Cope 1872, *Lepticthys agilis* Stewart 1900, and *Omosoma garretti* Bardack 1976. *Apateodus busseni* Fielitz and Shimada (2009) was the most recent addition to this list of smaller fishes in the chalk, based on a specimen (CMC VP 6941) collected from the Lower Campanian chalk of Logan County and donated by Jerome Bussen to the Cincinnati Museum Center. The authors estimated that the type specimen would have been about 40 cm (16 in) long in life. Another *Apateodus* sp. specimen (AMNH FF11560), as yet undescribed, was collected by George Sternberg from the Santonian chalk of Gove County and is in the collection of the American Museum of Natural History (Matt Friedman, pers. comm., 2010).

Discounting the large numbers of small fishes occasionally preserved inside inoceramids and the tiny jaws of minnow-size *Enchodus*, *Apsopelix* and *Stratodus* are the most common of this group. Just about all of them, however, would be considered rare in any collection. I have collected only one specimen of *Apsopelix* and three or four sets of fragmentary *Stratodus* remains in more than forty years. Other than a couple of occurrences of unidentified 'fish in shell,' I've never seen any of the other species in the field. Many of the smaller fishes documented from the Smoky Hill Chalk are single specimens or are the result of unusual preservation (e.g., inside inoceramid shells).

Other Fish Genera in the Smoky Hill Chalk

5.25. The photo shows only the second specimen of an eel (*Urenchelys abditus*; FHSM VP-17591) recovered from inside the shell of a *Platyceramus platinus* clam in the Smoky Hill Chalk. Scale bar in mm.

Small fishes are particularly rare in the chalk, either because they were completely consumed as prey by larger fishes or because they were literally too small to be preserved. That is, their remains (bones) were not large or solid enough to avoid being dissolved by the chemistry at work on the sea bottom. In some cases where whole schools of small fishes were preserved inside clam shells (see Stewart, 1990b), they simply are yet to be described and named. It appears likely that the fishes were sheltering inside the clam shells to avoid predators or possibly were even feeding on parasites or other tiny fauna also living there. When the clam died and the shell closed, they were trapped inside, where their skeletons had a better chance for preservation.

H. T. Martin (1920) reported the discovery of an eel from the chalk that he called *Anguillavus hackberryensis* (KUVP 927). That specimen has since been reidentified as a fish similar to *Stratodus*. The first actual eel to be described, *Urenchelys abditus* (KUVP 47241), was discovered along with other small fishes inside a giant inoceramid (Wiley and Stewart, 1981). A second specimen was collected by high school student Kris Super in 2010 from a inoceramid shell in southeast Gove County (FHSM VP-17591; Fig. 5.25). In both instances, the eels are small, less than 10 cm (4 in) in length.

In 1991 my wife discovered what was then the only known coelacanth remains (LACMNH 131958) from late Coniacian chalk of Gove County (Stewart, Everhart, and Everhart, 1991). Although the partial skull and lower jaws are identifiable as a small coelacanth, it is probably too fragmentary to be identified even to genus. In 2007 the partial skull of a giant coelacanth, *Megalocoelacanthus dobiei* (AMNH FF 20267), was collected from the early Santonian chalk of Lane County (Dutel et al., 2011, 2012). *M. dobiei* is a truly huge fish, estimated to be about

3.5 m (11.5 ft) by Schwimmer, Stewart, and Williams (1994:505), but it was previously known only from late Santonian to late Campanian strata of the Gulf and East coasts of North America. Unfortunately the authors of the 2012 paper stated the wrong stratigraphic level (early Campanian), an important difference of about 3 million years. A second specimen of a partial skull of *M. dobiei* (FHSM VP-18758) was collected by Anthony Maltese in 2007 from the chalk in Logan County and donated to the Sternberg Museum of Natural History. That specimen is of early Campanian age. A paper currently in press will correct the stratigraphic error in the first report and extend the known range of these giant coelacanths into the early Santonian.

Lastly, another strange fish, *Aethocephalichthys hyainarhinos*, was described by Fielitz, Stewart, and Wiffen (1999). While generally limited to North America, its enigmatic remains have been collected as far away as New Zealand. The known specimens are limited to an odd-shaped, solidly fused neurocranium (Fig. 5.26). The authors (ibid.) reported one specimen (KUVP 84901) from Gove County in the upper Smoky Hill Chalk, and in 2006 two additional specimens (FHSM VP-16465, 16466) were collected from southeastern Logan County. In life, the fish would have been narrow-bodied and no longer than about 30 cm (12 in) in length. According to the lead author in the paper describing the genus (Fielitz, pers. comm., 2016), the remains of these fishes are more common in the Pierre Shale. In 1995, my wife collected a small *Aethocephalichthys* skull (FHSM VP-15577; Fig. 5.26) from the Sharon Springs Member (middle Campanian) of the Pierre Shale, near McAllaster Butte in Logan County. At least 20 other specimens are in the collections of the University of Kansas, the Los Angeles County Museum of Natural History and the Smithsonian. Our friend Pete Bussen, who also collected and donated a specimen from the Pierre Shale, refers to them as the 'headlamp

5.26. The neurocranium of an unusual and poorly known little fish with big eyes called *Aethocephalichthys hyainarhinos* (FHSM VP-15577) in dorsal, right lateral, and ventral views. This specimen was collected from the Sharon Springs Member (Lower Campanian) of the Pierre Shale in Logan County by Pam Everhart. Similar specimens have been collected from the upper Smoky Hill Chalk, some as low as the Middle Santonian. Scale in mm.

fish' because of the large eye sockets that occupy most of the front of the skull. According to Fielitz, Stewart, and Wiffen (1999:97) the etymology of the name is *Aethocephalichthys* (Greek) 'strange-headed fish' and *hyainarhinos* (Greek) 'pig-nosed'—not exactly a complimentary name or description of a rare little fish.

The Smoky Hill Chalk has preserved a wide variety of fishes, large and small. It is likely that there are still new species to be discovered there, even after more than 145 years of collecting.

Turtles

Leatherback Giants

The giant turtle swam steadily through the sunlit water of the Inland Sea. Although it was early afternoon, the passing of time mattered little to her. She had been swimming almost continuously eastward now for three days on a journey that would take at least a week longer. Able to swim submerged for several minutes before surfacing to breathe, she was making good progress. Her large front limbs acted as underwater wings and enabled her streamlined body to 'fly' efficiently through the water. The hundred or so eggs in her lower abdomen were nearly ready to be laid, and she was returning to the warm sands of a beach where she herself had hatched more than 50 years before. She seldom swam continuously like this, preferring to move more slowly and feed leisurely among the sea jellies. Instinctively, however, she knew that the best time for laying her eggs was quickly approaching. The tides would soon be at their peak and would allow her to dig a nest that was safely above the high-water mark.

She was a mature adult *Protostega* sea turtle, with a 'leatherback' shell that measured more than 2 m (6 ft) across. She would continue to grow slowly throughout her long life, but she had already reached most of her adult size. Few predators were large enough or hungry enough to bother with something of her bulk, but it hadn't always been so. She was missing three toes on her left rear paddle from a bite by a shark when she was much younger. After the sudden attack, the shark had disappeared as quickly as it had come, and she had recovered from the wound without further problems. Life was harsh in the Western Interior Sea, and unlike nearly all of her nest mates, she had been lucky to survive.

She did not know or care that few, if any, of the fertilized eggs she carried would hatch and make it through the gauntlet of hungry predators as they made their way from the nest on the beach to the relative safety of the ocean. Once they were in the water, their chances of survival improved, but not by much. The ocean held many large fishes and mosasaurs that were quite capable of swallowing a baby sea turtle whole.

Unlike other marine reptiles, including mosasaurs and plesiosaurs, sea turtles had retained the ability to lay eggs, staying with a strategy of survival of the species by reproducing in overwhelming numbers. So every year at about the same time, she and thousands of other females of her species made the journey to the miles of beaches surrounding the great shallow sea to lay the millions of eggs that would ensure that a few young would always survive to reach adulthood.

When she surfaced to breathe the next time, there was a haze of fine dust in the air. Exhaling the stale air from her lungs and then inhaling, she noticed an acrid odor. A quick glance behind her revealed a dark cloud that stretched across the western sky. Already there were fine patches of dust floating on the water. Sensing only that something unusual was happening, she took several deep breaths and dived again, continuing to swim steadily to the east. As she swam, the water around her darkened as the black cloud covered the sun and turned day into night. The dust that had at first floated on the surface soon began to clump up, and then to sink, mixing with the seawater and clouding it with fine particles. Unknown to her, a giant volcano on the edge of the ocean more than a thousand miles to the west had exploded violently earlier in the day, sending millions of tons of volcanic ash high into the atmosphere, where it was picked up by the prevailing westerly winds. As the huge dust cloud was carried eastward over the Inland Sea, larger pieces of the ejecta began to fall out, pelting the surface of the water with a heavy rain of ash that covered and killed everything living immediately downwind of the volcano. Further east, the effects of the ash and the toxic gases that traveled with it were not immediately lethal to life, but they would have serious consequences on the local ecosystem for years to come. In the middle of the sea, the finer ash that had been carried by the wind would fall in much smaller but still significant amounts. This dust would slowly settle downward in the water and eventually completely cover the sea floor to a depth of an inch or more. Animals that depended on gills to extract oxygen from the water, such as fishes and most invertebrates, would have problems with damage caused by the gritty ash, and many would die. Enough would survive to repopulate the sea, but in the path of the heaviest plume of ash it would be a major natural disaster for sea life.

The turtle continued to swim through the now dark and muddied water between her and the distant shore. The sulfurous gases and the ash irritated her throat and eyes each time she surfaced to breathe, but she would survive. The eggs she laid eight days later would not be so lucky. Weather changes caused by the upsurge in volcanic activity would create a brief chilling of global temperatures, and her eggs would fail to develop and hatch. She would not know this, however, and would return each year for the next 48 years to lay more eggs at the place where she first entered the ocean.

Turtles

While a number of turtle species and specimens have been described from the Smoky Hill Chalk, surprisingly little is known of their occurrence in the Western Interior Sea during the Late Cretaceous. Part of the problem is the rarity of sea turtle remains as fossils, especially relatively complete specimens. We are fairly certain that turtles were never as numerous in the Western Interior Sea as they were in warmer waters along the Cretaceous Gulf Coast. Russell (1993) noted that turtles make up only

6.1. Fragment of the plastron of *Protostega gigas* (FHSM VP-2158) from the upper chalk of Graham County, Kansas, preserving multiple bite marks from the ginsu shark, *Cretoxyrhina mantelli*. Scale bar = 10 cm (4 in).

about 3 percent of vertebrate specimens from the Niobrara Formation in museum collections and are collected about as often as the remains of toothed birds (Chapter 11). Another issue is the generally smaller size of most marine turtles and the lack of speed or other defensive capabilities. While turtles might seem to be more likely than other animals to be preserved because of their bony shell, it turns out that most of the remains that have been discovered in the Smoky Hill Chalk are skulls and limbs that appear to have come from dismembered carcasses.

After discovering the type specimen of *Protostega gigas*, Cope (1872a) noted that the bones making up the shell of the large turtle were extremely thin and fragile. It seems likely that the larger predators (sharks, pliosaurs and mosasaurs) of the Cretaceous seas were well equipped to feed on the smaller sea turtles and must have done so on a regular basis. Even the relatively complete specimen of *Toxochelys latiremis* (ROM 28563) reported from the Kansas chalk by Nicholls (1988) was missing most of the readily detachable parts like the hind limbs and distal portions of the forelimbs, most likely as the result of scavenging by sharks. Turtle remains, including the complete skull of a small *Toxochelys*, were reported as stomach contents of *Squalicorax falcatus* (Druckenmiller et al., 1993; see Fig. 4.6), and pieces of *Protostega* shell and limb material have been described with embedded teeth and bite marks attributed to *Cretoxyrhina mantelli* (Shimada and Hooks, 2004; FHSM VP-2158, Fig. 6.1).

Cope (1872c) even named a new species of shark (*G. hartwelli* Cope = *Squalicorax falcatus*) from a single tooth located beneath the bones of his giant *Protostega gigas* specimen. The tooth is a possible indication of scavenging on that specimen by sharks. The humerus of a *Protostega* (1503.57) from Logan County in the collection of the University of

6.2. The skull (dorsal view), limb bones and fragments of the carapace and plastron of a small *Toxochelys* sp. turtle (FHSM VP-13449) from the Upper Coniacian chalk of Gove County, Kansas. Scale bar = 10 cm (4 in).

Wisconsin-Madison shows deep, serrated bite marks from *Squalicorax*. Schwimmer, Stewart, and Williams, (1997:78) reported bite marks from *Squalicorax* sp. on a *Toxochelys?* sp. humerus (LACMNH 50974), and two *Desmatochelys lowi* humeri (KUVP 32401 and 32405). A small *Toxochelys* specimen (FHSM VP-13449) that I collected in 1988 consisted of a complete skull, neck vertebrae, front limbs, and the front edge of the carapace (Fig. 6.2). The rest had apparently been bitten off.

Even the giant *Archelon* (YPM3000) specimen in the Yale Peabody Museum had apparently lost the lower portion of its right rear leg at some point during its life (Wieland, written communication in Williston, 1914:240). While turtles have not yet been documented as stomach contents of mosasaurs in the Western Interior Sea, Dollo (1887) noted turtle remains in association with a giant *Hainosaurus* (*Tylosaurus*) mosasaur in Europe. A fragment of a protostegid turtle's carapace (*Alloplueron hoffmanni*, Inv. No. 7431) in the Teylers Museum in the Netherlands includes a large, shattered depression from the bite of a predator, most likely a mosasaur with a crushing dentition like *Prognathodon*.

Further study of the distribution of turtle species in the Smoky Hill Chalk has certainly been hampered by this lack of complete, identifiable specimens and the absence of accurate stratigraphic data. Nicholls (1988:181) noted that toxochelyids, the most common turtles in the Western Interior Sea, are known mostly from skulls and that postcranial material is rare. Over the last 30 years of working in the chalk, my success in collecting turtle remains has been limited to finding a small but nicely articulated *Toxochelys* skull (FHSM VP-13449; Fig. 6.2) and a few specimens of isolated limb material or shell fragments, mostly from the lower chalk (Upper Coniacian). In 1994, my wife had the good fortune to discover

a nearly complete right plastron (half of the lower shell) of a *Protostega gigas* (FHSM VP-13448; Fig. 6.3) in the upper chalk (biozone of *Spinaptychus sternbergi*—Hattin's marker units 17–18). Unfortunately, most of the turtle material collected from Kansas during the last 130 years does not have good locality or stratigraphic data.

While Cope (1872b) gets credit for describing and naming the first turtle (*Protostega gigas*) from the Smoky Hill Chalk, the first specimen of this species was actually collected earlier in 1871 by the second Yale Scientific Expedition under O. C. Marsh. When I visited the Yale Peabody Museum in 2010, I discovered a box of bone fragments of a large turtle (YPM 1408). The curation information indicated that the specimen had been collected by Marsh on July 4, 1871, from the south side of the Smoky Hill River, 8 miles east of Fort Wallace. This means it was collected fully four months before Cope arrived in Kansas and discovered the type specimen. The bones were so weathered and broken, however, that they were not identified as a turtle until much later.

Cope (1872d:433) noted that his *Protostega gigas* had been discovered "near Fort Wallace" (upper chalk) and also named a new species (1872b:309) from several vertebrae of the tail of a small turtle (*Cynocercus incisus*—AMNH 1582) discovered by Sgt. William Gardner in "the yellow chalk near to Butte's Creek, south of Fort Wallace." Lane (1946:298) regarded the second specimen as "related to *Toxochelys*" but otherwise uninformative. Cope (1873) indicated that the type specimen of *Toxochelys latiremis* (AMNH 2362) had been collected by B. F. Mudge and later noted that the remains had been "found by Professor Mudge near the forks of the Smoky Hill River" (1875:99). This locality is somewhat confusing since exposures of both the Smoky Hill Chalk and the Pierre Shale are located in this vicinity. At the time of the discovery, Cope was unaware of the existence of the Pierre Shale and assumed that all fossils collected in this area were from the "Cretaceous chalk." The north fork of the Smoky Hill River branches off the main stream to the southeast of McAllaster Butte and just to the south of the abandoned townsite of

6.4. The skull of the type specimen of *Desmatochelys lowi* (KUVP 1200) in dorsal and ventral view, collected in 1893 from the Fairport Chalk, Jefferson County, Nebraska. This specimen is unusual for its uncrushed preservation. Scale bar = 10 cm (4 in).

Sheridan (See Fig. 1.7). Further examination (Schultze et al., 1985:25; Nicholls, 1988:185) of the material has determined that the type specimen of *Toxochelys* (AMNH 2362) most likely came from the Pierre Shale and not the chalk as Cope had assumed.

Two years later, Cope (1877) noted the receipt of two nearly complete skulls of *Toxochelys* from C. H. Sternberg (AMNH 1496 and 1497), but he failed to provide any locality information. Hay reported vaguely that both of these skulls had been "collected somewhere along the Smoky Hill River" (1908:169). In a similar situation, Williston noted in an 1876 letter to Marsh that Sternberg, who was working along the Smoky Hill River at the time, "got one or two large turtles that are good and some pretty good saurians" (Shor, 1971:77). At the time, Williston (ibid.) was "eight miles west of Monument [Rocks?]—four miles north of the river." This would have been in upper chalk of Logan County.

In 1894, S. W. Williston at the University of Kansas described a new species of protostegid turtle from the 'Benton Cretaceous' just across the state line near Fairbury, Nebraska. The specimen had been given to the university in 1893 by M. A. Low, and Williston named it *Desmatochelys lowi* in his honor. Although the specimen (KUVP 1200) had been damaged by "curiosity seekers" after its discovery, Williston noted that "the portions that were obtained are of the greatest importance" (1894a:5), including the skull, limbs, and most of the carapace. Hay (1908) indicated that portions of the back of the skull were damaged. When I examined the specimen in 2004, I was surprised to find that the skull was uncrushed, completely articulated, and in nearly perfect condition. It looked more like a modern turtle skull than one that was more than 90 million years old (Fig. 6.4). Although I could not locate any notes regarding the preparation of the skull, the lack of crushing most likely indicates that the specimen came from a concretion in the Fairport Chalk Member of the Carlile Shale (middle Turonian) and was prepared with acid.

Although the number of vertebrate remains occurring in the Fairport Chalk is small compared to the Smoky Hill Chalk, the preservation is somewhat better and there is less crushing compared to other Cretaceous

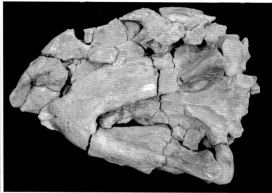

6.5. The crushed and fragmented skull and lower jaw of a *Desmatochelys lowi* turtle (FHSM VP-17470) in dorsal and ventral view, collected in 2008 from the Fairport Chalk, Russell County, Kansas. Scale bar = 10 cm (4 in).

strata in Kansas. Although there are older turtle remains in Kansas, *Desmatochelys lowi* probably represents the oldest turtle species yet to be officially described from the Western Interior Sea.

In 2008, Gail Pearson saw some strange bones eroding out of the Fairport Chalk from a roadside ditch in southern Mitchell County. He contacted me, and together we collected the best specimen of *Desmatochelys lowi* (FHSM VP-17470), including a crushed skull (Fig. 6.5), ever discovered in Kansas (Everhart and Pearson, 2009; Roth and Everhart, 2012). Referring to it as the 'best specimen ever' is a little misleading because almost all of the earlier remains associated with this species from Kansas are single bones or small pieces of the skeleton. In the case of VP-17470, we collected the front one-third of the skeleton (skull, front paddles, and anterior portions of the carapace and plastron) and suspect that the turtle had been cut in two by the bite of a large pliosaur called *Megacephalosaurus eulerti* (Chapter 8), the dominant marine reptile in the Western Interior sea at that time.

Beamon (1999) reported fragmentary remains of turtles (FHSM VP-13550, 13551, and 13553) from the Early Cretaceous (Albian) Kiowa Shale of McPherson County, but he noted that they could not be identified further than to family. Williston (1894b:2) reported the discovery of a "scapula-coracoid of a species as large as *Protostega*" from the Kiowa of Clark County and said that it was unlike "any turtle known to me." Unfortunately, I was unable to relocate the specimen in the collections at the University of Kansas.

In 1969, the nearly complete carapace and plastron of a medium-size turtle (KUVP 16370) were collected from the Kiowa Shale (Albian, Early Cretaceous) of Kiowa County, Kansas, by Orville Bonner. The specimen has not yet been identified or otherwise described, but it would certainly represent a much older species than *Desmatochelys lowi*.

Since the discovery of the type specimen of *Desmatochelys*, additional remains have been collected in South Dakota (Zangerl and Sloan, 1960), Arizona (Elliott, Irby, and Hutchinson, 1997), and possibly from Vancouver Island (Nicholls, 1992). Williston (1898:354) predicted, correctly, that *Desmatochelys* would also be discovered in Kansas. While a complete specimen has not yet been documented from the state, Schwimmer,

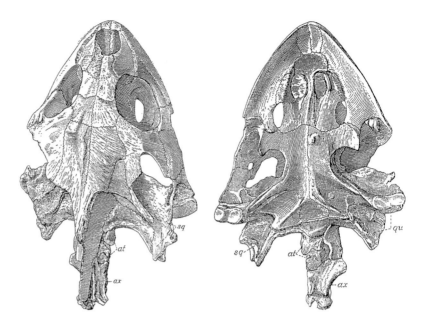

6.6. Dorsal and ventral views of the skull of *Toxochelys latiremis* (AMNH 1496) as figured by Hay (1908). *Toxochelys* is the most common turtle occurring in the Smoky Hill Chalk.

Stewart, and Williams (1997:78) reported two shark-bitten *Desmatochelys lowi* humeri (KUVP 32401 and 32405) from the Carlile Shale of Ellis County. More recently, Bruce Schumacher (pers. comm., 2003) located the remains of a large, as yet unidentified, turtle in the Fairport Chalk (Lower Turonian) of Russell County. The age of this specimen would be slightly younger than Williston's *Desmatochelys*. At this point, the relationship of this family (Desmatochelyidae) to other Late Cretaceous turtles remains uncertain. Elliott, Irby, and Hutchinson (1997) and Hooks (1998) consider it to be a separate family, while Hirayama (1997) includes it in the Protostegidae.

Much of the early work on turtles done by Cope was done hurriedly and on incomplete or otherwise damaged specimens, which resulted in significant errors that are still being corrected. Hay (1895) reexamined Cope's type specimen of *Protostega gigas* (AMNH FR 1503) and noted that Cope had mistakenly identified the left half of the plastron as being elements of the carapace. While noting that Cope's estimate of the size of the skull was wrong, Hay (ibid.:62) arrived at nearly the same overall length for *Protostega* specimen, 3.92 m (nearly 13 ft). Later, Hay wrote that the earlier estimates of the length of *Protostega* "made by Cope and Hay were too great" (1908:196).

In another paper, Hay (1896) also reviewed Cope's description of *Toxochelys latiremis* (Fig. 6.6) and compared it with a new skull that the Field Museum had acquired. Cope's description (1873) was based on a fragmentary specimen (part of the lower jaw, the coracoid, and some limb bones). Cope never completed a full description of the species even though he had received additional cranial material (AMNH 1476 and 1477) from C. H. Sternberg (Cope, 1877). Case followed Hay's work with a description of several new specimens and species of *Toxochelys* that were in the collection of the University of Kansas Museum of Natural History,

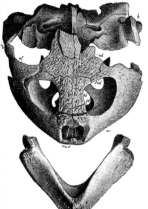
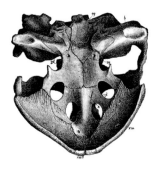

and noted that it was "the single well-established genus of the sea turtles from the Cretaceous of Kansas" (1898:370).

A total of four genera and six species of turtles were reported by Williston (1898) from the Upper Cretaceous of Kansas, including *Desmatochelys* from the 'Benton Cretaceous.' He further indicated that *Toxochelys* is the most common, "especially in the upper or yellow chalk" (ibid.:351). Williston (1901) described a new and unusual species of marine turtle (*Porthochelys laticeps*) from the lower chalk in Trego County near the Saline River. Although the type specimen is amazingly complete, *Porthochelys* is one of the rarest turtles described from the Smoky Hill Chalk. It was more than 110 years before another specimen (MCZ 4104, a lower jaw) was identified as *Porthochelys* by Densmore and Brinkman (2013).

The skull of this species (Fig. 6.7) is almost as wide as it is long, and the jaw is massive (Hay, 1908). The specimen was also unusual in that it included a nearly solid carapace. Unlike Cope's *Protostega* and most other marine turtles, the upper shell of the new species was completely ossified. The carapace of *Toxochelys latiremis* had not been discovered at that time, and Williston believed that it, too, would be completely ossified. He further noted that "the relationships of *Porthochelys* are clearly with *Toxochelys latiremis*" (1901:198). Later discoveries, however, made it readily apparent that *Toxochelys* and *Porthochelys* represent distinctly different genera.

Possibly as the result of his 1895 discovery of the giant *Archelon ischyros*, George Wieland (1896) became the next researcher to concentrate on the study of Late Cretaceous marine turtles. The nearly complete type specimen of *Archelon* (YPM 3000) that he collected from the Upper Campanian Pierre Shale of South Dakota, and thus it lived a million or so years after any of the turtles described from Kansas. Because of its completeness and because it was still mostly articulated, the specimen provided much information about protostegid turtles in general and served to deflate the exaggerated size attributed to *Protostega gigas* by Cope

6.7. Photograph of the exhibited skull, and dorsal and ventral views of the skull and jaws of *Porthochelys laticeps* (KUVP 1204) as figured by Hay (1908). This specimen, including a nearly complete carapace, is on exhibit in the University of Kansas Museum of Natural History.

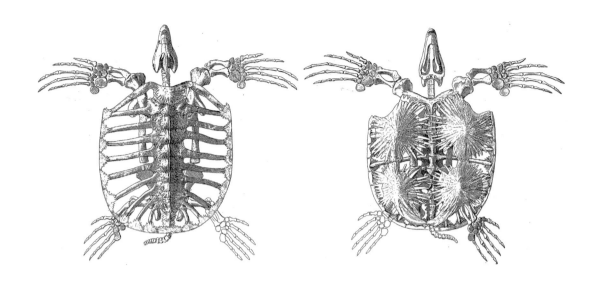

6.8. Dorsal and ventral views of the type specimen of *Archelon ischyros* (YPM 3000) as figured by Wieland (1909). Wieland noted that his drawings in print were 1/36 natural size. This means the distance across the outstretched front paddles is about 4.8 m (15.7 ft), and the length of the skeleton, front of skull to tip of tail, was about 4.3 m (14 f).

and others (Wieland, 1896:411). According to Wieland (ibid.:410), his field measurements of the type specimen of *Archelon ischyros* indicated that the length of the turtle was about 3.5 m (11.3 ft; Fig. 6.8). However, he indicated (Wieland, 1909:117) that the remains were not prepared until 1906. The specimen as mounted at the Yale Peabody Museum is somewhat larger, and the drawings in Wieland (1909:figs. 7 and 8; Fig. 6.8) indicate a length of 4.3 m (14 ft). Even so, Wieland was uncertain that his *Archelon* represented a distinct genus, and after reconstructing the plastron and noting that it was very similar to that of *Protostega*, he (1898) renamed it *Protostega ischyra*. After additional study, however, Wieland (1900) apparently determined that his original view was correct and resumed using the genus name *Archelon* when he described the skull and pelvis of the type specimen.

While *Protostega* was certainly a giant turtle, most of the specimens known at the time were incomplete. From the condition of the remains, it is likely that their carcasses were scavenged and torn apart by sharks or other large predators before reaching the sea bottom. Thirty years after Cope's 1871 discovery of the genus, the hind limb was still largely unknown. In looking through the collection at the University of Kansas, Williston (1902) came across a nearly complete hind limb of *Protostega* (KUVP 1201) that had been collected by Charles Sternberg near Russell Springs (Logan County) in 1902. In his brief description of the material, Williston commented that "the bones of the fore limb, moreover, are all much larger than those of the hind" (ibid.:276). In this case, the hind limb would have been 1.2 m (4 ft.) long and was certainly part of a very large turtle. Over a hundred years later, the specimen (Fig. 6.9) is still on exhibit at the University of Kansas Museum of Natural History.

Williston goes on to make an interesting comment regarding the stratigraphic occurrence of various Late Cretaceous genera that is worth repeating here. He noted that

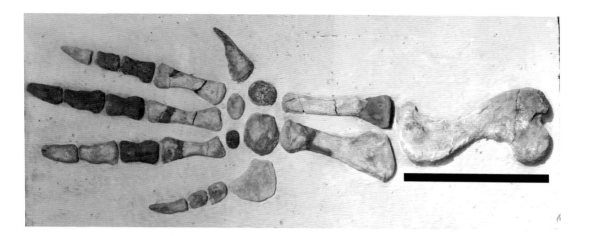

the characters separating *Archelon* Wieland from *Protostega* Cope, while not very important, would seem sufficient. Nevertheless, one can derive little justification from the different geological horizons in which the forms are found. The relations between the Niobrara and Fort Pierre vertebrates are for the most part very close. I have recognized in both horizons *Tylosaurus*, *Platecarpus* and *Mosasaurus* (*Clidastes*), as well as *Pteranodon* and *Hesperornis*, all very typical of the Niobrara deposits, and the existence of *Claosaurus* [Marsh's dinosaur discovery in Kansas] has been recently affirmed in the Fort Pierre. On lithological grounds, there is nothing separating the two groups of deposits and I protest against the names Colorado and Montana, as perpetuating a wrong impression. On paleontological and lithological grounds, there would be much better reasons for uniting the Niobrara with the Fort Pierre than with the Fort Benton. (1902:277)

6.9. The hind limb of *Protostega gigas* (KUVP 1201) collected by Charles H. Sternberg from the upper chalk in Logan County and described by Williston (1902). Scale bar = 36 cm (14 in).

In the same year, Wieland (1902) described and figured the front limb of a new specimen of *Toxochelys latiremis* in the Yale Peabody collection. He noted that this was probably "the first complete restoration of the fore flipper of an ancient marine Chelonian which has been given" (ibid.:96) and suggested that the earlier description by Cope was in error. In the same paper, Wieland also commented on the front limb of *Archelon* and (ibid.:107) proposed a general classification of marine turtles.

Wieland (1905) then turned his attention to the description of a new species (*Toxochelys bauri*) collected by C. H. Sternberg from near the middle of the Smoky Hill Chalk (Santonian) in western Gove County in December 1904. This was an important specimen according to Wieland because until then "no complete carapace of *Toxochelys* had been described" (ibid.:326). The figures of the carapace and plastron published in Wieland's paper demonstrated two important points: Williston's (1901) belief that *Toxochelys* would have a solid shell like *Porthochelys* was wrong; and the carapace of *Toxochelys* was similar in construction to those of *Protostega* and *Archelon*. The holotype specimen of *T. bauri* (YPM 1786) was reexamined by Zangerl (1953) and referred to *Ctenochelys stenopora*. A check of the online records at the Yale Peabody Museum in 2016 indicated that the name had not been changed.

Wieland (1906) was able to report on new *Protostega gigas* material when the Carnegie Museum of Natural History purchased two specimens that had been discovered by C. H. Sternberg. Wieland (ibid.:282) wrote that CMNH 1421 was collected from along Hackberry Creek in southern Gove County, Kansas, and that the two specimens were located fairly close together. If this locality had been correct, these specimens would be the earliest known (early Coniacian) examples of the species. However, Sternberg noted in his book *Life of a Fossil Hunter* that he "should like to correct this mistake. It was found about three miles northwest of Monument Rocks, in a ravine that opens into the Smoky [Hill River], east of where Elkader once stood" (1909:116). The chalk in this locality is likely to be early Santonian to early Campanian in age.

CMNH 1420 was the most complete of the two specimens, but Wieland wrote that Sternberg, in his "attempt to remove and separate the bones from their matrix of chalk, mismarked some of them, and also made it impossible to determine the outlines of any of the plastral elements with exactness" (1906:284–285). He then added, "however, none of the bones are broken, and Mr. Sternberg redeemed himself by discovering and securing in such excellent condition specimen No. 1421" (a skull, lower jaw, and limbs and plastron still in the chalk matrix).

Wieland's admonition to field collectors is well worth repeating here: "As will be evident to any student of the fossil vertebrates the removal of the fossil from its matrix in the absence of the necessary knowledge, training and equipment, was ill advised. Such work is difficult enough in the best equipped laboratories" (1906:284–285). Sternberg might have disagreed with Wieland that he lacked the knowledge necessary to collect and prepare specimens after working almost 30 years in the chalk, but apparently he took no offense. In fact, he repeated Wieland's remarks word for word (Sternberg, 1909:116–117). There is no doubt that Sternberg was pleased with the results of his efforts and those of the Carnegie Museum. The specimen (a composite of CMNH 1420 and 1421) is still one of the best examples of *Protostega gigas* currently on exhibit.

Williston provides a summary of what was then known about the "Ancient Sea Turtles" (1914:231) quoting extensively from his communications with Wieland. Regarding the general habits of the Protostegids, Wieland suggested that there was "no doubt that *Archelon* was strictly carnivorous" and was fast enough to pursue "other than relatively slow-moving prey (ibid.:239–240). Regarding the diet of *Protostega*, Williston added that the abundance of giant clams (inoceramids) would have "afforded an almost inexhaustible source of food for these large turtles" (1914:240). He then noted that "perhaps the formidable beak [Fig. 6.10] was used more in social quarrels than for food-getting" (ibid.).

While no evidence regarding feeding habits has been described since Williston's time, I think it is far more likely that *Protostega* fed near the surface, much like modern sea turtles. The toothless beak of these turtles appears far more useful for shearing flesh, like that of a modern snapping turtle, than for crushing hard-shelled invertebrates. In any case,

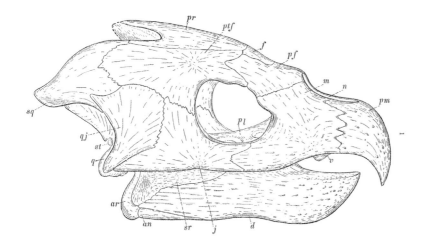

6.10. Skull of *Archelon ischyros* in right lateral view figured by Wieland (1900). Note the characteristic hooked beak (premaxilla) of this species. Skull is about 60 cm (24 in) in length (Wieland, 1909).

the giant inoceramids that Williston wrote about as possible turtle food were mostly gone from the central portions of the Western Interior Sea by the early Campanian when *Protostega* arrived. We may never know what they fed upon.

However, a specimen of *Toxochelys latiremis* from the Pierre Shale of Wyoming was associated with coprolites containing fish bones (Carpenter, 1996:46). Nicholls (1997:219) notes that modern sea turtles feed on a variety of food sources, including sea grasses, algae, mollusks, and jellyfish, but are not known to feed on fishes. Kear (2006) reported the discovery of fragments of small inoceramid shells as gut contents and coprolites associated with several specimens of the protostegid *Notochelone* sp. from the Lower Cretaceous of Australia. It is certainly possible that *Toxochelys*, *Protostega*, and *Archelon* could have scavenged dead fishes and other floating carrion.

Zangerl (1953) reviewed the occurrence of marine turtles, particularly those from the Gulf Coast Cretaceous. He described and named a new species, *Lophochelys natatrix* (YPM 3606), from a specimen collected in 1876 by B. F. Mudge and party from the Smoky Hill Chalk. However, he did not add further information about the occurrence of *Toxochelys*, *Protostega*, or *Archelon* in the Western Interior Sea. Stewart (1978) reported the discovery of the remains of small toxochelyid turtles in the early Turonian Fairport Chalk (Carlile Shale Formation) of Russell and Ellis counties at roughly the same time as *Desmatochelys*. Although it is likely that the Western Interior Sea had a population of marine turtles from its earliest days, it would be a long time before the truly giant Protostegids made their appearance in the early Campanian.

Gaffney and Zangerl (1968) revised the Chelonian genus *Bothremys*. Remains of this unusual marine turtle occur in the Cretaceous of New Jersey, Alabama, and Arkansas and the Miocene of Maryland. The turtle is notable for its solid carapace and plastron and for pits in its upper jaws that were possibly used to position round prey, like snails, so they could be crushed. During their research, the authors also discovered another shattered specimen (YPM 3608) in the Yale Peabody Museum that appeared to be similar to *Bothremys*. It had also been collected by B. F. Mudge and party in June 1876, apparently in what is now Logan County. Once the remains were reassembled, the specimen consisted of "a nearly complete plastron, . . . as well as about 60 percent of the carapace" and was identified as *Bothremys barberi* (ibid.:206). While similar to the New Jersey and Gulf forms, the authors noted several minor differences and designated it as "Sub-species C" (ibid.).

Following the collection of a nearly complete late Campanian specimen of *Toxochelys latiremis* from near Castle Rock in Gove County, Nicholls (1988) reviewed most of the known specimens and published a revision of the genus *Toxochelys*. She noted (ibid.:181) that the carapace was poorly known in existing specimens of this genus and that the new specimen (ROM 28563) was valuable because the carapace was complete. Whereas Zangerl (1953) noted that there were six species of *Toxochelys*, Nicholls (1988:186) determined that there were really only two species: *T. latiremis* from the Western Interior Sea and *T. moorevillensis* Zangerl 1953 from the Gulf Coast.

There is, however, a downside to the story of the collection and description of this otherwise excellent specimen (ROM 28563). The remains were discovered and collected from private property by a field crew (not including E. Nicholls) from the Royal Ontario Museum. While they had permission from the landowner to collect, there were some apparent misunderstandings that left the rancher feeling that the museum had not lived up to its side of the bargain. As a result, the landowner has not allowed any collecting of fossils on his property since 1988. When I talked with him in 2001, he was still adamant about refusing permission and indicated that he had never received further information about the disposition of the specimen or even a copy of the publication regarding it. I did make a point of providing him with a copy of the Nicholls (1988) paper.

Stewart (1990) noted that *Toxochelys latiremis* occurs throughout the deposition of the Smoky Hill Chalk and that the holotype of another species of toxochelyid (*Lophochelys natatrix* Zangerl 1953) occurred in the uppermost chalk (zone of *Spinaptychus sternbergi* or *Hesperornis*). He also indicated that the largest of the Niobrara turtles, *Protostega gigas*, had been collected only in the uppermost chalk (zone of *Hesperornis*). Curiously, he did not mention any of the other species that have been described from the chalk.

Hirayama (1997:234) discussed the distribution and diversity of marine turtles from the Cretaceous and suggested that the toxochelyids were bottom-dwellers and not necessarily good swimmers. While this may have been true for habitats along the margins of the Western Interior Sea, it doesn't explain the relatively large number of *Toxochelys* specimens collected from sediments that occurred near mid-ocean, hundreds of miles from the nearest shore, where the water was up to 200 m (650 ft) deep (Hattin, 1982).

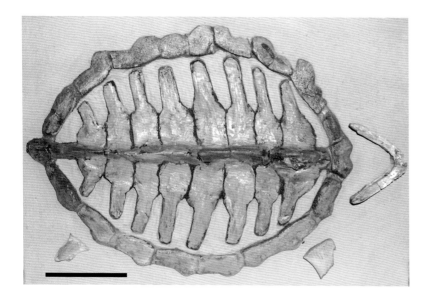

6.11. Carapace and lower jaw of a small marine turtle, *Ctenochelys stenoporus*, in the collection of the Keystone Gallery, Logan County, Kansas. The carapace is about 20 cm (8 in) in length. Scale bar = 5 cm (2 in).

Matzke published the most recent research on the smaller turtles from the Smoky Hill Chalk. His 2007 paper described a relatively complete (minus limbs) juvenile specimen of *Ctenochelys stenoporus* (Hay, 1905; USNM 357166) that had been collected by G.F. Sternberg in 1925. *Ctenochelys* (Fig. 6.11) is closely related to *Toxochelys*, but based on known specimens, it is somewhat smaller.

The following year, Matzke (2008) described a juvenile *Toxochelys* (USNM 12009–1) specimen and compared it with *Ctenochelys*. This is another G. F. Sternberg specimen collected from the Santonian of Gove or Logan County. Matzke (2009) then provided a very detailed description of the skull of *Toxochelys*.

About the same time, Rei Konuki, a graduate student at Fort Hays State University was studying bite marks and bone fractures on *Toxochelys* specimens to determine which predators were feeding on them. In his masters thesis, Konuki (2008) reported that the size and spacing of bite marks on a plastron, carapace, and limbs of a *Toxochelys* specimen (FHSM VP-790) were most likely from a large fish such as *Xiphactinus*, and not from a marine reptile. Since that time a number of very small and generally fragmentary specimens (FHSM VP-17572; Fig. 6.12) have been collected by Triebold Paleontology and by the author. We have a lot of work to do in order to make sense from these scattered remains.

Much of the current research on Late Cretaceous marine turtles is about *Protostega*. Hirayama also noted that while "the skull of the protostegids may have been modified to feed on hard-shelled animals" (1997:236), the limbs were well developed for swimming and may have enabled these turtles to hunt more mobile ammonites in the water column instead of the sessile, bottom-dwelling inoceramids. Again, with no evidence available as to what these turtles actually ate, I suspect that

6.12. Skeletal elements of a hatchling turtle (FHSM VP-17572), including a vertebra (circle), discovered by Keith Ewell in the Lower Santonian chalk of Gove County. The eroded condition of the bones suggests that the specimen was partially digested. The turtle would have been less than 10 cm (4 in) in length. Scale bar in mm.

catching ammonites near the surface would have been a more productive method of feeding than diving for the large inoceramids at the bottom of the sea. But without any fossil evidence of their feeding habits, I think the best bet is to look at the diet of modern large marine turtles like the leatherback (*Dermochelys coriacea*).

Hooks (1998) reviewed the Protostegidae and revised the systematics of the order based on a cladistic analysis. His cladogram indicates that *Toxochelys* is only a distant relative of the Protostegidae, sharing few characters with *Protostega* and *Archelon*. His cladogram also shows Williston's (1894a) *Desmatochelys* to be more closely related to *Toxochelys* than to the Protostegidae.

Based on the fragmentary remains of sea turtles from the earliest Cretaceous rocks in Kansas (Kiowa Shale—Albian), it is likely that some of these earlier species were about the size of a modern leatherback turtle. For comparison, the largest leatherback turtles (*Dermochelys coriacea*) can grow to a length of 2.4 m (8 ft) and weigh 900 kg (2000 lb), although most are smaller. By the Santonian, however, at least one turtle lineage in North America began to grow much larger. *Protostega gigas* was the largest marine turtle from the middle of the Santonian, which includes the 2 million or so years of deposition of the chalk.

The actual size of *Protostega* has been a matter of conjecture and confusion since it was discovered. Cope (1875:111), who collected and named the first specimen in 1871, estimated that *P. gigas* would have measured

3.9 m (12.8 ft) in length and 3.4 m (11.3 ft) across the outstretched paddles, with a skull that was about 0.50 m (24⅝ in) long. Note here that Cope (ibid.) repeated a math error in converting from meters to inches that he had made years earlier (Cope, 1872d:431). Half a meter is roughly 19⅝ inches, not 24⅝. When Hay reexamined Cope's specimen, he didn't catch Cope's conversion error but instead noted that Cope's estimated length of the skull was too high and revised the length downward to "18 inches or 45 cm" (1895:61). In his comparison of Cope's type specimen with two new specimens of *Protostega gigas* (CMNH 1420 and 1421) collected by C. H. Sternberg, Wieland saw Cope's mistake and noted "a palpable numerical error in the measurements of the cranium" (1906:280).

Hay (1895) indicated that Cope was in error in several interpretations of the skeleton, including his misidentification of the left plastron as part of the carapace. However, he agreed almost exactly with Cope's calculated length of the turtle (3.9 m). Ten years later, having just sold a nearly complete and articulated composite specimen to the Carnegie Museum of Natural History, C. H. Sternberg (1905) strongly disagreed with Cope's reconstruction of *P. gigas* but did not refute his measurements. The elder Sternberg had collected several reasonably complete specimens since the 1870s, including one that he sent to Cope in 1876. Wieland (1906) used the Carnegie specimens to establish a more accurate size for *Protostega*, including Cope's type specimen, noting that the actual length is far less than originally estimated. While not providing a measurement of the length of the specimens described in this paper, Wieland had noted earlier (1896:411) that "*Protostega gigas* must have been less than three meters in length" when compared to his newly discovered specimen of *Archelon ischyros*.

Archelon ischyros is the largest known marine turtle from the Cretaceous or any other time (Fig. 6.8). Its size is often compared with the size of a small car. Wieland (1896:410) gave the length of the type specimen as 3.52 m (11 ft 4 in). K. Derstler (written comm., 1999) noted that the length of the *Protostega gigas* in the Dallas Museum of Natural History is 3.4 m, that the Yale *Archelon* specimen (YPM 3000) was 3.0–3.1 m, and that he had worked with a specimen of *Archelon* that was 4.6 m long. Although the remains of *Archelon* have never been collected in the lower portion of the Pierre Shale in Kansas, we know that it lived in the Western Interior Sea during the latter part of the Campanian and may have survived well into the Maastrichtian. Rocks of that age in the Pierre Shale of western Kansas were probably exposed and eroded away many millions of years ago, along with the fossils that they contained. *Archelon*, however, is well known from the upper Pierre Shale of South Dakota.

The discovery of *Protostega* is an interesting story that was well documented by E. D. Cope. While Cope (1870:446) had already named a new species of Cretaceous marine turtle (*Pneumatoarthrus peloreus*, originally described by Cope as a dinosaur) from an isolated vertebra collected in the green sand deposits of New Jersey, he was apparently surprised by the peculiar characteristics of the much more complete specimen he discovered and collected during his only trip to the chalk of western Kansas in

1871. In a letter to Professor Leslie, dated October 9, 1871, and read at the Academy of Natural Science of Philadelphia on October 20 of that year, Cope described and named the first turtle from the Smoky Hill Chalk:

> On another occasion, we detected unusually attenuated bones projecting from the side of a low bluff of yellow chalk, and some pains were taken to uncover them. They were found to belong to a singular reptile, of affinities probably to the Testudinata, this point remaining uncertain. Instead of being expanded into a carapace, the ribs are slender and flat. The tubercular portion is expanded into a transverse shield to beyond the capitular articulation, which thus projects as it were in the midst of a flat plate. These plates have radiating lines of growth to the circumference, which is dentate [see Fig. 6.3]. Above each rib was a large flat ossification of much tenuity, and digitate on the margins, which appears to represent the dermoossification of the Tortoises. Two of these bones were recovered, each two feet across. The femur resembles in some measure that ascribed by Leidy to *Platecarpus tympaniticus* [a mosasaur specimen from Alabama], while the phalanges are of great size. Those of one series measured eight inches and a half in length, and are very stout, indicating a length of limb of seven feet at least. The whole expanse would thus be twenty feet if estimated on a Chelonian basis. The proper reference of this species cannot now be made, but both it and the genus are clearly new to science, and its affinities not very near to those known. Not the least of its peculiarities is the great tenuity [thinness] of all the bones. It may be called *Protostega gigas*. (1872a:175)

The remains of this huge turtle were dug up and packed as well as possible under the primitive conditions that existed in western Kansas at the time. This was several years before plaster and burlap jackets were used to protect fossil bones (Sternberg, 1884), and the techniques used to remove fossils from the matrix ("pick-axe and shovel," see below) were at best crude. In addition, the locality on Butte Creek was 15 miles or more south of the nearest railroad line, and the bones had to travel more than that distance in a horse- or mule-drawn wagon over the then roadless prairie. Once packed in crates and loaded into a freight car, the remains then traveled some 1,500 miles to Philadelphia, Pennsylvania, where they were transferred again to a horse-drawn wagon and transported across town to Cope's work place. Considering the rough handling involved, it was a wonder that any of these delicate bones ever arrived in good enough condition to be useful. Cope (Almy, 1987:189) was well aware of the problems involved with shipping fossils and had written to Dr. Turner two years earlier asking him to pack the next shipment of *Elasmosaurus* bones more carefully (see Chapter 7).

At the March 1, 1872, meeting of the American Philosophical Society, Cope (1872c:433) read his complete description (11 pages!) of the type specimen (AMNH FR 1503) collected from the chalk along Butte Creek, "near Fort Wallace," in Logan County. He also indicated that the remains were completely enclosed in the chalk when discovered and had to be excavated with "pick-axe and shovel." Apparently only the edge of the shell was visible when it was discovered. The bones were quite fragile, and Cope remarked that the specimen was "much fractured" by the time it had been shipped 1,500 miles to Philadelphia. Hay wrote:

> The specimen of this species which Cope originally described was found by him in 1871, in Niobrara deposits, near Butte Creek, south of Wallace, Kansas. This is now in the American Museum of Natural History and has the number 1503. At the time of its discovery the bones were much distorted and comprest [*sic*]; and most of them were much fractured in collecting them. Cope states that the portions described by him were reconstructed out of over 800 pieces. One of the large bony plates, described as overlying the ribs, had been broken into 108 pieces. Most of the bones are yet in the condition in which Cope left them; the large plates are, to a great extent, again broken up. (1908:191)

Cope (1872e:323–324) also described the discovery of *Protostega gigas* in his discussion regarding the geology and paleontology of western Kansas published in the *U.S. Geological Survey of Montana and Portions of the Adjacent Territories*. One has to wonder how Kansas was included with Montana, but it happened in this instance. Cope wrote:

> Tortoises were the boatmen of the Cretaceous waters of the eastern coast, but none had been known from the deposits of Kansas until very recently. But two species are on record; one large and strange enough to excite the attention of naturalists is the *Protostega gigas*, Cope. It is well known that the house or boat of the tortoise or turtle is formed by the expansion of the usual bones of the skeleton till they meet and unite, and thus become continuous. Thus the lower shell is formed of united ribs of the breast and of the breastbone, with bone deposited in the skin. In the same way the roof is formed by the union of the ribs with bone deposited in the skin. In the very young tortoise the ribs are separate as in other animals; as they grow older they begin to expand at the upper side of the upper end, and with increased age the expansion extends throughout the length. The ribs first come in contact, where the process commences, and, in the land-tortoise, they are united to the end. In the sea turtle, the union ceases a little above the ends. The fragments of the *Protostega* were seen by one of my party projecting from a ledge of a low bluff. Their thinness and the distance to which they were traced excited my curiosity, and I straightway attacked the bank with the pick. After several square feet of rock had been removed, we cleared up one floor, and found ourselves well repaid. Many long slender pieces of two inches in width lay upon the ledge. They were evidently ribs, with the usual heads, but behind each head was a plate like the flattened bowl of a huge spoon, placed crosswise. Beneath these stretched two broad plates, two feet in width, and no thicker than binder's board. The edges were fingered, and the surface hard and smooth. And this was quite new among full grown animals, and we at once determined that more ground must be explored for further light. After picking away the bank, and carving the soft rock, new masses of strange bones were disclosed. Some bones of a large paddle were recognized, and a leg-bone. The shoulder-blade of a huge tortoise came next, and further examination showed that we had stumbled on the burial-place of the largest species of sea-turtle yet known. The single bones of the paddle were eight inches long, giving the spread of the expanded

flippers as considerably over fifteen feet. But the ribs were those of an ordinary turtle just born, and the great plates represented the bony deposit in the skin, which, commencing independently in modern turtles, united with each other below at an early day. But it was incredible that the largest of known turtles should be but just hatched, and for this and other reasons it has been concluded that this "ancient mariner" is one of those forms not uncommon in old days, whose incompleteness in some respects points to the truth of the belief that animals have assumed their modern perfections by a process of growth from more simple beginnings. (1872e:334–335)

A brief description of the skeleton of *Protostega gigas* by Cope follows this section. Cope also mentions earlier discoveries of Cretaceous turtle material, including a humerus from Mississippi discovered by Dr. William Spillman (Manning, 1994:67). The specimen was originally figured in Leidy (1865:128, pl. VIII, figs. 1, 2) and described as "the humerus of *Mosasaurus*." While this might be considered a major mistake by today's standards, it is important to note that the limbs of both mosasaurs and giant turtles were unknown at the time. The specimen was reidentified as a turtle by Cope (1872c:433) and named *Protostega tuberosa*. Wieland (1900) examined the same bone and recognized that it was not a protostegid. He changed the genus name and called the turtle *Neptunochelys tuberosa*.

Leidy (1865:128, pl. VIII, figs. 3, 4, 5) also included three views of another "gigantic humerus" from the green sand of New Jersey on the same plate. Although Leidy didn't identify it directly, he did refer to the figure as "supposed to be of a humerus of the *Mosasaurus*" (ibid.:128). However, the same specimen had been previously identified by Agassiz (1849) as a turtle that he called *Atlantochelys Mortoni* (ANSP 9234) without further description. Unlike Leidy, Cope appeared to be aware of the name given by Agassiz but chose to ignore it because "it was unaccompanied with specific or generic description" (1875:113). As justification, Cope indicated that he considered it to be "not only a privilege but a duty to ignore names put forward in this manner" and renamed the protostegid turtle humerus *Protostega neptunia*. To add to the confusion, Wieland (1900) indicated that Leidy (1873) had "validly rehabilitated" (Wieland, 1900:419) the name and also indicated that Cope was in error when he considered it to be a protostegid. Wieland (1900) noted that the form of the humerus was actually closer to that of *Desmatochelys lowi* Williston and retained the name *Atlantochelys mortoni*.

In 2011, amateur fossil collectors Steven and Donna Loffer, and Susan Liebl discovered the paddle bones of a turtle eroding from the Smoky Hill Chalk in eastern Gove County. The exposed bones were collected, but they wisely decided not to dig for more. After Susan Liebl showed me the bones, I made arrangements to visit the locality and view the specimen. It only took a few minutes of careful digging to determine that they had discovered the posterior end (including the complete tail) of a medium-size turtle that was just beginning to erode from a gentle slope of chalk. We dug further and uncovered the carapace of a turtle that was about 0.9 m (34 in) across at its widest point. It appeared that the turtle had landed upright on the sea floor and been preserved there intact without any evidence of scavenging. The specimen appeared well preserved, but it was too large for us to jacket and remove without more help. Fortunately, I knew that Mike Triebold was in Kansas with his field crew at the time, and he agreed to come over in a couple of days and help us. In the meantime, I talked with the landowners and received permission to collect the specimen.

Once on the site, Mike was able to supply power tools (chainsaw, jack hammer, and air-powered chisels) that made clearing the overburden around the turtle much easier. When we had cleared the chalk around the bones and determined the extent of the specimen, Mike used his chainsaw to cut two deep grooves around the perimeter. The cuts enclosed an area about 1.2 m across by 1.5 m long. Then Anthony Maltese and Jacob Jett used the air-powered chisels to remove the chalk from the grooves, creating an edge to wrap the jacket around and space where we would be able to get under the specimen to drive long wedges under it to free it. Once these tasks were completed, the exposed bones were covered with tinfoil and Anthony and I began laying on the plaster-covered burlap strips. Hollow metal pipes were incorporated into the jacket as we built up the layers of plaster and burlap. They would be used later as handles to carry the jacket after we had rolled it. When the jacket was finished, we let it dry for about an hour and a half, just to make sure it was solid. While we waited, we began clearing space to accommodate the wedges and to provide an area to receive the jacket once it was rolled.

Although we were confident we had secured the specimen in a good jacket, some tension always builds up before it is rolled. Although most plaster-jacketed specimens are recovered successfully, we all knew of instances when the fossils fell out of the jacket as it was being lifted. Eventually the plaster was set to everyone's satisfaction. After pounding several wedges under the specimen and literally hearing when the layer of rock finally broke free at the base, three of us put our hands under the jacket to lift while two others stood on the other side to help lower it as the jacket rolled. On the count of three, we lifted and rolled the jacket over on its top. It came free cleanly and everything stayed inside the plaster bandages. The final task was to remove some of the 15 cm (6 in) of extra chalk that was attached to the specimen—that had been our safety margin to ensure that the bones stayed in place, but after the jacket was turned, the extra chalk was just added weight. At last we carried the jacket to the back of Triebold's truck and loaded it for transport to the prep lab.

The turtle turned out to be a beautiful specimen of a small to medium-size *Protostega*. It was mostly complete, especially the back two-thirds of the turtle, but it was apparent that this turtle had died a violent death. Half of the skull was missing, as were both front paddles, bitten off in mid-humerus, and there were more than a hundred puncture marks on the carapace and plastron. Most of the punctures were located on the lower shell (plastron), including the deepest wound on the left

6.13. Mosasaur bite marks on the left hyoplastron in ventral view of the *Protostega gigas* turtle (FHSM VP-17979) discovered in the Middle Santonian chalk of Gove County, Kansas. The base of the turtle's left front paddle would be where the scale is located in this photo. Scale bar = 10 cm (4 in).

hyoplastron (Fig. 6.13). Some of the bite marks were on the ridge or crest that covers the turtle's spine, more than 40 cm (16 in) from the edge of the carapace. The only predator in the Western Interior Sea with a mouth large enough to inflict that kind of damage to a turtle this size would have been a mosasaur called *Tylosaurus proriger* (Fig. 6.13). The bite marks were punctures, exactly what you would expect from the conical teeth of a large mosasaur. From the evidence, it appears that a large mosasaur had attacked and killed the turtle (Fig. 6.14), but then was unable to find a way to swallow it. The large number of bite marks probably means that the mosasaur turned the turtle around in its mouth, trying to find a way to swallow it. The turtle was simply too big. Eventually the mosasaur gave up and released the dead turtle, which then sank to the bottom without further damage from scavengers.

An interesting sideline to this story is that I collected a large (9 m) *Tylosaurus proriger* (FFHM 1997–10) from this same canyon in 1996 (Chapter 9). I'm still working on the description of this new *Protostega* specimen (FHSM VP-17979), but reported on it briefly in Everhart (2013).

The most commonly collected turtle remains throughout the Smoky Hill Chalk are those of *Toxochelys*, a much smaller species than *Protostega*. Cope (1872e:129) mentioned the discovery of a turtle "the size of

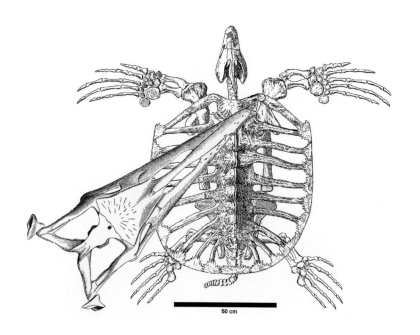

6.14. Skeletal reconstruction from above of a large *Tylosaurus proriger* attacking a protostegid turtle. The critical issue in the case of the FHSM VP-17979 *Protostega gigas* was that the predator had to be large enough to leave bite marks on the dorsal crest of the turtle's shell. The 1.2 m (48 in) skull of the *Tylosaurus* is large enough to be the source of the bites on the carapace. The total length of the *Tylosaurus* would have been about 9 m (30 ft). Scale = 50 cm (20 in).

some of the large *Cheloniidae* of recent seas" that was associated with a plesiosaur skeleton (*Plesiosaurus gulo*) from the Pierre Shale near Sheridan, Kansas. In a brief note, Cope (1873) indicated that B. F. Mudge had provided him with additional material from the specimen, which he named *Toxochelys latiremis* without further description. Although Cope spent considerable time and effort on the reconstruction (albeit in error) and description of *Protostega gigas*, he apparently did not have sufficient material or interest to do the same with *Toxochelys*. The type specimen is briefly noted and figured in his *The Vertebrata of the Cretaceous Formations of the West* (Cope, 1875:98–99, pl. VIII). His final note on the species (Cope, 1877) was a brief mention of two new skulls of *Toxochelys* discovered by C. H. Sternberg.

Like mosasaurs, marine turtles certainly reached their peak in terms of body size during the Late Cretaceous. The smaller varieties were apparently far more common in the warmer waters of the Gulf Coast than they were in the north. However, the fate of the toxochelyids, desmatochelyids, and protostegids was the same as that of the dinosaurs and other species at the end of the Cretaceous.

In retrospect, marine turtles were and are possibly the most successful of any of the reptiles that have returned to the sea, perhaps partly because they remained conservative in their evolution. Even as they evolved flippers in place of legs and feet, they still retained a measure of flexibility and skeletal integrity in the joints that was lost in the ichthyosaurs, plesiosaurs, and mosasaurs. Unlike these other highly adapted marine reptiles, marine turtles were still able to support their bodies to some extent, and to move about on land. Of the four major groups, only the turtles retained their connection with the land, returning each year to lay their eggs in the warm beach sand.

They also avoided the fatal situation of becoming apex predators. In their case, it is possible that *not* being at the top of the food chain was beneficial to their survival.

While *Archelon* and other closely related marine turtles did not survive much, if at all, beyond the end of the Cretaceous, it is readily apparent that other marine turtle species came through and prospered. If the extinction of marine reptiles at the end of the Cretaceous was largely due to a collapse of the marine ecosystem, it would be easier for a turtle feeding low on the food chain to survive than for a mosasaur or plesiosaur that depended on an abundance of fishes in its diet.

Where the Elasmosaurs Roamed

7

The warm, silt-laden water felt strange to the young long-necked plesiosaur as he slowly worked his way upstream against the strong current of the river. His streamlined body was less buoyant in the fresh water, which made raising his head above water to breathe a bit more difficult. Unlike in the ocean that was his home, he could not see very far in the murky water, so he moved cautiously along the sandy bottom, trying to stay in the deeper reaches of the main channel. Dimly he remembered being here several times before with a much larger member of his species. At the time, it was all he could do to stay close to her body and not be swept away by the current. Now sensory cells in his nasal passages 'tasted' strange new chemical odors as the water passed in through his nose and exited through his slightly opened mouth. Although he was hungry after his long journey, his instincts drove him up this river for something else.

Within a short distance from where the river entered the ocean, the sand bottom of the river had turned to coarser gravel. The plesiosaur lowered his head and began to probe the gravel with his lower jaw. Clouds of silt and other debris rose wherever his chin touched the bottom. The sensations he received through his lower jaw told him that this wasn't what he was looking for, and he began swimming again. Holding his head and neck stiffly in front of his body, he relied on the rhythmic movement of his wing-shaped front and rear flippers to provide the thrust he needed to 'fly' forward through the water. The river narrowed as he swam upstream, and the current became faster. Swimming against a current was not something he was used to doing. Such currents hardly ever occurred in the calm, shallow ocean where he had spent almost all his life. He soon began to encounter large rocks that formed eddies in the current. Around the base of a rock larger than he was, he finally located what he was looking for. Small round stones had collected on the downstream side of the boulder. Twisting his head slightly to the side, he saw that there were many more here than he would need. Slowly and delicately, he picked up one smooth rock after the other with his long, slender teeth. With each, he raised his head slightly and swallowed, feeling the hard stone move into his throat and start the slow trip down his 10 foot long esophagus to his stomach. He continued selecting and swallowing the round stones until he had satisfied his need. As he raised his head above the surface briefly to breathe, his sharp underwater vision saw only a dark green blur that was the dense forest that crowded over the riverbank. Submerging again, he turned and began his return journey back to the sea.

His movement flushed a small fish from its hiding place near the large rock. Before the fish could flee, the plesiosaur reacted and trapped it between jaws filled with long, interlocking teeth. Raising his head just above the surface, he swallowed his prey headfirst, then exhaled again and refilled his lungs. Vaguely satisfied, he submerged his head and began to swim downstream with the current with slow, deep strokes of his paddles.

Elasmosaurs

Although isolated fragments of Cretaceous plesiosaurs, mostly vertebrae, had been described earlier in the nineteenth century by Joseph Leidy (1823–1891) and others from the East Coast and Arkansas, the remains discovered by Dr. Theophilus Turner (1841–1869) near Fort Wallace in western Kansas in 1867 represented the first reasonably complete plesiosaur specimen from North America and the first to be described as an elasmosaur. Dr. Turner was an Army surgeon assigned to Fort Wallace in far western Kansas. According to the letters sent to his family (Almy, 1987), Turner had discovered the fossilized remains of the huge animal while exploring exposures of the Pierre Shale along the right-of-way of the approaching Kansas Pacific Railway near McAllaster Butte (see Fig. 1.7) in present-day western Logan County, Kansas. Dr. Turner gave several of the vertebrae to John LeConte, who was part of a survey crew for the railroad. In turn, LeConte delivered the bones to Edward D. Cope in Philadelphia for examination (LeConte, 1868). Cope had seen plesiosaur remains and their vertebrae during his studies in Europe and immediately recognized the bones as belonging to a large plesiosaur. He wrote back to Turner (Almy, 1987) asking him to secure the rest of the specimen and ship it to him at the expense of the Academy of Natural Sciences. Turner was reluctant to do so until the approaching railroad reached Fort Wallace. In February 1868, he finally agreed to Cope's request and loaded the specimen on a military wagon train that was headed 90 miles east to near Ellis, Kansas, where the railroad was being constructed.

After receiving the shipment in mid-March of 1868, Cope hurriedly unpacked and examined the remains. The specimen consisted of a nearly continuous string of over 100 vertebrae, and pectoral and pelvic girdles, but no limbs and only the front portion and other fragments of the skull. Some of the remains, including the limb girdles, were still enclosed in hard concretions. However, less than two weeks after first seeing the bones, Cope reported on the strange and heretofore unknown extinct animal at the March 24th meeting of the Academy of Natural Sciences of Philadelphia (Cope, 1868:93). Cope named the new plesiosaur *Elasmosaurus platyurus* (flat-tailed, thin-plate reptile) to describe what he saw as a long flat tail and the large pelvic and pectoral girdles. In August of the following year, Cope (1869), at his own expense, published a full description and figures of *E. platyurus* as a 'preprint' of the *Transactions of the American Philosophical Society*, a respected scientific journal, and sent copies to many of his friends and associates in the U.S. and Europe.

A.

B.

C.

Unfortunately, it appears that Cope did much of his work in secret and without a review by his peers. Following the early work on other American plesiosaur remains by his mentor, Joseph Leidy, Cope misidentified the cervical vertebrae as caudals and therefore believed the long neck was the tail (Fig. 7.1). As a result, he placed the head of his reconstruction on the wrong end, creating an animal with a short neck and an extremely long tail. For a recent reprinting of Cope's (1869) Plate II with the original reconstruction, see Davidson (2002:216). At the time it was a reasonable assumption, in part because Cope was an expert on lizards which have short necks and long tails.

After reading Cope's prepublication and eventually making his own examination of the specimen, Joseph Leidy, Cope's mentor, recognized that Cope had made a serious mistake. Nearly a year later, on March 8, 1870, Leidy presented his own view of the specimen, apparently without discussing it first with Cope. Leidy noted that "Prof. Cope has fallen into the error of describing the skeleton in a reversed position to the true one, and in that view has represented it in a restored condition in his recent 'Synopsis of the Extinct Batrachia, Reptilia, and Aves,' published in the Transactions of the American Philosophical Society" (1870:392).

In Cope's defense, while the half-round base of the skull (occipital condyle) is still firmly attached to the first vertebra of the neck (atlas/axis), the neck of this huge animal is so long and thin that even today (pers. obs., 2002, 2009) it looks like it should be the tail. The centra of the most anterior cervical vertebrae are only 2.5 cm (1 in) in diameter, about the width of a broom handle, and the small fragment from the base of the skull is not obvious. Such a long, thin neck would not have been expected on an animal that was over 12 m (40 ft) in length. It is also possible that the atlas/axis vertebra was not fully prepared initially because it was just

A.

B.

7.2. A, the muzzle of *Elasmosaurus platyurus* Cope (ANSP 10081) in right oblique view; the anterior ends of the upper and lower jaws were preserved together and represent the largest fragment of the skull that was recovered; the muzzle was figured by Cope (1869:pl. II, fig. 8); scale bar = 5 cm (2 in); B, the upper and lower jaws have since been separated; scale bar = cm (2 in).

assumed to be the end of the tail and of less taxonomic value or interest than other parts of the skeleton.

As might be expected, Cope was embarrassed by what now appears to have been an obvious oversight on his part. His immediate reaction was to try to minimize the damage by printing a notice and offering to replace all copies of the preprint (Storrs, 1984). The text of the notice was reprinted in O. C. Marsh's comments more than 20 years later in the *New York Herald* (W. H. Ballou, New York Herald, January 19, 1890, 11): "An error having been detected in the letter press of the 'Synopsis of the Extinct Batrachia and Reptilia of North America,' by Edward D. Cope, it will be necessary to cancel and replace one of the forms. The author therefore requests that the recipient of this notice would please return his copy of said work to the author's address, at his expense. The volume will be returned, postpaid, with Part II, of the same work, which will be sent to those who have received the corrected Part I".

Cope's mention of an "error" in the preprint is certainly understated, but understandable from his point of view. Working quickly with the specimen, and with his printer, he wrote and published a revised description (Cope, 1870) in April of that year, which included a partially corrected drawing showing the head at the end of an extremely long neck. As might be expected in the rush, however, the second version was not completely updated. Beside the many errors remaining in the text and the reconstruction, the paper was reprinted using the original cover page from the first version, and thus is still dated August 1869. This created the perception, justified or not, that Cope was trying to cover up his original mistakes (Storrs, 1984). Although the arguments over the orientation of the vertebrae between Cope and Leidy died down quickly, this incident would continue to haunt Cope for years.

It is worth noting here that there is no record of any published criticism by O. C. Marsh during this time (1868–1870). The only comment by Marsh appeared several years later in the appendix of the June 1873 issue of the *American Naturalist*, on the last page of a petty nine-page attack on Cope. Marsh states, "After investigating a very perfect specimen

7.3. The gastroliths associated with a fragmentary elasmosaur specimen (KUVP 129744) from the Sharon Springs Member of the Pierre Shale in Logan County, Kansas. These stones are the largest gastroliths documented from any animal, including those of sauropod dinosaurs (Everhart, 2000). Scale bar = 10 cm (4 in).

for months, he placed *the head on the end of the tail* [emphasis added by Marsh], and restored the animal in this position as the type of a new order, *Streptosauria*" (1873:ix).

There may be more to the story of *Elasmosaurus platyurus*. Cope (Almy, 1987:189) noted that when Cope received the specimen from Turner, it was missing most of the skull (Fig. 7.2), some dorsal and cervical vertebrae, and all of the gastralia (belly ribs in plesiosaurs that support the abdomen). Although Turner did recover and send some additional remains, many of the dorsal vertebrae and the gastralia (and the limbs) were never recovered. It appears likely that portions of the floating carcass may have been removed by scavengers or simply dropped off before it reached its final resting place on the sea bottom.

In 1991, two dorsal vertebrae and a number of very large gastroliths (stomach stones) were discovered by Pete Bussen, a local rancher, at a site about a mile northwest of where Turner dug up the remains of *E. platyurus*. In fact, the gastroliths were the largest ever associated with any elasmosaur remains (Everhart, 2000; Fig. 7.3). Larry Martin and John Chorn from the University of Kansas Museum of Natural History conducted a limited dig on the site and collected the vertebrae, rib fragments, and gastroliths (KUVP 129744). Several years later, in 1998, I participated in another dig on the same site with Glenn Storrs of the Cincinnati Museum Center (CMC), when the last of the remains (CMC VP6865) were recovered. They consisted of a few more gastroliths, several complete ribs, and many gastralia. Nothing was articulated, and the remains appeared to have literally been dropped in a pile on the sea floor.

That could have been the end of the story of just another incomplete plesiosaur specimen, but, while going through the collection at

Specimen #	Taxon	Collected	Locality	Collected by:	Described by:
YPM 1130	Elasmosauridae	1876	Niobrara, Wallace Co.?	H. A. Brous	
YPM 1640	*Elasmosaurus nobilis***	1874	Ft. Hays Lm., Jewell Co.	B. F. Mudge	Williston, 1906
YPM 1644	*Styxosaurus snowii*	1874	Gove Co., KS	Mudge, Williston	
YPM 1645	*Thalassonomosaurus marshii*	1889	Logan Co., KS	H. T. Martin	Williston, 1906
KUVP 434	*Styxosaurus snowii*	1874	Logan Co., KS	Mudge, Williston	
KUVP 1301	*Styxosaurus snowii* (Holotype)	1890	Logan Co., KS	E. P. West	Williston, 1890
KUVP 1302	Elasmosaur indet.	1895	Logan Co., KSC.	H. Sternberg	
KUVP 1312	*Elasmosaurus sternbergi***	1895	Logan Co., KS	C. H. Sternberg	Williston, 1906
KUVP 434	*Thalassiosaurus ischiadicus***	1874	Logan Co., KS	B. F. Mudge	Williston, 1903
USNM 11910	*Styxosaurus snowii*	1927	Logan Co. near Oakley, KS	G. F. Sternberg.	

* Carpenter (1999) made many of Williston's species names synonymous with *Styxosaurus snowii*.
** Smoky Hill Chalk Member unless otherwise noted.
*** Considered to be Elasmosauridae indeterminate by Storrs (1999).

Table 7.1. Elasmosaurs* collected from the Niobrara Formation** of Kansas: by date of collection, locality and collector(s). Currently, most of these specimens are considered to be either Elasmosauridae indet. (Storrs, 1999) or *Styxosaurus snowii* (Carpenter, 1999). Abbreviations: FFHM—Fick Fossil and History Museum; KUVP; University of Kansas Museum of Natural History; YPM—Yale Peabody Museum; USNM—United States National Museum; indet.—indeterminate/ unknown because of the fragmentary nature of the specimen.

the Sternberg Museum in 2002, I came across seven plesiosaur dorsal vertebrae (FHSM VP-398) that were almost identical in size and preservation to the ones discovered by Pete Bussen in 1991 (KUVP 129744). They had been donated to the museum in 1954 by a Logan County rancher. More importantly, George Sternberg's handwritten note accompanying the specimen indicated that the site was at the same locality where the University of Kansas and the Cincinnati Museum crews had recovered the other plesiosaur remains. Taken as one specimen, the three sets of remains contain most of the same bones Cope noted to be missing from *E. platyurus*. They also include gastroliths that are now known to be present in almost all elasmosaurs (Cicimurri and Everhart, 2001), but were not part of the original material collected by Turner in 1867–1868. Whether or not it can be proved that they are one and the same as the holotype specimen of *E. platyurus* remains to be seen; my paper (Everhart, 2005a) suggesting that connection was not well received. I readily admit it is a long shot, but then again, I am the only person in the world who has seen all four of the specimens and both localities.

As one suggested scenario, I believe that the bloated carcass of the long-dead and decomposing elasmosaur floated along in the current at or near the surface until it finally ruptured, releasing a large number of loose bones and gastroliths from the center of the body. Then the remaining portion of the carcass, including the head, neck, and tail, lost buoyancy and drifted farther to the southeast until it sank to the bottom. There it was covered with hundreds of feet of shale until millions of years of erosion exposed the remains that were discovered in 1867 by Dr. Turner. See Reisdorf et al. (2012) for a recent and more detailed discussion of the decomposition of the carcasses of ichthyosaurs and other air-breathing vertebrates in the ocean.

The original remains of *Elasmosaurus platyurus* were located in what later became known as the Sharon Springs Member of the Pierre Shale. At the time of their discovery, however, Cope would have assumed they were from the "Niobrara Cretaceous," just as he did in regard to the type specimen of *Toxochelys latiremis* (Chapter 6).

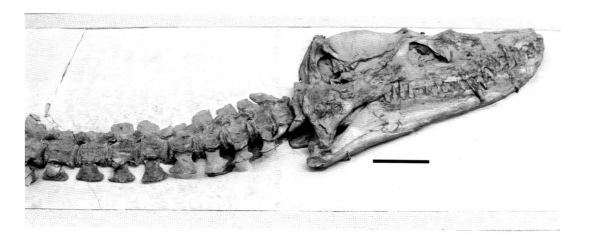

Besides *Elasmosaurus platyurus*, there are relatively few remains of elasmosaurs known from the Smoky Hill Chalk (Table 7.1) and Pierre Shale in Kansas. Most in Kansas have occurred in the Pierre Shale, and few are as complete as Turner's *Elasmosaurus* discovery. A specimen of *Styxosaurus snowii* (NJSM 15435; Cicimurri and Everhart, 2001) also discovered by Pete Bussen, is the most complete collected to date, but it is also missing the skull. See Carpenter (1999) and Everhart (2006) for more information about the most recent summaries of the elasmosaur specimens from the Smoky Hill Chalk.

Perhaps the best known of the other Kansas elasmosaur specimens is the skull (Fig. 7.4) and cervical vertebrae of *Styxosaurus snowii* (KUVP 1301—holotype) that was collected by 70-year-old retired Judge E. P. West (1820–1892) in 1890 from the upper chalk of Logan County (Williston, 1890; Everhart, 2006, 2015). An article in the *Lawrence Daily Journal* (October 13, 1890, 4) noted that "Dr. Williston, the 'new Prof' of geology and paleontology appears as pleased over the find as if he had discovered the end of a rainbow, and says that many an eastern museum would gladly pay $500 for it." When Cope visited the University of Kansas in 1893 (Lawrence Daily Journal, August 17, 1893, 4), he used the opportunity to examine the skull and subsequently described the structure in greater detail (Cope 1894). The specimen represents the only known complete skull of an elasmosaur ever collected in Kansas. Note that the skull of the *Elasmosaurus platyurus* specimen consisted only of the muzzle (Fig. 7.2) and a few tooth-bearing fragments.

Carpenter (1999) lists several other Kansas specimens that he considers synonymous with *S. snowii*: KUVP 434, USNM 11910, YPM 1130, YPM 1644, and YPM 1645. All except YPM 1130 are from the upper chalk of Logan County. According to the records of the Yale Peabody Museum, YPM 1130 was collected a little farther west in the upper chalk of Wallace County. This distinction, however, is based on a political boundary and is meaningless in terms of the actual occurrence of the specimen. Since Logan County was the eastern half of a much larger Wallace County until 1881 and called St. John County until 1885 (Bennett, 2000), it's likely that

7.4. The skull of the type specimen of *Styxosaurus snowii* (KUVP 1301) in right lateral view. The specimen was collected by Judge E. P. West in the upper chalk of Logan County in 1890 and represents the only elasmosaur skull known from the Smoky Hill Chalk. Scale bar = 10 cm (4 in).

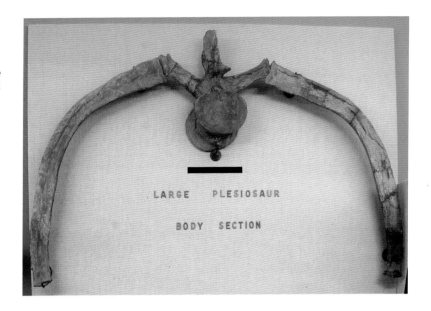

7.5. Reconstruction of the cross section of a very large elasmosaur (KUVP 1312) at the University of Kansas. The diameter of the body would be about 115 cm (4 ft), somewhat larger than *Elasmosaurus platyurus*. Scale bar (diameter of a dorsal vertebra) = 16.5 cm (6.5 in).

LARGE   PLESIOSAUR

BODY   SECTION

some locality information on the earlier discoveries is confused. All of these specimens, however, are from the upper Smoky Hill Chalk (early Campanian). Considered together, these specimens may represent the return of elasmosaurs to the middle portion of the Western Interior Sea as it became narrower and shallower.

In 1893, Charles H. Sternberg collected two very large dorsal vertebrae of an elasmosaur (KUVP 1312; Fig. 7.5) from the chalk at an unknown locality, originally reported to be in Gove County. In his description of the vertebrae, Williston (1906) noted that they were the largest he had ever seen and named a new species in honor of Sternberg: *Elasmosaurus sternbergi*. The two vertebral centra measured by Williston (ibid.) were 165 and 155 mm (6.5 and 6.1 in) across, suggesting an enormous elasmosaur "that could not have been less than 60 feet in length." In comparison, by my measurements, the largest vertebrae of *Elasmosaurus platyurus* are 130 and 120 mm (5.1 and 4.7 in) across. Jim Martin (pers. comm., 1999) collected a single very large elasmosaur vertebra from Vega Island, off the coast of Antarctica, that is an estimated 180 mm (7.1 in) across.

Welles (1952) disagreed with Williston's identification of *E. sternbergi* and believed the vertebrae were more likely those of a large pliosaurid. Since that time, however, it has been determined that pliosaurids were long extinct by the time that the Smoky Hill Chalk was deposited (Chapter 8). Storrs (1999) noted, contrary to Williston, that the specimen actually consisted of one dorsal and two cervical vertebrae, but agreed with Williston (1906) that the remains were from an elasmosaur. The story could have ended there, but it didn't.

In 2004, I came across additional details in a book written by the man who had collected the specimen, Charles H. Sternberg. In *Hunting Dinosaurs on the Red Deer River, Alberta, Canada*, he described the discovery of the remains of a huge elasmosaur that had been uncovered and nearly destroyed by a farmer during an excavation for a building on his homestead. Sternberg wrote:

> I told Maud [his daughter] of a complete skeleton that had once been found by a farmer in the Kansas Chalk of Butte Creek, Logan County. He started to excavate a place for a stable, when he uncovered some huge vertebrae, and ribs over five feet long. He supposed they were elephant bones, and as they were broken, he thought they could not be saved, and so he dug up the bones with the chalk. They were dumped into a cow yard and beaten to powder under their feet, and could never be restored. I grieved much over the loss to science of that splendid specimen that has never been duplicated. Dr. S. W. Williston [1906], the oldest living American vertebrate paleontologist, described the few bones I was able to save from the general wreck. He did me the honor of naming it after me. (1917:176)

The discovery of this specimen in the upper chalk (early Campanian) of Logan County makes much more sense to me than Williston's note that it came from Gove County (generally lower chalk). To date, more than 110 years later, nothing quite like this specimen of a truly huge elasmosaur has been discovered in Kansas.

Plesiosaurs have also been discovered in Kansas from below the Niobrara. In 1931, a large and fairly complete elasmosaur was discovered on an oil lease in southwestern Ellsworth County by Joe Purzer, an oil company geologist. According to George F. Sternberg's notes, the specimen was collected by M. V. Walker and himself from the Lincoln Limestone Member of the Greenhorn Limestone (middle Cenomanian) in September of that year. The remains (UNSM 1195; new number UNSM 50134) were sold, unprepared, to the University of Nebraska State Museum around 1935 per UNSM records. Sternberg's original description notes that the specimen included many vertebrae and ribs, a complete right front paddle, portions of the pectoral and pelvic girdles including at least one ilium, both femora, most of both rear paddles, and a large assemblage of gastroliths.

When I examined UNSM 1195 in 2000, the remains included 248 gastroliths (Fig. 7.6), weighing 2.5 kg (5.5 lb). The paddle currently on display in the UNSM was originally prepared and mounted by George Sternberg as indicated by archival photographs at the Sternberg Museum of Natural History (Everhart, 2007:fig. 1). The records also included a series of black and white photographs taken during the dig near Holyrood in Ellsworth County. The pictures show both George Sternberg and his assistant, Myrl Walker, uncovering and jacketing the bones from a rocky stream channel (ibid.:fig. 2). In one photo, Myrl Walker is holding a handful of gastroliths from the specimen (ibid.:fig. 3). Other photos show the crowd of interested spectators that came by to watch the dig.

I was able to use George Sternberg's photographs of the dig for UNSM to locate the place where it was recovered. On a hunch, I contacted the town clerk in Holyrood, Kansas, and asked her there was anyone still alive in the town who might remember the dig. She did some checking

7.6. At least 248 gastroliths from an elasmosaur (UNSM 1195) were collected by George Sternberg from the Greenhorn Formation near Holyrood, Kansas (Ellsworth County) in 1931. Scale.bar = 10 cm (4 in).

and located a man who remembered hearing about the elasmosaur. As it turned out, Harold Ehler, the grandson of the property owner at the time, was born in 1933, two years after the fact, but remembered hearing stories about the fossil. He was able to take me to the site just west of town. At the time of the discovery, the land was being used as a picnic and recreational area by the townspeople, complete with a baseball diamond and other facilities. We were able to match the pictures from 1931 with the place he took me to. As with many other specimens George Sternberg collected, he had neglected to record good locality information, but in this rare case, his photographs nailed it down (Everhart, 2007).

Although the remains of UNSM 1195 are nearly contemporaneous (middle Cenomanian) with two large specimens of *Thalassomedon haningtoni* discovered in the Graneros Shale of Colorado (DMNH 1588) and Nebraska (UNSM 50132), the characteristics of the paddle suggest that UNSM 1195 (UNSM 50134) represents an undescribed taxon (Schumacher, pers. comm., 2004, 2016). In that regard, another specimen (FHSM VP-15576) consists of an articulated series of nine cervical vertebrae from an as yet unidentified elasmosaur collected from the Graneros Shale of Ellis County that may be similar to the Colorado and Nebraska *Thalassomedon* specimens.

A bit of history of the discovery of plesiosaurs provides a better understanding of the occurrence of these long-necked marine reptiles in the Late Cretaceous rocks of western Kansas. Although their bones and teeth had probably been picked up as curiosities much earlier, the first nearly complete plesiosaur was discovered in the Jurassic rocks of Lyme Regis, England, by Mary Anning in the winter of 1820–1821 (McGowan, 2001:23). The name 'plesiosaur' was given by the Reverend William Conybeare

(De la Beche and Conybeare, 1821) and means 'near-reptile,' a reference to the belief at the time that plesiosaurs were closer to reptiles than were the more 'fishlike' ichthyosaurs. Thus, plesiosaurs were well known to science, and several complete specimens were already in European museums by the time the American West was being settled. Some of these Jurassic plesiosaurs, such as *Plesiosaurus* and *Cryptoclidus*, had relatively long necks and small heads and would have looked very much like their Late Cretaceous elasmosaurid cousins. However, there is some evidence that many of these early plesiosaur lineages became extinct at the end of the Jurassic (Bakker, 1993), and the exact origin of the elasmosaurs is unknown. Whatever the reason, the number and diversity of plesiosaur species appears to have declined steadily from the Late Jurassic through the end of the Cretaceous.

Elasmosaurus platyurus was the first of several very long necked plesiosaurs (elasmosaurs) to be discovered (1867) and described from the late Cretaceous deposits in the Midwest, and elsewhere in North America (Carpenter, 1999; Storrs, 1999). In May of 1874, B. F. Mudge collected the fragmentary remains of a large elasmosaur from the Fort Hays Limestone of Jewell County, in north-central Kansas. Later, Mudge (1876:214) would describe the Fort Hays Member of the Niobrara Formation, saying, "Its fossils are *Inocerami*, fragments of *Haploscapha*, *Ostrea*, with occasional remains of fishes and Saurians. The vertebrates are so rare that we never wasted our time in hunting them in this stratum; still, our largest Saurian, *Brimosaurus* of Leidy, was found in it in Jewell County." The specimen was later sent to the Yale Peabody Museum (YPM 1640) and described as *Elasmosaurus nobilis* by Williston (1906).

Although *Elasmosaurus nobilis* is considered by Storrs (1999) to be 'Elasmosauridae indeterminate,' it does have some lasting value. Williston wrote, "The specimen, notwithstanding what [damage] it has suffered, is of much interest since it is the only vertebrate I have any knowledge of from the [Fort] Hays limestone" (1906:233, see also pl. IV), Almost a hundred years later, there are still very few vertebrate remains known from the early Coniacian-age Fort Hays Limestone Formation, and for that matter, very few elasmosaurs known from the Coniacian anywhere.

Later in 1874 Mudge and his assistant at the time, S. W. Williston, collected part of a very large plesiosaur (YPM 1644; *Styxosaurus snowii*) from the Smoky Hill Chalk in Logan County, Kansas. Williston, who was 23 years old at the time of the discovery, wrote, "It was the first specimen of a plesiosaur that I ever saw" (1906:232–233). This was apparently the first time Mudge had seen gastroliths in association with plesiosaur remains. Mudge noted that "in the Plesiosauri, we found another interesting feature, showing an aid to digestion similar to many living reptiles and some birds. This consisted of well-worn siliceous pebbles, from one-fourth to one-half inch in diameter. They were the more curious, as we never found such pebbles in the chalk or shales of the Niobrara" (1877:286). In his discussion of *Elasmosaurus platyurus*, Williston indicated that "it was with a specimen of an elasmosaur (*'Elasmosaurus' snowii*) that Mudge first noticed the occurrence of the peculiar siliceous pebbles which he described; and it was also with another, a large species yet unnamed, from the Benton Cretaceous, that the like specimens were found described by me in 1892" (1906:226–227; see also Williston, 1893).

Eventually, elasmosaurs would be collected from Cretaceous marine deposits on almost every continent, including Europe (Mulder et al., 2003), South America (Welles, 1962), Australia (Long, 1998), Japan (Nakaya, 1989), New Zealand (Hector, 1874), and even Antarctica (Chatterjee and Zinsmeister, 1982:66; Gasparini et al., 2003; Martin et al., 2007; Thompson, Martin, and Reguero, 2007; O'Gorman et al., 2014). The localities where most plesiosaur fossils have occurred seem to indicate that they preferred cooler waters of the higher latitudes over those of warmer equatorial climates. In North America, they occur more often in the Cretaceous of Canada (Nicholls, 1988; Russell, 1993; Sato, 2003; Druckenmiller and Russell, 2006; Kubo, Mitchell, and Henderson, 2012) than in the central United States, and they are extremely rare in the Gulf states.

Plesiosaurs are fascinating creatures, for both what we know and what we don't know about them. By the early part of the Late Cretaceous, they had evolved (or been reduced) to two distinct groups that are generally referred to as short-necked (polycotylids—Chapter 8) and long-necked (elasmosaurs). The giant pliosaurids (in Kansas, *Brachauchenius lucasi* [USNM 4989] and *Megacephalosaurus eulerti* [FHSM VP-321]) became extinct earlier in the Late Cretaceous and were not a part of the fauna of the Smoky Hill Chalk. The short-necked polycotylids were initially small, generally reaching lengths of no more than 3 m (10 ft) during the deposition of the chalk. A much larger specimen of *Dolichorhynchops* from the Campanian Pierre Shale in South Dakota now at the University of Kansas (KUVP 40001; see Carpenter, 1996; O'Keefe, 2008) suggests that they may have been as large as 6–7 m (20–23 ft). They had long, narrow jaws filled with slender, seizing teeth (Massare, 1987) that were apparently well adapted for catching small prey. It is likely that they were capable of swimming rapidly in pursuit of small fishes and cephalopods that were their preferred prey (Adams, 1997).

Elasmosaurs, on the other hand, had shorter, more compact heads set on extremely long necks. They reached lengths of 14 m (45 ft) or so, and weighed several tons as adults. Based on work by Alexander (1989) in determining the weight of dinosaurs by the amount of water that is displaced by an accurate model, my estimate of the weight of a 12 m (40 ft) long elasmosaur is about 6,600 kg (more than 7 tons). That is certainly much less than a modern whale would weigh (a grey whale of that length weighs about 31,000 kg, or 34 tons), but it would have been a very large animal nevertheless. Among other physical limitations, their size, weight, and limb structure meant they could not have crawled ashore to rest or to lay eggs. Also, due to their unusual body plan, they were probably slow moving and not much of a threat to anything larger than a small fish.

For various reasons, plesiosaurs, and elasmosaurs in particular, seem to have captured the imagination of almost anyone interested in ancient creatures. In spite of their extinction more than 65 Ma, their mystique lives on in Nessie (the Loch Ness monster), Champ (something seen in Lake Champlain), other sea and lake 'serpents' from around the world, basking shark carcasses netted by Japanese fishing boats, and other unexplained sightings (O'Neill, 1999). All that actually remains of these plesiosaurs, however, are their bones. In this chapter, I will try to put some flesh on these old bones.

Williston credits the Reverend William Buckland (1784–1856) with describing a plesiosaur as "a snake drawn through the shell of a turtle" (1914:77), although the exact source of this quote appears to have been lost (Ellis, 2003:119), or at least misplaced. Over a space of several years, I asked various plesiosaur workers on the Internet about the quote. Although we were unable to verify the source for certain, it appears that Gideon Mantell was the first author to describe a plesiosaur in print, stating "that the *Plesiosaurus* might be compared to a serpent threaded through the shell of a turtle" (1854:715). However, Mantell notes that the description was "a simile from an eloquent professor." This left us with two possibilities: that Mantell was referring to his friend Buckland or that Mantell was referring to himself as the "eloquent professor."

The body of elasmosaurs, like other plesiosaurs, was almost totally enclosed by long ribs along the top and sides, and by large, flat pectoral and pelvic girdles (Fig. 7.7) connected by a continuous basket of gastralia (belly ribs) across the chest and abdomen. Four winglike paddles projected at nearly right angles, and a long neck was attached to the front of a teardrop-shaped body. Instead of lengthening the neck by increasing the length of the vertebrae as is the case in the giraffe, elasmosaurs simply (and greatly!) increased the number of vertebrae.

In his revised description of *Elasmosaurus*, Cope (1870:49) had originally counted 68½ cervicals, but remarked that some were likely to be missing (note that he had reported it as a half 'caudal vertebra' in the earlier printing). Other paleontologists since Cope have reported that the neck had more than 70 vertebrae (Leidy, 1870; Williston, 1906; Welles, 1952; Sachs, 2005). However, no one else mentioned Cope's 'half' vertebra. What happened to it?

When I examined the type specimen in 2002, I noticed that there was a fragment of a cervical vertebra mixed in with other bone scraps. While all the other vertebrae have been arranged and numbered at least three times (Cope, 1869; Welles, 1952; and most recently Sachs, pers. comm., 2003; Fig. 7.1A), the cervical fragment was unmarked and had apparently never officially been counted with the rest after Cope. I mentioned the half cervical vertebra in my paper (Everhart, 2005a) on the possible additional remains of *Elasmosaurus*. Apparently, it went pretty much unnoticed there, too, except by a German friend of mine, Sven Sachs. After some discussion, we then included the overlooked fragment in a complete redescription of the neck (Sachs, Kear, and Everhart, 2013). So it took nearly 145 years for

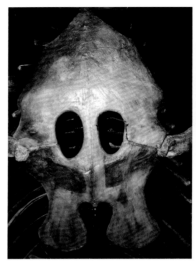 

7.7. Ventral views of the pectoral (left) and pelvic (center) girdles of *Elasmosaurus platyurus* Cope (ANSP 10081) in the exhibit at the Fort Wallace Museum, Wallace, Kansas. This exhibit is a copy of the original display at the Academy of Natural Sciences of Philadelphia by Triebold Paleontology. The girdles were reconstructed from Cope's (1869) original drawings.

Cope's original half cervical vertebra (Fig. 7.8) to officially become part of the type specimen of *Elasmosaurus platyurus* (ANSP 10081).

Other species of elasmosaurs, such as *Thalassomedon*, had fewer vertebrae in the neck, but their cervicals tended to be proportionately longer. In all elasmosaurs that are known from relatively complete specimens, the length of the neck is usually equal to or slightly longer than half the total length of the animal.

Since the discovery of plesiosaurs and elasmosaurs, this unusually long neck has produced a number of incorrect descriptions of its 'snake-like' flexibility. Descriptions and figures in earlier European publications often showed plesiosaurs on the surface of the ocean with their heads raised high in the air. These caricatures had an obvious influence on how North American plesiosaurs would be visualized.

E. D. Cope was certainly among the most influential when he described the appearance of swimming plesiosaurs:

> Far out on the expanse of this ancient sea might be seen a huge, snake-like form which rose above the surface and stood erect, with tapering throat and arrow-shaped head, or swayed about describing a circle of twenty feet radius above the water. Then plunging into the depths, naught would be visible but the foam caused by the disappearing mass of the life. Should several have appeared together, we can easily imagine tall, flexible forms rising to a height of the masts of a fishing-fleet, or like snakes, twisting and knotting themselves together. This extraordinary neck—for such it was—rose from a body of elephantine proportions, and a tail of the serpent-pattern balanced it behind. (1872:320)

(See also Webb, 1872, for another, illustrated, version of Cope's reconstruction of *Elasmosaurus*; Fig. 13.1.) Later, Cope added that the "snake-like neck was raised high in the air, or depressed at the will of the animal, now arched swanlike, preparatory to a plunge after a fish, now stretched in repose on the water, or deflexed in exploring the depths below" (1875:44).

7.8. The 'missing' half vertebra from the cervical series in *Elasmosaurus platyurus* (ANSP 10081). Based on size, the broken vertebra probably fits between cervicals C-14 and C-15 as numbered by Sachs (2005). Scale bar = 10 cm (4 in).

Cope's hand-drawn sketch of two plesiosaurs with necks intertwined appears in Osborn (1931:fig. 17).

Cope's mentor, Joseph Leidy, considered by some (Warren, 1998) as the "last man who knew everything," also weighed in on the appearance of elasmosaurs and got it wrong in his description of the elasmosaur *Discosaurus*. Leidy wrote: "We may imagine this extraordinary creature, with its turtle-like body, paddling about, at one moment darting its head a distance of upwards of twenty feet into the depths of the sea after its fish prey, at another into the air after some feathered or other winged reptile, or perhaps, when near shore, even reaching so far as to seize by the throat some biped dinosaur" (1870:10).

Many paleo-artists, including Charles Knight, have drawn popular but anatomically impossible representations of these fanciful (and fictional) elasmosaurs. These ideas still persist in spite of publications by Williston (1914:91), Shuler (1950:24–31), Storrs (1993:74), and others that have shown that the cervical vertebrae of elasmosaurs had limited movement up and down and only slightly more in a horizontal plane. For a contrasting opinion, however, see Welles (1943:147–152). Some of the earlier work on flexibility of the neck in elasmosaurs assumed a 1 cm space between the vertebrae (ibid.) for cartilage. Sato (2003) suggested a 0.5 cm thickness for the intervertebral disks in a newly discovered elasmosaur from Canada (*Terminonatator ponteixensis*). An inspection of these vertebrae in articulated specimens, however, indicates that there was probably little, if any, space between them in life. The close fit, the relatively flat surface of the articulation between the vertebrae, and the lack of space between the neural spines and transverse processes on adjoining vertebrae probably meant that there was little opportunity for movement between the individual vertebrae. Several good reasons for this limited flexibility are discussed below.

Another form of this misconception is the view of elasmosaurs shown cruising along on the surface with head and neck arched, swanlike, high above the water as they search for prey. Two issues present themselves. First, since the eyes of a plesiosaur are located on top of the skull and are generally directed upward, the plesiosaur would almost have to turn its head upside down to look down at the water in search of prey from above. In addition, their eyes were more than likely adapted for underwater vision and may have been unable to focus on prey from above the surface of the water.

A second problem is even more challenging. Unless the laws of physics were suspended on behalf of these extinct creatures, it would have been physically impossible for them to lift much more than their head above water to breathe. To prove this for yourself, try lifting a heavy pole from one end while floating in water and not touching the bottom. As you try to raise the object, your legs and lower body will also rotate toward the surface to counterbalance any weight above the water. In elasmosaurs, the weight of the long neck (as much as a ton or more in a full-grown adult) dictated that the center of gravity was just behind the front flippers. Holding the head and entire neck above the surface could never happen unless the rest of the plesiosaur's body were sitting on a firm bottom in shallow water (a possibly fatal situation for the elasmosaur). Even then, the limited movement between the vertebrae, the limited musculature of the neck, and the sheer weight of the long neck would greatly limit the height to which the head could have been raised (Henderson, 2006).

Still another problem that becomes evident regarding the long neck of elasmosaurs is what happens when the head is moved to one side or the other while the animal is moving. Located 3–6 m (10–20 ft) or more in front of the anterior set of paddles, the head (and neck) would act as a highly effective rudder whenever it was moved away from the central axis of the animal as it swam. In other words, moving the head to the right while swimming would cause the whole body of the plesiosaur to turn in that direction. While it is certainly possible that plesiosaurs used their head as a rudder to change directions while swimming, it would also be nearly impossible for them to swim in a straight line with their heads darting from side to side, or up and down, in search of prey as shown in many paleo-life reconstructions. In any case, it seems likely that the tiny brain of an elasmosaur would have been busy trying to maintain the position of the head even during stationary feeding activities through fine movements of the two pairs of large paddles located up to 8–11 m (25–35 ft) to the rear. No disrespect for the mental capacity of these creatures is intended here; plesiosaurs, and elasmosaurs in particular, were very successful marine predators for millions of years, even if we brainy humans don't yet understand quite how they did it.

The length of the neck and the number of cervical vertebrae in elasmosaurs appears to have increased through time, conveying some adaptive advantage for this group that we do not fully understand. *Thalassomedon haningtoni* (Welles, 1943) from the late Cretaceous (Cenomanian) Graneros Shale of Colorado and Nebraska, *Elasmosaurus platyurus* Cope 1868 from the Campanian Sharon Springs Member of the Pierre Shale of Kansas, and *Albertonectes vanderveldei* (Kubo, Mitchell, and Henderson, 2012), a recent discovery from the upper Bearpaw Formation in Canada, are the longest elasmosaurs presently known. All three probably grew to 14 m (45 ft) or more in length. The necks of these three species are about the same length, but *Elasmosaurus* has about ten more cervical vertebrae than *Thalassomedon* (Welles, 1943). Elasmosaurus and *Albertonectes* appear to have about the same number of cervical vertebrae, but a lot depends on defining exactly where the cervicals transition into pectoral vertebrae. No matter where that point is, the number of cervicals is more than 70. Unlike long-necked mammals and birds that evolved longer vertebrae (usually seven) through time, the neck of elasmosaurs became longer by adding more vertebrae (as many as 73 in *Elasmosaurus* and 76 in *Albertonectes*).

While the other large marine predators of the Late Cretaceous, including the short-necked polycotylids, snakelike mosasaurs, giant fishes such as *Xiphactinus*, and huge ginsu sharks (*Cretoxyrhina mantelli*), used speed to chase down or ambush their prey, it is more likely that elasmosaurs used stealth to stalk the schools of small fishes that were their primary food source. A long neck would allow the large body of the plesiosaur to remain concealed in the darkness far below a school of fishes while the small head moved slowly at the end of its long neck to stealthily approach the unlucky victims from below.

That they had their eyes on top of their heads, directed generally upward, as is the case in most elasmosaurs, seems to support this method of feeding. The eyes were not unusually large but were placed so that they may have been capable of stereoscopic vision, something that would be useful for accurately locating small prey (Shuler, 1950:8). Because of limited movement capability in the neck, it is likely that the head was moved fairly close to the prey before striking. Precise and rapid coordination between the eyes and the distant paddles of the elasmosaur would have been essential for successful hunting of highly mobile fishes and other marine prey. Approaching from below may have also been used as a strategy to silhouette the prey of an elasmosaur against the sunlit surface while using the darker, deeper waters for its own concealment (Fig. 7.9).

We know very little about the other senses of elasmosaurs. Cruikshank, Small, and Taylor (1991) suggested that the nostrils of plesiosaurs were well positioned to passively direct a continuous flow of water through the nasal cavity, where the senses of taste and smell may have been useful in detecting prey. Williston (1914:88) notes that the inner ear (semicircular canals) of plesiosaurs was large and well developed. This area of the brain maintains equilibrium and coordinates muscular control, and its relatively large size implies that plesiosaurs were probably quite graceful in their movements, although not necessarily fast. While there is no evidence of external ears in plesiosaurs, they may also have evolved a sensory system to locate nearby prey from the vibrations generated by their movement. It appears likely to me, however, that elasmosaurs relied primarily on their eyesight to find food.

Evidence from stomach contents preserved in plesiosaur remains indicates that the prey they ate was limited to small fishes and invertebrates. Earlier plesiosaurs from the Jurassic and early Cretaceous appear to have preferred squid and other cephalopods, while the later and longer-necked elasmosaurs fed on small fishes such as *Enchodus* (Cicimurri and

Everhart, 2001). While most elasmosaurs were huge by any measure, their heads and their prey remained relatively small in relation to the size of their bodies. A 12 m (40 ft) elasmosaur had a head length of about 0.5 m (20 in), or about 4 percent of the total length of the animal. Unlike the kinetic (flexible) skulls of most mosasaurs (Russell, 1967), the skulls of plesiosaurs were rigidly constructed (Fig. 7.10). This means that the cross-sectional size of the prey they could swallow was limited by the distance between the fixed hinge points (quadrates) of their jaws at the back of the skull. Even in the largest of the elasmosaurs, like *Thalassomedon*, this distance was no more than about 18 cm (7 in), and the prey they fed on was probably no longer than about 0.45 m (18 in) in length (Shuler, 1950; Cicimurri and Everhart, 2001). Their long, slender teeth also argue for trapping small prey. In all cases, elasmosaur teeth appear to have been used to seize small fishes or invertebrates, and not to tear flesh or otherwise dismember larger prey. Once captured, the prey would have to be swallowed whole.

Because of their unusual body plan, elasmosaurs were most likely slow-moving animals, using their paddles as wings or foils to generate the lift/

7.10. The reconstructed skull of *Styxosaurus snowii* (SDSMT 451) at the South Dakota School of Mines and Technology. The rigid construction of the skull and jaws in elasmosaurs limits the size of prey that could be swallowed whole. This specimen was collected from the Pierre Shale of western South Dakota in 1945.

thrust necessary to 'fly' through the water. This type of swimming is more efficient than the rowing or paddling (e.g., like a duck) methods that are sometimes depicted in reconstructions. The front and rear limbs of plesiosaurs are similarly constructed to function as winglike hydrofoils. Whether both were used in that manner is still open to question. While modern underwater 'fliers' such as penguins use only their front limbs, many researchers now believe that plesiosaurs used a coordinated movement of both pairs of limbs. Others think the rear limbs did not have either the range of movement or the musculature necessary to effectively generate the lifting forces necessary for this method of swimming (Lingham-Soliar, 2000). Since the hind limbs were located well behind the center of gravity, they may have been more effectively used for steering. In that case, the rear limbs may have been rudders and/or stabilizers, especially to maintain the position of the body and position the head and neck while feeding.

In elasmosaurs, both pairs of limbs are about equal in size and are highly modified into a rigid paddle (Fig. 7.11). The upper bone of the paddle, called the propodial (humerus/femur) is attached to the body at the shoulder and hip. The lower limb bones (radius/ulna and tibia/fibula) are reduced to flat, polygonal shapes that interlock with the smaller but similar-shaped bones of the wrist and ankle. The finger and toe bones (phalanges) are hourglass-shaped. Bones of the adjoining fingers fitted closely together and were probably held tightly within a sheath of muscles, tendons, and ligaments. Although plesiosaurs had five 'fingers,' they exhibited a condition called hyperphalangy, which means they have many more individual finger bones than usual (humans have 3 bones in each

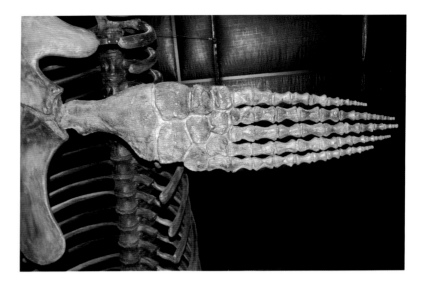

7.11. Reconstructed left front limb of *Elasmosaurus platyurus* cast at the Fort Wallace Museum, Wallace, Kansas. Approximately 1.5 m (5 ft) in length.

finger; some plesiosaurs had as many as 24). The effect of the compacted lower arm/leg and wrist/ankle structures and the extra finger/toe bones, was to modify what had originally been a walking limb into a long, tapering paddle that had limited flexibility. In cross-section, the limb was also thicker on the leading edge than the trailing edge. It had evolved into a wing or hydrofoil for flying through the water. Movement of the paddles in a manner similar to that of a bird's wing in flight created lift and pulled the animal through the water. Unlike in birds, however, the wrist was effectively 'locked' and the paddles could not be folded or otherwise flexed far. At rest, a plesiosaur's limbs probably were held at right angles to the body. Extensive computer simulations of plesiosaur swimming by Liu et al. (2015; see also Carpenter, et al., 2010) concluded that they mainly used their forelimbs to propel themselves like modern turtles and penguins.

Another peculiarity that has been noted among long-necked plesiosaurs since their discovery are masses of gastroliths (stomach stones) that are commonly associated with their remains (Whittle and Everhart, 2000; Everhart, 2000; Cicimurri and Everhart, 2001; Everhart, 2004b; Thompson, Martin, and Reguero, 2007; Kubo, Mitchell, and Henderson, 2012; O'Gorman et al., 2014). These rounded and usually highly polished rocks are normally located inside the abdomen of complete specimens (Figs. 7.3 and 7.6). There have been as many as several hundred stones of various sizes, generally from hen-egg-size down to pea-size or smaller, reported to be associated with plesiosaur remains. In the case of a recently discovered specimen from Antarctica, there were thousands of small quartzite pebbles 1 cm (0.1 in) or less in the stomach area (James Martin, pers. comm., 1999; Thompson, Martin, and Reguero, 2007). Typically, most gastroliths are siliceous rocks such as quartz, quartzite, or chert. Occasionally they are granite, basalt, or other igneous rocks. Until recently, the largest gastroliths known were about as big as a tennis ball, but in two Kansas specimens collected in the early 1990s, they were much larger. Everhart (2000) noted that the largest stones in a University of Kansas specimen (KUVP 129744) weighed about 1.4 kg (3 lb) each and were nearly 17 cm (7 in) in maximum diameter (roughly the size of a softball). These are the largest individual sizes and the greatest total weight of gastroliths ever documented from a plesiosaur—or, for that matter, from any animal (Fig. 7.3).

The total weight of the stomach stones from this plesiosaur was about 13.1 kg (29 lb), but considering the weight of the 14 m (45.9 ft) plesiosaur that carried them (estimated to weigh 6,600 kg [14,500 lb]), their relative weight is insignificant. In fact, the weight of the gastroliths is probably much less than the daily food intake of the plesiosaur. The gastroliths from the recently named new genus and species of an elasmosaur from Canada, *Albertonectes vanderveldei* (Kubo, Mitchell, and Henderson, 2012), were also fairly large, but had an estimated mass of only 9 kg (20 lb). More commonly, however, the total weight of gastroliths documented from individual plesiosaurs is less than 5 kg (10 lb) (Everhart, 2000; Thompson, Martin, and Reguero, 2007). It appears highly unlikely that these stones would have been effective in changing the buoyancy of large animals that probably had to consume large quantities of food daily in order to maintain their elevated body temperature and sustain themselves while living in mid-ocean. Even buoyancy changes caused by inhalation and exhalation through a 6 m (20 ft) long trachea would have been more than could have been offset by the small amount of weight represented by the gastroliths.

This leads us to the question of what gastroliths were actually used for. While early fossil collectors such as Benjamin Mudge (1877:268) concluded from plesiosaur remains he had seen in Kansas in the 1870s that the stomach stones were used as "an aid to digestion similar to many living reptiles and birds," Samuel Williston (1893) and others initially believed that gastroliths were swallowed for other reasons, such as the animal craving food or mistaking the stones for prey.

Many researchers believe they were used in some manner for buoyancy control or for adjusting attitude/longitudinal balance (Taylor, 1981). However, a nearly complete specimen of *Styxosaurus snowii* (NJSM 15435) discovered in the Sharon Springs Member of the Pierre Shale of western Kansas provides a rather convincing counterargument (Cicimurri and Everhart, 2001). These remains included about a hundred gastroliths, fish bones, scales, and teeth commingled in a crop or gizzard near the front of the body. Unlike the skeletal material preserved as stomach contents of mosasaurs and fishes that appear to have been partially digested (dissolved) by stomach acids, the remains of the fishes in the stomach of this elasmosaur were broken up into nearly unrecognizable pieces. Most of the fish bones appeared to have been ground up into very small (2–3 mm; 0.1 in or less) fragments. This specimen suggests that whatever use gastroliths may have had for buoyancy control, they were definitely useful in grinding up prey as an aid to digestion. Another elasmosaur specimen (UNSM 1111–002) discovered in the Mobridge Chalk Member of the Pierre Shale in northeastern Nebraska in 2002 had more than 600 gastroliths (Fig. 7.12) in its abdomen (pers. obs.; unpublished data), some of which had been reduced to the size of large sand grains (less than 4 mm [0.2 in]).

7.12. Twenty-four of the more than 600 gastroliths recovered in 2002–3 from an elasmosaur (UNSM 1111–002) in the Mobridge Chalk Member (Maastrichtian) of the Pierre Shale in northeast Nebraska. This is the largest number of gastroliths ever recovered from any plesiosaur in North America. Scale bar = 10 cm (4 in).

So where did these gastroliths come from? Egg-size quartz or other igneous rocks did not occur normally in the soft limey mud of the Cretaceous sea bottoms in Kansas. Mudge noted that "how far the Saurians wandered to collect them is a perplexing problem" (1877:286). Williston (1893:122) noted that some of the quartzite stones were similar to boulders he had seen in northwestern Iowa, or in the Black Hills of South Dakota. Each of these potential sources is 400–500 miles from the localities where the plesiosaur remains that contained them were collected in western Kansas. Cobbles associated with a recently discovered Nebraska elasmosaur specimen (UNSM 1111–02) may have come from nearby granite exposures in southeastern South Dakota, or further south from the Ouachita Mountains in western Arkansas. Wherever their origin, they certainly did not come from the soft mud bottom of the Western Interior Sea that covered Kansas or Nebraska at that time. This in itself implies that some plesiosaurs traveled long distances to the sources of these stones. That the stones are always rounded and smooth suggests that they were eventually worn down to sizes small enough to pass though from the crop or stomach into the gut, and that they had to be replaced periodically throughout the life of the plesiosaur.

One feature I have observed on gastroliths from a number of elasmosaur specimens from Texas to South Dakota is that the gray or black chert stones are covered with numerous tiny arc-shaped surface markings. The age of the elasmosaurs in which they were preserved ranges from the Cenomanian through the Upper Maastrichtian, a time span of more than 30 million years. Closer examination indicates that these markings were the result of small (2–5 mm) conchoidal (bowl-shaped) fractures of the chert (Fig. 7.13). The fractures generally cover the surface of the stone, exhibit varying degrees of wear, and often cross other fractures. Similar fractures can occur naturally due to stone-on-stone impacts in river or beach gravels, but they do not occur on non-gastroliths in the numbers

7.13. Markings (conchoidal fracture scars) are seen frequently on elasmosaur gastroliths. These markings are probably the result of the crushing of the stones against each other in the muscular crop or stomach of the plesiosaur. Scale bar in mm.

observed on the gastroliths that I have examined. The more frequent occurrence of conchoidal fractures on the edges of angular-shaped stones in an experiment using a rock tumbler indicates that such damage is an important part of the mechanism for rounding and smoothing gastroliths.

Gastroliths with these same arc-shaped markings were discovered within a plesiosaur specimen (Cicimurri and Everhart, 2001) from Kansas in intimate association with the finely comminuted bones of small fishes. These markings suggest that the conchoidal fractures occurred as the stones were ground against one another by peristaltic contractions within the plesiosaur's digestive tract and most likely within a crop- or gizzard-like structure ahead of the stomach. I believe that the more frequent occurrence of these markings on chert gastroliths in plesiosaurs compared to those seen on similar stones from river and shore deposits is evidence of their continuous use in processing food (Everhart, 2005b).

Similar tractures were observed on smooth black stones scattered about in the lower Kiowa Formation (Early Cretaceous, Albian) discovered in Kiowa County, Kansas. These probable gastroliths were not associated with skeletal material, but the most common vertebrate remains at the same stratigraphic level are the bones of plesiosaurs (Everhart, 2005b). A plesiosaur specimen ('*Plesiosaurus mudgei*' Cragin; KUVP 1305) figured by Williston (1903;plate XXIX; Fig. 7.14) from the Kiowa Formation included more than 200 gastroliths. Unfortunately, I was not able to relocate this specimen.

A final subject is the issue of plesiosaurs crawling ashore to lay their eggs. This method of reproduction has been long assumed because they are reptiles, or possibly because they appear superficially similar to marine turtles. Consequently, shore-dwelling plesiosaurs have been

7.14. Photograph of the remains of *'Plesiosaurus mudgei'* and more than 200 gastroliths from Williston (1903:pl. XXIX). Williston noted that "all the pebbles were dark in color and none were quartzite" (ibid.:76).

"Stomach Pebbles" of Plesiosaur.

the subject of imaginative works by paleo-artists for more than 150 years. While many modern reptiles do lay eggs, it is highly unlikely that the larger marine reptiles of the Mesozoic could have reproduced in that manner. Turtles (Chapter 6) appear to be the only major group of marine reptiles that still come ashore to lay their eggs. Unlike turtle paddles, the limbs of plesiosaurs were so modified for use in underwater flying that they would have been too rigid to be used effectively for movement on land, much less for scooping out a nest for eggs in beach sand. In addition, the length and weight of the long neck of an elasmosaur would have effectively counterbalanced the back half of the animal, lifting it off the ground and probably making the rear flippers useless for movement on land, or anything else, under the best of circumstances.

Ichthyosaurs have long been known to have given live birth to their young, as evidenced by well-preserved remains in the Jurassic shales of Germany. Live birth in the other marine reptiles (mosasaurs and plesiosaurs) has not been so easy to prove. A mosasaur (*Plioplatecarpus*) specimen collected from the Pierre Shale of South Dakota included the remains of several young of the same species in the pelvic region (Bell et al., 1996). More recently, Caldwell and Lee (2001) reported the remains of at least four embryos preserved inside the abdomen of a mosasauroid called an aigialosaur. There is also at least one specimen of a short-necked polycotylid associated with presumed fetal material in the abdomen (Rothschild and Martin, 1993; pers. obs.; O'Keefe and Chiappe, 2011; see Chapter 8). It appears likely that live birth is one of the necessary adaptations that marine reptiles had to make in order to successfully return to life in the ocean.

If so, this raises additional questions as to how marine reptiles nurtured a fetus inside the mother's body for a relatively long period of gestation. Did they simply retain the developing embryos and yolk sacs inside the mother's body until they were ready to hatch, as do some modern snakes? Or had they developed some other method, such as a placenta, for nurturing the growth of plesiosaur embryos? We may never know. Plesiosaurs were highly successful animals that clearly had to have evolved some efficient method for reproduction in the marine environment.

Giving live birth to their young probably also meant that plesiosaurs provided some form of parental care, or at least some minimal form of protection. Turtles and other reptiles that lay many eggs at one time depend on safety in numbers for the survival of a few of their young. In most cases, ichthyosaurs and mosasaurs appear to have given birth to six or fewer young. The remains of the smallest *Tylosaurus* 'babies' that have been collected in the Smoky Hill Chalk are about 2 m in length (Everhart, 2002). Modern animals that invest their energy in birthing a few larger babies, such as mammals, generally provide some form of protection for those young as a means of improving their chances of surviving to adulthood. Like some dinosaurs, plesiosaurs may have traveled together in small groups for that purpose. Otherwise, the survival rate for any young animals in the middle of an ocean populated with giant predatory fishes like *Xiphactinus*, great white shark–size ginsu sharks, and 10 m (30 ft) long hungry mosasaurs would have been close to zero.

Plesiosaurs may have been driven to the edge of extinction by the 'mosasaur explosion' during the late Cretaceous. The greatest number of Late Cretaceous plesiosaur fossils in Kansas occurs in the early middle Turonian (Bruce Schumacher, pers. comm., 2003; Schumacher and Everhart, 2005). This is roughly the same time that mosasaurs make their first appearance (Martin and Stewart, 1977; Schumacher, 2011; Everhart and Pearson, 2014) in the Western Interior Sea. As mosasaurs became more numerous and larger during the Coniacian, Santonian, and early Campanian time, plesiosaurs (as judged by the number of their remains discovered to date) essentially disappeared from the middle of the Interior Sea. Their preferred habitat may have been closer to shore, or they, especially their young, may have been the prey of large mosasaurs such as *Tylosaurus*. One juvenile polycotylid specimen from Kansas was discovered as stomach contents of a large *Tylosaurus* (Sternberg, 1922; Everhart, 2004a). We do know from the bite marks on plesiosaur bones (Fig. 4.2) that their carcasses were scavenged by sharks and that the more easily detachable parts, such a limbs, tails, heads, and necks, were often carried away (Everhart, 2003; Everhart, 2005c).

For all their size and amazing adaptations to life in the ocean, it appears likely that plesiosaurs were an evolutionary dead end and were already on their way out by the end of the Cretaceous. Their greatest diversity and numbers probably occurred during the Jurassic or Early Cretaceous. By early Maastrichtian time (72 Ma) there were only a few species of elasmosaurs and short-necked polycotylids left in Earth's oceans. Like the ichthyosaur 'fish-lizards' that became extinct in the early part of the Late Cretaceous (Lingham-Soliar, 2003), plesiosaurs were probably losers of an evolutionary arms race that saw the rise of larger, faster teleost fishes as competitors for the same prey during the Early Cretaceous and the explosive entry of the highly adaptable mosasaurs, who were both competitors and predators, in the Late Cretaceous.

Generalized Geologic Map of Kansas

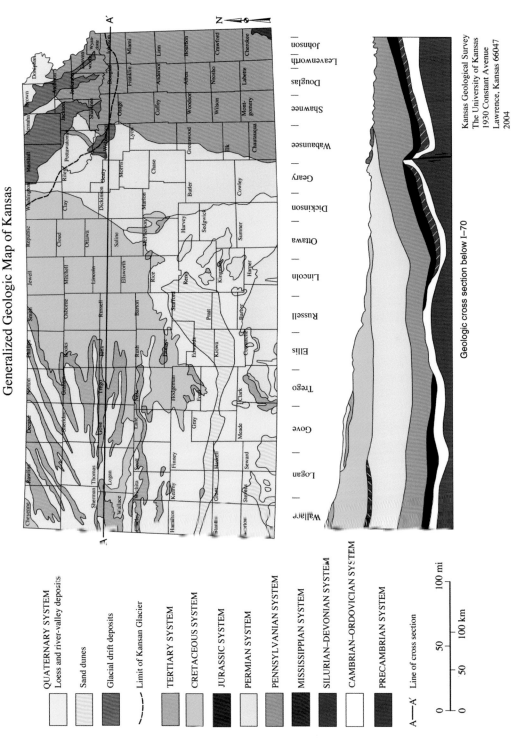

QUATERNARY SYSTEM
Loess and river-valley deposits

Sand dunes

Glacial drift deposits

Limit of Kansan Glacier

TERTIARY SYSTEM

CRETACEOUS SYSTEM

JURASSIC SYSTEM

PERMIAN SYSTEM

PENNSYLVANIAN SYSTEM

MISSISSIPPIAN SYSTEM

SILURIAN–DEVONIAN SYSTEM

CAMBRIAN–ORDOVICIAN SYSTEM

PRECAMBRIAN SYSTEM

A——A'   Line of cross section

0   50   100 mi
0   50   100 km

Geologic cross section below I–70

Kansas Geological Survey
The University of Kansas
1930 Constant Avenue
Lawrence, Kansas 66047
2004

Plate 1. Generalized surface and cross-section geologic maps of Kansas. Courtesy of the Kansas Geological Survey, Lawrence, Kansas.

Plate 2. The hunter becomes the hunted, as a giant pliosaur called *Megacephalosaurus eulerti* attacks an early mosasaur. Although pliosaurs became extinct during Turonian time, they were still present in the Western Interior Sea when the first mosasaurs arrived. Adult mosasaurs were top predators, but their young were preyed on by sharks and other species of mosasaurs. Life could be short for the unwary. Painting © Dan Varner.

Plate 3. A lone polycotylid plesiosaur (*Trinacromerum osborni*) cruises in the shallow water of the Western Interior Sea during Turonian time as a group of ammonites and a much smaller *Ptychodus* shark move out of its path. Painting © by Dan Varner.

Plate 4. Squid, such as this early relative of *Tusoteuthis longa*, were often the favored prey of plesiosaurs, including the pliosaur *Brachauchenius lucasi*. This large predator was one of the last pliosaurs and made its final appearance in the waters over Kansas during the deposition of the Blue Hill Shale Member (upper Middle Turonian) of the Carlile Formation. Painting © by Dan Varner.

Plate 5. During the last few minutes of its life, a giant *Xiphactinus audax* tries to swallow its prey, a smaller but still struggling *Gillicus*. Then, for unknown reasons, the *Xiphactinus* died before the smaller fish could be digested. The fossilized remains were collected by George F. Sternberg from the Smoky Hill Chalk in 1952, and are currently displayed in the Sternberg Museum of Natural History as the famous fish-within-a-fish specimen. Painting © by Dan Varner.

Plate 6. A giant ginsu shark (*Cretoxyrhina mantelli*) at the moment of impact in a high-speed attack on a subadult mosasaur (*Tylosaurus kansasensis*). While we are unsure whether these sharks attacked living mosasaurs or simply scavenged dead ones (or both), the partially digested bones of mosasaurs are frequently discovered in the Smoky Hill Chalk and sometimes include the broken tips of ginsu shark teeth. Painting © by Dan Varner.

Plate 7. In this painting, a *Clidastes propython* (one of the smaller mosasaurs, about 3.6 m (12 ft) long, is about to put the bite on a spiky marine turtle called *Calcarichelys* somewhere off the ancient Gulf coast of what is now Alabama. While mosasaurs fed primarily on fishes, some of them ate anything small enough to swallow. And sometimes they made mistakes. Painting © by Dan Varner.

Plate 8. It is hard to imagine the scale of this painting by Varner. The swimming birds (*Hesperornis regalis*) are about 1.5 m (5 ft) in length and the *Tylosaurus* is, well, huge. Modeled after one of the largest specimens on exhibit, the 'Bunker Tylosaur' (KUVP 5033) at the University of Kansas Museum of Natural History was at least 12 m (40 ft) long, including a massive skull that was nearly 1.8 m (6 ft) in length. Painting © by Dan Varner.

Plate 9. This painting is a recreation of the attack of large *Tylosaurus proriger* on a much smaller *Clidastes propython*. The largest mosasaur in the Western Interior Sea, *Tylosaurus* occasionally killed and ate other species of mosasaurs, as shown by preserved gut contents discovered in the Pierre Shale of South Dakota. Painting © by Dan Varner.

© Dan Varner

Plate 10. Two *Globidens dakotensis* mosasaurs are shown feeding on clams and other shellfish that lived on the bottom of the shallow sea that covered South Dakota and much of the middle of North America during Campanian time. *Globidens* was a very specialized mosasaur with round, ball-shaped teeth and a short, heavily built skull. Painting © by Dan Varner.

Plate 11. A giant *Mosasaurus hoffmanni* just misses in an attack on a marine crocodile (*Thoracosaurus*) in the Late Cretaceous seas that covered present-day New Jersey. These Maastrichtian-age mosasaurs are known from North America and Europe. The much smaller North Atlantic Ocean of about 68 million years ago did not present a barrier to these and other mosasaurs. Although first discovered in the Netherlands, *M. hoffmanni* has now been identified from many Late Cretaceous localities, including Texas. *Thoracosaurus* survived for some time after the K/T (K/Pg) event that drove mosasaurs, plesiosaurs, dinosaurs, and many other groups of animals to extinction. Painting © by Dan Varner.

Plate 12. *Mosasaurus hobetsuensis*, a close relative of *M. hoffmanni*, cruises the underwater shoreline of the Japanese archipelago, looking for prey. During the Late Cretaceous, the western rim of the Pacific Ocean, from Japan to New Zealand, was inhabited by many of the same genera of marine reptiles (mosasaurs and plesiosaurs) as were living in the Western Interior Sea. Painting © by Dan Varner.

Pliosaurs and Polycotylids

8

Just below the surface, the pod of four adult female and three smaller juvenile short-necked plesiosaurs moved steadily eastward as they migrated across the expanse of open sea toward shallower coastal waters. Their narrow, toothy heads were held stiffly in front of them as they moved their long paddles up and down rapidly like the wings of a bird. With little apparent effort, they were flying through the clear, warm water. At regular intervals, almost in unison, they would break through the surface and come completely out of the water as they quickly exhaled and inhaled. Their smooth, scaleless skin, blue-black on top and a pale cream color underneath, was briefly visible when they were above the surface.

The bright midday sun illuminated the water around them but did not penetrate far into the depths. There was only a gentle breeze, and the surface of the dark blue waters around them was almost calm. Normally, they would have avoided the deeper waters near the center of the sea, preferring the relative safety nearer to shore. It was late in the season, however, and the females were about to give birth to their young. The leader of the group, an older pregnant female, was taking the most direct route across the sea to a sheltered nursery area near the mouth of a large river. The stubby but well-streamlined bodies of the polycotylids were making little splash or other noise as the animals came out of the water periodically to breathe.

That slight noise was, however, enough to draw the attention of a large tylosaur hunting nearby. His acute hearing picked up the sounds from nearly a mile away and alerted him to the passage of the short-necked polycotylids. Normally slower than these fish-eating plesiosaurs, this time the tylosaur swam rapidly with wide sweeps of his long tail at nearly right angles to the pod on a course that would intercept them. He would have only one chance to attack from below before they outdistanced him.

As the pod approached, the tylosaur surfaced, inhaled, and dived again. This time his descent took him under the path of the pod. Nearly motionless now, he looked up and watched as they approached, silhouetted against the sunlit surface. Then, using his front paddles to orient his body upward, he lashed his tail suddenly and accelerated toward the surface. His target was one of the smaller juveniles in the center of the group, and he timed his approach perfectly. Coming up underneath the polycotylid, he opened his jaws at the last moment, then closed them savagely around the head and neck of his prey. Bones crunched as the plesiosaur's lightly built skull was crushed. The momentum of the tylosaur carried his upper body and that of his prey out of the water, then he fell over sideways with a great splash. The remaining plesiosaurs scattered quickly in all directions, then reformed, moving quickly away from the scene of the ambush.

Still holding the much smaller polycotylid tightly between his jaws, the tylosaur began to position his prey so that it was pointed headfirst into his throat. Attracted to the noise of the attack and the blood of the plesiosaur that was now spreading into the water, a group of small sharks swam in wary circles around the big tylosaur. They remained at a safe distance, and the tylosaur ignored them for the moment. Satisfied with the position of his prey, the tylosaur opened his mouth wider and surged forward, lodging the narrow head of the short-necked plesiosaur into his throat. Then he began using his double-jointed lower jaws to slowly ratchet the body into his gullet. Each time the prey was moved backward, the sharp, hooked teeth on the roof of the tylosaur's mouth grabbed hold of the plesiosaur's smooth scaleless skin, allowing the lower jaw teeth to be released and moved forward. Slowly the plesiosaur's body was swallowed, much like a large rat being eaten by a snake. The tylosaur rested briefly while floating at the surface as it began to digest its large meal.

Pliosaurs and Polycotylids

Cretaceous plesiosaurs have traditionally been divided into two groups: the long-necked, small-headed variety, called elasmosaurids (i.e., *Elasmosaurus*); and the short-necked, large-headed variety, called pliosaurids. Pliosaurids such as *Pliosaurus* (Owen, 1842) and *Liopleurodon* from the Jurassic, and Early Cretaceous varieties such as *Kronosaurus* (Australia), *Brachauchenius* (Kansas / Texas / Utah), and the recently described *Megacephalosaurus* (Kansas; Schumacher, Carpenter, and Everhart, 2013), were true 'sea monsters' of their day (Ellis, 2003). Some of these were huge, with skulls as long as 3 m (9 ft). Smaller 'short-necks,' or polycotylids, from the Late Cretaceous, such as *Trinacromerum*, *Dolichorhynchops*, and *Polycotylus*, have been frequently lumped into the pliosaurid group because of the obvious resemblance of their body plan to that of their extinct cousins. Carpenter (1996), however, determined that these Late Cretaceous short-necks were actually more closely related to the longer-necked elasmosaurids than to the extinct pliosaurids, on the basis of similarities in their skulls. Assuming this determination is correct, the pliosaurid lineage became extinct for unknown reasons during the early part of the Late Cretaceous, along with the last of the ichthyosaurs.

In a review of stomach contents preserved with specimens, Cicimurri and Everhart (2001) noted that the food preferences of plesiosaurs in general appeared to change from cephalopods (squid, baculites, etc.) to bony fishes through time from the middle of the Jurassic to the end of the Cretaceous. The polycotylids, including *Trinacromerum* and *Dolichorhynchops*, may have evolved as a smaller, faster version of the elasmosaur lineage, possibly to fill the ecological niche left vacant by the extinction of the ichthyosaurs and pliosaurs.

A more plausible ending to the story at the beginning of this chapter would have been the polycotylids sensing the approach of the giant tylosaur and dodging at the last moment. Then, as now, predators probably failed in capturing prey more often than they succeeded. In either case, while the story is fiction, the fact is that a large *Tylosaurus proriger* mosasaur was discovered by Charles Sternberg in 1918 with the remains of a partially digested polycotylid within its ribcage. Did adult tylosaurs hunt and kill plesiosaurs as the story suggests? Or did they simply scavenge the bloated remains of dead plesiosaurs that happened to become available from time to time, competing with sharks as scavengers? We don't know the answer to those questions; about all we are certain of is that a juvenile polycotylid was eaten by a 9 m (29.5 ft) long *Tylosaurus proriger* and that the *Tylosaurus* died before completely digesting its meal.

In the summer of 1918, Charles Sternberg and his sons were prospecting for fossils in the Smoky Hill Chalk along Butte Creek in Logan County, Kansas. According to Sternberg:

> I was so fortunate as to find a fine tylosaur skeleton the second day in the field. There were twenty-one feet of the skeleton present in fine chalk. The complete skull was crushed laterally, nearly the complete front arches and limbs were present, as was also the pelvic bones and both femora. All the vertebrae to well into the caudal region beyond the lateral spines were continuous, with the ribs in the dorsal region. Between the ribs was a large part of a huge plesiosaur with many half-digested bones, including the large humeri, part of the coracoscapula, phalanges, vertebrae, and, strangest of all, the stomach stones, showing that this huge tylosaur, that was about twenty-nine feet long, had swallowed this plesiosaur in large enough chunks to include the stomach. How powerful the gastric juice that could dissolve these big bones! This specimen I sent to the United States National Museum. (1922:119–120)

After reading Sternberg's account in 2001, I was amazed that the intimate association of these two marine reptile specimens had never been mentioned again in the literature. I contacted Bob Purdy and Michael Brett-Surman at the Smithsonian and learned that the *Tylosaurus* (USNM 8898) was, in fact, their exhibit specimen (Fig. 8.1), and had been so since it was obtained from Charles Sternberg in 1919. A photograph of the preparation of the *Tylosaurus* exhibit was even featured in the *Scientific American* magazine (see Gilmore, 1921). The plesiosaur material (USNM 9468), however, had been stored safely away in the collection, and more or less forgotten. That is where I saw it, in the drawer where it had been gathering dust for more than 80 years. A review of the Smithsonian's curation records clearly showed the association, but otherwise the plesiosaur remains had been essentially ignored by paleontologists since their discovery.

I made arrangements to visit the collection, and early on the morning of September 11, 2001, my wife and I drove up the Kansas Turnpike

toward Kansas City for a flight to Washington, D.C. It became one of the longest days of our lives as we listened to reports of the terrorist attacks on the World Trade Center and the Pentagon on National Public Radio. We made it as far as Lawrence, Kansas, before realizing that we weren't going anywhere, and like the contrails of the commercial airliners in the clear blue sky overhead, we turned around and went home.

Later that year, in November, I visited the Smithsonian, where I was able to examine the plesiosaur specimen. Unfortunately, after reading and rereading Sternberg's glowing note, I had set my expectations a little too high. The plesiosaur remains were contained in a single drawer, and there wasn't much to see. The bones that Sternberg had discovered were badly corroded, apparently by the mosasaur's stomach acid as he had indicated. Unlike the mosasaur specimen in the exhibit, much of the chalk matrix had not been entirely removed from the bones of the little plesiosaur (Fig. 8.2). I have described the specimen more fully elsewhere (Everhart, 2004b), but it wasn't a "huge plesiosaur" as indicated by Sternberg (1922:119). Rather, the remains appeared to be those of a juvenile polycotylid, probably no more than 2–2.5 m (6–7 ft) long. Most of the plesiosaur's skeleton was missing, and it is likely that some of the bones had been

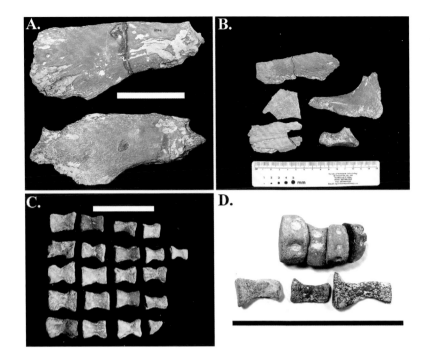

8.2. These are some of the juvenile polycotylid plesiosaur remains (USNM 9468) recovered as gut contents from the *Tylosaurus proriger* (USNM 8898) in the collection of the United States National Museum (Smithsonian) in Washington, D.C. A, two of the badly corroded upper limb bones (propodials—humeri or femora); B, partially digested scraps of plesiosaur bone including girdle and possible skull elements; C, finger bones (epipodials) from one or more of the plesiosaur's paddles; D, caudal vertebrae and caudal ribs. Scale bars = 10 cm (4 in).

removed by sharks that had scavenged the mosasaur carcass (there was a single *Squalicorax kaupi* shark tooth included in the box of plesiosaur bones). As noted by Sternberg (1922), the fact that the plesiosaur bones were located inside the heavy ribs of the mosasaur probably meant that the much smaller *Squalicorax* sharks were unable to reach them. In any case, the evidence shows that mosasaurs did eat plesiosaurs, and for that matter, probably anything else they could swallow.

Sternberg (1922) also reported stomach stones (gastroliths) that were included with the plesiosaur remains. I was dubious of his observation, but when I looked at the specimen, I saw that there were eleven small stones and two masses of what appeared to be compacted sand or a sandy conglomerate. The is very unusual because gastroliths rarely occur in association with polycotylids or in plesiosaurs for that matter. Since then, I have recovered one gastrolith from a new Fort Hays Limestone polycotylid (FHSM VP-16459) and have also learned of the discovery of a new species of polycotylid in Utah that contained at least 289 gastroliths (Schmeisser and Gillette, 2009; *Dolichorhynchops tropicensis*, Schmeisser McKean, 2012). One or a few gastroliths can be explained away, like Williston (1893:122) suggested, as accidental, but finding nearly 300 of them in a polycotylid is difficult to understand.

As a postscript to the story of the plesiosaur eaten by the *Tylosaurus*, I also talked to the staff at the National Geographic headquarters in Washington, D.C., and suggested that this would make a great documentary for TV or at least a feature article in their monthly magazine. Neither of those options happened, but what did happen was so much better. After several years of false starts, the National Geographic IMAX 3-D movie *Sea*

Monsters hit the big screen in October 2007, featuring a family of *Dolichorhynchops* plesiosaurs ('Dollies') as the central figures. I was fortunate to serve as a science advisor on the movie, along with Ken Carpenter, Glenn Storrs, and Larry Martin. One of our major challenges as scientists was to convince the CGI animators that the plesiosaurs they were creating for the movie had bony, internal skeletons and did not flex while swimming like they were made of rubber. Eventually we were happy with the results, and I was pleased recently to see that computer simulations of plesiosaur swimming by Liu et al. (2015) supported our view.

It was a thrill for me to introduce the movie in big-screen theaters in Portland, Oregon, and Galveston, Texas. As a bit of movie trivia, it is my hand holding the pen in the close-up when 'Charlie Sternberg' is drawing the mosasaur in his field notebook. I also wrote the book *Sea Monsters* (2007b) for National Geographic to accompany the movie. Almost completely filmed in Kansas, with local actors, the movie was a huge success and showed in theaters around the world.

Unlike the other kinds of marine reptiles such as ichthyosaurs, plesiosaurs, and mosasaurs that were relatively well known from European specimens by the 1860s, polycotylids (and elasmosaurs) were discovered first in Kansas. It was shortly after the discovery of *Elasmosaurus platyurus* by Dr. Theophilus Turner (Cope, 1868) that the bones of another 'new' kind of plesiosaur were collected from the upper Smoky Hill Chalk "about five miles west" of Fort Wallace (Cope, 1871:386). A railroad land agent and part-time adventurer and fossil collector named William E. Webb from Topeka (Everhart, 2016) had obtained the fragmentary remains of a plesiosaur that Cope (1869) called *Polycotylus latipinnis* (USNM 27678 *and* AMNH 1735). The locality given by Cope (and probably provided to him by the collector) is suspect because there are no exposures of "yellow cretaceous limestone" to the west of Fort Wallace. Most likely the locality was along the Smoky Hill River five or so miles to the east. Note that the specimen has two numbers because portions of it are curated in both the Smithsonian (United States National Museum) in Washington, D.C., and the American Museum of Natural History in New York City (Storrs, 1999).

In his review of science discoveries in the early days of Kansas, Peterson wrote:

> Although there were no organized scientific expeditions into the area [western Kansas] in 1868, the Kansas Pacific Railway promoted several well attended excursions to the end of the line at Sheridan near Fort Wallace [Fig. 1.7]. One small group organized as a hunting and adventure party by William E. Webb, on reaching the end of the line in September was told of a large fossil near the fort [for a fictionalized account of this expedition, see Webb, 1872; also Davidson, 2003; Everhart, 2016]. Webb, with the help of a 'professor' in the group, collected the fossil and shipped it to Cope who placed it in a new genus and described it as the "first true Plesiosauroid found in America." (1987:228)

While Cope (1871:34) did use those exact words in his publication, the fragmentary remains of other plesiosaurs had been discovered in North

America prior to *Polycotylus*. It was, however, the first specimen of an unknown (at the time) group of Late Cretaceous plesiosaurs that would be called polycotylids. A number of plesiosaur specimens, including *Polycotylus*, were also collected by O. C. Marsh and the Yale College Scientific Expeditions from the same area between 1870 and 1875, but were not reported or described at the time.

Polycotylus latipinnis was described originally from a pelvic arch and 21 vertebrae. The genus name (*Polycotylus*) refers to the deep 'cupping' of the anterior and posterior surfaces of the vertebrae, a characteristic that was quite different from other plesiosaur vertebrae seen by Cope until then. According to Carpenter (1996), however, "the holotype material is scrappy and of questionable value." That could be said of many of the first specimens from Kansas described by Cope and Marsh, but in most cases better specimens were discovered after the first hectic years of collecting in Kansas.

Williston (1906) apparently held the same opinion of the type specimen, and he redescribed *Polycotylus* from a second, more complete set of remains (YPM 1125) that was collected from the chalk of Logan County personally by O. C. Marsh in November of 1870. Considering the growing rivalry between Cope and Marsh during the 1870s, it is interesting that Marsh did nothing with this much more complete specimen and did not even mention it in his report (1871) of the fossils collected by the 1870 Yale College Scientific Expedition. Perhaps even more interesting in view of Marsh's belated criticism (Ballou, 1890) of Cope's mistake regarding the reconstruction of *Elasmosaurus platyurus*, Marsh did not publish a single paper on Cretaceous plesiosaurs during his long career. I think this is odd considering that the Yale Peabody Museum has at least 11 plesiosaur specimens collected from the chalk of western Kansas from the 1870s. *Polycotylus latipinnis* has since been redescribed from a nearly complete specimen from South Dakota (Schumacher and Martin, 2016).

While most of the original specimens of *Polycotylus latipinnis* were fragmentary (Carpenter, 1996:268), two other species within the family were better represented in the fossil record. In the past the genus names *Trinacromerum* and *Dolichorhynchops* have often been used interchangeably, most recently by Adams (1997), who mistakenly named a new species of a large *Dolichorhynchops* from the Campanian of Wyoming as *Trinacromerum bonneri*. The type (KUVP 40002) and paratype specimen (KUVP 40001; Fig. 8.3) were subsequently redescribed by O'Keefe (2008), who correctly revised the genus to *Dolichorhynchops*.

In his discussion of the occurrence of polycotylids (e.g., short-necked plesiosaurs of the Cretaceous), Carpenter (1996:284) determined that *Trinacromerum bentonianum* (Cragin, 1888) has a 3.3-million-year range in the fossil record, from Upper Cenomanian into the Turonian, and *Dolichorhynchops osborni* (Williston, 1902) has a 4-million-year range, beginning in the earliest Campanian. This suggests that there is an 8- to 9-million-year gap in the Western Interior Sea between the middle Turonian and the beginning of the Campanian that is unoccupied by

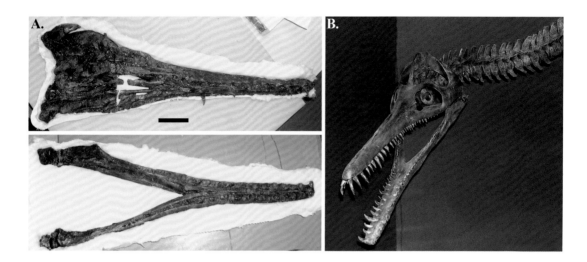

8.3. A, the paratype skull and jaws of *Dolichorhynchops bonneri* (KUVP 40001) from the Pierre Shale of South Dakota; this specimen represents the largest known individual of the species; scale bar = 10 cm (4 in); B, the skull of *Dolichorhynchops bonneri* as reconstructed by Triebold Paleontology from the KUVP 40001 specimen; the skull is nearly 1 m (39 in) in length.

polycotylids. In their review of the occurrence of pliosaurids, polycotylids, and elasmosaurids in the old 'Benton Formation' in Kansas (Graneros Shale, Greenhorn Limestone, and Carlile Shale), Schumacher and Everhart (2005) documented that *Trinacromerum* was relatively abundant through the middle of the Fairport Chalk Member (middle Turonian) of the Carlile Shale.

Within the Smoky Hill Chalk, the gap noted by Carpenter (1996) until the first occurrence of *Dolichorhynchops* and *Polycotylus* extended from the beginning of the chalk (late Coniacian) through the end of the Santonian, a period of almost 3.5 million years. This interval also corresponds roughly to the greatest expansion of the Western Interior Sea (Hattin, 1982:59), and the beginnings of the subsequent regression, a period in which nearshore, shallow-water environments, possibly the preferred habitats of polycotylids, were the farthest away from the depositional area of the Smoky Hill Chalk in western Kansas. Subsequently, Everhart (2003:130) reported the discovery of several specimens of fragmentary plesiosaur remains from the lower chalk (late Coniacian to early Santonian) and suggested that polycotylids were present throughout the deposition of the chalk, "albeit in small numbers."

The discovery of a polycotylid by Ramo and Pam Decker from the base of the Fort Hays Limestone in Jewell County has further closed the gap between *Trinacromerum* and *Dolichorhynchops / Polycotylus*. The headless specimen (FHSM VP-16459) was donated to the Sternberg Museum of Natural History where it has been reported (Everhart, Decker, and Decker, 2006), but is not yet identified or fully described. Whether it turns out to be *Trinacromerum* or *Dolichorhynchops* (most likely), it essentially closes the time gap between these lineages of polycotylids. Although not previously reported, the discovery of such rare specimens should not be considered unusual in light of the fact that the polycotylid lineage would not have disappeared and suddenly reappeared. The most likely explanation is that these fish-eating polycotylids

SHORT NECKED CRETACEOUS PLESIOSAUR
Dolichorhynchops osborni

preferred a different environment somewhere away from the deeper water in the middle of the Western Interior Sea.

The holotype of *Dolichorhynchops osborni* (KUVP 1300; Fig. 8.4) was discovered by a young George F. Sternberg in 1900 and collected by his father, Charles H. Sternberg (Williston, 1903). The skull of KUVP 1300 is crushed laterally but is well preserved and complete (Fig. 8.5). Storrs (1999:9) indicated that the specimen came from the Smoky Hill Chalk (Campanian) of Logan County, Kansas. Carpenter (1996:271) concurred, noting that KUVP 1300 was collected "east of Wallace" in Stewart's (1990:23) zone of *Hesperornis*. Exposures in this area are all in the upper chalk.

Bonner (1964:41) reported that the specimen of '*Trinacromerum*' *osborni* (FHSM VP-404; now *Dolichorhynchops osborni*, per Carpenter, 1997:193) on exhibit in the Sternberg Museum of Natural History at Fort Hays State University was collected from Logan County, a mile southwest of Russell Springs (Fig. 8.6). A third, partial *D. osborni* specimen (MCZ 1064) collected by G. F. Sternberg in 1926 (Sternberg and Walker, 1957; Everhart, 2004a) was also discovered in Logan County (Carpenter, 1996). The partially digested bones of the plesiosaur (USNM 9468) reported by Charles Sternberg (1922; Everhart, 2004b) as gut contents within a large *Tylosaurus proriger* (USNM 8898) mosasaur and the subject of the story at the beginning of this chapter, came from an exposure along (Twin) Butte Creek in Logan County. The remains of this partially digested plesiosaur were tentatively identified as a juvenile polycotylid (O'Keefe, pers. comm., 2001). Although their exact stratigraphic occurrence is unknown, the horizon of all these Logan County localities is in the upper one-third (Upper Santonian through Lower Campanian) of the Smoky Hill Chalk.

One of the fragmentary late Coniacian plesiosaur specimens (FHSM VP-13966) that I discovered in 1992 consisted of the partially digested remains of the back of a skull and lower jaws of *Dolichorhynchops* (Everhart, 2003). The bones were scattered across a wide area of the chalk at one of our favorite sites in Gove County. The specimen now consists of more than 20 small fragments of bone, all of which appeared initially to be badly weathered. I have since determined that they were actually partially digested, most likely by a large shark (see Chapter 4 in regard to bones regurgitated by sharks). At the time of discovery, however, I

8.4. The mounted type specimen of *Dolichorhynchops osborni* (KUVP 1300) in the collection of the University of Kansas Museum of Natural History. Note that the skull is a model constructed by H. T. Martin because the original was crushed and regarded as too fragile to place in the exhibit. It has now been added to the exhibit (Fig. 8.5).

Plesiosaur
Dolichorhynchops osborni
Niobrara Chalk, Kansas
Cretaceous

8.5. A left lateral view of the crushed skull of *Dolichorhynchops osborni* (KUVP 1300). This is the same view as published in Williston (1903:pl. II). The slender teeth and delicate construction of the skull suggest that *Dolichorhynchops* was feeding only on small prey.

was unable to identify the remains. When J. D. Stewart examined the material in 1992, he concluded that the bone fragments were from the skull of a plesiosaur, noting that two of the larger fragments appeared to be the hinge joints (right and left articular, surangular, and angular) of the lower jaws.

This was really good news as far as I was concerned because previously we had collected few plesiosaur specimens from the lower chalk. Two years earlier, Stewart had written that he knew of no "lower chalk specimens" other than a "few juvenile propodials" (1990:25). Few things can ever be said to be certain in paleontology. As luck would have it, I happened to be with J. D. Stewart in May of 1990 when he discovered a fairly complete hind limb of a *Dolichorhynchops* (LACMNH 148920) in the lower chalk. Unfortunately, it was too late to add it to his paper (Stewart, 1990) in the *Niobrara Chalk Excursion Guidebook* that was published for the fiftieth anniversary meeting of the Society of Vertebrate Paleontology (SVP) in Lawrence, Kansas, in October.

The plesiosaur skull fragments and our other plesiosaur specimens sat around for almost 10 years before I got around to working with them. In 2001, I sent the skull fragments (FHSM VP-13966) to Ken Carpenter at the Denver Museum of Nature and Science. He had recently published on his studies of short-necked plesiosaurs (Carpenter, 1996) from the Western Interior Sea and is familiar with polycotylids. Ken was able to identify the fragments, with a reasonable amount of certainty, as elements of the skull, braincase, and lower jaws of a plesiosaur, most likely *Dolichorhynchops*. After talking to Ken and rereading his 1996 paper, I realized that this specimen and five other sets of remains from the low chalk represented the first real evidence that plesiosaurs were living in the Western Interior Sea during late Coniacian time. While this was certainly not unexpected, it was the first time it could be actually documented from this portion of the chalk. Ugly and uninformative as the specimens were, I published a short paper describing them (Everhart, 2003) and added them to the record of plesiosaur remains from the Smoky Hill Chalk.

8.6. Skull of *Dolichorhynchops osborni* (FHSM VP-404) in dorsal (top) and ventral views, collected from the upper chalk of Logan County, Kansas, by Marion Bonner and donated to the Sternberg Museum of Natural History. Scale bars = 10 cm (4 in).

In the process of describing these specimens, I learned that the upper limb bones (propodials) of immature plesiosaurs had been picked up for many years, more or less as curiosities, and occasionally mentioned in the literature, but without stratigraphic data. Williston was the first to mention and figure the "propodial bones of young plesiosaurs" based on four specimens in the collection of the Field Museum of Natural History in Chicago (1903:73, pl. XXIII). Several years later, Moodie described and figured seven similar specimens in the University of Kansas Museum of Natural History and referred to the smallest of them as "an embryonic plesiosaurian prodigal" (1911:95; see also Moodie, 1908). The Sternberg Museum of Natural History also has six specimens collected by George F. Sternberg and several more that have been added since his retirement. With one notable exception (FHSM VP-16868), these specimens are all single limb bones from small plesiosaurs that appear to have been partially digested, most likely by large sharks, and then regurgitated. One of Sternberg's specimens (FHSM VP-649) includes a very obvious serrated bite mark that is attributable to *Squalicorax*. The FHSM VP-16868 specimen is unusual because it includes the propodial, 17 elements from the lower limb including 8 phalanges, and a caudal vertebrae. The partially digested specimen was collected as a jumbled pile of bones from the middle chalk in Gove County by Kenneth Jenkins in 2006 and almost certainly indicates feeding on a juvenile plesiosaur by a large shark such as *Cretoxyrhina mantelli*. Taken as a group, these specimens certainly indicate the presence of plesiosaurs throughout the deposition of the Smoky Hill Chalk.

Several notable polycotylid specimens occur outside the Smoky Hill Chalk formation in Kansas. Certainly one of the first discoveries was 10 articulated vertebrae collected from Russell County by B. F. Mudge in 1872 in the 'Benton Cretaceous.' The stratigraphic occurrence was determined by Schumacher and Everhart (2005) from Mudge's locality information to be the middle of the Fairport Chalk (middle Turonian). Mudge's handwritten note is still affixed to the specimen and indicates that he believed he had discovered the remains (KUVP 1325) of an *Ichthyosaurus*. Later, the specimen was described as the type of *Trinacromerum anonymum* by Williston (1903) and then reidentified as *T. bentonianum* by Carpenter (1996). A more complete specimen of *T. anonymum / bentonianum* (YPM 1129) collected a year later (1873) in Osborne County by Joseph Savage, an amateur collector, is now in the Yale Peabody Museum.

The type specimen of *Trinacromerum bentonianum* (Plate 3) was described by Cragin (1888) from two skulls (USNM 10945 and 10946) that were collected from the upper part of the Fairport Chalk (middle Turonian) in Osborne County, Kansas. Riggs (1944) described another specimen (KUVP 5070) that had been discovered in 1936 on a road cut through the Hartland Shale Member of the Greenhorn Formation (Upper Cenomanian) along U.S. Highway 81 a few miles south of Concordia in Cloud County. The construction workers had apparently taken great care to protect the specimen, and the skull was recovered intact along with much of the rest of the skeleton. Riggs (1944) named the plesiosaur *Trinacromerum willistoni* in honor of S. W. Williston, but it was later determined by Carpenter (1996) to be yet another example of *T. bentonianum.*

In talking to Dr. Paul Johnston (pers. comm., 2003), who retired from the geology department at Emporia State University, I learned of the discovery of another, fairly complete *Trinacromerum bentonianum* specimen that turned into a paleo-crime story. According to Dr. Johnston, in the summer of 1971 he noticed plesiosaur bones eroding from a road cut at the top of the Greenhorn Limestone, about a mile south of Wilson Lake in eastern Russell County. He covered up the bones and brought help back with him the next week. They uncovered the skeleton of a large plesiosaur coming out of the hillside headfirst at an angle. The skull and most of the cervical vertebrae were already gone, probably taken off by heavy equipment when the road was built. He remembers seeing one of the front paddles, the pectoral girdle, an articulated vertebral column, ribs, and at least one of the rear paddles in their excavation. They were on a public road and had quite a bit of traffic past the dig, including two men on motorcycles who stopped and looked things over pretty closely. He remembered that they asked a lot of questions. Near the end of the excavation, Dr. Johnston and his crew covered the remains and returned to Emporia for a few days. When they returned, the plesiosaur bones had been hacked out of the ground. All that was left were a few bone fragments and one complete front paddle that had still been buried in the hillside. The paddle is currently on exhibit (Fig. 8.7) in the Paul Johnston Museum of Geology

8.7. A forepaddle of *Trinacromerum bentonianum* (ESU 5000) in the collection of the Paul Johnston Museum of Geology at Emporia State University, Emporia, Kansas. The specimen was discovered in the Pfeifer Shale of Russell County in 1971.

at Emporia State University and has been identified as *T. bentonianum* (B. Schumacher, pers. comm., 2003). So far as I am aware, the rest of the specimen has never been seen again.

In 2009, Brian Baalmann was looking for his golf ball when he saw fossil bones in the ditch next to the greens just south of La Crosse, Kansas. Later he brought them to the Sternberg Museum for identification. I eventually saw the remains and realized that they were plesiosaur bones. I made arrangements to meet him early on a cold day in February 2010. Together we walked to the site in the ditch between the highway and the golf course, and he showed me where he had picked up the bones. I was a bit concerned when I saw golf cart tracks running through that portion of the ditch. Fortunately there were other bones visible, and I was able to collect additional vertebrae and limb material from what appeared to be the back half of a juvenile *Trinacromerum* plesiosaur. Similar to the specimen described above, the remains had been discovered and partially destroyed by heavy equipment during the recent widening of the highway. Among the bones and fragments was a section of one of the plesiosaur's ribs that had been broken and had healed before it died (Fig. 8.8). Life was tough in the Late Cretaceous! I continued to visit the site over the course of several months and recovered hundreds of small bone fragments, mostly from the pelvic girdle and ribs. Stratigraphically, they occurred in the basal Fairport Chalk (Middle Turonian), just above the Fencepost Limestone layer. The 'Golf Course Plesiosaur' (FHSM VP-17543) was donated to the Sternberg Museum; it represents the smallest (youngest) individual of this species collected to date.

Two important pliosaurid specimens (not polycotylids) that came from below the chalk in central Kansas should be mentioned here to avoid any further confusion. Williston (1903) described and named a new species of pliosaur, *Brachauchenius lucasi* (Plate 4), from a 90 cm (nearly 3 ft) long skull and 37 vertebrae in the Smithsonian collection (USNM 4989) collected from the 'Benton Formation' in Ottawa County, Kansas. According to Robert Purdy (pers. comm., 2003), the Smithsonian purchased the specimen from Charles Sternberg in the spring of 1884, but the story is actually more complicated than that. The specimen was collected in 1884 by Charles Sternberg and one of his assistants, but they did so as employees of O. C. Marsh at Yale (Everhart, 2007a). Although Sternberg shipped the type specimen of *Brachauchenius lucasi* to O. C. Marsh at Yale, Marsh was working for the United States Geological Survey at the time and had used federal funds to pay

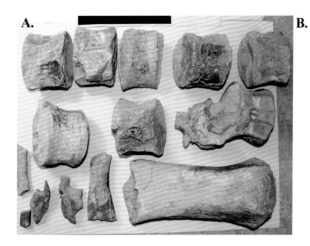

8.8. A, some of the vertebrae and the proximal end of a podial (femur) from the Golf Course Plesiosaur (FHSM VP-17543); although only a partial specimen, it is significant because of its small size; scale bar = 10 cm (4 in); B, upper and lower views of a rib fragment from the same specimen preserving a healed break; scale bar = 5 cm (2 in).

for the acquisition. Apparently Marsh did nothing with the specimen while it was in his possession. The remains (USNM 4989) were eventually transferred to the United States National Museum (Smithsonian), where they were examined by S. W. Williston. By that time, the skull had been prepared and exhibited upside down, along with the articulated cervical and dorsal vertebrae. It has since been removed from the original mounting so that the damaged upper portions of the skull can be examined (McHenry, pers. comm., 2004; Schumacher, Carpenter, and Everhart, 2013).

The skull of a second specimen (FHSM VP-321) that was assumed to be *Brachauchenius lucasi* is much larger, more complete, and better preserved than the specimen described by Williston (1903; Fig. 8.9). The remains were discovered in 1950 near the town of Fairport in Russell County by Robert and Frank Jennrich while they were looking for shark teeth. Although the skull and lower jaws were collected by George Sternberg in October 1950, and placed on exhibit soon afterward in the Sternberg Museum, the discovery was not officially reported until it was briefly described by Carpenter (1996). Williston (1907) and Carpenter (1996) both agreed that *Brachauchenius* was closely related to the Jurassic pliosaur *Liopleurodon ferox*. After his examination of the Sternberg specimen, McHenry (pers. comm., 2004) suggested that there were also many similarities between *B. lucasi* and an Early Cretaceous pliosaur from Australia called *Kronosaurus queenslandicus*. A cast of the FHSM VP-321 skull is currently on exhibit in the Sternberg Museum of Natural History at Fort Hays State University, and an earlier cast of the skull is on display at the University of Kansas Museum of Natural History, Lawrence, Kansas.

The FHSM VP-321 skull itself is huge, over 1.5 m (5 ft) in length. Once Sternberg had prepared the bones, he mounted them in a plaster slab so that only the top (dorsal) surface of the skull and the left lower jaw were visible. Although that made a spectacular mount as an exhibit, the plaster completely obscured the underside (palate) of the skull, hiding its structure and limiting the usefulness of the specimen to science.

8.9. The skull of a giant pliosaur (*Megacephalosaurus eulerti*—FHSM VP-321) in the collection of the Sternberg Museum of Natural History. The specimen was collected in 1950 from the Fairport Chalk Member (middle Turonian) of the Carlile Shale in Russell County, Kansas; FHSM VP-321 represents one of the last known occurrences of a pliosaur in North America. Scale bar = 100 cm (39 in).

While searching through the Sternberg Museum's records in 2006, I discovered a black and white photograph of the underside of the skull taken by G. F. Sternberg just before he entombed it in plaster. When I shared the photo with Bruce Schumacher, we decided that there was something different about the bones as compared to Williston's description of the type specimen. By then, Ken Carpenter was involved in the project, and we received permission from the museum to remove the skull from the plaster mounting. The specimen was transferred to the Denver Museum of Nature and Science, where Ken and his volunteers went through laborious process of removing George Sternberg's 60-year-old plaster. When that was done, we were able to see the palate of the skull for the first time since 1950. Ken was also able to mold the skull and make casts for display and study purposes. Once the newly prepped skull was back at the Sternberg Museum, we (mostly Bruce) were able to examine the bones and understand the differences that became apparent between it and the type specimen. We concluded that the specimen was a new genus and species of Late Cretaceous pliosaur and gave it the name of *Megacephalosaurus eulerti* (Schumacher, Carpenter, and Everhart, 2013). The genus name means 'big headed lizard,' and the species name honors Otto C. Eulert, the landowner who generously donated the specimen.

Large as the VP-321 skull is, fragments of an even larger *Megacephalosaurus* skull (UNSM 50136) are in the collection of the University of Nebraska State Museum. While visiting the UNSM in 2005 on another plesiosaur project, I came across a drawer full of large fragments of a pliosaur skull. Although I recognized some of the bones, I couldn't make much sense out of them. I did take pictures, and later I showed them to Bruce Schumacher. Bruce was intrigued with what he saw and visited the UNSM to examine the specimen. He was able to reassemble the fragments of the skull into the muzzle of what turned out to be the largest pliosaur ever discovered in North America. Compared to the 1.5 m (5 ft) length of the type specimen (VP-321), the skull of the new specimen would have been about 1.75 m (5.7 ft) in length (Schumacher, 2008; Schumacher, Carpenter, and Everhart, 2013).

8.10. The skull of *Megacephalosaurus eulerti* (FHSM VP-321) as reconstructed by Triebold Paleontology, on display at the 2012 Annual Meeting of the Society of Vertebrate Paleontology. The skull is 1.5 m (5 ft) in length.

Casts of the FHSM VP-321 *Megacephalosaurus* skull were sent to Triebold Paleontology to be used in 3-D reconstruction of the skull. The first copy of the skull (Fig. 8.10) was completed in time to be exhibited at the 2012 Annual Meeting of the Society of Vertebrate Paleontology in Raleigh, North Carolina, where Mike Triebold generously donated the cast for the SVP auction at the end of the meeting.

Liggett et al. (1997) reported the discovery of a partial paddle (FHSM VP-13997; Fig. 8.11) of a large pliosaur from the base of the Lincoln Limestone Member of the Greenhorn Limestone in western Russell County. If complete, the paddle would have measured more than 2 m (6.5 ft) in length (Schumacher, pers. comm., 2003). The specimen was tentatively identified as the limb of *Brachauchenius lucasi*, but based on size it might actually be from *Megacephalosaurus*. Conservatively, this paddle would represent a pliosaur that was about 5 m (16 ft) wide from paddle tip to paddle tip, and 7 m (22 ft) long from nose to tail. The Sternberg Museum also houses two other fragmentary specimens that have been previously attributed to *B. lucasi* (a partial propodial, FHSM VP-2149; vertebrae, VP-2150).

Additional remains of a big pliosaur were discovered in the Blue Hill Shale of Russell County (J. D. Stewart, pers. comm., 1999), but the specimen was taken to California by the collector without permission of the landowner. Efforts to have it returned to Kansas have not been successful up to now. Most of what we know about it is contained in copies of a few grainy black and white photographs. The locality was recorded, however, and in 2010 Bruce Schumacher and I visited the site. We were able to locate enough fragments of bone and teeth to verify

8.11. The middle portion of the paddle of a large pliosaur (probably *Megacephalosaurus eulerti*) from the Upper Cenomanian Lincoln Limestone of Russell County, Kansas. Scale bar = 10 cm (4 in).

that pliosaur remains had been there and assign a specimen number (FHSM VP-17469). This specimen and fragments of a large rib (FHSU VP-17299) that I collected from another Blue Hill Shale site (Everhart, 2009), represent some of the last of the pliosaurs living in the Western Interior Sea prior to their extinction.

Even with the discovery of three nearly complete specimens of *Dolichorhynchops osborni* in the past century or so, plesiosaur remains of any kind are rare occurrences in the Smoky Hill Chalk. *Dolichorhynchops* persisted past the end of the deposition of the chalk (about 82 Ma) and lived well into the deposition of the Pierre Shale during Campanian time. Although rarely seen in the Pierre Shale of Kansas, its remains have been collected more often farther north in South Dakota and Wyoming. Two *Dolichorhynchops* specimens in the University of Kansas collection (KUVP 40001 and KUVP 40002; from the Sharon Springs Member of the Pierre Shale, Fall River County, South Dakota) appear to show that the species was getting much larger (Carpenter, 1996; O'Keefe, 2008). The skull length of this specimen is 98 cm (38 in) as compared with 71 cm (28 in) for the specimen (FHSM VP-404) in the Sternberg Museum. Scaling up, suggested a pliosaur that was nearly 6 m (20 ft) long, very close to the size of the much earlier *Brachauchenius* specimens from Kansas. This also approximates the body size (not including the neck and head) of *Elasmosaurus platyurus*.

In 1992, we participated in a dig by the New Jersey State Museum on a nearly complete *Styxosaurus snowii* in the Sharon Springs Member of the Pierre Shale of Logan County that had been discovered by our friend Pete Bussen. At the time, my wife, Pamela, was having trouble 'seeing' vertebrate remains in the selenite-filled dark-gray shale. She asked Pete for help, and he went up the hill with her to show her what to look for. Within 30 minutes, she had picked up a podial (finger bone of a paddle) and asked him if it was from a plesiosaur. It was, and she discovered the remains of one of the largest specimens of *Dolichorhynchops osborni* currently known from Kansas. Most of the remains were contained in a

8.12. Bones of a large polycotylid, most likely *Dolichorhynchops* (CMC VP-7055), recovered from a concretion in the upper Sharon Springs Member of the Pierre Shale, Wallace County, Kansas. A, the proximal end of a femur at right and a fragment of the pelvic girdle (pubis or ischium); the brown material between the bones is matrix from the concretion; scale bar = 10 cm (4 in); B, a complete femur from the same specimen, damaged by the concretion; scale bar = 10 cm (4 in).

A.

B.

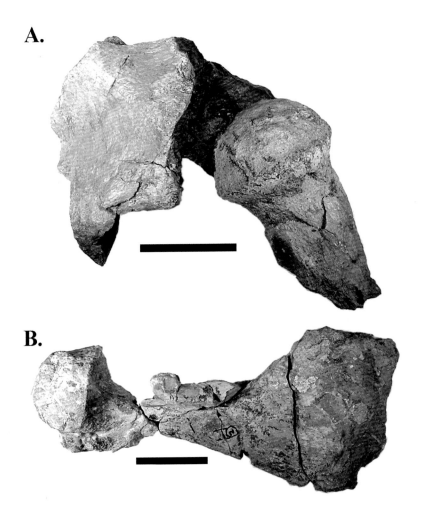

concretion, badly damaged and far from complete. Elements recovered included both rear paddles, part of the pelvis, and a number of caudal vertebrae. One of the femora was about 40 cm (16 in) long, compared with the 33 cm (13 in) femur of FHSM VP-404 (Bonner, 1964), and much more massive. The material was identified by Ken Carpenter in 1994 and later donated to the Cincinnati Museum Center (CMC VP-7055; Fig. 8.12).

The remains of another *Dolichorhynchops osborni* were collected about 1987 by Charles (Chuck) Bonner in the Sharon Springs Member of the Pierre Shale in Logan County, just above the contact with the Smoky Hill Chalk. It was headless and otherwise unremarkable, except for what was preserved inside the abdominal area. In their book on paleopathology, Rothschild and Martin cited a personal communication with O. (Orville) Bonner in 1991 which alluded to the discovery of "young within the body" of "short-necked, polycotylid plesiosaurs" (1993:294). The specimen (LACMNH 129639) was obtained by the Los Angeles County Museum of Natural History in 1988 (S. McLeod, pers. comm., 2000).

In September 2000, Chuck Bonner showed me many photographs of the plesiosaur dig and the partially prepared specimen and gave me some additional information. The scattered remains were removed in several jackets. A photograph of one of the partially prepared jackets showed adult-size limb material next to what appeared to be the remains of at least one smaller (baby?) plesiosaur. The dorsal processes of the small vertebrae were not fused to the centra, and the limb girdles appeared to be incompletely formed. The remains appear to represent a small plesiosaur but one that was much too large to have been consumed by the adult *Dolichorhynchops*. This, and the fact that the bones of the smaller individual do not appear to have been damaged by stomach acids, pretty much rules out cannibalism. That was about all I could say about the remains in the first edition of this book (2005).

The specimen has now been formally described, and identified as a female *Polycotylus latipinnis* that was associated with a near-term fetus (O'Keefe and Chiappe, 2011). It is partially reconstructed, including the addition of the missing skull and cervical vertebrae on the adult, and exhibited in a plaster mounting on the wall of the LACMNH (ibid.:fig. 1A). It is still my opinion, based on what I saw in the original photographs, that the adult plesiosaur is *Dolichorhynchops*, but that is simply my view, not a scientific fact. The specimen will remain as identified by O'Keefe and Chiappe until someone else examines it and can prove that it is something else. Part of my concern in this case is that the adult plesiosaur was missing its skull and at least half of its cervical vertebrae. Only 10 actual cervical vertebrae are included in the mounted specimen. The number of cervical vertebrae is important to the diagnosis of *Polycotylus* because it has a noticeably longer neck than *Dolichorhynchops*. Williston (1902, 1903) stated that there are 19 cervical vertebrae in *Dolichorhynchops* and 26 cervicals in *Polycotylus* (see Schumacher and Martin [2016] for an extended discussion of this specimen and the cervical vertebrae counts in polycotylids).

The remains of the smaller individual are incomplete, also missing a skull, but sufficient for O'Keefe and Chiappe (2011) to estimate the length of the fetus at 1.5 m. This is about one-third of the estimated length of the adult, leading the authors to conclude that the fetus was near term and that *Polycotylus* gave birth to only a single offspring at time. This may well be, but what they did not mention was the probability that both the mother and the fetus had been scavenged before burial. Besides the missing neck and skull of the adult, most of her ribs and gastralia are missing, portions of the body that sharks would readily remove to gain access to organs inside the abdomen. Similar preservation, including severed or missing ribs, was noted in my description of a mosasaur carcass that had been scavenged by both *Cretoxyrhina* and *Squalicorax* (Everhart, 2004c). In spite of my concerns as to the identification, however, this specimen appears to prove that plesiosaurs did give live birth to their young, just like ichthyosaurs and mosasaurs.

The earliest (oldest) remains of plesiosaurs occurring in Kansas are from the Kiowa Shale (Early Cretaceous–middle Albian). The Kiowa

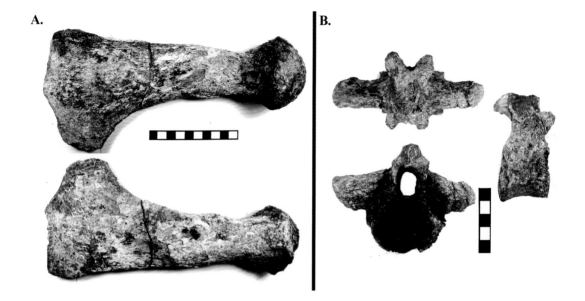

8.13. Plesiosaur bones from the Early Cretaceous (Albian) Kiowa Formation of south-central Kansas. A, two views of an isolated plesiosaur propodial (humerus or femur; FHSM VP-17708); scale bar = 10 cm (4 in); B, a plesiosaur dorsal vertebra in three views (FHSM VP-17710); scale bar = 10 cm (4 in).

Shale is exposed along the southern border of Kansas in Clark, Kiowa, and Comanche counties, where it overlies rocks of Permian age. The Kiowa also crops out in the central part of Kansas (mainly in Saline, McPherson, and Ellsworth counties). According to Scott (1970), much of Kansas was on the edge of a shallow sea at the time, and these interbedded shale and sandstone deposits represent a nearshore, high-energy environment. In this case, high-energy environment means shallow water that is subject to tidal flows and wave action, both of which are detrimental to preserving complete fossils.

Consequently, the remains of the plesiosaurs and other vertebrates recovered from the Kiowa are generally disarticulated and incomplete. Although the Kiowa has been collected since the 1870s, little is known about the plesiosaurs or other animals of that time in Kansas. F. W. Cragin, who discovered the type specimen of *Trinacromerum bentonianum*, was among the first collectors of plesiosaur material from the Kiowa Shale. About 1893, the remains of a possible elasmosaur (*Plesiosaurus mudgei*—KUVP 1305; Fig. 7.14), consisting of nine vertebrae, a fragmentary femur, and about 200 gastroliths, were collected and named by Cragin (1894) from the Kiowa Shale in Clark County (see also Schultze et al., 1985). The specimen is briefly mentioned in the text by Williston (1903), who apparently believed it was synonymous with another fragmentary specimen (*Plesiosaurus gouldii*). An excellent black and white photograph (ibid.:pl. 29) shows the remains, including the many gastroliths. Unfortunately, most of the specimen was missing when I checked on it at the University of Kansas.

Both Sternberg Museum and the University of Kansas have many fragmentary specimens of plesiosaurs from the Kiowa Shale in their collections. I was able to add a plesiosaur ischium in 2008 (FHSM VP-17302) and a propodial and dorsal vertebrae in 2011 (FHSM VP-17708 and VP-17710; Fig. 8.13) to the Sternberg collection. Eventually I'd like to

8.14. The short-necked polycotylids, *Polycotylus latipinnis* and *Dolichorhynchops osborni*, depended on their speed and long, toothy jaws to capture small prey. As teleosts evolved into larger and faster forms during the Late Cretaceous, the competition for food increased. The young of these plesiosaurs were occasionally prey for mosasaurs. Drawing by Russell Hawley, used with permission.

describe both collections because it appears that they include material from elasmosaurs, polycotylids, and pliosaurs.

Kansas has produced the remains of many plesiosaurs since the discovery of *Elasmosaurus* and *Polycotylus* (Fig. 8.14) in the late 1860s. They occur in every Cretaceous rock unit in the state, including the Dakota Sandstone (FHSM VP-2151 from Stanton County), and are especially common in the Kiowa Shale, the Greenhorn Limestone, the Fairport Member of the Carlile Shale, and the Sharon Springs Member of the Pierre Shale. I have no doubt that discoveries of major new specimens lie ahead.

9

Enter the Mosasaurs

The female mosasaur swam slowly through the calm, warm water of the Inland Sea. Three of her week-old young swam cautiously on either side and slightly above her rear paddles. A fourth baby, the smallest and weakest of the litter, had not been able to keep up with the steady pace of the mother. When it dropped behind two days ago, it was quickly swallowed whole by a large fish. The three surviving babies were instinctively alert to any signs of predators and kept as close as possible to their mother's scaly side. They also stayed well behind her head and out of the reach of her toothy jaws. Even though she was their mother, her feeding instinct was very strong. A momentary mistake in the recognition of her offspring could be instantly fatal to them.

For the past several days, the mother mosasaur had guided her young through areas of the ocean that teemed with swarms of small, soft-bodied prey. The young mosasaurs were accomplished hunters almost from the moment of birth and had easily caught enough of the little squid to keep their bellies full. If they were to survive, it was essential that they eat as much and as often as they could. They were growing quickly but still would be vulnerable to attacks by other larger predators in the Western Interior Sea for many months to come.

The mother mosasaur paused briefly in the water to allow her babies to rest, floating motionless in the calm sea. Her young slithered partway up on her narrow back to take advantage of the relative safety of her large body. Few other animals in the ocean, except mosasaurs of her own species, could match her 9 m (30 ft) length.

For her young, it was a rare chance to absorb warmth from the sun that shone hotly in the clear, blue-white sky. Just maintaining their body temperatures in the cooler water required a lot of energy.

While the mother mosasaur rested with her eyes and nose barely above water, her predatory senses were active. Hunger gnawed at her belly. She was used to eating often and well, and her recent pregnancy had all but depleted her body's reserves. Dimly aware that she could not hunt her usual prey without endangering her offspring, she had curbed her urge to stalk the schools of large fish that she sensed were nearby. She had eaten only twice in the last three days, first the bloated carcass of a small shark and then a large, swimming bird that had blundered across her path.

As she watched and waited, her senses detected the commotion caused by several large fish feeding on a trapped school of smaller fish.

She could feel vibrations that carried through the water as the smaller fish leaped into the air to avoid the jaws of their attackers. For a few moments, she didn't move, but the urge was too great. With a powerful flick of her long, sinuous tail, she started her large body moving in the direction of the feeding frenzy.

Startled by her sudden movement, her young splashed awkwardly into the water and then swam swiftly to catch up and return to her side. As she swam beneath the surface, her movements were efficient and silent. Years of experience had taught her that hunting was more successful when her prey was unaware of her presence until it was too late to get away. The noise created by the smaller fish in their panic to escape would conceal her approach from the larger fish. They were her intended prey. One would fill her belly and satisfy her hunger for a day or so.

As she got closer to the commotion, she sensed rather than saw the larger fish darting into the turbulence caused by the panic of their smaller prey that had been corralled into a bait ball. Moments later, a long, silver torpedo shape emerged from the cloud of bubbles and almost ran into her open mouth. Too late, the fish recognized the danger and tried to turn away. With a swift lunge, the mosasaur's jaws snapped closed just behind the head of the fish. The water turned red with blood, and a shower of scales glittered in the sunlit water as the fish struggled briefly and then went limp.

Waiting a moment to make certain that the fish was dead, the mosasaur then opened and closed her jaws in rapid movements to position her victim to be swallowed headfirst. When the head of the fish was inside the mosasaur's mouth, two rows of sharp teeth on the roof of her mouth helped hold the fish in place as her lower jaw flexed and pulled it deeper inside. Once the fish was securely started down the mosasaur's throat, she raised her head out of the water and used gravity to help her swallow her large meal. When the fish's large bony tail was the only part remaining outside her jaws, she closed her mouth and shook her head sharply. With a snap, the tail broke off at the base and skipped across the water.

Again, the mosasaur floated almost motionless in the water as she finished swallowing the large fish. Her young milled nervously around her flanks, uncertain what to do in all the confusion. She had almost forgotten about them in her drive to satisfy her hunger.

The feeding frenzy had also attracted other predators to the area. A large shark moved swiftly toward the fray, sensing the same signals that the mosasaur had reacted to. As it got closer, the shark detected the faint traces of blood in the water. Seeing three small objects that struggled at the side of the larger stationary one, it raced upward from the depths.

By the time the mother mosasaur and her young sensed the pressure wave generated by the approaching shark, it was too late to react. As the babies turned to flee, the shark's jaws closed viciously across the muzzle and neck of the largest one. The shark's sharp, bladelike teeth sliced easily through the smaller animal's skull and cervical vertebrae, killing it instantly.

With the body of the young mosasaur in its mouth, the momentum of the shark carried it into the side of the mother mosasaur. The shark was less than half the length of the mosasaur and no match for the larger animal. As the mosasaur turned almost double on herself, slashing with open jaws at the flank of the shark, the shark flicked its tail and swam swiftly away with its victim held securely in its mouth. The mother mosasaur pursued the shark for a short distance, then slowed so that her two remaining young could catch up. Swimming out of danger, the shark wolfed down the carcass of the little mosasaur. Hours later, it would regurgitate a few of the indigestible bones from the mosasaur's skull.

Mosasaurs

We know virtually nothing about the social or parenting behavior of mosasaurs. Their modern relatives (snakes and monitor lizards) mostly do not display the sort of parental involvement that I have imagined for this mosasaur mother and her young. Mosasaurs appear to have been solitary animals for the most part, at least on the basis of thousands of individual mosasaur specimens that have been collected to date. In fact, the only associations of two or more mosasaurs that I was aware of were: (1) the *Plioplatecarpus* specimen with at least four embryonic young reported by Bell et al. (1996) and Bell and Sheldon (2004); (2) the *Tylosaurus proriger* specimen described by Martin and Bjork (1987) that contained a much smaller *Platecarpus* as stomach contents; and (3) the commingled and partially digested bones of two juvenile *Platecarpus* skulls (FHSM VP-14846 and VP-14847; Amy Sheldon, pers. comm., 1994) that I discovered in 1990, which were probably the remains of a shark's meal. Then, at the Second Mosasaur Meeting in 2007, Gorden Bell reported on a Texas specimen of *Tylosaurus nepaeolicus* that included three *Platecarpus* young as stomach contents (see also Bell, Barnes, and Polcyn, 2013). His 2007 talk also included two other specimens, a second *Tylosaurus proriger*, and a *Hainosaurus* [*Tylosaurus*] sp., both with mosasaur remains as stomach contents.

Both Mudge (1876) and Williston (1898a) mentioned damage to mosasaur remains (bitten skulls, mangled paddles) that were almost certainly caused by another mosasaur. Other examples of interaction (fighting) between mosasaurs include a *Mosasaurus conodon* skull from the Pierre Shale with bite marks and the tooth of another *M. conodon* embedded in its left quadrate (Bell and Martin, 1995) and a *Tylosaurus kansasensis* skull (FHSM VP-2295) from the lower (late Coniacian) chalk with deep bite marks across the top of the skull and right lower jaw that could have only been caused by a larger mosasaur (Everhart, 2008a; Plate 9). A *Platecarpus tympaniticus* specimen (FHSM VP-322) on exhibit at the Sternberg Museum has a healed bite mark on the left lower jaw and a pair of paddle bones that are fused together because of infection caused by a bite at the joint (ibid.:fig. 1E). Beyond these few examples of mosasaur interactions, the fossil record is largely mute regarding questions of whether mosasaurs may or may not have lived in close association or cared for their young.

While the above story is fiction, it is based on a fragment from the front of the skull of a young mosasaur (FHSM VP-13748) discovered in 1997 by Tom Caggiano in Gove County (see Fig. 4.3). It had been severed cleanly across the muzzle, behind the third tooth in both maxillae, and partially digested. The premaxillae and the anterior ends of both maxillae were still firmly joined together. I knew it was partially digested because of the eroded condition of the surface of the bone (Everhart, 1999; Varricchio, 2001) and because the teeth were completely dissolved down into their sockets. Most likely this was the work of a large ginsu shark (*Cretoxyrhina mantelli*). Other partial remains of mosasaurs, including several with the broken tips of *Cretoxyrhina* teeth still embedded in them, are common occurrences in the lower (late Coniacian) chalk. One such specimen that I discovered, FHSM VP-13283 (Shimada, 1997: fig. 4; Everhart, 1999), is an articulated series of five vertebrae from the lower back of a large (est. 6 m) mosasaur that has two shark teeth embedded in it (Fig. 4.1). The anterior and posterior vertebrae (approximate diameter 5 cm; 2 in) in the series are both severed across the centra, and most of the surface of all the vertebrae has an eroded appearance. When I discovered this specimen in 1995, it was still mostly in the chalk, so the erosion of the bone surface is not due to weathering. There is no way to tell, however, whether or not these specimens were the result of a shark attack on a living mosasaur or from scavenging of a carcass. I suspect that most of these fragmentary specimens resulted from sharks feeding on already dead mosasaurs, much like the behavior of modern sharks. However, there is no reason to think that Cretaceous sharks would have missed an opportunity for an easy meal on an injured or otherwise vulnerable mosasaur (Chapter 4).

In the first edition, I asked the question, what is a mosasaur? Almost this exact question was addressed by Michael Caldwell in the proceedings of the 2010 Third Mosasaur Meeting, held in Paris, France. Caldwell (2012) thoroughly discussed the various attempts to define the term, along with the various misconceptions and misrepresentations that have occurred in the nearly 200 years since it was first used to describe the remains of a giant reptile discovered near the Meuse River in the Netherlands. Caldwell's point was that while those of us who work with these marine reptiles 'know' what they are, there is no concise or scientifically accurate definition of what the term actually means (Fig. 9.1).

For my purposes here, mosasaurs were marine lizards that lived during the last 35 million years or so of the Late Cretaceous. They were the top predators of the Earth's oceans at roughly the same time as *Tyrannosaurus rex* and *Triceratops* lived in western North America. In the Western Interior Sea, *Tylosaurus proriger* was the largest predator around, with a skull that was 1.2 m (4 ft) or more in length (Fig. 9.2). The closest modern relatives of mosasaurs are snakes (Caldwell, 1999), and probably monitor lizards (varanids) like the Komodo dragon, although the exact relationships are still a matter of debate among mosasaur workers. Molecular evidence reported by Lee (2005) indicates that mosasaurs are actually

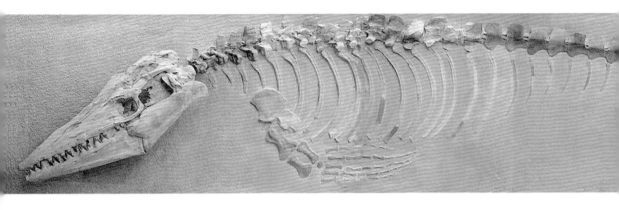

9.1. The exhibit specimen of *Tylosaurus proriger* (FHSM VP-3) at the Sternberg Museum. This 9 m (29 ft) specimen was discovered in Logan County by the son of George F. Sternberg in 1927. The specimen is complete except for some missing vertebrae in the lower back where a gully had cut through the remains.

more closely related to snakes than to varanid lizards. The common ancestor of both mosasaurs and snakes was a small lizard that successfully transitioned from terrestrial to marine life. In adapting to life in the ocean, mosasaur legs evolved into webbed paddles, and the early snakes eventually lost their limbs completely. At some later time, the ancestors of modern terrestrial snakes then made a second, and apparently legless transition back to the land.

The ancestors of mosasaurs were probably related to small, shore-dwelling lizards called aigialosaurs that lived close to the ocean from the Late Jurassic through the Cretaceous. They may have been very much like modern marine iguanas in size and lifestyle. Carroll and Debraga (1992) reported three mosasaur-like aigialosaur specimens from Cenomanian-Turonian (93 Ma) deposits in Yugoslavia. As they evolved, they became better adapted to life at sea, becoming an intermediate form that we refer to as mosasauroids—not quite mosasaurs.

About 100 million years ago, just into the second half of the Cretaceous, the shore-dwelling ancestors of mosasaurs entered the ocean and began to evolve rapidly. Recent discoveries in North America suggest a small mosasauroid ancestral form was in Kansas (Everhart and Pearson, 2014) and Utah (Michael Polcyn, pers. comm., 2015) around 95 million years ago (Upper Cenomanian). Within a relatively short time, geologically speaking, they had adapted to life in the sea to the point that they could no longer leave the water. Mosasaurs were not the first terrestrial reptiles to return to the sea, but they were probably the most successful in terms of their diversity, large numbers, and eventual dominance of the marine environment (Table 9.1). Then, for reasons we don't yet fully understand, they suddenly became extinct along with plesiosaurs, pterosaurs, dinosaurs, and many other groups of animals and plants at the end of the Cretaceous.

Mosasaurs seem to suddenly appear, fully developed, in the fossil record of the Western Interior Sea during the second half of the Cretaceous period (early Turonian), several million years prior to deposition of the Smoky Hill Chalk (Polcyn et al., 2008; Bice and Shimada, 2016; Everhart, 2016b). However, compared to the number of specimens

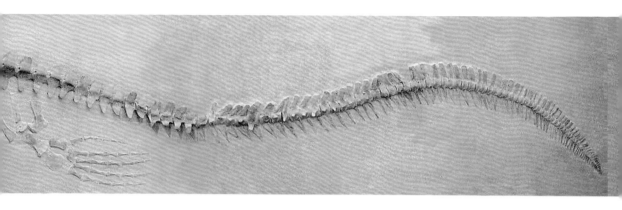

occurring in the Smoky Hill Chalk, very few mosasaur fossils have been collected anywhere prior to the beginning of Coniacian time (89 Ma). In fact, some fragmentary specimens of "early mosasaur remains" that had been reported have turned out to be the jaws and teeth of a large predatory fish called *Pachyrhizodus* (Thurmond, 1969; Stewart and Bell, 1994; see Chapter 4).

Martin and Stewart (1977) described two sets of vertebrae and a jaw fragment from the middle Turonian Fairport Chalk Member of the Carlile Shale Formation in Kansas, and noted their affinities with *Clidastes*. Another skull element (a *Platecarpus*-like frontal; KUVP 97200) is also in the KUVP collection (Bell, pers. comm., 2004) from the same strata in Ellis County, Kansas (Fig. 9.3A). Schumacher (2011) commented on these University of Kansas specimens in his description of mosasaur vertebrae (FHSM VP-17564; Fig. 9.3B) that he collected from the lower Middle Turonian Fairport Chalk, and noted that the first mosasaurs were living in the Western Interior Sea for at least a million years with the last of the pliosaurs (Chapter 7). It is likely that these early mosasaurs were subject to predation by much larger pliosaurs such as *Megacephalosaurus eulerti* (Plate 2).

Lingham-Soliar (1994) reviewed early mosasaur remains from the upper Turonian of Angola in western Africa. Bell and VonLoh (1998; see also and Bell et al., 2013) reported on new records of mosasauroids from the Greenhorn Formation (lower Turonian) of South Dakota and the Boquillas Formation of western Texas. A more detailed discussion of the stratigraphic occurrence of mosasaurs was subsequently provided by Bell (1997b).

New discoveries kept coming and pushing the occurrence of mosasaurs further back in time. Following the 2007 Second Mosasaur Meeting at the Sternberg Museum in Hays, Kansas, Polcyn et al. (2008) described the known mosasaur remains from the middle Turonian of Texas and Kansas. Much of this work was based on discoveries by Ramo and Pam Decker, who discovered scattered mosasaur material in the Codell Sandstone of Jewell County, Kansas. About the same time, Gail Pearson, a friend of mine who is a talented amateur collector, showed me a mosasaur-like vertebra (Fig. 9.4) that he had collected from the upper part

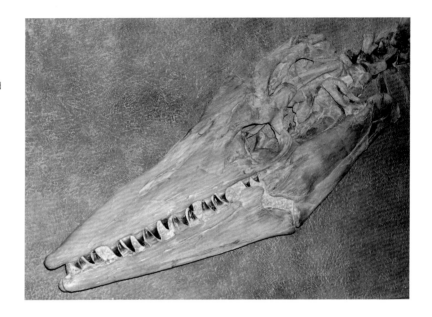

9.2. The skull of this *Tylosaurus proriger* (FHSM VP-3) is about 1.2 m (4 ft) in length. Other specimens in Kansas have skulls that are even larger. Note the extended toothless rostrum in front of the front teeth in the upper jaw.

of the Greenhorn Formation (upper Cenomanian) in Mitchell County more than 20 years earlier. It was only 3.8 cm long (1.5 in) and unlike any other mosasaur vertebra I had ever seen. But the age was the most interesting part. Until I saw the vertebra, there were no mosasaur specimens from that time in Kansas. When I told Gail I thought it was important, he was generous and donated the specimen to the Sternberg Museum. We took a shot at describing the vertebra (FHSM VP-17246) in 2012 as the earliest mosasauroid in North America, but the paper was criticized by skeptical reviewers who did not believe that the latest Cenomanian age we assigned to the specimen was accurate or that it was from an early mosasaur.

So we went back to the drawing board and considered what to do next. On a hunch, I looked at the opening in the vertebra where the spinal cord passes through (neural canal) and saw that it was blocked with the same limestone matrix that had once enclosed the bone. Working under a microscope, I was able to extract a small amount of the material. I hoped that this bit, less than the size of an aspirin, would contain tiny invertebrate remains (nannofossils) that could be used for positive dating. I sent the matrix off to David Watkins at the University of Nebraska and a few weeks later received a note confirming what we had reported earlier. Based on the remains of the tiny forams and other nannofossils that had lived at the same time, the bone had come from the upper part of the Jetmore Chalk Member of the Greenhorn Formation and was upper Cenomanian in age. So we were then able to resubmit the paper confirming the age of the specimen. In order the get around the reviewer's objection that the vertebra was not from a mosasaur or even a mosasauroid, we referred to it as a 'squamate' vertebra, a higher classification that includes coniasaur lizards and snakes (Everhart and Pearson, 2014). After publication, I learned of a more complete specimen of a mosasauroid from the Cenomanian rocks

Genus *Clidastes*	Cope, 1868b
Clidastes liodontus	Merriam, 1894
Clidastes propython	Cope, 1869b
Clidastes sp.	Informally named *Clidastes moorevillensis* (see Bell, 1997b)
Genus *Globidens*	Gilmore, 1912
Globidens alabamaensis	Gilmore, 1912
Globidens dakotensis	Russell, 1975
Globidens sp. "Kansas"	(Right dentary; see Everhart, 2008b)
Globidens schurmanni	Martin, 2007
Genus *Halisaurus*	Marsh, 1869
Eonatator sternbergii	(Russell, 1970); see *Clidastes sternbergi* Wiman 1920
Genus *Mosasaurus*	Conybeare, in Parkinson (1822:298)
Mosasaurus hoffmanni	Mantell, 1829
Mosasaurus conodon	Cope, 1881
Mosasaurus ivoensis	Persson, 1863 (KUVP 1024)
Mosasaurus missouriensis	Harlan, 1834 (*Mosasaurus Maximiliana* Goldfuss 1845)
Mosasaurus maximus	Cope, 1869d
Mosasaurus sp.	(Undescribed specimens from the Weskan Shale, Kansas)
Genus *Platecarpus*	Cope, 1869d
Platecarpus tympaniticus	Cope, 1869d; includes *P. ictericus* and *P. coryphaeus*
Plesioplatecarpus planifrons	Cope, 1874 (originally *Clidastes planifrons* Cope)
Latoplatecarpus willistoni	Konishi and Caldwell, 2011
Latoplatecarpus nichollsae	(Cuthbertson et al., 2007)--See Konishi and Caldwell, 2011
Genus *Ectenosaurus*	Russell, 1967
Ectenosaurus clidastoides	Merriam, 1894 (Originally *Platecarpus clidastoides*)
Genus *Plioplatecarpus*	Dollo, 1882
Plioplatecarpus primaevus	Russell, 1967
Plioplatecarpus depressus	Cope, 1869d
Genus *Prognathodon*	Dollo, 1889 (*Prognathodon solvayi* Dollo)
Prognathodon rapax	(Cope, 1872a) See Williston, 1898a:180
Prognathodon overtoni	Williston, 1897
Prognathodon stadtmani	Kass, 1999
Genus *Selmasaurus*	Wright and Shannon, 1988
Selmasaurus johnsoni	Polcyn and Everhart, 2008
Genus *Tylosaurus*	Marsh, 1872b
Tylosaurus proriger	Cope, 1869c
Tylosaurus nepaeolicus	Cope, 1874
Tylosaurus kansasensis	Everhart, 2005a
Tylosaurus bernardi	(*Hainosaurus* Dollo, 1885; see also Nicholls, 1988)

A.

B.

9.3. The earliest mosasaur remains in Kansas occur in the Fairport Chalk Member of the Carlile Formation. A, dorsal view of a '*Platecarpus*-like' mosasaur frontal (top of the skull; KUVP 97200); scale bar = 5 cm (2 in); B, seven caudal vertebrae from a plioplatecarpine mosasaur from the Fairport Chalk (FHSM VP-17564); scale bar = 5 cm (2 in).

of Utah that has almost exactly the same dorsal vertebrae (Polcyn, pers. comm., 2015). Our original identification had been correct.

Ichthyosaurs and plesiosaurs had inhabited the oceans for millions of years, beginning in the Triassic, evolving into many forms and surviving several major extinction events (Ellis, 2003). For unknown reasons, the ichthyosaurs declined significantly after the beginning of the Cretaceous (Bakker, 1993; Russell, 1993; Lingham-Soliar, 2003) and are thought to have been extinct by the time that the earliest mosasaurs reentered the water. Shimada (1996) reported the only known specimen of an ichthyosaur from Kansas, a single vertebra (FHSM VP-2169) from the Early Cretaceous (Albian) Kiowa Shale.

Russell (1993) suggested that mosasaurs were able to fill many of the ecological niches left vacant by the demise of ichthyosaurs. About the

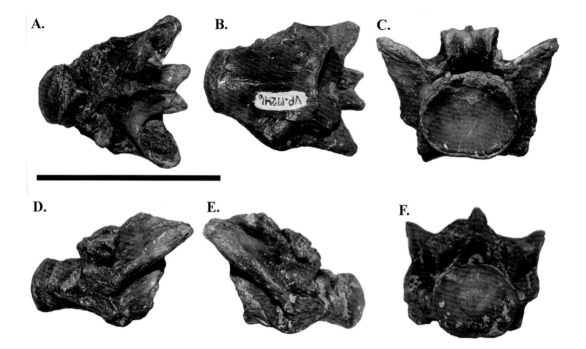

A. B. C.

D. E. F.

9.4. Mosasauroid dorsal vertebrae (FHSM VP-17246) from the Jetmore Chalk Member of the Greenhorn Formation (Lower Turonian) in: A, dorsal; B. ventral; C, anterior; D, right lateral; E, left lateral; and F, posterior view. Scale bar = 5 cm (2 in).

same time, Bakker (1993) suggested that plesiosaurs replaced the ichthyosaurs and mosasaurs may have replaced the marine crocodilians that had also disappeared by the beginning of the Coniacian in the Western Interior Sea.

In Kansas, we find a few crocodile remains (teeth and scutes) in the late Early Cretaceous Kiowa Shale (Beamon, 1999), and a few more specimens in the Cenomanian Dakota Formation, including the type specimen of *Hyposaurus vebbii*, described by Cope (1875) from a single vertebra recovered from a hand-dug well in Saline County. A more complete crocodile specimen in a concretion was described by Mehl (1941) as *Dakotasuchus kingi*, but then it was somehow misplaced. In 2014 I was able locate the type specimen at Kansas Wesleyan University, where it is currently on display after having been in storage for several years. Since then, a second, larger specimen of *D. kingi* has been discovered in the Cedar Mountain Formation (Cenomanian) of Emery County, Utah. The new material is in the process of being described (J. Frederickson, pers. comm., 2016).

In 1959 a partial skull and limb bones of a crocodyliform called *Terminonaris* were discovered in Russell County (FHSM VP-2079). There is a second, more unusual occurrence of *Terminonaris sp.* (FHSM VP-4387) in the middle Turonian Fairport of Russell County (Shimada and Parris, 2007). This specimen appears to have floated into the middle of the WIS from somewhere else. In addition, the giant crocodilian *Deinosuchus rugosus* lived along the coast of the Western Interior Sea during the Campanian (Schwimmer, 2002), but probably didn't venture far out to sea.

It is quite possible that both the ichthyosaurs and the plesiosaurs had been losing the evolutionary battle of "who eats what" to competition from faster, larger, and more advanced varieties of bony fish such as *Xiphactinus audax* (Massare, 1987; Lingham-Soliar, 1999a) and the giant ginsu sharks (*Cretoxyrhina mantelli*). Other groups of reptiles, including marine crocodiles, teleosaurs, placodonts, and turtles (Chapter 5), had also enjoyed limited successes in the oceans of the Mesozoic, but none approached the worldwide domination of the seas that mosasaurs would attain in the Late Cretaceous.

Plesiosaurs also appear to be less numerous in the Late Cretaceous than during the Jurassic (there is no Jurassic record in Kansas) and had evolved into specialized forms like the slow-moving, long-necked *Elasmosaurus* and *Styxosaurus*, and the fast-swimming, short-necked *Trinacromerum*, *Polycotylus*, and *Dolichorhynchops* (Carpenter, 1996; Adams, 1997; O'Keefe, 2008). Initially, the short-necked plesiosaurs (polycotylids) that lived in the ocean over Kansas were much smaller than their distant Jurassic cousin, *Liopleurodon*; an Early Cretaceous relative, *Kronosaurus*, from Australia and South America; and the Late Cretaceous pliosaurs from Kansas, *Brachauchenius lucasi* and *Megacephalosaurus eulerti*. Similar to the trend shown in nearly all marine reptiles from the Mesozoic, polycotylids would grow much larger during the Campanian.

During deposition of the chalk in the Late Cretaceous, however, mosasaurs outnumbered the remaining elasmosaurs and polycotylids by more than ten to one, at least in terms of the number of museum specimens recorded from the Niobrara (Russell, 1988). In support of Russell's data, my wife and I collected 70 mosasaur specimens and just five fragmentary plesiosaur specimens between 1988 and 1995, mostly in the lower chalk of Gove County. The likely bias here is that we collected fossils preserved in the middle of the Western Interior Sea, and plesiosaurs may have preferred to live close to shore.

The early ancestors of mosasaurs probably fed in shallow coastal waters and returned to land to rest and breed much like the marine iguanas that inhabit the Galapagos Islands today, and to a lesser extent like modern sea snakes. Several species of modern monitor lizards (varanids) spend much of their time in the water, but are otherwise terrestrial.

Over the space of a few million years, these ancestral mosasaurs became much larger and more specialized, evolving rapidly into a diverse assortment of highly successful marine predators (Table 9.2). By the middle of Coniacian time (87 Ma), the 'first wave' of mosasaurs (*Tylosaurus*, *Platecarpus*, and *Clidastes*) was well established in the Western Interior Sea that covered Kansas and most of the central portions of North America (Williston, 1898a; Russell, 1967; Everhart, 2001, 2005b; Polcyn, et al., 2008). As mosasaurs continued to evolve, growing larger and diversifying rapidly, a 'second wave' of genera and species showed up about the beginning of the Campanian (83.5 Ma). Following a possible near-extinction event near the middle of the Campanian reported by Lindgren

Table 9.2. Approximate duration of ages that comprise the Upper Cretaceous in the Western Interior Sea of North America, with remarks regarding mosasaurs.

Age	Time Span	Remarks
Maastrichtian	71.3–65.4 Ma (5.9 m.y.)	Greatest diversity and distribution; invasion of freshwater habitats. Beginning of the 'third wave' of mosasaurs.
Campanian	83.5–71.3 Ma (12.2 m.y.)	'Second wave' mosasaurs; *Tylosaurus bernardi, Mosasaurus, Globidens, Plioplatecarpus, Prognathodon*
Santonian	85.8–83.5 Ma (2.3 m.y.)	Mosasaurs get much larger; worldwide distribution
Coniacian	89.0–85.8 Ma (3.2 m.y.)	'First wave' mosasaurs; *Tylosaurus, Clidastes,* and *Plesioplatecarpus*
Turonian	93.5–89.0 Ma (4.5 m.y.)	Early mosasaurs; *Clidastes* and *Tylosaurus/Platecarpus* precursors
Cenomanian	99.0–93.5 Ma (5.5 m.y.)	Ancestral mosasaurs (aigialosaurs?) return to the ocean, mosasauroids in Kansas, Texas, Utah

and Siverson (2004), mosasaurs rebounded, and a very diverse 'third wave' was just getting underway during the final years of the Cretaceous, shortly before their final extinction.

Tylosaurus proriger (Cope 1869a; Figs. 9.1, 9.2) was the third new species of mosasaur to be named from North America, after *Clidastes* Cope (1868b) and *Platecarpus* Cope (1869d), and the second to be named from the Western Interior Sea. Most other discoveries up to that point, including the skull from South Dakota described by Goldfuss (1845) had been referred to the European genus, *Mosasaurus*. And actually, *Tylosaurus* didn't start out to be the genus name of the first mosasaur from Kansas. Unfortunately for Cope, his rival, O. C. Marsh, gets credit for coining the genus name *Tylosaurus*.

Tylosaurus

The type specimen of *Tylosaurus* (MCZ 4374) was originally named *Macrosaurus proriger* by Cope in 1869. The description of the new species by Cope (1869a) in the minutes of the Academy of Natural Sciences of Philadelphia meeting of June 1 was brief by modern standards, referring the species to a European genus, *Macrosaurus* Owen. The following year Cope (1870.pl. 12, figs. 22, 23; Fig. 9.5) described the same specimen more completely, figured it, and then, with little explanation, referred it to another European genus, *Liodon* Owen. Two years later, O. C. Marsh, apparently recognizing significant differences between the two American and European mosasaurs, proposed a new genus (*Rhinosaurus*: meaning 'nose-lizard') from a more complete specimen (YPM 1268—*Rhinosaurus micromus*) that he had collected "on the south side of the Smoky Hill River" in 1871 (Marsh, 1872b: 462). However, that genus name had already been used, and Cope (1872b) proposed the genus name *Rhamphosaurus* as a substitute.

In a brief note, Marsh responded that "as this name [*Rhinosaurus*] proves to be preoccupied, it may be replaced with *Tylosaurus*. The name *Rhamphosaurus*, since suggested by Prof. Cope, cannot be retained, as it was given to a genus of lizards in 1843 by Fitzinger" (1872c:147). Leidy

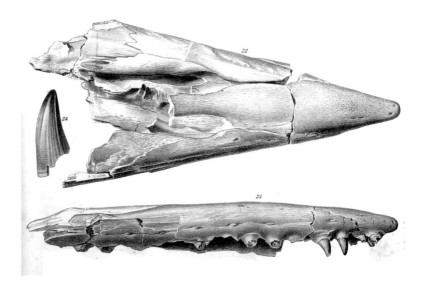

9.5. The type specimen of *Tylosaurus proriger* (MCZ 4374) as figured by Cope (1870:pl. XII). According to Cope, the skull fragment is about 78 cm (31 in) in length.

(1873:274) was the first to publish Marsh's generic name, *Tylosaurus proriger*, as the correct genus for '*Macrosaurus*' *proriger* Cope, 1869. That said, Cope didn't give up; in a paper he wrote the following disclaimer as part of his description of a another new species, *Rhamphosaurus (Tylosaurus) nepaeolicus*: "This name [*Rhamphosaurus*] was applied by Fitzinger to two species of lizards, which had already received several generic names, and hence became at once a synonym. Further, he did not characterize it; for these reasons the name was not preoccupied at the time I employed it as above; hence there is no necessity for Prof. Marsh's subsequent name *Tylosaurus*, given on the supposition of preoccupation" (1874:36). By then, however, Marsh's name had been accepted, and is unlikely to be changed. (There is more to the story later in this chapter.)

The meanings of mosasaur names are of some interest because they give some insight into certain characteristics that early workers saw in the first fossil remains. *Tylosaurus* (Marsh, 1872b) means 'knob (or snout) lizard' after the elongated bony muzzle or rostrum that projected beyond the anterior-most teeth of the upper and lower jaws (Fig. 9.6). Three species of *Tylosaurus* are known from the chalk: *T. proriger* (Cope 1869b—'prow-bearing'); *T. nepaeolicus* (Cope 1874; from the Nepaholla River, an old name for the Solomon River in north-central Kansas; Everhart, 2002a); and *T. kansasensis*, a third species that I named in 2005 (Stewart, 1990; Bell, 1997b; Everhart, 2004b, 2005a). The species name honors the Kansa (Kaw) Indians who once inhabited Kansas.

Tylosaurus kansasensis Everhart (2005a) is the most common mosasaur in the lower chalk (Late Coniacian). One of the unusual things that I've noticed about this species is that there are a large number of nicely articulated skulls in various collections around the country, including the type specimen (FHSM VP-2295) at the Sternberg Museum. The first known complete skull (YPM 1336) was collected by B. F. Mudge from Trego County in 1875. Originally identified as *Tylosaurus proriger*, it was

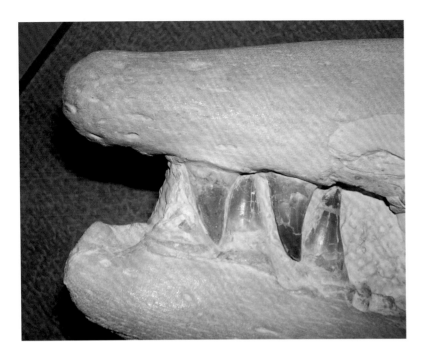

9.6. The toothless muzzle or rostrum that extends in front of the anterior teeth in *Tylosaurus proriger* (FHSM VP-3) and other tylosaurine mosasaurs. Note that the anterior ends of the lower jaws are also toothless.

still sitting in a plaster jacket when I visited the Yale Peabody Museum in 2010. Besides having the characteristics described for *T. kansasensis*, the fact that *T. proriger* does not occur in the lower chalk, or Trego County, makes the original identification suspect. G. F. Sternberg collected a complete skull (MCZ 1031) in 1924 from the low chalk of northwest Ellis County; it was sold to the Harvard Museum of Comparative Zoology. A note by George Sternberg on his photographs of the MCZ 1031 skull says, "*Tylosaurus*, species not determined" (Everhart, 2008c:fig. 9). He collected yet another specimen (FM VP43; Fig. 9.7) in 1939 and sold it to the Fryxell Museum of Geology at Augustana College, Rock Island, Illinois. Unfortunately, Sternberg did not provide a record of the locality for this specimen.

J. D. Stewart collected the paratype specimen (LACM 127815) of the species from the low chalk of Rooks County in 1966. The type specimen (FHSM VP-2295) was collected from northwest Ellis County by Myrl Walker in 1968, not too far above the contact between the Smoky Hill Chalk and the underlying Fort Hays Limestone. Coincidentally, I collected my first mosasaur, a larger *T. kansasensis* (FHSM VP-13742), in 1979 from within a couple of miles of the type specimen.

The type specimen (FHSM VP-2295) was originally identified in museum records as "*Clidastes* or *Tylosaurus*" when first placed on exhibit in the original Sternberg Museum, was later changed to *Platecarpus*, then to *Tylosaurus nepaeolicus*, before being relegated to *Tylosaurus* n. sp. (new species by Stewart (1990) and *Tylosaurus* sp. novum by Bell (1993). During the 1990 Society of Vertebrate Paleontology field trip, Stewart, Bell, and I looked at the specimen at the Sternberg Museum and agreed it was something different, but it still didn't get described. Years went by,

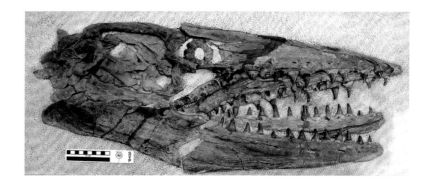

9.7. The skull of *Tylosaurus kansasensis* (FM VP43) in right lateral view in the Fryxell Museum of Geology, Augustana College, Rock Island, Illinois. The specimen was collected by George F. Sternberg in 1939. The lower jaw measures 82.5 cm (32 in).

and I did a presentation (Everhart, 2004b) at the First Mosasaur Meeting in Maastricht. I ended with the suggestion that it was time to give the new species a name, and that's when Lou Jacobs (pers. comm., 2004) said, "Well, get it done."

My paper naming *Tylosaurus kansasensis* was published in 2005, along with others from the First Mosasaur Meeting. At the Second Mosasaur Meeting in Hays, Mike Caldwell (pers. comm., 2007) suggested that I had not adequately diagnosed the species and said I should do so in a follow-up paper. At the annual meeting of the Society of Vertebrate Paleontology in 2013, he and a student coauthored an abstract (Jiménez-Huidobro and Caldwell, 2013) questioning whether or not *T. kansasensis* could be separated from *T. nepaeolicus*. Then, as this book was being revised, a paper was published that basically said *T. kansasensis* was simply a juvenile form of *T. nepaeolicus* (Jiménez-Huidobro, Simões, and Caldwell, 2016). Although I strongly disagree with their hypothesis and believe the 2016 paper to be poorly researched and written, any comments I would make at this point would be considered to be sour grapes. *Tylosaurus kansasensis* remains a valid species until proven otherwise.

On a positive note, *Tylosaurus proriger* and *Pteranodon longiceps* were named as the official State Fossils of Kansas in 2014. I was able to testify before both House and Senate committees as part of the citizen campaign promoting these two iconic species, discovered first and occurring mostly in Kansas, and was pleased that the legislature and the governor agreed with House Bill HB2595.

Platecarpus

Platecarpus Cope 1869 means 'oar (or paddle) wrist' for the flatness and relative immobility of the wrist joint. Several species of *Platecarpus* were named initially, some from fragmentary remains, but current workers (Russell, 1967; Stewart, 1990; Bell, 1997b; Everhart, 2001) now agree that most of the earlier names (*P. coryphaeus* Cope and *P. ictericus* Cope) are synonymous with *Platecarpus tympaniticus* Cope 1869 (referring to the difference in the form of the tympanic bone or quadrate). *P. tympaniticus* was originally discovered in Mississippi by Dr. William Spillman (Manning, 1994) and was later recognized in the fauna of the Western Interior Sea (Russell, 1967).

The fragmentary type specimen (AMNH 1491) of *Platecarpus plani-frons* (Cope 1874) (meaning 'flat forehead/frontal bone') was collected in 1873 from Trego County by Professor B. F. Mudge. It was originally identified and named as *Clidastes planifrons* by Cope (1874). In the first comprehensive review of mosasaur species, Williston (1898a) determined that the remains were not those of a *Clidastes* and renamed it *Platecarpus planifrons*. However, because the type specimen was so fragmentary, Russell (1967:176) considered the specimen to be of "uncertain taxonomic position." *P. planifrons* was again recognized in the Smoky Hill Chalk by Bell (report of pers. comm. by Schumacher, 1993; Bell, 1997b).

All of this history is leading up to the naming of a new mosasaur species in Kansas discovered by an amateur collector. In 1996 Steve Johnson of Wichita, Kansas, collected several small mosasaur vertebrae from the surface of the chalk while on a family outing. When he showed them to me, he told there were more bones in the chalk. We made arrangements to meet him and his family at the locality, and as I had hoped, we collected a nearly complete but disarticulated skull and other remains from just below marker unit 7 (lower Santonian). Steve Johnson was pleased with his discovery and generously donated the specimen (FHSM VP-13910) to the Sternberg Museum. After preparing the bones, I realized that it was something I hadn't seen before, and I could not identify it for several months.

However, the quadrates were very much like the drawing published by Cope (1875:pl. XXXV, fig. 25) for *Platecarpus (Clidastes) planifrons*. So we reported the discovery (Everhart and Johnson, 2001) at the annual meeting of the Society of Vertebrate Paleontology. By then, Triebold Paleontology had made casts of the bones and assembled a 3-D cast of the skull that I took one along for 'show and tell' during the SVP poster session in Bozeman, Montana. I even had a T-shirt made for presenters at the 2002 Paleontology Symposium at the Kansas Academy of Science annual meeting to celebrate the discovery.

At the 2007 Second Mosasaur Meeting at the Sternberg Museum in Hays, Kansas, Mike Polcyn looked at the actual specimen and said, "No, this is something really different." So, making this long story short, we described a new species of mosasaur from Kansas and named it *Selmasaurus johnsoni* (Fig. 9.8) in honor Steve Johnson. The type specimen of the genus (*S. russelli*) was originally described from the Selma Formation of Alabama by Wright and Shannon (1988). Konishi (2008) reported on a second specimen from Alabama, but until Steve Johnson discovered the Kansas specimen, the genus was unknown outside the Gulf. Currently *S. johnsoni*, a plioplatecarpine mosasaur, is considered to be most closely related to *Ectenosaurus*. The species is characterized in part by a shortened skull and the fewest number of teeth in the maxilla (10) and dentary (11) known in any mosasaur (Table 9.3). Note that Burnham (1991) reported that an as yet unnamed species of *Plioplatecarpus* (UNO 8611–2) from the Demopolis Formation in Alabama had 11 teeth in one dentary and 12 in the other.

9.8. A cast of the reconstructed skull (length 38 cm; 15 in) of *Selmasaurus johnsoni* (FHSM VP-13910) in the collection of the Sternberg Museum of Natural History. This specimen was discovered in Gove County by Steve Johnson of Wichita, Kansas, in 1996, and is the only known complete skull of this rare species.

The foregoing story highlights the difficulty in identifying mosasaur remains based on descriptions from more than a century ago. Although many mosasaur workers have agreed that Cope's original descriptions are inadequate for separating the various species of *Platecarpus*, there had been little effort until recently to straighten out the problems. Fortunately, there are now many excellent specimens available for study.

Konishi and Caldwell (2007) examined new material from various collections and determined that *Platecarpus planifrons* was a valid species in the Western Interior Sea along with *Platecarpus tympaniticus*. In a follow-up paper, Konishi and Caldwell (2011) named a new species of plioplatecarpine mosasaur, *Latoplatecarpus willistoni*, from the Bearpaw Shale (Campanian) of Montana, and moved *P. planifrons* into a new genus, *Plesioplatecarpus planifrons*. *Platecarpus tympaniticus* Cope was redescribed by Konishi, Caldwell, and Bell (2010), and then discussed in greater detail by Konishi et al. (2012) on the basis of their examination of one of the best-preserved Kansas specimens (LACMNH 128319).

Bottom line for those of us working in the Smoky Hill Chalk is that *P. planifrons* occurs in the lower chalk (Upper Coniacian–Middle Santonian) and is replaced by *P. tympaniticus* in the upper chalk (Middle Santonian–Lower Campanian). *P. tympaniticus* appears to be then replaced by *Plioplatecarpus primaevus* in the Sharon Springs Member of the Pierre Shale.

Clidastes

Clidastes Cope 1868 roughly translates as 'one who locks up' and describes the bony processes (called the zygosphene and zygantrum) on the vertebrae of this genus, which restrict movement and limit the bending of the dorsal portion of the vertebral column (Russell, 1967). This was a necessary adaptation to stiffen the back of the animal for more efficient swimming using its broad tail. *Clidastes propython* Cope

TABLE **9.3** MOSASAUR
TOOTH COUNTS

	Maxilla	Dentary	Pterygoid	Citation/Specimen #
Mosasauridae				
Eonatator sternbergii	unk	22	unk	RMM 6890
Russellosaurinae				
Tylosaurus kansasensis	12	13–14	10–11	Everhart, 2005a
Tylosaurus nepaeolicus	13	13	10	FHSM VP-7262
Tylosaurus proriger	13(12)	13	10	Russell 1967
Ectenosaurus clidastoides	17	16?	9–11	FHSM VP-401
Selmasaurus johnsoni	11–12	11	11	Polcyn and Everhart, 2008
Plesioplatecarpus planifrons	10	11	9	FHSM VP-13910
Platecarpus tympaniticus	12	12(11 rarely)	10–12	Russell, 1967
Plioplatecarpus primaevus	11	12	13	Holmes, 1996
Plioplatecarpus UNO 8611-2	12	11–12	10–13	Burnham, 1991
Mosasaurinae				
Clidastes liodontus	14–15	16	15	Russell, 1967
Clidastes propython	18	17–18	14	Russell, 1967
Globidens alabamaensis	12–13?	unk	unk	Gilmore, 1912
Globidens dakotensis	13	unk	0	Russell, 1975
Globidens "Kansas"	unk	13–14	unk	Everhart, 2008b
Globidens schurmanni	12–13	13–14	unk	Martin, 2007
Mosasaurus missouriensis	14	15	8–10	Russell, 1967
Mosasaurus conodon	15	17	10	Russell, 1967
Prognathodon overtoni	12	14	6	SDSMT 3393
Prognathodon stadtmani	11–12	14	6	Kass, 1999
Plesiotylosaurus crassidens	13	16–17 unk		Camp, 1942
Plotosaurini				
Plotosaurus bennisoni	18	17	15	Camp, 1942
Plotosaurus tuckeri	18	17+/–	12–13	Camp, 1942
European Species				
Mosasaurus hoffmanni	14	14	8	Lingham-Soliar, 1995
Mosasaurus maximus	13–14	14	8	Russell, 1967
Tylosaurus bernardi	12	13	8?	Lingham-Soliar, 1992
Prognathodon solvayi	12	13	8	L.-Soliar and Nolf, 1989
Plioplatecarpus marshi	12/13	1?	11	Lingham-Soliar, 1994
Plioplatecarpus houzeaui	12	11	?	Lingham-Soliar, 1994
African Species				
Goronyosaurus nigeriensis	11	12	8	Lingham-Soliar, 1999a
Pluridens walkeri	unk	30	unk	Lingham-Soliar, 1998

Abbreviations: FHSM—Fort Hays Sternberg Museum, Hays, KS; RMM—Red Mountain Museum, Birmingham, AL; SDSMT—South Dakota School of Mines and Technology, Rapid City, SD; UNO—University of New Orleans, New Orleans, LA; unk—unknown.

1869 ('before python,' refers to the snakelike affinities Cope saw when he classified mosasaurs into a separate order called Pythonomorpha) was a relatively late arrival (Middle Santonian) in the Western Interior Sea and was far more common along the Gulf Coast (Plate 7). *Clidastes liodontus* Merriam 1894 ('smooth-toothed') first appears in the Smoky Hill Chalk during late Coniacian time (FHSM VP-13909; Everhart, 2001).

Clidastes sternbergii was described by Wiman (1920) but was later transferred to the genus *Halisaurus* (Marsh 1869—'sea lizard') by Russell (1970) on the basis of similar material occurring in Alabama. The only reasonably complete Kansas specimen was discovered in Logan County in 1918 by Levi Sternberg (Sternberg, 1922), and purchased for the Uppsala University in Sweden. Bardet and Suberbiola (2001) reexamined the Kansas remains (UPI R 163) and concluded that it is the oldest representative of the genus *Halisaurus* and the most primitive mosasaurid (see also Bell, 1997b; Lindgren and Siverson, 2004). Then Bardet et al. (2005) revised their conclusion, noting that it really wasn't like other halisaurine mosasaurs, and moved it into a new genus (*Eonatator*—'dawn swimmer'). The following year, Lindgren and Siverson (2005:764) considered Eonatator to be an "invalid junior synonym" and retained the genus *Halisaurus*. At the Fifth Mosasaur Meeting in May 2016, *Eonatator* was again use in a presentation by Polcyn and Bell (2016:31). So whatever you call it, at this point *Clidastes-Halisaurus-Eonatator sternbergii* remains one of the most primitive and enigmatic mosasaurs known.

Another unusual and apparently rare species in the Smoky Hill Chalk is *Ectenosaurus clidastoides* Merriam 1894. Merriam (1894:30) originally named the species *Platecarpus clidastoides* because of its *Clidastes*-like appearance. However, Russell (1967:156–158) disagreed with its placement within *Platecarpus* and erected a new genus (*Ectenosaurus*—'drawn-out lizard') to describe the unusually long, narrow snout. The type specimen was probably destroyed in WWII (ibid.). The most complete specimen known (FHSM VP-401; Fig. 9.9), missing only the tail and hind limbs and including lots of preserved cartilage and scale impressions, was collected by George Sternberg northwest of WaKeeney in Trego County in 1953 (ibid.). A reconstruction of this skull is figured by Russell (ibid.:fig. 86). One of the distinguishing features of this species is its long, narrow skull and relatively large number of teeth, especially when compared to *Platecarpus* (17 versus 12 in the maxilla of *Platecarpus*—see Table 9.3).

When George Sternberg was preparing the specimen, he noted that there were impressions of the scale-covered skin under in the throat region of the specimen. Sternberg was careful to preserve the scale impressions in separate storage when he mounted the rest of the mosasaur in plaster. Although they were mentioned by Russell (1967:70), no one had actually examined the scale impressions. After Johan Lindgren saw the specimen at the Second Mosasaur Meeting in 2008, he and I began to work together on describing the material. Our examination of my initial close-up photographs (Fig. 9.10) indicated that we were looking at something more than just the impression of the scales. Subsequent

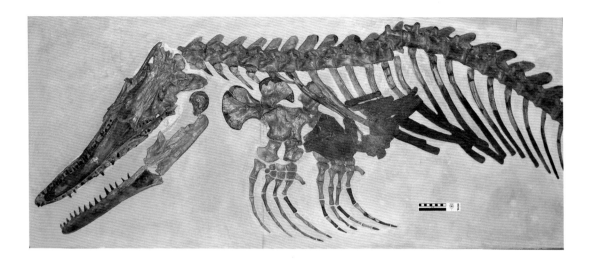

testing by Lindgren under a scanning electron microscope (SEM) confirmed that what we were seeing was the preserved (phosphatized) skin of the mosasaur, complete with layers of preserved fibers. This was the first time that the detailed structure of the skin of a mosasaur was described (Lindgren, Everhart, and Caldwell, 2011). See Lindgren et al. (2009) for an earlier description of the scales of a California mosasaur (*Plotosaurus*).

Tylosaurus was the largest of the early mosasaurs, arriving long before *Mosasaurus* (Fig. 9.11). *Tylosaurus* reached about 9 m (29.5 ft) in length by Santonian time (Everhart and Everhart, 1997; Everhart, 2001) and were even larger by the early Campanian. They were the 'heavyweights' of the first wave of mosasaurs, with robust, conical teeth capable of capturing a variety of prey, including birds and other mosasaurs (Martin and Bjork, 1987) and plesiosaurs (Sternberg, 1922; Everhart, 2004a). *Platecarpus* was probably the most common in terms of the sheer number of individuals in the first wave and reached lengths of about 7 m (23 ft). Judging by its slender, backward-curving teeth (Massare, 1987; Russell, 1970), it is likely that *Platecarpus* fed on smaller, soft-bodied prey including squid and other cephalopods that were abundant at the time. *Clidastes*, the smallest and probably the most primitive of the three major groups, was generally less than 5 m (16 ft) in length when full grown and fed on fish and other small marine creatures. Recently, several specimens of an unusually large (6–7 m) *Clidastes* sp. have been discovered in the Smoky Hill Chalk and also noted to be in the collection of the Yale Peabody Museum from the 1870s (Ott, Behlke, and Kelly, 2002). Although they are nominally assigned to *C. moorevillensis* (an informal name), a formal description has not yet been published.

By the early Campanian, *Tylosaurus* had reached lengths of 13–14 m (42–45 ft), as evidenced by the 'Bunker Mosasaur' specimen on exhibit in the Natural History Museum at the University of Kansas. The remains of this huge tylosaur (KUVP 5033) were discovered just above the contact

9.9. The skeleton of *Ectenosaurus clidastoides* (FHSM VP-401) in the collection of the Sternberg Museum of Natural History. The remains were collected from the middle Santonian chalk of Trego County by George F. Sternberg in 1953. This is the most complete specimen known of this rare species. The specimen also includes preserved scales and skin fibers (Lindgren, Everhart, and Caldwell, 2011). Scale bar = 10 cm (4 in).

9.10. Preserved scales and scale impressions from the neck region of *Ectenosaurus clidastoides* (FHSM VP-401). Scale bars in mm.

between the Smoky Hill Chalk and the Pierre Shale near Wallace, Kansas, in 1911. The group, led by Charles D. Bunker, was collecting birds for the university's ornithology collection, but the huge mosasaur specimen was too important to pass up. The rest of the KUVP 5033 specimen was eventually collected by H. T. Martin and students from the University of Kansas. The exhibit version by Triebold Paleontology at KU represents one of the largest mounted mosasaur skeletons in the United States. The Sternberg Museum has a similar-sized but incomplete *Tylosaurus* specimen (FHSM VP-2496) in its collection, which is also from the Sharon Springs Member of the Pierre Shale. The bases of the teeth of the VP-2496 specimen are almost as large as a man's clenched fist (Schumacher, 1993; Fig. 9.12). Similar very large *Tylosaurus* remains have also been discovered in South Dakota and Texas, including a nearly complete specimen ('Sophie') collected by Triebold Paleontology from near Dallas.

Later in the Campanian, a 'second wave' of new species began making their appearance as mosasaurs diversified and dispersed into the Cretaceous oceans around the world (Table 9.1). *Prognathodon* and *Globidens* (Plate 10) evolved more robust skulls and heavy, crushing teeth, most likely for eating hard-shelled prey. *Plioplatecarpus* replaced *Platecarpus*, and *Mosasaurus* appears to have replaced *Clidastes*. Tylosaurine mosasaurs such as '*Hainosaurus*' (U.S. and Europe), *Taniwhasaurus oweni* (New Zealand), and a new species (*Lakumasaurus antarcticus*) from Antarctica (Novas et al., 2002) are evidence of the worldwide distribution of this group. Note that *Hainosaurus* has since been determined to be a junior synonym of *Tylosaurus* (Bullard and Caldwell,

9.11. *Tylosaurus proriger* was the largest of the mosasaurs in the Western Interior Sea during the deposition of the Smoky Hill Chalk, reaching 10 m or more in length. The larger mosasaurs, such as *Tylosaurus*, preyed on a variety of marine vertebrates, including fishes, plesiosaurs, birds, and even other smaller mosasaurs. Drawing by Russell Hawley.

2010; Jiménez-Huidobro and Caldwell, 2016) and that *Lakumasaurus* is synonymous with *Taniwhasaurus* (Martin and Fernández, 2007). In the Western Interior Sea, this means that the very large tylosaurine from Canada, *Hainosaurus pembinensis* described by Elizabeth (Betsy) Nicholls, is now officially *Tylosaurus pembinensis* (Nicholls, 1988).

Highly derived mosasaurs such as the ichthyosaur-like *Plotosaurus* and *Plesiotylosaurus* that lived in the Pacific Ocean off the coast of California (Camp, 1942), and a number of poorly known species from Africa suggest that some mosasaurs were also evolving in other directions as isolated populations in these areas. Within the space of a few more million years, by Maastrichtian time (71 my), many mosasaur species were truly huge, with several lineages (*Mosasaurus* and *Tylosaurus*) reaching more than 15 m (50 ft) in length. By then *Mosasaurus* was widely distributed around the northern hemisphere, from western Europe and Russia, and across North America. Most recently, *Mosasaurus hobetsuensis* (Plate 12) was described from northern Japan by Suzuki (1985).

9.12. Fragment of the jaw of a very large *Tylosaurus proriger* (FHSM VP-2496) in dorsal view, showing the bases of four teeth. The specimen was collected from the lower Sharon Springs Member of the Pierre Shale in Logan County. The roots of the teeth are about 6 cm (2.5 in) in diameter. The *Tylosaurus* would have been about 14 m (45 ft) long in life. Scale bar = 10 cm (4 in).

The giant specimen of *Mosasaurus hoffmanni* from Europe described by Lingham-Soliar (1999b; see also Meijer, 1971), from a partial dentary (NHMM 009002) with 13 tooth positions that is 0.9 m (36 in) in length, in the collection of the Natuurhistorisch Museum Maastricht, was 17 m (almost 56 ft) in length. This specimen was revisited in our talk ("Mosasaurs—How Large Did They Really Get?") at the Fifth Mosasaur Meeting in Uppsala, Sweden (Everhart, et al. 2016), along with a truly huge quadrate (NHMM 003892) that suggests an even larger *Mosasaurus* (18 m—59 ft). The *Mosasaurus hoffmanni* specimen reported by Grigoriev (2014) from Penza, Russia, is also in the 17 m size range, as is the *Tylosaurus (Hainosaurus) bernardi* from Belgium. Note that a recent paper by Street and Caldwell (2016) provides a new description of the type specimen of *Mosasaurus hoffmanni* and its relationship to other species of *Mosasaurus*. Specimens of *Prognathodon* from Utah (Kass, 1999), the Netherlands (Dortangs et al., 2002), and Israel (Christiansen and Bonde, 2002) suggest that this genus was also becoming larger and more heavily built. The fossil record indicates that mosasaurs were becoming very diverse and were able to adapt to a variety of available prey.

During this age of the last great marine reptiles, there was no doubt which were the biggest and most dangerous predators in the oceans of the Late Cretaceous. Mosasaurs continued to grow throughout their lives, although much more slowly as they got older, and we don't know just how large they could have been. It is highly unlikely that we will ever find the largest member of any species; the odds against preservation and subsequent discovery of a specific individual are just too great. That said, 20 m (65 ft) is probably about the absolute maximum body length of extremely old individuals of *Mosasaurus* and *Tylosaurus*, regardless of what is portrayed in current and very popular dinosaur movies.

A complete listing of mosasaur species in the Smoky Hill Chalk is still being worked out. Even though they have been collected in Kansas for more than 130 years, only recently has much attention been paid to their biostratigraphy (when and where they occurred). Williston (1897) briefly discussed the stratigraphic occurrence of mosasaurs in the Pteranodon Beds (upper Smoky Hill Chalk) for the first time. He also made the observation that *Clidastes* does not occur in the Rudistes Beds (lower chalk), indicating that other genera probably occurred within 100 feet of the contact of the chalk with the underlying Fort Hays Limestone. To date, only one set of mosasaur remains, a *Tylosaurus* sp., has been collected from the Fort Hays Limestone (FHSM VP-2297; Everhart, 2005b).

In the years that followed, S. W. Williston, Charles H. Sternberg and his sons, H. T. Martin, and others continued to collect spectacular examples of mosasaurs and other marine species from the Smoky Hill Chalk without adding significantly to the knowledge of the stratigraphic record of these creatures. Williston (1898a) published the first comprehensive description of the systematics and comparative anatomy of mosasaurs from the Smoky Hill Chalk, and discussed their range and distribution in comparison with specimens discovered earlier in New Zealand and Europe. He commented that *Tylosaurus*, "so far as was known, begins near the lower part of the Niobrara [Smoky Hill Chalk] and terminates at its close or in the beginning of the Fort Pierre [Pierre Shale]" (ibid.:88). Of *Platecarpus*, he stated that the species on which the genus is based are "known nowhere outside of Kansas and Colorado, and are here restricted exclusively to the Niobrara" (ibid.). He again concluded that the lowest horizon of *Clidastes* "is the upper part of the Niobrara in Kansas" (ibid.:89).

Russell (1967) reviewed the specimens collected by the 1870s Yale scientific expeditions and postulated that the Smoky Hill Chalk could be divided into a lower, *Clidastes liodontus*—*Platecarpus coryphaeus*—*Tylosaurus nepaeolicus* zone and an upper, *Clidastes propython*—*Platecarpus ictericus*—*Tylosaurus proriger* zone. He also suggested that the increased abundance of *Clidastes* specimens in the upper portion of the chalk was an indication of a gradual change from a mid-ocean to a nearshore environment. Russell (1970) noted significant differences between the distribution of mosasaur species in the Smoky Hill Chalk compared to Gulf Coast species occurring in the Selma Formation of Alabama. In his initial work concerning the biostratigraphy of the Smoky Hill Chalk, Stewart stated that he was aware of several exceptions to Russell's 1967 stratigraphic distribution of mosasaurs in the Smoky Hill Chalk that caused him to regard it with "a degree of skepticism" (Stewart, 1988:81). Field work since that time has shown that Stewart was correct, but I give Russell a lot of credit for the conclusions he was able to draw from old specimens and incomplete collection information. Even after 50 years, his *Systematics and Morphology of American Mosasaurs* (1967) is still the best reference available about mosasaurs.

It was not until Hattin (1982) published his composite measured section of the Smoky Hill Chalk that significant progress could be made in documenting the vertebrate biostratigraphy of this formation, including mosasaurs. Hattin used bentonites (layers of volcanic ash) and other physical features of the chalk to delineate his 23 lithologic marker units and divided the chalk into five biostratigraphic zones based on the occurrence of invertebrate species. In doing so, he provided field workers with the first dependable method of determining their stratigraphic location in the section.

Stewart (1990) incorporated Hattin's marker units as upper and lower boundaries for his six proposed biostratigraphic zones (Table 2.1). He also provided the first comprehensive description of the distribution of known invertebrate and vertebrate species in the Niobrara Formation and made the first attempt to assign specific stratigraphic ranges for mosasaur species within the Smoky Hill Chalk.

Schumacher (1993) and Sheldon (1996) reviewed existing collections of mosasaur material in the Sternberg Museum of Natural History, the Yale Peabody Museum, and other institutions and further refined the occurrence of mosasaurs in the Smoky Hill Chalk by building upon the stratigraphic methodology provided by Hattin (1982) and Stewart (1990). Since that time, additional discoveries (Everhart, 2001; Everhart and Johnson, 2001; Everhart, 2002a; Everhart, 2005a, 2005b; Polcyn, et al., 2008; Polcyn and Everhart, 2008) of mosasaur remains with associated stratigraphic data have aided in the refinement of the known temporal ranges of the most common species: *Tylosaurus proriger, T. nepaeolicus, T. kansasensis, Platecarpus tympaniticus, Plesioplatecarpus planifrons, Clidastes liodontus,* and *C. propython.*

Based on the latest discoveries, we can now say that the genus *Tylosaurus* occurs throughout the chalk, with *T. nepaeolicus* and *T. kansasensis* (Everhart, 2005a) living together during the late Coniacian and being replaced by (or evolving into) *T. proriger* by the early Santonian. Tylosaurs appear to continue well into the Campanian, and possibly the Maastrichtian, as several species, including the very large *Tylosaurus bernardi* and *T. pembinensis,* which evolved during the Campanian.

Plesioplatecarpus planifrons (Konishi and Caldwell, 2011) first appears in the lower chalk (late Coniacian) and is replaced by *Platecarpus tympaniticus* in the middle Smoky Hill Chalk. Note that *P. tympaniticus* is the senior synonym of both *P. coryphaeus* and *P. ictericus* (Russell, 1967). Sometime in the early Campanian (Pierre Shale) *P. tympaniticus* is then replaced by one or more species of *Plioplatecarpus.*

The genus *Clidastes* may have first appeared during the deposition of the Fairport Chalk in middle Turonian time (Martin and Stewart, 1977), with *C. liodontus* being recognized from the late Coniacian (Everhart, Everhart, and Bourdon, 1997; Everhart, 2001) and *C. propython* arriving in the late Santonian. *Clidastes* is then replaced by or evolves into a variety of *Mosasaurus* species during the early Campanian.

In Kansas, the first identifiable remains of *Mosasaurus* occur in the middle to upper Campanian Weskan Member of the Pierre Shale (Everhart et al., 2015). Two fairly complete specimens were collected by Triebold Paleontology in 2013 and 2014 (Fig. 9.13).

Interestingly, the first *Mosasaurus* remains (KUVP 155838) from western Kansas were actually collected by a geological survey field party from the University of Kansas in the summer of 1930 (Elias, 1931), but the specimen went unrecognized until David Burnham and I relocated them in 2015 in the museum collection. The remains were enclosed in a concretion and included the premaxilla, jaw fragments with teeth (Fig. 9.14), and vertebrae of a large mosasaur. They were originally identified by museum curator H. T. Martin as a new species of *Platecarpus* (ibid.), but he died in 1931 before describing the specimen. It was also about this time that the shell-crushing mosasaur, *Globidens*, first appeared in Kansas (Everhart and Everhart, 1996; Everhart, 2008b) in the early Campanian Sharon Springs Member of the Pierre Shale, apparently as the genus moved northward from the Gulf Coast (location of the type specimen, *G. alabamaensis* Gilmore, 1912) on its way to South Dakota (Russell, 1975).

As yet, there are not enough specimens of *Ectenosaurus*, *Selmasaurus*, *Halisaurus*, or several other rare mosasaur occurrences to say more than that they appear in the fossil record during the late Santonian/early Campanian. While this scenario is a considerable improvement over that envisioned by Williston (1897) or Russell (1967), future discoveries will make it even better. Enlarging a bit on Williston's (1897) plea, I would reiterate that additional collecting is required to improve the accuracy of these ranges, and it is essential that good stratigraphic information be obtained for all specimens from the Smoky Hill Chalk.

As mosasaurs increased in size and diversity during the last 25 million years before the end of the Cretaceous, they also spread around the world. While they have been collected frequently in Kansas, their remains are known from every continent, and have even been recovered from Late Cretaceous rocks exposed on islands off the coast of Antarctica (Chatterjee and Zinsmeister, 1982; Case et al., 2000; Novas et al., 2002). Recent discoveries in Israel (Christianson and Bonde, 2002); Turkey (Bardet and Tunoglu, 2002); Angola, Africa (Jacobs, et al., 2006), Hungary (Makádi, Caldwell, and Osi, 2012); and Japan (Sakurai, Chitoku and Shibuya, 1999; Konishi et al., 2015) further document a worldwide distribution during the Late Cretaceous.

History of Discovery

The history of the discovery of mosasaurs goes back more than 200 years. The word 'mosasaur' means 'Meuse River lizard,' a tribute to the locality where the first reported specimen of a mosasaur was discovered, in an underground mine near Maastricht in the Netherlands about 1780 (Saint-Fond, 1799; Williston, 1898a; Russell, 1967; Mulder, 2003). Another,

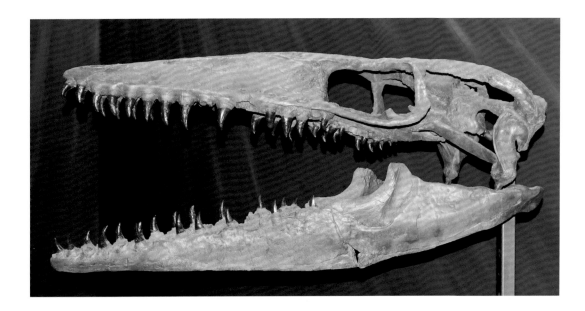

less complete specimen (No. 7424; see Saint-Fond, 1799:pl. 5) currently exhibited in the Teylers Museum (Haarlem, the Netherlands) had been discovered several years earlier but did not generate the same interest (Mulder, 2003). However, J. L. Hoffman, a military doctor, recognized the scientific value of the more complete Meuse River specimen and brought it to the attention of several noted scientists of the day.

At the time, nearly 50 years before the discovery of dinosaurs, the identity of this strange creature was a genuine mystery, and it was believed to be everything from a large fish to a giant crocodile. In 1808, Baron Cuvier wrote about "le grand animal fossile de Maastricht" (Bardet and Jagt, 1996:570) and agreed with Adrian Camper (1800; see also Williston, 1898a) that the animal was closely related to modern monitor lizards. Fourteen years later, the name *Mosasaurus* was coined by W. D. Conybeare, and mosasaurs began to be recognized as a group of extinct marine creatures in their own right. Mosasaur remains were soon discovered in England, on the east coast of North America, and as far away as New Zealand. It was not until 1829, however, that Gideon Mantell (1829:207) authored the species name *Hoffmannii* (Plate 11) in honor of the man who was responsible for bringing "le grand animal fossile" to the attention of scientists.

Mitchell (1818:pl. VIII, fig. 4) published the first North American record of mosasaur remains (a tooth and jaw fragment of *Mosasaurus dekayi*) from the Navesink Formation of New Jersey, and Emmons (1858) described mosasaur specimens from the green sands of North Carolina. It is possible, however, that the first mention of the discovery of mosasaurs in the American West came from notes kept by the Lewis and Clark Expedition of 1804–1806. Mitchell (1818:406) recounts the discovery of a "petrified skeleton of a very large fish, seen in Sioux country, on the Big Bend of the Missouri River," noting that "Sergeant Gass, who had

9.14. Three teeth on a fragment of the dentary in a concretion of the first Kansas *Mosasaurus* sp. specimen (KUVP 155838), collected in 1930. Scale bar = 10 cm (4 in).

discovered the skeleton in 1804, wrote in his journal that it was forty-five feet long, and lay on the top of a high cliff." Harlan wrote:

> It is not improbable that Lewis and Clark, in their Expedition up the Missouri, allude to the remains of a similar animal in the following extracts: "Monday, September 10th, 1804, we reached an island (not far from the grand detour, between Shannon creek and Poncarrar River), extending for two miles in the middle of the river, covered with red cedar, from which it takes the name of *Cedar Island*; just below this island, on a hill, to the south, is the back-bone of a fish forty-five feet long, tapering towards the tail, and in a perfect state of petrifaction, fragments of which were collected and sent to Washington." (1834b:408)

This place on the river is several miles below the confluence with the White River and is probably in the northwest corner of present-day Gregory County, South Dakota.

Four members of the expedition wrote slightly different accounts of the discovery in their journals. (See Moulton, 1983–1997, for entries of September 10, 1804). A portion of the specimen was reported to be collected and sent back to Washington, D.C., but was subsequently lost (Simpson, 1942). While the remains were probably the articulated vertebrae of a large mosasaur, we will likely never know for certain.

Almost 30 years later, another mosasaur fossil was discovered in the same general area along the Missouri River in what is now South Dakota. A nearly complete skull with associated vertebrae and limb material was collected and brought back to St. Louis by a retired Indian agent named

Major Benjamin O'Fallon (Goldfuss, 1845; see also Goldfuss, 2013—English translation), where it was displayed in the formal garden of his home. The bones eventually came to the attention of Prince Maximilian zu Wied during his travels in the American West and were acquired by him. The material was shipped to Bonn, Germany, where it was examined by Dr. August Goldfuss (1782–1848), a well-known naturalist. Dr. Goldfuss prepared the specimen and named it *Mosasaurus Maximiliana* in honor of the man who brought it back to Germany. His paper, "Der Schädelbau des *Mosasaurus*" (The structure of the skull of *Mosasaurus*) was originally presented at a scientific meeting in Mainz, Germany, in the fall of 1842 and later published in 1845 (Fig. 2.3). The historic specimen is still exhibited in the collection of the Goldfuss Museum (University of Bonn), Bonn, Germany, along with other specimens described by Dr. Goldfuss.

This story has further twists. The skull described and figured by Goldfuss had the end of the snout and the tips of the lower jaws missing.. A paper published by Dr. Richard Harlan (1834b), a Philadelphia surgeon and naturalist (1796–1843), had described and named a new species of "Ichthyosaur" from a bone fragment (which later turned out to be the anterior end [premaxilla] of a mosasaur snout). Harlan reported that his specimen was obtained by a beaver trapper from "Missouri" (probably from along the Missouri River in South Dakota) about 1832. He gave it the name of *Ichthyosaurus missouriensis*. Soon, however, it became clear that the fragment was part of a mosasaur and not an ichthyosaur as suggested initially by Harlan. In fact, it came from the same species of mosasaur as the one Goldfuss had described more than ten years later (Fig. 9.15). In spite of Harlan's erroneous identification, the name of *Mosasaurus Maximiliana* Goldfuss 1845 was short-lived and eventually became a junior synonym of *Mosasaurus missouriensis* (Harlan 1834b).

Not only was it from the same species of *Mosasaurus*, it appeared likely that Harlan's 'ichthyosaur' fragment was actually the missing premaxilla of the Goldfuss mosasaur skull. The 'mystery' of the missing piece was noticed soon after the Goldfuss paper was published. Although Russell (1967) indicated that the missing piece could not be located, he also credits an 1845 letter from Hermann von Meyer to Professor Bronn, published in the *Neues Jahrbuch für Mineralogie, Geognosie, Geologie und Petrefaktenkunde* (308–313) as the first mention that Harlan's 'ichthyosaur' fragment might be a part of the Goldfuss mosasaur. Camp (1942) provides a more detailed explanation of the circumstances surrounding this mosasaur mix-up.

However, the story does have a happy ending. In May 2004, just before the First Mosasaur Meeting at the Natural History Museum of Maastricht in the Netherlands, Gordon Bell and Mike Caldwell were examining the mosasaur collection in the Muséum National d'Histoire Naturelle in Paris and came across a fragment of a mosasaur snout (No. 1314). When they looked closer, they could see that Harlan's name had been written in ink on the bone (Fig. 9.16). Photographs taken the following week of

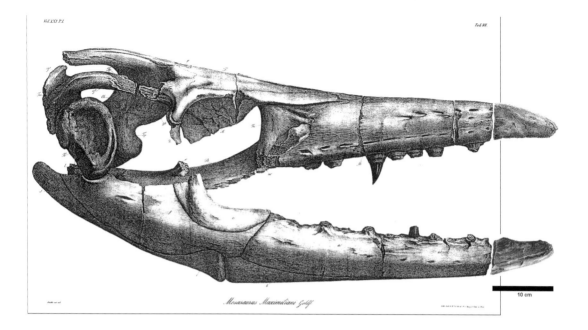

Mosasaurus Maximiliana Goldf.

10 cm

9.15. Drawing of the skull of *Mosasaurus maximiliana* from Goldfuss (1845) with the missing premaxilla and tip of the dentary added from Harlan (*Ichthyosaurus missouriensis*, 1834). Adapted from the English translation of Goldfuss (2013, edited by M. J. Everhart). Scale bar = 10 cm (4 in).

this specimen and the rest of the skull in the Goldfuss Museum in Bonn by my friend Takehito Ikejiri confirmed that Harlan's 'Ichthyosaurus' was the missing piece of the skull of *Mosasaurus Maximiliana.*

Since 1845, mosasaur remains have been discovered in many parts of the United States, from New Jersey to California, from Texas and the Gulf Coast to North Dakota, and even from Alaska. In fact, they are known from just about everywhere in the world where a Late Cretaceous marine fossil record is preserved. They are, however, most numerous and best preserved from the chalks and shale of the Western Interior Sea and especially in western Kansas. According to Williston, "Kansas, *par excellence*, has been the great collecting ground of the world for these reptiles" (1898a:83).

Mosasaur fossils occur in Late Cretaceous rocks deposited as near-shore or shallow marine sediments in many places around the world. Although mosasaurs probably preferred the shallower coastal waters where their prey was most abundant, they apparently also were excellent open-water swimmers. Mosasaurs were certainly capable of traveling across large bodies of water, as evidenced by the numerous examples of their remains recovered from the Smoky Hill Chalk. These rocks were deposited near the middle of the Western Interior Sea, hundreds of miles from the nearest land. From this evidence, it appears likely that mosasaurs were living continuously in mid-ocean and not just migrating through it. Fossils discovered in Canada and in Africa also suggest that near the end of the Cretaceous at least two mosasaur species, *Plioplatecarpus* (Holmes, Caldwell, and Cumbaa, 1999) and *Goronyosaurus* (Lingham-Soliar, 1999a), were also adapting to estuarine and freshwater habitats that had been the primary domain of alligators and crocodiles for millions of

9.16. Photograph of a replica of the premaxilla of *Mosasaurus missouriensis* (Harlan) from the Muséum National d'Histoire Naturelle in Paris, France. Harlan's name is written above the green circle. Scale bar = 10 cm (4 in).

years. More recently, a large number of mosasaur bones were recovered from freshwater sediments deposited by an ancient river in Hungary (Makádi, Caldwell, and Osi, 2012).

It would have been interesting to see how mosasaurs might have evolved had they survived the extinction at the end of the Cretaceous. Having mosasaurs as apex predators would have severely hindered the return of mammals to the sea. The fact that some primitive whales (e.g., *Basilosaurus*) first adapted a 'mosasauroid' body plan also suggests that the long, sinuous form and swimming style of mosasaurs continued to be successful for several million years after their extinction. As a historical note and for completeness: Harlan (1834a) also misidentified and named the type specimen of *Basilosaurus*. The name means king reptile or lizard, a genus Harlan believed to be most closely related to plesiosaurs, instead of an early whale.

Discovery of Mosasaurs in Kansas

As the pioneers and the railroads moved into the Midwest after the Civil War, military personnel, railroad surveyors, geologists, and other explorers came across new sources of fossils. One of the first and most famous Kansas fossils (*Elasmosaurus platyurus*; Cope, 1868a) was discovered in 1867 by an Army doctor named Theophilus H. Turner in the Pierre Shale near McAllaster Butte in Logan County (Chapter 7). The news of the discovery of fossils in western Kansas created a sensation in the East. *Elasmosaurus*, *Tylosaurus*, and the other fossils that had been shipped from Kansas by Turner and two other military doctors, by William E. Webb, and by Prof. B. F. Mudge, sparked a 'fossil rush' by Marsh, Cope, and other paleontologists of the day.

The first Kansas mosasaur, '*Macrosaurus*' (*Tylosaurus*) *proriger* (MCZ 4374), was named by Cope (1869a:123) from a specimen obtained by Dr. Louis Agassiz from "Western Kansas." Cope later reported it had been discovered by "Col. Connyngham [*sic*] and Mr. Minor near Monument Station, Kansas" (1875:166) but said no more. How Agassiz received the type specimen, however, had been a paleo-mystery until recently, when at least part of it was solved by Everhart (2016a).

Even as the Kansas Pacific Railway was being built, travel and commerce between Kansas City and Denver was still being conducted by stage coach and wagon along the Butterfield Overland Despatch (BOD) trail along the Smoky Hill River, about 20–25 miles south of the route planned for the railroad. Way stations were maintained every 10–15 miles along the route for changing horses and caring for passengers. Because of the threat of attack by Indians, detachments of Army troops were stationed at some of those stage stops. One such location was near Monument Rocks in western Gove County. The soldiers of Company I, 38th Infantry, who served there referred to it as Fort Monument, but it was little more than one wooden building and several sod-covered dugouts that served as rifle pits in case of an attack (Wetzel, 1960).

As the railroad neared completion, the Butterfield Overland Despatch went out of business, and the stagecoach way stations and the military presence along the trail were no longer needed. In June 1868, Captain John B. Conyngham (1827–1871) marched his troops 25 miles northwest to a military camp where the present-day town of Monument was being established along the railroad (Wetzel, 1960). Apparently he also brought an important discovery with him, the partial skull and some vertebrae of the first mosasaur, *Tylosaurus proriger*, ever to be described from Kansas. (Dr. George M. Sternberg, surgeon, U.S. Army, was probably the first to collect mosasaur bones from western Kansas [Fig. 9.17], but they were initially sent to the U.S. Army Medical Museum and not readily available at the time for study or description).

Two months later a 'fact-finding' party of federal government officials, including two United States senators, traveled on the Kansas Pacific Railway through Monument, Kansas, on their way to Denver. The party also included Louis Agassiz from Harvard University (Leavenworth Conservative, August 28, 1868, 1) as a scientific advisor. Taft (1953) believed that William E. Webb, the president of the National Land Company, a subsidiary of the Kansas Pacific Railway, with his knowledge and contacts in western Kansas, would have also been included in the group as the representative of the railroad.

It is still unclear how or where Webb received the type specimen of *Tylosaurus proriger* from Conyngham during the summer of 1868. It is possible that Agassiz may have seen the bones in August, or at least indicated his interest to Webb in regard to receiving fossils collected in Kansas. What is known is that later that same year, Webb had the specimen in his possession, because various Kansas newspapers reported that he had discovered an "immense saurian. The bones were petrified

9.17. This *Tylosaurus proriger* premaxilla (USNM 3894) was collected from the Smoky Hill Chalk by Dr. George M. Sternberg about 1868 and sent to the United States Army Medical Museum in Washington, D.C. From there it was transferred to the United States National Museum (Smithsonian) and subsequently figured by Leidy (1873; pl. 36, fig.13). The date of collection is uncertain, but it may represent the first *Tylosaurus* specimen ever collected in Kansas. Note Sternberg's signature on the bone. Scale = 15 cm (6 in).

and consisted of the jaws, teeth, and vertebrae, etc. . . . The gigantic remains were exhumed near Sheridan on the plains, the last town built by the National Land Co. . . . The jaw bones were over five feet long and full of teeth. . . . It is understood that Professor Agassiz will visit in a few days to examine the remains" (Wyandotte Commercial Gazette, November 21, 1868, 2). The *White Cloud Kansas Chief* newspaper reported, "We examined, while in Topeka the other day, some of the petrified bones of the monster found recently on the plains. The entire remains are in the possession of W. E. Webb, Esq., of that city, who had the same collected and brought down the road at great expense" (December 3, 1868, 1).

At the December 22, 1868, meeting of the Academy of Natural Sciences of Philadelphia (ANSP), Joseph Leidy "exhibited some photographs of fossil bones, received from Mr. W. E. Webb, Sec. of the National Land Co., at Topeka, Kansas. They represent vertebrae, and fragments of the jaws with teeth of a skeleton of *Mosasaurus*, reported by Mr. Webb to be about 70 feet in length, recently discovered on the great plains of Kansas, near Fort Wallace" (1868:316). In the preface of Webb's 1872 book *Buffalo Land*, he mentions that "whenever it was possible, we had photographs taken" (1872a:xii). However, the photos of the fossil bones reported by Leidy were never mentioned again and have not be located.

While the length and completeness of this specimen (Leidy, 1868; Webb, 1872b:326; and various newspaper reports) are certainly exaggerated, the actual identity of the unusual specimen and what became of it was unresolved in the paleontological record. Webb wrote that the fossil had been "sent to Agassiz" (1872b:129) and that it "now rests in the museum at Cambridge, Massachusetts" (1872a:326).

A critical issue that made identification of Webb's specimen uncertain up to this point was the misinformation provided by Webb as to the locality where it was collected. Webb (1872b) said that the remains were discovered near Sheridan, Kansas (western Logan County, Pierre Shale) while the records at Harvard's Museum of Comparative Zoology (MCZ) indicate that the type specimen of *Tylosaurus proriger* came from near Monument Rocks (Smoky Hill Chalk) in western Gove County. These localities are about 35 miles apart and quite different geologically. It is also unclear why Webb failed to mention that it had been discovered by Conyngham, although his version makes for a more interesting fictional tale in *Buffalo Land*. Why he deceived the newspapers, the readers of his 1872 book, and Joseph Leidy (photographs), but not Louis Agassiz, will never be known.

In fact, the mosasaur specimen was delivered to Agassiz at Harvard University's Museum of Comparative Zoology, where it is curated as MCZ 4374. A printed form receipt is preserved in the museum's Ernst Mayr Library, which reads: "April 6, 1869. Received of Professor Agassiz five hundred dollars for a fossil saurian, etc." It is signed by W. E. Webb. Agassiz, in turn, donated the specimen to the MCZ as included in "boxes of fossils from the Smoky Hill route, Rocky Mountains" (Shaler, 1869:44). Even so, Williston, based on Cope's 1875 report, repeated the story that the specimen had been "collected by Colonel Cunningham [*sic*] and Mr. Minor in the vicinity of Monument station, and sent by them to Prof. Louis Agassiz" (1898a:28).

E. D. Cope almost certainly saw Webb's photos of the specimen at the December 21, 1868, ANSP meeting. Then, in an early oral report to the January 20, 1869, meeting of the Boston Natural History Society, Cope said that the specimen was "exhumed by W. E. Webb, near the town of Topeka in Kansas. My friend Professor J. Parker, of Lincoln College, of that place [Topeka] informs me that it is seventy-five feet in length, and the gentleman who discovered it, that it measures eighty feet . . . as stated on photographs by my friend W. E. Webb" (1869b:263–264).

In any case, Cope borrowed the specimen from Agassiz at Harvard, then briefly described and named it orally at the June 1, 1869, meeting of the ANSP. Cope:

made some remarks on a fine fragment of the muzzle of a large Mosasauroid, which pertained to a cranium of near five feet in length. The pterygoid bones were separated from each other, and support nine teeth. A peculiarity of physiognomy was produced by the cylindric prolongation of the premaxillary bone beyond the teeth, and a similar flat prolongation of the extremity of the dentary. He referred the species to *Macrosaurus* Owen, under the name *M. proriger*. The specimen he stated belonged to Prof. Agassiz, who obtained it from Western Kansas, probably from the No. 3 of the Upper Cretaceous [Niobrara Fm.] of Hayden. (1869c:123)

The following year Cope (1870:202, pl. XII, figs. 22–24; see Fig. 9.5) described the specimen in more detail, illustrated the anterior portion of the skull and a single tooth, and stated that Agassiz had lent it to him. He also referred the specimen to a different genus, *Liodon proriger* Cope. After a three-year back and forth published dispute with O. C. Marsh regarding the genus name, the new mosasaur officially became *Tylosaurus proriger* (Leidy 1873:274).

Marsh was certainly interested in the new fossil discoveries from western Kansas, and in 1870 he led the first the Yale College Scientific Expedition, collecting fossils in many places in the West, including Kansas. Late in November of that year, Marsh and his students made a brief visit to the Smoky Hill River valley east of Fort Wallace. As might be expected for Kansas, the weather was already turning cold in November. Accompanied by a military escort, they were able to collect for several days in the Smoky Hill Chalk along the Smoky Hill River. One of the students later wrote that they spent four days digging out the nearly complete skeleton of a mosasaur "allied with the genus *Mosasaurus*" (Betts, 1871:671). His note regarding the length of the specimen ("not been less than sixty feet") was certainly an exaggeration. The find was important, however, because it provided the first evidence that mosasaurs had hind limbs, something that was otherwise unknown to Cope and others at the time.

Marsh (1871a) reported in a brief note shortly after their return to Yale that "some interesting reptilian and fish remains" were collected during their visit to Kansas. Later, Marsh (1871b) published a more detailed account of the mosasaurs collected by the 1870 expedition, including the nearly complete *Clidastes* ('*Mosasaurus*') specimen mentioned by Betts (1871).

For the next several years, Cope and Marsh competed to see who could collect and name the most new animals from Kansas and elsewhere. While Cope and Marsh were both in the Kansas chalk during 1871, they usually employed others, including professional collectors, to find and secure most of the fossils they described. Over a four-year period (1870–1873), the Yale expeditions collected more than a thousand mosasaur specimens from Kansas for the Peabody Museum. They also collected fragmentary remains of the first fossil bird (*Hesperornis*) and the first *Pteranodon* in the chalk. After B. F. Mudge sent him the first bird with teeth (*Ichthyornis*) in 1872, Marsh became preoccupied with the pursuit of the remains of toothed birds from the chalk and hired Mudge to locate as many specimens as possible (Shor, 1971). While this strategy established Marsh as the expert on these primitive birds, it also allowed Cope to collect and describe many new species of fish and marine reptiles.

In 1876 Cope hired a young Charles Sternberg to collect for him, and Cope seemed to always be one step ahead of Marsh in naming new species of mosasaurs from Kansas (Bell, 1997a). In retrospect, Marsh may simply have been simply overwhelmed by the sheer volume of new and undescribed material that was being returned to Yale by his more numerous employees. Cope was certainly the more agile of the two, producing many more papers over the course of his career than Marsh. One adverse result of the competition between Cope and Marsh, however, was a unrealistically large number of new species of mosasaurs from the Niobrara of Kansas, cited as 17 by Cope (1872a) early in the exploration of the chalk.

Although other workers (Baur, 1892; Williston and Case, 1892; Williston, 1893; Merriam, 1894) subsequently published papers on mosasaurs, it was Williston (1898a) who first described the occurrence of mosasaurs as a group. He suggested that many of the differences observed by Cope and Marsh were due to distortion of the bones during preservation and that fully "four-fifths of all the described species must be abandoned." Russell (1967) published an extensive review of the systematics and morphology of American mosasaurs and significantly reduced the number of valid taxa. Bell (1997b) applied cladistic analysis to mosasaurs as a group for the first time in his phylogenetic revision of North American and Adriatic Mosasauridae. Since then, however, new species of mosasaurs continue to be discovered and described. We are still some years away from being able to say that we fully understand their evolution and diversity on a worldwide basis, or even in Kansas.

We are currently trying to understand the adaptations of mosasaurs to life in the ocean, and their success in the relatively short time (35 million years or so) that they existed at the end of the Late Cretaceous. Even the way they swam was somewhat of a departure from the time proven method used by the earlier and very successful ichthyosaurs and plesiosaurs. The long, muscular tails of mosasaurs were flattened from side to side, and they used a side-to-side undulating motion (like a snake or a varanid lizard) of their tails to propel their bodies through the water. Because this method of swimming is inefficient compared to that used by more streamlined fish, the shortened body and highly developed caudal fin of some ichthyosaurs, and the winglike paddles of plesiosaurs, mosasaurs were probably not capable of traveling long distances at high rates of speed (Massare, 1987). Rather, they swam rather slowly and probably hunted from 'ambush,' surprising and outrunning their prey over a short distance. Even though mosasaurs were far less streamlined than the 'fish-lizard' ichthyosaurs, their method of hunting conserved much of the energy ichthyosaurs were required to expend to swim constantly in pursuit of prey. This adaptation, probably more than anything else, enabled mosasaurs to compete successfully with the bony fish (*Xiphactinus*, *Pachyrhizodus*, and *Ichthyodectes*) and sharks (*Cretoxyrhina*) that were also becoming much larger and more common during the Late Cretaceous.

The tail accounts for half or more of the body length in most mosasaurs and is composed of about 100 relatively short (compared to their diameter) vertebrae. Williston (1898a:pl. LXXII, Restoration of Kansas Mosasaurs) published skeletal drawings of *Clidastes*, *Platecarpus*, and *Tylosaurus* with vertebral columns that were drawn out perfectly straight. Although not accurate as living animals in that regard, these drawings have been used since then by artists and museum preparators in their reconstructions. However, it is likely that the plate was prepared that way to save space, given that three dissimilar-sized mosasaurs had to be shown in the same space. An expert on reptiles, Williston (1893:pl. III) certainly

What about Mosasaur Tails?

recognized the actual curvature of the vertebral column because he had published a much more lifelike skeletal drawing of *Clidastes* years earlier. It is worth noting here the expanded dorsal processes on the vertebrae near the end of the tail in *Clidastes*, a feature that would have significantly expanded the vertical depth of the tail and increased the thrust.

Osborn (1899:178, fig. 13) was the first to report a tail bend in the specimen of a *Tylosaurus* [*dyspelor*] *proriger* (AMNH FR 221) that is now on exhibit in the American Museum of Natural History, noting that his measurements "tend to show that the vertebral centra were slightly longer above than below and thus produced the [downward] curve" (ibid.). What he did not report was the fact that the inclination of the dorsal processes in area of the curve changes from inclined posteriorly to vertical to inclined anteriorly. This is a clear indication that something is going on in that portion of the tail. The question remains just what is happening? Note that the famous artist Charles Knight provided a horribly inaccurate reconstruction of *Tylosaurus* in the same article (Osborn, 1899:fig. 14), complete with Williston's erroneous 'dorsal fringe,' a sagging throat pouch, a big belly, strange looking paddles, and the tail bend with an expanded dorsal fin (Fig. 9.18).

Wiman (1920:11) also noted a "tail-curve" in specimens of *Platecarpus tympaniticus* and *Halisaurus* (*Eonatator*) *sternbergii* he had received from Charles Sternberg in Kansas, and noted a similarity to the modifications of the caudal vertebrae of ichthyosaurs. In mosasaurs, as in ichthyosaurs, this adaptation would have most likely served to deepen the end the tail from top to bottom and provide a greater area for producing thrust, as well as produce a slight downward thrust to assist in overcoming buoyancy and raising the head above water for breathing. He illustrated an enlarged dorsal fin on the tail of his *Platecarpus* specimen (ibid.:fig. 2), a truly large dorsal fluke on his modified version of Knight's 1899 *Tylosaurus* drawing in Osborn (ibid.:fig. 3), and an enlarged dorsal fin on his type specimen of *Clidastes* [*Halisaurus*, *Eonatator*] *sternbergii* (ibid.:fig. 7). Wiman (ibid.:12) wrote, "If I am correct in my interpretation, then, at least, certain Mosasaurians would have developed a caudal fin similar to that of the Ichthyosaurians and Thalattosuchians [crocodile-like marine animals from the Early Jurassic]" (ibid.:12).

Although the ugly mosasaur fringe popularized by Charles Knight has persisted, the work by Osborn and Wiman has been largely ignored in reconstructions of mosasaurs over the years. Schumacher and Varner (1996, 2007) measured the same tail bend in *Clidastes*, *Platecarpus*, and *Tylosaurus*, and observed the slight wedge shape in each of the vertebrae in the area of the tail bend as reported by Osborn (1899). They also noted that the transverse processes of the caudal vertebrae are reduced in size and that the neural spines become taller (Schumacher, pers. comm., 2004). This means the tail becomes laterally compressed at the same time as it gets taller, forming a more flexible and efficient paddle. Schumacher and Varner noted that the "hypocercal tails of mosasaurs

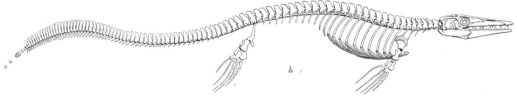

Fig. 13. *Tylosaurus dyspelor*. Restoration after drawings by W. D. Matthew and Bruce Horsfall, under direction of the author. ⅟₁₀ nat. size.

Fig. 14. Restoration of the Ram-nosed Tylosaur, by Charles Knight.

bear a gentle sinuous curve throughout their length, with a slight upward flexure anterior to a more significant downward flexure. This shape, likely lacking a pronounced dorsal caudal fin, would produce downward thrust to compensate for positive buoyancy. Mosasaurs retain elongate neural and haemal spines in the area of the downward bend, heightened in the manner of primitive ichthyosaurs, which coupled with elongate body plans indicates a similar from of eel-like swimming in early members of both groups" (2007:24).

This long-winded discussion concerns a relatively important, current issue about mosasaur tails: did they or did they not have an enlarged dorsal fluke of some sort? Lindgren, Jagt, and Caldwell (2007) published their description of the tail of *Plotosaurus*, a highly derived mosasaur from California. They interpreted the remains to include a fleshy dorsal fluke in the area of the tail bend (ibid.:fig. 3B). Lindgren et al. (2010) then described an exceptionally well preserved specimen of *Platecarpus tympaniticus* (LACM 128319), including soft tissues, from the upper chalk in Kansas, and proposed that the observed bend in the vertebral column (ibid.:fig. 8B; originally reported by Stewart, 1993) also supported a large dorsal fluke. In neither specimen, however, was there any indication of soft tissue preservation in the area of the suggested tail fluke. The following year, Lindgren, Polcyn, and Young (2011) published a well-researched paper about the evolution of swimming in mosasaurine mosasaurs. In that paper they suggested that the vertebral column and especially the caudal vertebrae became organized through time into regions supporting the use of a progressively larger dorsal fluke (ibid.:fig. 12A–B). There was still, however, no physical evidence of a mosasaur dorsal fluke. I have remained skeptical (and critical) over the years throughout this process as the circumstantial observations piled up. Unfortunately, in the meantime, the paleo-art community was putting big dorsal flukes on nearly every mosasaur reconstruction.

9.18. Figures 13 and 14 from Osborn (1899) showing the skeletal reconstruction, including the tail bend, of a specimen of *Tylosaurus proriger* (AMNH FR 221) and a life reconstruction by artist Charles Knight. Total length of the skeleton is 8.3 m (27 ft, 4 in).

Half a world away from Kansas in 2008, the remains of a small juvenile mosasaur (1.5 m) were collected and described from Maastrichtian-age chalk of Jordan. Although originally described as a new genus in 2009, the specimen most likely represents a juvenile *Prognathodon* (Lindgren, Kaddumi, and Polcyn, 2013). In any case, subsequent preparation of the mostly complete specimen reviewed a considerable amount of scale impressions AND the outline of a dorsal fluke in the area of the tail bend, almost exactly where earlier reconstructions predicted it would be. It is a wonderful discovery, and I congratulate the authors on the paper. It is the best of all evidence for the presence of a dorsal fluke in some mosasaurs.

That said, I still do not believe that every mosasaur had a dorsal fluke or an especially large one. We know from the fossil record that not all ichthyosaurs, especially the early ones, had a fishlike tail. The mosasaurs that I work with in Kansas are some of the earliest representatives of group, only a few million years into their evolution away from the land. It is certainly possible that they developed dorsal flukes early in their adaption to life in the sea, but we have yet to find the fossil evidence. I would like to be the one who discovers the preserved soft tissue outline of a *Tylosaurus* dorsal fluke. Until then, I will remain the loyal opposition.

One final note of caution here. Williston (1898b, 1899), an expert on many things, including mosasaurs, made a mistake in the interpretation of soft tissues (tracheal cartilages) that led to the erroneous belief that mosasaurs had a dorsal fringe running down the middle of the back. In his report regarding a wonderfully preserved specimen of *Platecarpus tympaniticus* (KUVP 1001—collected by Albin Stewart in 1898), Williston noted that it included scales, preserved sternal cartilage, and "a row of dermal processes." The specimen was first reported by Williston (1898b) in a brief editorial note in the *Kansas University Quarterly* and then followed by a short article (Williston, 1899). Charles Knight, a well-known artist of the day who specialized in recreations of prehistoric animals, then painted a mosasaur with a frilly mane along its back (as well as the figure done for Osborn, 1899), and the mistake has been repeated time and again ever since. About the same time, H. F. Osborn (1899) observed the same structures preserved in a large *Tylosaurus proriger* (AMNH FR 221) from western Kansas and interpreted the structures correctly as cartilaginous tracheal rings. However, he also credited Williston for his discovery of the "nuchal fringe" (ibid.:167) and published the figure by Charles Knight that showed the fringe (Fig 9.18).

Russell (1967:88) briefly discussed "cartilaginous structures in the thoracic region of mosasaurs" and diplomatically stated that both Williston and Osborn had seen similar "tracheal structures" in these "two excellent skeletons from the Niobrara chalk." To Williston's credit, he did recognize his error and made the following statement: "In conclusion, I wish to correct an error made by myself. That which I considered to be the nuchal fringe in the mosasaurs is evidently only the slender cartilaginous rings of the trachea, first described and figured by Professor Osborn.

I have no excuses to make for the mistake, which I recognized when too late to correct" (1902:254). Indeed, it was too late to correct, and for more than 115 years, images of mosasaurs have continued to be plagued with an ugly accessory that they do not deserve. I hope the same thing is not happening with mosasaur tails.

Even after 50 years, Dale Russell's (1967) *Systematics of a American Mosasaurs* remains the single best reference regarding the skeletons of the various mosasaur genera. Updates are certainly needed to address new discoveries and new interpretations, but for now, the most common mosasaurs in Kansas (*Tylosaurus*, *Platecarpus*, and *Clidastes*) are adequately covered by Russell's work.

The tail of *Clidastes* is shorter in proportion to its body (about 2/5 of total length) than that of other mosasaurs. *Clidastes* also has elongated dorsal processes on the caudal vertebrae that would have supported a proportionately deeper tail. This may indicate that *Clidastes* was somewhat faster than other mosasaurs. Swimming faster would have been a useful adaptation when your larger cousin (*Tylosaurus*) saw you as his next meal. However, speed wasn't always enough, as demonstrated by preserved stomach contents of a large *Tylosaurus proriger* discovered in South Dakota, which included the remains of a smaller mosasaur, quite possibly *Clidastes* (Martin and Bjork, 1987).

In addition to differences in size and proportion between the three most common genera (Fig. 9.19), mosasaurs also differed in the number of vertebrae that made up their extremely long bodies (Russell, 1967). *Tylosaurus* had the most vertebrae: 7 cervicals, 23 dorsals, and 95–119 caudals, including the pygals, for an average of about 140. *Platecarpus* had between 125 and 130 vertebrae, and *Clidastes* had the fewest, with 108.

Mosasaur vertebrae were procoelous in form; that is, the anterior face of the centrum (cotyle) was concave and the posterior face (condyle) was convex. The round end of one vertebra was set securely in the cupped centrum of the following vertebra with little or no space for cushioning between the vertebrae, forming a long column that had little vertical flexibility. Lateral movement of the vertebral column was also limited in the anterior dorsal vertebrae but increased toward the hips and tail. While swimming at cruising speeds, the head and forward portion of the body of a mosasaur would have been held fairly stiff, with the paddles probably folded against the body, while the hips and tail undulated from side to side. In most cases, the paddles were probably used only for steering, although the limbs of some mosasaurs, including *Plioplatecarpus*, *Mosasaurus*, and *Plotosaurus*, were much more heavily constructed than those of *Tylosaurus* and *Platecarpus*. There has been some discussion regarding the possible use of such paddles by *Plioplatecarpus* as an adaptation to 'subaqueous flying' similar to that used by plesiosaurs (Lingham-Soliar, 1992; Nicholls and Godfrey, 1994; see also Lindgren, Caldwell, and Jagt,

The Mosasaur Skeleton

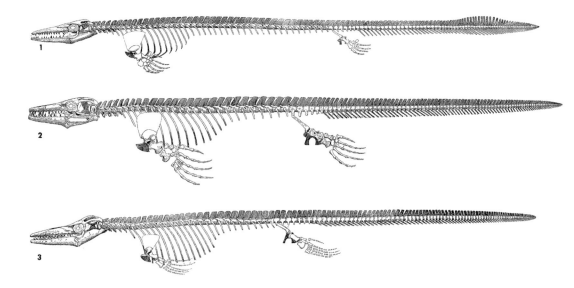

9.19. Plate LXXII from Williston (1898) illustrating the three most common genera of mosasaurs (not to scale) in the Smoky Hill Chalk: 1. *Clidastes*; 2. *Platecarpus*; 3. *Tylosaurus*. In life, an adult *Tylosaurus proriger* could be as much as three times the size of an adult *Clidastes propython*.

2008). While anything was possible as mosasaurs evolved and spread into a variety of different environments, it is likely that almost all of them continued to use their tails as their primary means of propulsion.

Because mosasaurs had adapted so well to living in the ocean, it is improbable that they ever voluntarily returned to land. As they adapted to life in the ocean, their upper limb bones became shortened and their feet became modified into broad, flat paddles. Bell and Polcyn (2005) discussed the transition from 'plesiopedal' limbs to 'hydropedal' limbs in their description of the primitive (with feet) mosasauroid *Dallasaurus turneri* from the Turonian of Texas. In contrast to the extreme hyperphalangy (multiple finger bones) seen in plesiosaurs and ichthyosaurs, the number of bones in each of the fingers of a mosasaur was fairly small, ranging from 3 to 5 in *Clidastes* to 5 to 11 in *Tylosaurus* to a maximum of 16 in *Plotosaurus*. The five digits of each paddle were widely separated and loosely webbed together in most species. Mosasaurs probably also retained more flexibility in their limb joints compared to the rigid, bony flippers of the plesiosaurs and ichthyosaurs, although genera like *Mosasaurus*, *Prognathodon*, and *Plotosaurus* (Lindgren, Jagt, and Caldwell, 2007) show a trend toward more robust limb bones and rigidity.

The shoulder and hip girdles of mosasaurs were not solidly connected to their rib cages and backbones. Over time, much of what had been bone in their limbs was replaced by cartilage since the excess strength and rigidity of bone was not needed as much in the weightless environment of the ocean. This meant, however, that mosasaur limbs were basically useless out of the water and could not support the weight of their bodies on land. While mosasaurs may have been able to slither through shallow water much like a snake, they would have been nearly

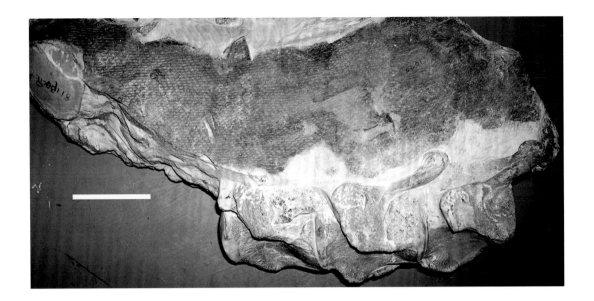

9.20. Small diamond-shaped scale impressions preserved on the vertebrae of a *Tylosaurus proriger* specimen (KUVP 1075) and described by Snow (1878). Snow reported that there were nearly 90 scales per square inch in this specimen. Scale bar = 10 cm (4 in).

helpless on dry land. Once beached, they would have died either from suffocation due to the collapse of their lungs or from overheating, much like a beached whale. And the larger the mosasaurs got, the more problems they would have had out of the water.

Ichthyosaurs and plesiosaurs had a much longer history in the Mesozoic than mosasaurs and appear to have been scaleless. It is possible that a smooth skin was much better adapted for a life in the ocean than was one with scales because it offered less resistance in the water or because it provided fewer opportunities for the attachment of parasites, or maybe scales simply were no longer needed to conserve moisture. Mosasaurs, still relative newcomers, were covered with small scales, the shape and arrangement of which resembles those of modern snakes and monitor lizards (Williston, 1898a). The scales of *Tylosaurus* (Fig. 9.20) were first described by Snow (1878) from impressions preserved with mosasaur remains (KUVP 1075) discovered near Hackberry Creek in southern Gove County, Kansas. They are rhomboidal and quite small, roughly 3.3 mm in length by 2.5 mm in width, and have a raised keel along the long axis. Snow noted that a "comparison with the scales of the living rattlesnake of the plains [Prairie rattlesnake *Crotalus viridis*], indicates that the scales of the saurian were somewhat smaller than those of the snake, which, in a full grown 'rattler,' average eighty to the square inch, instead of ninety" (1878:56). The mosasaur (KUVP 1075) is estimated to be about 5 m (16 ft) in length (Lindgren, Everhart and Caldwell, 2011).

The scales of *Platecarpus* (LACM 128319; Lindgren et al., 2010) and *Ectenosaurus* (FHSM VP-401; Lindgren, Polcyn, and Young, 2011) also appear to be relatively small and about the same size. Lindgren et al. (2009) reported a specimen of *Plotosaurus* (middle Maastrichtian age) from California that included preserved scales. In addition, as mentioned earlier in this chapter, the juvenile *Prognathodon* specimen from the middle Maastrichtian chalk of Jordan with the preserved out-line of a dorsal fluke included scale impressions that averaged about 1.25 mm in height by 1.75 mm in length (Lindgren, Kaddumi, and Polcyn, 2013:5).

In the first edition of this book (2005), I suggested that we may find that Maastrichtian mosasaurs were generally smooth-skinned, like the ichthyosaurs and plesiosaurs. My remark was justifiably criticized by Lindgren et al. (2009) on the basis of scale preservation in the middle Maastrichtian *Plotosaurus* specimen they described from California. At this point, I'll just acknowledge the new information and look forward to the discovery of soft tissue preservation in later occurring specimens of *Tylosaurus* and *Plioplatecarpus*.

Another early misinterpretation of mosasaur remains should be mentioned here. The large "dermal scutes" described by Marsh (1872a:291) were reexamined by Williston (1891) and identified as fragments of the thin sclerotic ring that covered much of the outer surface of the eye of the mosasaur (Fig. 9.21). Contrary to some reconstructions, mosasaurs do not have 'dermal scutes.' Instead, they have a complicated arrangement of thin bony plates that circles in the cornea of their eye, very similar to those seen in modern birds such as owls and in some reptiles. In a recent analysis of these sclerotic rings in *Tylosaurus*, *Platecarpus*, *Mosasaurus*, and *Clidastes* by Yamashita, Konishi, and Sato (2015) and Yamashita, Konishi, and Everhart (2015), it was discovered that the specific arrangement of these thin plates differs between plioplatecarpine and mosasaurine mosasaurs. The arrangement of sclerotic plates that occurred in *Tylosaurus* and *Platecarpus* is still seen in a number of extant lizards, including iguanas, while the mosasaurine arrangement is extinct. I was gratified to learn that the *Tylosaurus proriger* that I had collected and prepared (FFHM 1997-10; Fig. 9.21) preserves one of the best known examples of a plioplatecarpine sclerotic ring.

Reproduction

The structure of their skeletons shows that mosasaurs were certainly closely related to reptiles, but their bodies had changed in many other ways as they adapted to life in the ocean. The discovery of a *Plioplatecarpus* 'mother mosasaur' in South Dakota with the remains of five unborn young preserved inside her abdomen demonstrates conclusively that these animals bore their babies alive (viviparity; Bell et al., 1996; Bell and Sheldon, 2004). While similar specimens have not yet been discovered in the Smoky Hill Chalk of Kansas, some remains, like the fragmentary skull of a very small mosasaur (FHSM VP-14845; Fig. 9.22) in the Sternberg collection, suggest a near-term fetus or a newly born individual (Everhart, 2007). Like the ichthyosaurs, mosasaurs probably gave birth to a maximum of four or five fairly large babies at a time. This number is small compared to other reptiles such as crocodiles, alligators, and turtles, which may lay dozens of small eggs and depend on sheer numbers for the survival of their young. Note here that live birth is not that unusual

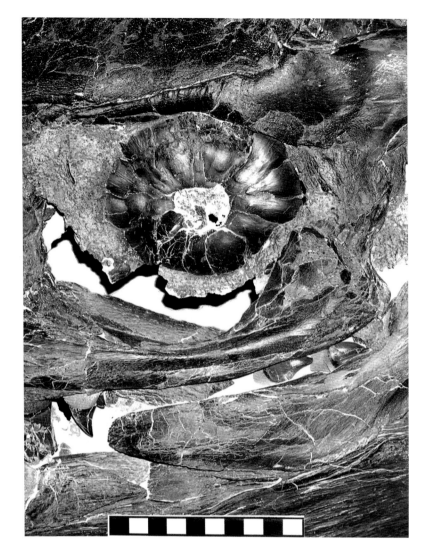

9.21. The sclerotic ring in the left orbit of a large *Tylosaurus proriger* (FFHM 1997–10) skull collected from Smoky Hill Chalk (middle Santonian) of Gove County in 1996. Anterior is to the left. The sclerotic ring of this specimen and other mosasaurs was described by Yamashita, Konishi, and Sato (2015). Scale bar = 10 cm (4 in).

among modern squamate reptiles and is something that has evolved many times within the lineage (Pyron and Burbrink, 2014).

Reproduction has been a controversial topic among mosasaur scientists for almost as long as mosasaurs have been studied. While acknowledging that mosasaurs would have been "practically helpless" on land, Williston (1898a:214) and other paleontologists firmly believed that mosasaurs must have somehow crawled up on the shore to lay their eggs. They also concluded that young mosasaurs lived in estuaries or other nearshore, protected environments because their remains had not been located along with those of adult mosasaurs. However, a recent review of the collections at the Yale Peabody Museum located a number of neonate mosasaur jaw fragments that had been misidentified as the remains of toothed birds (Field et al., 2015). Whether or not those remains were simply overlooked or ignored by Williston, Cope, Marsh, Sternberg, and others in the rush for bigger, more spectacular specimens, we now know

9.22. Skull fragments of a small (neonate) mosasaur (FHSM VP-14845) from the middle chalk of Gove County, south of Castle Rock. The premaxilla (ventral view) and a partial quadrate are shown at upper left. Scale bar in mm.

that mosasaurs of all ages lived and died in the middle of the Western Interior Sea (Sheldon, 1996; Everhart, 2002b, 2007).

The fairly common presence of the remains of young individuals in the Smoky Hill Chalk of Kansas suggests to me that mosasaurs were giving birth in mid-ocean, 200 miles or more from the nearest land. While there is no evidence for parental care or involvement, the fact that very young animals occur in an open-water environment indicates that female mosasaurs, out of necessity, may have lived in groups for the protection of their young. There were simply too many other hungry predators around, including 6 m (19.7 ft) sharks (*Cretoxyrhina mantelli*), 5 m (16.4 ft) giant predatory fish (*Xiphactinus*), and other species of mosasaurs for young mosasaurs to have survived long without some kind of parental care or protection. While modern monitor lizards do not care for their young, the female American alligator (*Alligator mississippiensis*) is known to move newly hatched babies from their nest to the water, and to protect them to some extent from predators. Unlike alligators and monitor lizards, mosasaurs were not egg-layers, and they were well adapted in many other ways for their life in the ocean. That said, as with most other animals in the wild, the mortality rate among young mosasaurs must have been high. However, from the number of adult specimens that have been collected so far, and their success in rapidly spreading around the world, it appears that many of them survived and successfully reproduced.

While it is also possible that the poorly circulated Western Interior Sea had masses of floating seaweed, like the Sargasso Sea, that young mosasaurs could hide in, there is no fossil evidence to support this idea. The presence of many fast-swimming predators, such as the ichthyodectid fish *Xiphactinus audax* and short-necked, wing-flying poly-cotylids like *Polycotylus* and *Dolichorhynchops*, seems to argue for large areas of unobstructed open water. The other marine reptiles, including the giant marine turtles (*Protostega* and *Archelon*) and the elasmosaurs, also moved through the water with long, outstretched limbs. One variety of primitive swordfish (*Protosphyraena perniciosa*) had long pectoral fins that extended 0.6–0.9 m (2–3 ft) outward on either side of its body, hardly a good design for efficient hunting in a stalked kelp forest, but living with small patches of floating *Sargassum*-like sea weed would be a different matter. The snakelike form of the mosasaurs, on the other hand, might have been well suited for such an environment if it existed.

Lizards are cold blooded (ectothermic) and mosasaurs are marine liz-ards, so they were cold blooded, right? Well, yes and no. Yes, mosasaurs did evolve from a group of advanced reptiles called squamates (scaled lizards). This group includes modern snakes and lizards such as geckos, iguanas, and varanids, a diverse family that includes the largest living squamate, the Komodo dragon (*Varanus komodoensis*). Mosasaurs were, in fact, the largest ever known squamates, reaching lengths of 18 m (59 ft) or larger (Lingham-Soliar, 1995; Grigoriev, 2014; Everhart et al., 2016). However, mosasaurs branched off the squamate family tree relatively early and evolved into fully marine forms during Turonian time.

Water temperatures of the Western Interior Sea were subtropical to tropical (Hattin, 1982) and probably about the same as today's Gulf of Mexico. Even so, water that is 24–29°C (75–85°F) is cooler than the opti-mum body temperature of an active large lizard like a Komodo dragon (33.9–36.1°C; 93–97°F) or a larger mosasaur. Unlike modern marine mammals, mosasaurs did not have a layer of insulating fatty tissue (blub-ber) to help them retain body heat. The cooler water would have drained body heat continuously, and fully marine mosasaurs had no way to warm themselves in the sun like a terrestrial lizard (or a marine iguana). While a large adult mosasaur like *Tylosaurus* might have conserved internal heat by its sheer size and bulk (gigantothermy), newly born and juvenile mosasaurs would not have been able to stay active in cooler water unless they were warm blooded (endothermic) from birth. Just the fact that mosasaurs were so successful worldwide implies, in my view, that they were also active warm-blooded marine reptiles. However, that idea does require some sort of proof.

At the First Mosasaur Meeting in Maastricht, the Netherlands, Lou Jacobs presented data (see also Jacobs et al., 2005) indicating that modern marine (ectothermic) squamates (sea snakes and marine iguanas) are geographically limited to environments where the water temperature is 20–30°C (68–86°F). However, analysis of oxygen iso-topes in the teeth of marine reptiles by Bernard et al. (2010; see also review by Motani, 2010) indicated that ichthyosaurs and mosasaurs were able to maintain body temperatures well above the temperature of the

Warm Blooded or Cold Blooded?

surrounding water, while cautioning that some form of gigantism might be involved in mosasaurs. Most recently, oxygen isotope analysis of teeth of fish, mosasaurs, and birds living in the Gulf of Mexico during the Late Cretaceous by Harrell et al. (2016) indicated that mosasaurs of all sizes and species tested had body temperatures well above the temperature of the water they were living in. While there is certainly more research to be done, I am confident at this point that mosasaurs were active, 'warm-blooded' marine predators and very different from modern marine squamates.

What Mosasaurs Ate

Mosasaurs probably fed primarily on fish and cephalopods such as squid and belemnites, but the fossil record shows that the larger species would have eaten just about anything they could swallow. The preserved stomach of a large *Tylosaurus proriger* (SDSMT 10439) on exhibit in the Museum of Geology at the South Dakota School of Mines and Technology contains the bones of a smaller mosasaur (*Clidastes*), a toothed, swimming bird (*Hesperornis*; Plate 8), and a fish (*Bananogmius*), and several shark teeth (Martin and Bjork, 1987). Bell and Barnes (2007) reported that the gut contents of a large *T. nepaeolicus* included three young mosasaurs (*Platecarpus* sp.) and a *Ptychodus* shark. Fish remains were observed in a specimen of *Plotosaurus* from California (Camp, 1942), a *Platecarpus* from Kansas (Lindgren et al., 2010), and a *Tylosaurus proriger* from Kansas (FFHM 1997–10; Everhart, pers. obs.). Konishi, Newbrey, and Caldwell (2014) documented fish as the stomach contents of a young *Mosasaurus missouriensis* from the Bearpaw Formation (Campanian) of Canada. The remains of a turtle were reported in a 15 m specimen of *Hainosaurus* [now *Tylosaurus*] from Belgium: "Quoi qu'il en soit, le Hainosaure se nourrissait assurément de tortues marines, car nous en avons trouvé des restes dans sa carcasse" (Dollo, 1887:520). Translated, he said, "However, *Hainosaurus* undoubtedly fed upon marine tortoises, because their remains have been found in its carcass."

One of the recently 'rediscovered' items in the *Tylosaurus* diet was a polycotylid plesiosaur (Chapter 8). A large (10 m; 30 ft) *Tylosaurus* specimen discovered in Logan County in 1918 by Charles H. Sternberg included the remains of a juvenile polycotylid plesiosaur as stomach contents (Sternberg, 1922; Everhart, 2004a; Fig. 8.2). While the *Tylosaurus* specimen was acquired and exhibited by the USNM (Gilmore, 1921), the plesiosaur remains were largely ignored. Sternberg's brief report in the *Transactions of the Kansas Academy of Science* had been overlooked by mosasaur workers until 2001, when I came across it while searching for something else. It appears that these mosasaurs, especially the larger individuals, were generalists when it came to suitable prey and ate almost anything they could get in their mouths. Similarly, the 2011 discovery of a medium-sized (about 1 m in length) *Protostega* specimen (FHSM VP-17979; Fig. 9.23) that had been bitten and killed, but not eaten, by a large mosasaur (most likely *Tylosaurus*) indicates that there

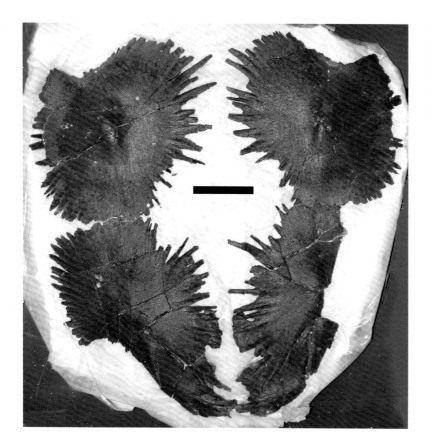

9.23. Ventral view of the plastron of a medium-sized *Protostega gigas* (FHSM VP-17979) that exhibits more than 90 punctures (bite marks) from a large predator, most likely *Tylosaurus proriger*. Scale bar = 10 cm (4 in).

were limits to the size of the prey that these predators could swallow (Everhart, 2013).

A *Globidens* specimen (SDSM 74764) from the Pierre Shale of South Dakota contains fragments of bivalve shells in the abdominal area (Martin and Fox, 2004, 2007). This specimen is the youngest and most complete known, and has been designated as a new species, *Globidens schurmanni* (Martin, 2007). Ammonite shells have been discovered with what appear to be mosasaur bite marks (Kauffman and Kesling, 1960), but there is some question as to whether these holes were caused by mosasaur teeth or by shell-boring invertebrates (Kase et al., 1998). To me, the question now is not so much whether mosasaurs fed on ammonites (Chapter 1, Introduction), but rather whether their teeth were responsible for the marks left on ammonite shells. My opinion is that it is likely they did prey on ammonites, but I think such predatory attacks would have resulted in the shells being crushed, not simply punctured. That said, there is simply no evidence of mosasaurs preying on ammonites in the Smoky Hill Chalk because ammonites are rarely preserved.

Tooth count (Table 9.3) is one method of distinguishing among the various taxa of mosasaurs, and it suggests a general trend toward a reduction in the number of jaw and pterygoid teeth over time. Most mosasaur teeth are conical and slightly recurved (Fig. 9.24). They may be smooth or distinctly striated or faceted, slender or robust, but usually do not have

9.24. Anterior portion of the skull of the FHSM VP-3 *Tylosaurus proriger,* showing the edentulous rostrum and robust conical teeth typical of tylosaurs. Scale = 30 cm (12 in).

the sharp, flesh-cutting edges and serrations typical of shark or theropod dinosaur teeth. They were also replaced continuously by new, slightly larger teeth throughout the life of the mosasaur, with the new tooth dissolving away the root of the older tooth and pushing it out as it grew (Leidy, 1858; Caldwell, Budney, and Lamoureux, 2003).

With the exception of a few specialized late-evolving species, mosasaurs used their jaws and teeth for seizing, killing, and holding their prey until it could be swallowed, and not for tearing it apart. It is unlikely that after capturing large prey in open, sometimes deep, water, a mosasaur would have had the luxury of tearing the meal apart into manageable pieces without risking it sinking to the bottom or being lost to other predators. Short of dismembering its prey by the sheer force of its bite, a mosasaur essentially had to swallow whatever it caught whole and head-first, much as a modern snake does.

Mosasaurs and snakes are also similar in the way their skulls are constructed (Caldwell, 1999; Lee, Bell, and Caldwell, 1999), including a flexible joint between the tooth-bearing dentary and the surangular (and other bones) that make up the back half of the lower jaw. In addition, the dentarys were not fused together at the anterior end of the lower jaw, allowing several degrees of movement at that point. This flexibility allowed the mosasaur to 'ratchet' prey into the throat with the aid of two rows of sharp 'holding' teeth on the pterygoid bones in the roof of the mouth.

A simple demonstration devised by Cope (1872c) can help to visualize the mechanics of how the mosasaur's lower jaw worked. Hold your arms straight out in front of you, hands extended, with your fingertips touching. Now try to pull your hands toward your face. Notice how

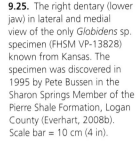

9.25. The right dentary (lower jaw) in lateral and medial view of the only *Globidens* sp. specimen (FHSM VP-13828) known from Kansas. The specimen was discovered in 1995 by Pete Bussen in the Sharon Springs Member of the Pierre Shale Formation, Logan County (Everhart, 2008b). Scale bar = 10 cm (4 in).

your elbows have to flex outward and that the angle between your fingers widens. This is the same process that occurred repeatedly as a mosasaur swallowed large prey. Each time the prey was pulled further into the mouth, the pterygoid teeth on the roof of the mouth caught the prey again and prevented it from escaping. Such feeding adaptations were necessary for survival in the Cretaceous oceans, and mosasaurs were certainly winners in that regard. Russell (1975) noted that the type specimen of *Globidens dakotensis* is the only mosasaur species known that does not have pterygoid teeth, but Bell (pers. comm., 2002) suggested the teeth of that specimen had been removed accidentally during the collection or preparation of the specimen. A fragmentary *Globidens* specimen from Texas was reported by Polcyn and Bell (2005) to have 10 teeth on the pterygoid. However, the tooth-bearing portion of the pterygoid was obscured in the newest specimen, *G. schurmanni*, described by Martin (2007).

Only a few of the later genera (Campanian–Maastrichtian) of mosasaurs developed specialized teeth for cutting or tearing flesh (*Leiodon*), feeding on crustaceans (*Carinodens*; Schulp, 2005), or crushing clams and other hard-shelled invertebrates (*Globidens* and possibly *Prognathodon*). The only specimen of *Globidens* (FHSM VP-13828) known from Kansas was discovered in 1995 by Pete Bussen near the top of the Sharon Springs Member (early Campanian) of the Pierre Shale in Logan County (Figs. 9.25, 9.26). Several other specimens, including the type specimen of *G. dakotensis*, are known from the Pierre Shale in South Dakota. An unusual and poorly known species from West Africa, *Pluridens walkeri*, appears to have reversed a trend toward shorter, heavier skulls and fewer teeth and became more 'ichthyosaur-like,' with 30 small, closely set teeth in its long narrow jaws (Lingham-Soliar, 1998). *Goronyosaurus nigeriensis*, another African species, had unusually large 'canine-like' teeth that interlocked and fitted into deep

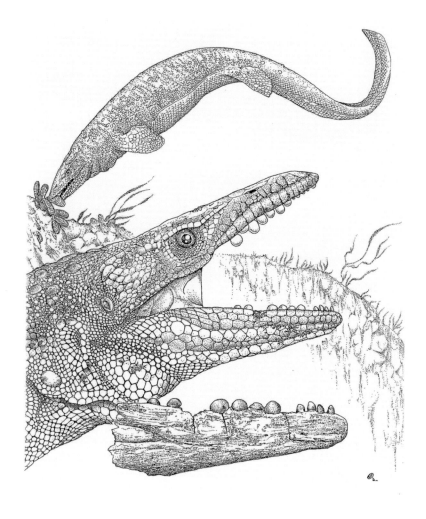

9.26. *Globidens* was a late arrival in the Western Interior Sea, moving north from the warmer waters of the Gulf Coast during the Campanian. The ball-shaped teeth in its jaws were well adapted for crushing the hard shells of clams and other shellfish. Drawing by Russell Hawley; Used with permission.

sockets in the upper and lower jaws, very similar to the appearance and function of the dentition of crocodiles (Lingham-Soliar 1999a; 2002).

Life in the oceans of the Cretaceous was dangerous at best, and mosasaurs didn't always have things their way. Early workers in the chalk, Cope (1872c), Mudge (1876), and Williston (1898a), noted signs of sharks scavenging on mosasaur carcasses. The serrated teeth marks of *Squalicorax falcatus* and *S. kaupi* are clearly visible on the scavenged bones of many fish and marine reptile specimens from the oceans of the Late Cretaceous (Bardet et al., 1998; Schwimmer, Stewart, and Williams, 1997). Recent evidence (Shimada, 1997; Everhart, 1999), including shark teeth embedded in partially digested mosasaur bones (FHSM VP-13283) from the late Coniacian (86 my) Smoky Hill Chalk of Kansas, indicates that mosasaurs were frequently fed upon by the large ginsu sharks (*Cretoxyrhina mantelli*). While it is not possible to determine whether this feeding activity was the result of an attack on a live mosasaur or the scavenging of a carcass, it is probable that a 6 m (20 ft) ginsu shark would have behaved much like a modern, similarly sized great white shark. The painting on the cover the first edition of this book (Plate 6), by Dan Varner, depicts a shark attack on a living mosasaur that could have produced the fossil evidence described from the specimen mentioned above. Most likely, ginsu sharks attacked small, injured, or sickly prey, including mosasaurs, because they were vulnerable, but they probably would not have passed up a free meal on a dead mosasaur.

Sharks weren't always successful in their attacks on living mosasaurs, as shown by the evidence of infections (exostosial bone growth) caused by shark bites on mosasaur tails that resulted in fusion of vertebrae (Rothschild and Everhart, 2015). In some cases the broken tips of shark teeth remained in the bone as the wound healed (Martin and Rothschild, 1989; Shimada, 1997; Everhart, 2004c; Rothschild and Everhart, 2015).

The fossil record, however, also indicates that these giant *Cretoxyrhina* sharks became extinct a few million years later (early Campanian) at roughly the same time that mosasaurs were becoming larger, more numerous, and more widespread. Did the success of mosasaurs cause the extinction of these large sharks? No one knows for sure, but we do know that modern sharks suffer greatly from overfishing by humans because they do not reproduce rapidly. It seems reasonable that bite-sized juvenile sharks were on the menu for the expanding population of large, hungry mosasaurs.

Mosasaurs ruled the oceans of the Late Cretaceous and were apparently beginning to invade freshwater environments such as estuaries, swamps, and rivers when the Age of Dinosaurs ended. At that point, there were more than 45 species of mosasaurs, many of which are known only from the Maastrichtian (Lingham-Soliar, 1999a). Like the dinosaurs, it is unlikely that any mosasaur species survived much past the Cretaceous–Tertiary boundary. Did they all die suddenly because of the catastrophic effects of an asteroid impact in the Yucatan, or was their extinction more gradual, following a general collapse of the marine ecosystem? We may never know.

Addendum

The study of mosasaurs is moving along at a more rapid pace, as more specimens are discovered worldwide and more researchers become interested in these fascinating animals. In May of 2004, I attended the First Mosasaur Meeting in the Natural History Museum of Maastricht in the Netherlands, along with about 20 other paleontologists from around the world (the Netherlands, Sweden, Denmark, Germany, France, Bulgaria, Japan, Canada, and the United States). It was the first time that such a meeting had been held on mosasaurs only. We were able to tour the historic limestone quarries around Maastricht, where the first mosasaur remains were collected, and to visit the Teylers Museum in Haarlem, where other specimens that were discovered before the type specimen of *Mosasaurus hoffmanni* have been on exhibit for more than 200 years. In between, we spent two days listening to papers and sharing information on mosasaurs. The meeting was an overwhelming success, thanks in large part to the work done by the staff at the museum (Anne Schulp, John Jagt, Eric Mulder, and others) and the enthusiastic support of the museum director, Douwe Th. De Graaf. The proceedings of the first meeting was published in 2005.

The Second Mosasaur Meeting was held in the Sternberg Museum of Natural History at Hays, Kansas. Given the success of the first meeting and the publicity from the resulting publications, this event was even larger. Although I had warned participants that springtime weather in Kansas can be stormy, the F5 tornado that almost totally destroyed the town of Greensburg, Kansas, just 90 miles south of Hays, on the first night was unexpected. We had two and half days of excellent presentations and ended the meeting with a field trip to my favorite chalk locality in Gove County. I served as the editor for the Proceedings of the Second Mosasaur Meeting (Everhart, 2008d).

Nathalie Bardet hosted the Third Mosasaur Meeting at the Muséum National d'Histoire Naturelle in Paris, France, May 18–22, 2010. Although I was unable to attend the meeting, I know that it also turned out well.

The Fourth Mosasaur Meeting was hosted by Mike Polcyn on the campus of Southern Methodist University in Dallas, Texas, and literally as I am writing this, the Fifth Mosasaur Meeting is taking place in Uppsala, Sweden, hosted by Johan Lindgren.

Nearly 200 years after mosasaurs were first named, mosasaur research is now proceeding at a rapid pace. Many interesting discoveries lie ahead of us.

Pteranodons

Rulers of the Air

10

Soaring above the seemingly endless stretch of bright blue water on thin, kitelike wings, the young male pteranodon effortlessly rode the sea breeze upward as he searched for signs of a school of fishes feeding near the surface. From his vantage point, he could look down on a large area of the water's surface, seeking the shifting dark shadows made by huge schools of small fishes as they migrated across the shallow sea. With swift movements of his head, he also kept track of other members of his flock, watching them for a sudden dive toward the surface that indicated prey had been sighted.

It was hatching time and there were many new babies to be fed. This flock of males had left the rookery late the previous afternoon, catching the rising air currents and prevailing winds that carried them far out to sea during the hours of darkness. Their instinctive navigation for long distances over water was necessary in order for the pteranodons to reach the most productive feeding grounds, and where there was the least competition.

Now the sun was again rising in the sky, and his sharp eyes were searching as he soared in large, counterclockwise circles. Off to one side, he saw something dark moving slowly in the water. Adjusting his wings slightly, he dipped and turned to the right, spiraling downward toward the object. He soon recognized the long dark object as it flew slowly through the water on wings of its own. It was a plesiosaur, a huge, long-necked marine reptile that fed on the same small fishes the pteranodon was seeking. Still, it was only one predator, and it was no threat to the pteranodons. Having other predators in the area sometimes meant that schools of fishes would be frightened and forced to the surface, creating a feeding opportunity for the flock. Keeping an eye on the plesiosaur, the pteranodon turned into the wind, climbed higher, and resumed his aerial search pattern.

The hours wore on without the flock sighting any schools of migrating fishes. If necessary, the pteranodons were capable of staying in the air for days at a time, but that would not feed their hungry young or the females waiting back at the rookery. About noon, he saw one of his flock mates turn and rapidly lose altitude, followed almost immediately by several others. As they descended, he could see a large shadow forming near the surface. He was the furthest away from the feeding opportunity that was developing and would be the last to arrive. As he turned and descended toward what was certainly a school of fishes, he saw other dark, sinuous forms circling the shoal. A pod of snakelike mosasaurs had surrounded the fishes and were moving around them in an ever-tightening circle. In the center of the trapped school, the closely crowded fishes began to jump out of the water, causing the surface to erupt into a foaming melee.

The first group of pteranodons skimmed quickly across the surface, dipping their beaks into the water and scooping up mouthfuls of small fishes. They were quickly joined by a flock of small white birds with toothed beaks that dove into the water and took their prey one at time from the school. The first group of pteranodons wheeled about and started another run across the now boiling school of fishes as a second group completed its pass.

Gauging his approach and watching out for his flock mates, the young male pteranodon adjusted his wings to slow his forward speed as much as possible as he descended in a shallow dive. Flying low across the school of fishes, just inches from the surface, he dropped the sharp tip of his lower jaw into the water and immediately felt the impact of the small prey as they were scooped up and forced back into his mouth and throat. Lifting his beak out of the water, he began flapping his wings to gain altitude as he swallowed the wriggling fishes. Rising slowly at first, then more rapidly, he turned to make another pass at the school. This time he saw that the mosasaurs were now actively feeding along the outer edges of the bait ball, but too far away to endanger the feeding pteranodons.

Again and again, he skimmed the water's surface, each time adding a number of small fishes to his catch. Soon his gullet was full and it took much more effort to gain altitude at the end of a pass. Finally, as he turned into the wind and labored upward, he saw that the rest of the flock was gathering above him. Below him, the school of small fishes had been dispersed by the attacking mosasaurs.

He could also see that a number of larger fishes, including sharks, had joined the feeding frenzy. Small groups of prey fishes scattered and then reformed into larger groups, only to be scattered again. Even as the well-fed pteranodons gathered to leave the area, another flock of pterosaurs arrived and began to skim the surface. They were a smaller species, although much larger than the birds that still wheeled and screeched over the dwindling bounty.

Once the pteranodon flock was back together, an older male took the lead and headed back toward the distant shore where the rookery had been established. The flock followed the leader, flying one behind the other in a relatively close V-formation that took advantage of 'slipstreaming' to reduce wind resistance and energy use. They would have to fly for many hours to reach the rookery on the shore. Ahead, however, a fast-moving cool front was building storms across their flight path. Within an hour, they began to run into turbulence. For the lightly built pteranodons, this was a dangerous situation. The flight leader instinctively moved between the rising thunderheads, seeking calmer air. Still the flock was buffeted about, breaking up into smaller and smaller groups as flight conditions became more difficult. As darkness fell, flashes of lightning allowed the young male to adjust and maintain his position near the center of a group of five. The other four pteranodons were older and stronger than he was, but they were still having trouble with the winds. The effort of flying in this weather was wearing him down more quickly, however, and he was having a difficult time keeping up. Then a sudden downdraft scattered the small group and he was alone, fighting wind gusts near the wave crests of the storm-swept sea. Moments later his wing tip caught the crest of a wave and he crashed heavily into the water. Dazed by the impact, he tried to untangle his wings, but they only wrapped more tightly around him. Struggling to breathe, he was submerged and trapped beneath the dark water. Mercifully, blackness soon enveloped him.

The story above, especially the long flight far out to sea to feed, is fiction. We don't know very much about these giant pterosaurs even though an estimated 2,000 specimens have been collected from the Smoky Hill Chalk since the 1870s. We know they ate fishes, but we don't know if they fed while on the wing or while sitting in the water, or grabbed like a heron walking along the sea shore. For that matter, we don't know if they soared for long periods over the Western Interior Sea like a modern albatross, or only periodically migrated across the sea. We have never discovered evidence of where or how they nested, and we certainly know nothing about their social behavior. We do not know why so many of their remains occur hundreds of miles from what would have been the nearest shore, or how they died, or, for that matter, how they lived.

Pteranodon

The Smoky Hill Chalk has produced a wealth of *Pteranodon* (meaning 'wing without tooth') specimens. Note that the word 'pteranodon' is both a scientific name that should be capitalized and italicized (*Pteranodon*) when referring to the genus, and a common name that refers generally to the winged and toothless flying reptiles of the Late Cretaceous (i.e., *Quetzalcoatlus northropi*, the largest known flying reptile from North America, is also a pteranodon).

Years after his initial discovery of the first *Pteranodon* wing bones in 1870, Marsh (1884:424) reported that the Yale Peabody Museum had the remains of about 600 individuals in its collection. From a survey of the Yale collection and others Russell (1988) calculated that a total of 878 specimens were known, or roughly 12 percent of all vertebrates in museum collections from the Smoky Hill Chalk, placing them a distant third behind fishes (58 percent) and mosasaurs (25 percent). Four years later, Bennett (1992) examined more than 1,100 specimens in his study of size differences between male and female specimens of *Pteranodon*.

In the years I have spent collecting in the chalk, I have collected only a dozen sets of *Pteranodon* remains, most of which were very fragmentary. I discovered my first one in 1990 when I came across the ends of several wing bones eroding from the chalk in northeastern Lane County. Once they were uncovered, it was easy to see that the thin-walled bones had been crushed. Working with them was like trying to pick up two layers of broken eggshell. Even so, I was able to collect the lower jaw and most of one wing of a medium-size *Pteranodon* (FHSM VP-17702; wingspan 4.2 m; 14 ft). All of the bones were severely damaged by 'root rot' (a condition in which the bones are literally enclosed by a web of plant roots that eventually destroy the bone in a search for nutrients), and there was nothing left of the large, shield-shaped sternum except an empty space filled with roots. The nearly complete lower jaw of the specimen was quite interesting. The narrow, V-shaped mandible was 63 cm (25 in) long and narrowed to a point that was as small as a pencil lead (about 2 mm). I still have to wonder how something that small and delicate could survive flying low over the water and grabbing fishes near the surface (Brower, 1983). If that was how *Pteranodons* fed, I suspect they weren't taking very big fishes. A study by Humphries et al. (2007) comparing pterosaurs to modern skim-feeding birds concluded that none of the pterosaurs, including *Pteranodon*, were adapted for skim-feeding.

Our next *Pteranodon* discovery was a bit more spectacular. It happened late in the afternoon of June 1, 1996. We had been collecting all day and were about ready to call it quits and head for home. As I was walking back to the van, my wife, Pam, called me over to where she had located the ends of two bones eroding out of the chalk. It was clear from their flattened appearance that they were wing bones from a *Pteranodon*. They were about halfway up the east side of an 2.5 m (8 ft) high, nearly vertical wall of a narrow gully in the middle chalk (early Santonian) of southeastern Gove County. The bones had apparently been exposed fairly recently but were already bleached to a light blue-gray color that contrasted well with the pale tan of the chalk. After digging back into the soft chalk about 15 cm (6 in), I uncovered most of the still articulated 'elbow' (the joint in the wing between the humerus and the radius/ulna) of a fairly large *Pteranodon*. The 'fresh,' unweathered material was a reddish brown color and was extremely fragile (again, the comparison to working with two layers of broken eggshell held together with a soft, chalky matrix is apt). The bones were oriented in a V-shape, with the open end of the V pointing into the side of the gully. That they were from a large individual (humerus about 24 cm in length; 9.4 in), and were still articulated, indicated to me that more of the *Pteranodon* might still be in the chalk. Finding them certainly made it easy for us to plan another trip to the site. I completely exposed and removed the humerus. The two other bones (radius and ulna) were already broken near the middle, so I carefully removed what I could and covered the rest. Since the length of the humerus is roughly 4.5 percent of the length of the wingspread (Bennett, 1992:431), I figured we had discovered the remains of a fairly large *Pteranodon* (wingspread about 5.3 m or 17.4 ft).

On June 29, we returned to the site about 8 a.m. to take advantage of as much of the cool morning temperature as possible. The first order of business was to start taking the overburden off the area where I expected the rest of the fossil to be. About a meter of hard tan chalk interbedded with

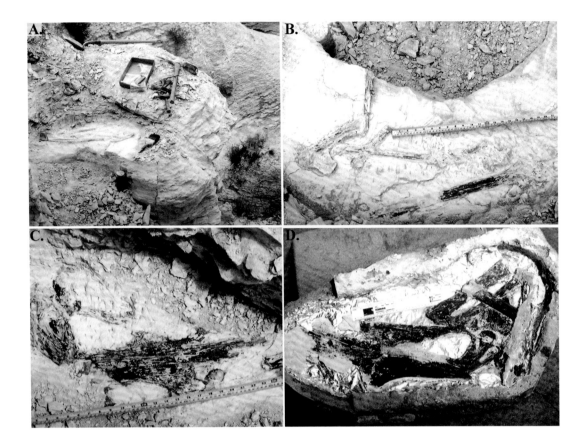

calcite seams, covering an area about 2 m (6.6 ft) on a side, needed to be removed. Part of this was done right away in order to get a minimum working area and to expand the initial entry point to locate additional remains. Additional chalk was removed as needed over the next day and a half. All the initial overburden removal was done with a heavy pick and flat-bladed shovel (and lots of sweat!). The chalk was fairly hard (for chalk) and partially cross-bedded with calcite seams, causing it to break up in mostly small, uneven pieces. We did this hot, somewhat tedious work in small increments.

When I finished digging with the pick and shovel, about 10 cm (4 in) of chalk remained on top of the fossil. This layer was removed slowly with a sharp ice pick and a brush and occasional careful use of my rock hammer. As soon as I started working at the level where the original material occurred, I encountered additional bone, including the distal portions of the radius/ulna that I had been unable to take out initially. This led to additional bones that were scattered randomly to the right (southwest) of the initial find. These included the other humerus, a scapulo-coracoid, two cervical vertebrae, and additional wing bones, including most of the fourth metacarpal (MC IV—the long wrist bone of the pteranodon wing), which was 56 cm in length (Fig. 10.1). These were exposed, coated with a preservative, photographed, and then removed. The preservative hardens the bone but can be removed with acetone if necessary during final preparation.

10.1. Collecting a *Pteranodon* (CMC VP-7203) skull in 1996: A, a view of the site from above, taken after the additional wing bones had been exposed; B, close-up of the wing bones; the second humerus is at far left; yellow scale is 53 cm (21 in); C, skull and lower jaw in right lateral view; orbit is at far left; frame is about 51 cm (20 in) wide; D, partially prepared skull in left lateral view, still in isocyanate foam jacket; several wing bones were lying under and around the crest of the skull.

By early afternoon, another ridge of bone was encountered at the back (east side) of the excavation. As this was uncovered, it became apparent by the size and shape that we had located the lower jaws, lying together with the right jaw on top of the left. Further excavation at the posterior end of the jaw showed additional bone above the upper surface of the jaw; eventually the entire skull was seen to be in place, still articulated with the lower jaw. The skull was lying on its left side, with the lower jaw closed. Working around the skull, I encountered several other wing bones, including one that went directly under where the crest should have been.

By the time that I had the skull exposed, it was almost 5 p.m., and I knew we were not going to get the remains out that day. I covered up the specimen with a tarp, and we checked back into the Q-Motel at Quinter, the nearest small town, for the night after a 25-mile drive to Interstate 70.

Sunday was supposed to be even hotter, so we were back in the field by 7:45 a.m. The *Pteranodon* skull was cleaned further in preparation for the application of more preservative and jacketing. At this point, the last of the overburden to the east and south of the skull had to be removed in order to make room for jacketing of the skull and turning of the jacket. Several of the wing bones around the skull were removed individually in order to cut down on the size (and weight) of the final jacket. (This was important to me because I was going to have to carry the skull and jacket back to our van.) In the process, several small pieces of bone were isolated at the back of the skull. Field examination showed one wing claw mixed in with other material. Two more unarticulated cervical vertebrae (third and fourth?) were also discovered near the back of the skull and removed. From what I could see without too much digging around, it appeared that the atlas/axis vertebrae were still connected to the base of the skull.

At this point, the block containing the skull was isolated by digging a 10 cm deep trench completely around it. Once this was done, the block was undercut slightly to allow the jacket material to fill in and support as much of the chalk as possible and to find a seam in the chalk that could be used to separate the block from the underlying matrix. Once the undercut was completed, the exposed bones of the skull were again painted with the preservative and allowed to dry.

When the preservative had dried, aluminum foil was placed over the block to prevent the jacketing foam from contacting the specimen. A narrow piece of thin plywood was added along the length of the skull to reinforce the jacket. Then a temporary form for the jacket was made of cardboard, placed around the block containing the skull, and supported with pieces of broken chalk. A two-part mixture of isocyanate foam was prepared and poured into the form. The foam expands and cures in about half an hour, producing a rigid yet extremely light jacket for the specimen. I could have used a standard plaster and burlap jacket, but I was concerned about being able to carry the specimen once I had it safely jacketed. A comparable plaster jacket would have weighed nearly as much as the piece of chalk containing the skull. In this case, high-tech modern chemistry won the day.

When the jacket had cured and the form was removed, thin chisels and metal blades were driven under the block of chalk to free it from the matrix. Once it was loosened, the block was carefully but quickly turned over so that it was now upside down and resting safely inside the 'top' of the foam jacket. This is always the moment of truth, when you have something inside a jacket that needs to be turned over. Occasionally, if you don't do it right, the matrix and the fossil can fall out of the jacket as it is being rolled. In this case, everything stayed in place. I removed some of the excess chalk to lighten the load, and then carried it about 50 yards up the hill to our van for the trip home. We left the field about 1 p.m. on Sunday, feeling pretty good about our efforts.

In May 1998, almost two years after a long-delayed preparation of the skull, Chris Bennett, then at the University of Kansas, identified our discovery as a remains of a young male *Pteranodon sternbergi*. Interestingly, the only bones from below the shoulders that have been identified so far are several tarsals and a toe claw that Chris noted had been lying under the beak of the animal. It's possible that other remains were still at the site, or had eroded out and been destroyed earlier, or that the lower body of the *Pteranodon* had been severed by scavengers before burial. As noted in the fictional story above, the wings were literally wrapped around the skull. It is possible that they helped protect and preserve the skull itself.

In June 1998, we made a final trip to the site, hoping to collect additional postcranial material. After removing a lot of overburden, however, we were able to find only one more bone, the other scapulo-coracoid. In 1999, the specimen was donated to the Cincinnati Museum Center, where the preparation was completed. The skull is currently on exhibit (CMC VP-7203). It turned out that, according to Bennett (1992), our specimen was almost exactly the average size he calculated for a male *Pteranodon* from the Smoky Hill Chalk (wingspan 5.6 m; 18 ft).

Pterosaurs were flying reptiles (not dinosaurs) that lived alongside dinosaurs through most of the Mesozoic. The first pterosaurs were small, bird-size creatures, but through time they evolved into giant fliers with wingspans as big as a small airplane. They were superbly adapted for flight, with hollow, air-filled bones, a relatively large, birdlike brain (Seeley, 1871; Wellnhofer, 1991), and membranous wings that were supported by the elongated fourth finger of each hand (Fig. 10.2). In *Pteranodon* and other larger species from the Cretaceous, their upper bodies were stiffened by rigidly binding the fused dorsal vertebrae and ribs together into a solid structure (notarium) that supported the large muscles needed to power their wings. Fossil evidence shows that some of the smaller pterosaurs were covered with fur-like fibers, and it is likely that they were 'warm blooded' to some extent. Most had three clawed fingers on each hand and four clawed toes on each flat, narrow foot. Contrary to the reconstructions shown in science fiction movies, the toes were not opposable like those of an eagle, and it would have been impossible to them to pick up and carry off prey with their feet (see also Williston, 1891:1126).

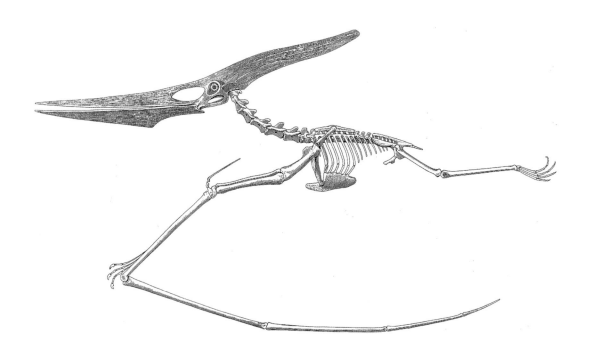

10.2. A drawing of the skeleton of a male *Pteranodon longiceps* from Eaton (1910:pl. XXI). Note the size of the skull compared to the rest of the body. No scale.

Pterodactylus from the Jurassic of Europe was one of the smallest and was about the size of an American robin. The largest pterosaurs, like *Quetzalcoatlus*, had a wingspread as large (11–12 m; 36–39 ft) as a light airplane. The largest species (*Pteranodon*) that occurred in Kansas had a wingspread of about 8 m (26 ft) but would probably have weighed no more than 11 kg (25 lb). Compared to their heads and their wings, their bodies are almost absurdly small. One interesting comparison provided by Hankin and Watson was that "with a body little larger than that of a cat, they had a span of wing asserted in some cases to have reached 21 feet [6.4 m] or more" (1914:324).

Although pterosaurs (wing-lizards) had been known from Europe since the early 1800s, they were not discovered in North America prior to 1870. The much older European specimens from the Jurassic period were generally small with long tails and jaws filled with sharp teeth. They looked much like a reptilian version of a bat and were often depicted as something similar in the literature of the day. The whole concept of these flying lizards was about to be radically changed by a chance discovery in western Kansas.

During the summer and fall of 1870, O. C. Marsh and his first Yale College Scientific Expedition had collected fossils from as far west as Wyoming and Utah. On their way back to Yale in late November, they made a brief visit to western Kansas in the vicinity of Fort Wallace. The weather was cold, but they were still able to collect for several days along the Smoky Hill River in what are now Wallace and Logan counties. Charles W. Betts (1871), one of Marsh's students from Yale, wrote a detailed and well-illustrated popular account of the expedition's adventures. Their time in Kansas, however, was only a small part of their journey and was allocated only a half-page of the eight-page essay. Betts did indicate that they spent four days digging out the nearly complete skeleton of a "sea-serpent . . . allied with the genus *Mosasaurus*" (ibid.:671). The story was published in the popular *Harper's New Monthly Magazine* in October of the following year.

In a short note published after returning to Yale in mid-December, Marsh (1871a:143) noted only that "some interesting reptilian and fish remains" were collected during their two-week stay in Kansas. A few months later, Marsh (1871b) published a more detailed account of the mosasaurs discovered by the 1870 expedition in Kansas, including the nearly complete *Clidastes* specimen mentioned by Betts (see Chapter 9).

This rather long introduction to the discovery of the first *Pteranodon* in Kansas makes the point that finding a giant pterosaur wasn't expected, even by Marsh, nor was it readily apparent what he had. At the time, larger, more complete specimens of strange fishes and marine reptiles were being collected, and mosasaurs were among the most popular of the "new" discoveries.

In his first mention of the *Pteranodon*, Marsh (1871c) reported that the distal ends of two long bones (two metacarpals from the wings of two individuals, YPM 1160 and 1161) had been collected, and he noted that they were not unlike those figured by Richard Owen (1851:tab. XXXII). Marsh (1871c:472) also noted that the bones were thin-walled and hollow. From these few fragments, he named the new species *Pterodactylus Oweni*, "in honor of Professor Richard Owen of London." Marsh also estimated the size of the creature from the fragments and noted that the outstretched wings would have measured "not less than twenty feet!" (ibid.).

Surprisingly, and as a complete fabrication since no skull material was reported to be present, Marsh (1871c:472) also indicated that "the teeth are smooth, and compressed." In his much more complete description of additional remains of this and two other species collected by the 1871 expedition, Marsh again noted the presence of teeth in *Pteranodon*: "The teeth found with remains of this species, and supposed to belong to them, are very similar to the teeth of Pterodactyls from the Cretaceous of England. They are smooth, compressed, elliptical in transverse outline, pointed at the apex and somewhat curved" (1872:244). Of the teeth of a second species (*Pterodactylus ingens*), Marsh wrote that "the dental characters of this species are at present only known from a single crown of a tooth; found with one series of the specimens and from two larger and very perfect teeth found by themselves, which agree so closely with the former that they deserve notice in this connection. These specimens are less curved and less compressed than the teeth referred to *Pt. occidentalis*, but in other respects they are nearly identical" (ibid.:247). Apparently Marsh assumed that the newly discovered American 'Pterodactyle' would have teeth like its European cousins and was hedging his bet on their discovery when the first skull was collected. Although he was going out on a limb, at the time there were no known pterosaurs without teeth.

In that regard, he was not the only one who would be playing 'fast and loose' with this unsupported conclusion. His rival E. D. Cope also indicated, more conservatively, that *Pteranodon* skulls "were slender and the teeth indicated carnivorous habits" (1872a:337). As shown in an illustration (see Fig. 13.1) in a book called *Buffalo Land* (Webb, 1872:facing 357), however, it is clear that Cope believed that at least one of the species he had named (*Ornithochirus umbrosus*) had teeth.

In the summer of 1871, Marsh and the second Yale College Scientific Expedition returned to Kansas. Marsh was able to return to the spot where he had collected the first remains of a *Pteranodon*, and he located additional pieces of the same bone. Joseph Savage (Lawrence Daily Journal, July 25, 1871, 2), who was at Fort Wallace with Professor Mudge at the time, reported that Marsh "says he has not obtained anything yet very remarkable, except a pterodactyl, . . . which has wings twenty feet long from tip to tip; it is web footed, and in appearance some like a large bat."

By the time Marsh (1872) published more complete descriptions, however, the giant *Pteranodon* specimens were already taking a back seat to the recent discovery of toothed birds in western Kansas (Chapter 11). Marsh (ibid.:241) did note that the name given to the first specimens (*Pterodactylus Oweni*) was preoccupied by a specimen described by Seeley and replaced it with the name *Pterodactylus occidentalis*.

Besides this species, Marsh collected several specimens of "the most gigantic of Pterosaurs," which he called *P. ingens* (YPM 1169 and 1170) and estimated had a wingspread "of nearly 22 feet!" (1872:246–247). Marsh also named another, smaller species (*P. velox*; YPM 1176) discovered in 1871 on the basis of what he believed were differences he saw in the wing bones. Bennett (1994a) examined the Marsh collection at the Yale Peabody Museum and indicated that because of the stratigraphic level (upper chalk) in which the pterosaur remains collected by Marsh in 1871 and 1872 occurred, all of Marsh's first specimens were probably *Pteranodon longiceps*. In that regard, Bennett (ibid.:14) considered all the early names given by Marsh and Cope (below) to be nomina dubia because the material that was collected does not exhibit any species-specific characters and was too fragmentary to be accurately identified beyond the genus *Pteranodon*.

Cope apparently collected at least two sets of *Pteranodon* remains during his trip to Kansas in late 1871. In a short note, Cope (1872a:337) named two species, *Ornithochirus umbrosus* (AMNH 1571) and *O. harpyia* (AMNH 1572), that were apparently distinguished from one another only on the basis of size. In doing so, he accepted Seeley's name for the genus and apparently ignored Marsh's *Pterodactylus* without further comment.

In a narrative that preceded the listing of the two new species, Cope described them in their natural habitat:

> The flying saurians are pretty well known from the descriptions of
> European authors. Our Mesozoic periods had been thought to have lacked

these singular forms until Professor Marsh and the writer discovered remains of species in the Kansas chalk. Though these are not numerous, their size was formidable. One of them, *Ornithochirus harpyia* Cope, spread eighteen feet between the tips of its wings, while the *O. umbrosus,* Cope, covered nearly twenty-five feet with his expanse. These strange creatures flapped their leathery wings over the waves, and often plunging, seized many an unsuspecting fish; or, soaring, at a safe distance, viewed the sports and combats of the more powerful saurians of the sea. At night-fall, we may imagine them trooping to the shore, and suspending themselves to the cliffs by the claw-bearing fingers of their wing-limbs. (1872a:323)

While this image of pteranodons hanging from rocks along the seashore has been shown in numerous illustrations over the years, it is probably just a fantasy. As noted by Stein (1975), it would have been impossible for *Pteranodon* to land on all fours on level ground without collapsing its wings first and thus losing lift. Performing such a landing against the vertical wall of a cliff would seem to be a death-defying act. Furthermore, none of the cores of the wing claws I have collected or examined show any damage on the tips as might be expected from a daily routine of hanging from one's fingertips.

In a follow-up paper read before the American Philosophical Society, Cope (1872b) provided a more complete description of the specimens (both consisting of the fragments of wing bones). Cope noted that some of the phalanges (finger bones) bore large claws that were similar to Seeley's genus *Ornithochirus*. He also stated that "as it is not likely on other grounds that the species of the Niobrara cretaceous strata belong to the genus *Pterodactylus* of Cuvier, which is chiefly known from the Jurassic period, I place the Kansas species for the present in *Ornithochirus*, as established by Seeley" (ibid.:420–421). In addition, Cope suggested that "this species is the largest Pterodactyle as yet known from our continent, the end of the wing metacarpal exceeding the diameter of that of the species described by Professor Marsh" (ibid.).

Clearly the challenge had been given, and a response was not long in coming. In a report regarding Cope's paper published in the *American Journal of Science,* an 'anonymous reviewer' noted that "*Ornithochirus umbrosus* Cope = *Pterodactylus ingens* Marsh, and *Ornithochirus harpyia* Cope = *Pterodactylus occidentalis* Marsh. As separate copies of Prof. Marsh's article were distributed March 7th, while the paper of Prof. Cope was not issued before March 12th, the names given by Prof. Marsh have priority" (Anonymous, 1872:374–375). Again, however, there were no diagnostic characters associated with the specimens collected by Marsh or Cope to justify even two species, let alone four.

Apparently Cope had lost the race to name his species by five days. Not one to give up easily, Cope noted that the species had been first "described by the writer in 1872 under the name of *P. harpyia;* but a fire occurring in the establishment printing the paper, its publication was delayed until two days after Professor Marsh had republished his species as *P. occidentalis*" (1875:67).

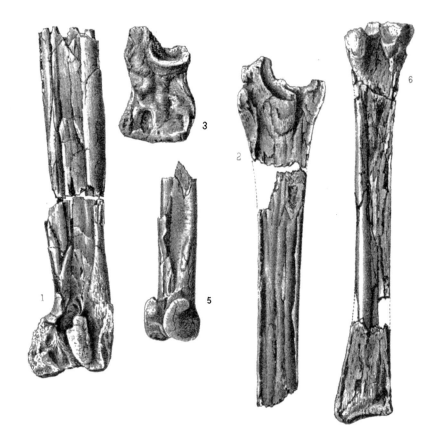

10.3. Figure adapted from Cope (1875:pl. VII) of the *Pteranodon* wing bone fragments that he collected in 1871 from the chalk of western Kansas, from which he named two species, *Ornithochirus umbrosus* and *O. harpyia*. No scale.

Cope (1874:26) included a list of Pterosauria from the Niobrara Cretaceous and conceded *Pterodactylus ingens:occidentalis*, and *P. velox* to Marsh. However, he kept *Pterodactylus umbrosus* Cope and placed it at the top of the list. Keeping things a bit confused, Cope on one page and on a plate (1875:65 and pl. VII) refers to the winged reptiles as Ornithosauria (Seeley), but lists them as Pterosauria on another page (ibid.:249). When all was said and done, Cope (ibid.:pl. VII) was the first to illustrate *Pteranodon* wing bones, in this case both the AMNH 1571 and 1572 specimens (Fig. 10.3).

By 1874, Marsh had the definite advantage of having B. F. Mudge, Joseph Savage, Harry Brous, S. W. Williston, and other experienced men collecting for him in Kansas. Cope, on the other hand, had hired a relatively young and inexperienced Charles H. Sternberg to work for him in 1876 (Sternberg, 1909:32–34). Sternberg, at the age of 25, and his young helpers were still learning about collecting fossils in general, while Marsh's better-organized, more experienced field crew was collecting birds and pteranodons.

Sternberg (1889) wrote a semifictionalized account in a popular magazine about his early experiences called "The Young Fossil-Hunters: A True Story of Western Exploration and Adventure." The serialized story closely follows the early years of his collecting fossils for Cope (Davidson and Everhart, 2014). Sternberg mentioned the discovery of a pterodactyl that "proved to be the largest of its kind that had ever been discovered, with a wing span of twenty-five feet" (1889:74). The same specimen also had been mentioned earlier by Sternberg when he wrote "the Pterodactyls of the Niobrara differed very much from those of Europe; they were much larger; I have found them with an expanse of wing of twenty-five feet; they were also without teeth, and Marsh (1876) has made the new genus 'Pterodon' [sic] for them" (1881:3).

The fossil-collecting crews watched each other's activities closely. In 1876, Williston wrote, "I do not think they will have any success in pterodactyls or small birds. One or two excellent ones they had struck carelessly with a pick and abandoned" (Shor, 1971:77). Later he noted that Sternberg was "down on the Smoky. . . . He doesn't know what a pterodactyl looks like and hardly what a saurian is. He has directions from Cope to collect all vertebrates—and we will take pains to leave him plenty of fishes" (ibid.:78). Williston was correct. Sternberg would collect many fish specimens, but the outcome was probably not quite what he or Marsh expected.

In a published June 2, 1877, letter to Dr. E. W. Seymour, Charles Sternberg wrote that he had "collected one hundred and forty-nine fish" during his first field season working for Cope (Junction City Weekly Union, June 16, 1877, 4). Cope would eventually name most of the fish species that were originally described from the Smoky Hill Chalk. Marsh would name none.

In the case of the American pterodactyls, however, Marsh had the final words on the subject between the two men. Apparently believing that American species were different enough to deserve their own order, Marsh created a new order (Pteranodontia) based on "the absence of teeth" (1876a:507), and a new genus (Pteranodon). He also described a new species, *Pteranodon longiceps*, which was the first to include a complete skull (YPM 1177). The specimen had been collected by S. W. Williston in May 1876. Marsh noted that "there are no teeth, or sockets for teeth, in any part of the upper jaws, and the premaxillary shows some indications of having been encased in a horny covering. The lower jaws, also, are long and pointed in front, and entirely edentulous [toothless]" (Marsh, 1876a:507). It should be recognized here that H. G. Seeley had reached the same conclusion about a new genus of flying reptiles discovered in the Cambridge Greensand several years earlier. Seeley wrote that "they have the ordinary dagger shaped snout, but appear to be entirely destitute of teeth. I provisionally name the genus *Ornithostoma*" (1871:35). Seeley originated the term 'ornithosaur,' which means 'bird-reptile.'

In a later note, Marsh reported that the remains from the Upper Cretaceous of Kansas "differ widely from the Pterodactyls of the old world, especially in the *absence of teeth* [italics by Marsh]" (1876b:479). It had been six years since the first discovery of a *Pteranodon*, and Marsh had finally understood that the fish teeth that had been collected with the earlier specimens did not belong to the specimens. Bennett reported that the remains collected in 1870 "consisted of the distal ends of two right metacarpals and a tooth of a *Xiphactinus* [a large fish] that Marsh

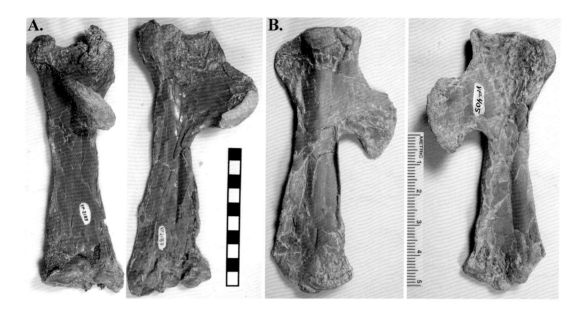

10.4. The humeri of *Pteranodon* and *Nyctosaurus* can be differentiated by the difference in size and the shape of the deltoid crest: A, left and right humeri of a *Pteranodon longiceps* (FHSM VP-2183) collected by George Sternberg in 1964; scale = 10 cm (4 in).; B, front and back views of the humerus of *Nyctosaurus gracilis* (FHSM VP-405) collected by Marion and Orville Bonner in 1956 from the upper chalk of Logan County; scale = 5 cm (2 in).

believed came from the pterosaur" (1994a:4). When I asked Chris Bennett for more information, he provided the following comment: "I found the actual teeth referred to by Marsh in the YPM collections. They were not kept with the *Pteranodon* specimens, but rather were separated out after it was realized that they did not belong with the *Pteranodon* specimens. However, they still had the appropriate YPM numbers on them. The teeth are teeth of *Xiphactinus* and perhaps *Ichthyodectes* as well, and so they were hidden away in the YPM fish collection" (pers. comm., 2004). Three teeth currently catalogued in the Yale Peabody collection were collected by Marsh in 1870 and later identified as *Xiphactinus audax* (two teeth as YPM 1163A and one as YPM 1171A).

Two other new and noticeably smaller *Pteranodons* were briefly mentioned for the first time and named by Marsh (1876a). One of those, '*Pteranodon*' *gracilis* Marsh, actually represented a new genus (*Nyctosaurus*), which would be recognized as such in a subsequent paper. The second new species, *Pteranodon comptus* (YPM 2335), was collected by Mudge in 1875, but is considered to be a nomen dubium by Bennett (1994a:48). It was Eaton (1910:3) who first noted that Marsh had misidentified the lower leg bones (tibiae) of a *Nyctosaurus* as wing bones of the new species. Marsh (1876b:480) renamed '*Pteranodon*' *gracilis* as *Nyctosaurus gracilis* (*Nyctosaurus* means 'night-lizard'). Although he still considered it to be closely related to *Pteranodon*, he noted that it could be distinguished from the larger species because the coracoid was not "coossified with the scapula" (ibid.). In the larger and heavier *Pteranodon*, the coracoid and the scapula are fused together into a single, U-shaped element. The ends of this bone are connected to the sternum and the fused dorsal vertebrae (notarium) and form a strong support for the attachment of the wing. Marsh believed *Nyctosaurus* would have had a wingspread of

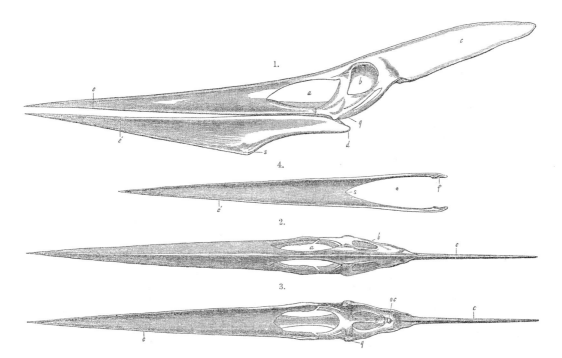

about 2.5 to 3.0 m (8 to 10 ft). Note also that, besides being much smaller, the humerus of *Nyctosaurus* also has a distinctive "hatchet" shape that is readily distinguishable from that of *Pteranodon* (Fig. 10.4).

Adding a bit of confusion to the name, Marsh noted later that "the name *Nyctosaurus* . . . appears to have been preoccupied, and hence may be replaced by *Nyctodactylus*" (1881:343). However, Williston stated that "the name has never been used otherwise for a genus of animals. Doubtless he [Marsh] thought the term conflicted with Nyctisauria, used for a group of sauria [geckos]. It does not, however, according to the accepted canons of nomenclature, and the original name should not be displaced" (1903:125). Williston's comment ended the discussion of the new genus name, and the name was not changed.

The collection of *Pteranodon* specimens from the Smoky Hill Chalk continued at a steady pace during the 1870s and 1880s. Marsh noted that "the remains of more than six hundred individuals of these reptiles have been secured . . . and now are in the museum of Yale College" (1884:423). Marsh (ibid.) further described the skull of a specimen of *Pteranodon longiceps* that he first reported in 1876 (YPM 1177). In addition, he figured the skull (ibid.:pl. 15) and noted that it was about 760 mm (30 in) long from the tip of the upper jaw to the back of the sagittal crest (Fig. 10.5). The illustration was the first ever to show four views of a three-dimensional reconstruction of a *Pteranodon* skull and was quite well done, although Williston (1892) would question the length of the added crest.

In Marsh's last paper on *Pteranodon*, he wrote that "there was apparently no ring of bony sclerotic plates, since in the best preserved specimens no traces of this have been found" (1884:425). Several years later, Williston (1891:1124, 1892:4) reported that the remains of a sclerotic ring

10.5. The first published drawing of the skull of *Pteranodon longiceps* (YPM 1177) by Marsh (1884:pl. 15). Top to bottom: left lateral view of the complete skull; dorsal view of the lower jaw; dorsal view of skull; ventral view of skull, lower jaw removed.

had been discovered in a new specimen of *Pteranodon longiceps* (KUVP 2212). In two brief papers written largely in defense of Marsh's uncompleted work on *Pteranodon*, Eaton (1903, 1904) attacked Williston's conclusions repeatedly, sometimes without justification. In one instance, Eaton erred in citing a reference to the sclerotic ring of *Nyctosaurus* (Williston, 1902a:528) instead of Williston's decade-earlier original comments (Williston, 1891:4) on *P. longiceps*, stating that "oddly enough, in this revision, it is *Pteranodon* that is credited with a sclerotic circle, and not *Nyctosaurus* in which Williston observed the structure" (Eaton, 1904:318). Eaton apparently missed the point that Williston had identified the sclerotic ring in both *Pteranodon* and *Nyctosaurus*.

After Marsh, Williston was the next to concentrate on the study of pteranodons of the Smoky Hill Chalk. He had worked for Marsh for eleven years, from 1874 to 1885 (Shor, 1971), and had himself collected many of the Yale Peabody specimens from Kansas. However, Williston and others were not allowed to publish papers on them so long as they were working for Marsh. Soon after he joined the faculty at the University of Kansas in 1890, Williston began collecting specimens from the chalk, including *Pteranodon* and *Nyctosaurus*, for the KU Museum of Natural History.

A year or so later, Williston provided probably the first complete description of a *Pteranodon*:

> About five or six species are known, varying in size, when alive, from four feet to not over twenty feet in expanse of wing. The head (in all the larger species, at least) was elongate and slender, with a well-developed occipital crest, and without teeth. The jaws may have been encased in horn, but I have never seen any evidence whatever that such was the case. . . . The body was short, the pelvis of moderate size, the hind legs comparatively small, with great freedom of movement, the tail short, and the feet without much, if any, prehensile power. Their food probably consisted of fishes. (1891:1126)

In the footnotes, Williston added (ibid.:n.4) that although the wingspread of *Pteranodon* had earlier been reported as being larger, he had measured the largest specimen and determined that the distance could not have exceeded twenty feet. He also indicated that "several coprolites found within the above described pelvis, ellipsoidal in shape, and about the size of an almond, showed bones so finely comminuted that their precise character could not be made out" (ibid.:n.5).

Barnum Brown mentioned a specimen at the American Museum of Natural History that included the "backbones of two species of fish and the joint of a crustacean, lying in the position of the throat pouch when death overtook the animal" (1943:106). This nearly complete lower jaw (AMNH 5098) was collected by Charles H. Sternberg in 1877 in Lane County (Mehling, pers. comm., 2004). A small mass (3 cm; 1.2 in) of fish vertebrae located behind the posterior margin of the symphysis of the jaw was interpreted by Bennett (pers. comm., 2004) as a bolus of food regurgitated at the time of death. Carpenter reported that "fish bones

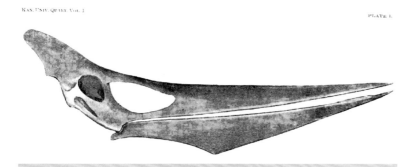

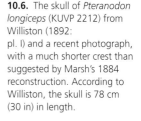

10.6. The skull of *Pteranodon longiceps* (KUVP 2212) from Williston (1892: pl. I) and a recent photograph, with a much shorter crest than suggested by Marsh's 1884 reconstruction. According to Williston, the skull is 78 cm (30 in) in length.

were found in two small coprolites associated with a *Pteranodon* sp. skeleton" (1996:44–46) from the Pierre Shale. Hargrave (2007) described 19 specimens of *Pteranodon* from the Pierre Shale of South Dakota and noted that two specimens included fish vertebrae (*Enchodus* sp.; ibid.:fig. 9).

Williston (1892) redescribed the skull of *Pteranodon longiceps* from a new, more complete specimen (KUVP 2212) that had been discovered by E. G. Case during the previous field season. He also criticized Marsh's reconstruction of a specimen he himself had collected in 1876 for Marsh (*P. longiceps*, YPM 1177; above). Williston noted that the type specimen was essentially complete when collected, except for the crest. Then he noted that Marsh had restored the crest from "indications presented by the basal portion, but without indicating that such a conjectural restoration had been made. The result is unfortunate" (ibid.:1124). According to Williston, the new specimen indicated that the crest was half as long as Marsh's reconstruction, and with a different shape. The specimen cited by Williston (KUVP 2212) is still on display as a part of a composite specimen in the University of Kansas Museum of Natural History. When Marsh's figure (1884:pl. XV) is compared to Williston's (1892:pl. I), it is readily apparent that the reconstructed crest is much larger in comparison to the skull (Fig. 10.6). As measured from the back of the orbit, the crest on the YPM 1177 skull represents about 32 percent of the total length compared to 26 percent of the slightly larger KUVP 2212 specimen.

However, the argument did not end there. Eaton (1903:pl. VI) reported and figured the crest of another specimen (YPM 2473, collected by Brous and Williston in 1876) that was being prepared at the Yale Peabody Museum. When the outline of the YPM 2473 crest was overlaid on Marsh's drawing of the YPM 1177 skull, the new crest was

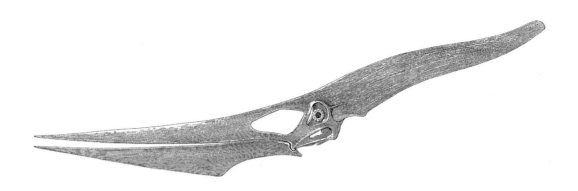

10.7. Composite drawing of the skull of *Pteranodon longiceps* (YPM 1177) overlaid with the crest of a partial skull (YPM 2473) as published in Eaton (1910:pl. IV, fig. 1). This crest is substantially longer than the one reconstructed by Marsh in 1884. Note the presence of the sclerotic bones around the eye in this figure.

substantially longer (43 percent of the skull length) than what Marsh had reconstructed (see also Eaton, 1910:pl. IV; Fig. 10.7). However, all three men apparently failed to consider the differences in the size and age of the animal, or sexual dimorphism, as reasons why the crests might be of different sizes.

In Eaton's case, he probably had a few other things on his mind at the time. The new Yale Peabody restoration of *Pteranodon longiceps* was being prepared under Eaton's direction "as the contribution of the Department of Vertebrate Paleontology of the Yale Museum to the University's exhibit at the Louisiana Purchase Exposition" (1904:318) at the 1904 World's Fair in St. Louis, Missouri. That same plaster reconstruction of a *Pteranodon* with a 3.0–3.6 (10–13 ft) wingspan is currently on display in the Great Hall of the Yale Peabody Museum (C. Bennett, pers. comm., 2004; pers. obser., 2010).

Returning to Williston's work in the early 1890s: it is apparent that Williston had trouble initially with the toothless character of *Pteranodon*. His 1892 paper was devoted mostly to the description of a new specimen of *Nyctodactylus (Nyctosaurus) gracilis* Marsh that was essentially complete except for the skull. The specimen had been collected during the previous field season near Monument Rocks (Williston, 1892:6). Williston noted that "not a single character has been given to distinguish this genus from *Pterodactylus*, and it is not at all impossible that it may prove to be the same; its location in *Pteranodon* rests solely on the assumed absence of teeth, and that is a character that is yet wholly unknown. . . . It seems very probable that the genus *Nyctodactylus* has no teeth in the jaws; it agrees in *every other respect* with the genus *Pterodactylus*, so far as is known. If the genus has teeth it must be united with *Pterodactylus*" (ibid.:11). Until the first skull of *Nyctosaurus* was collected, however, Williston was simply unable to distinguish its remains from those of *Pterodactylus*.

In his conclusion, Williston suggested that *Pteranodons* would also be discovered "in Europe, and if so, it is probable that the name *Pteranodon* must eventually be given up" (1892:12). In this case, he was referring to toothless remains of pterodactyls that had been described by Seeley in 1871 and named *Ornithostoma*. This was the same genus to which Cope had originally referred his discoveries from Kansas. Williston, however, had not seen Seeley's material and was careful not to make a decision at that time. By the following year, however, Williston had received copies of Seeley's papers and was convinced that "there can no longer be any reasonable doubt of the congenerousness [membership in the same genus] of our species with those included in the genus *Ornithostoma*" (1893:79). He also noted, and credited Seeley for recognizing, that it was Cope "whose acumen led him to refer his fragmentary material to the genus *Ornithocheirus*; an acumen all the more noteworthy in contrast with the total incomprehension of their affinities displayed by Marsh, notwithstanding his wealth of material upon which to base an opinion" (ibid.). The comparison here is that Marsh had hundreds of specimens to work with while Cope had only a few. It seems likely that Williston was expressing some anger regarding the oppressive scientific and management methods of his former employer. In a later note, Williston quoted Seeley (1871) as saying, "There is, so far as I can discern, no evidence of generic difference between *Ornithostoma* and *Pteranodon*" (Williston, 1897:35).

Williston wrote a number of papers on *Pteranodon* between 1891 and 1904. He described the lower jaw (Williston, 1895) from a specimen collected by H. T. Martin. Previously, the only remains of the skull and lower jaw had been crushed from side to side. In this specimen, the lower jaw was preserved more or less upright, allowing the true shape and other features to be seen for the first time. Williston (1896) described a new skull of *Ornithostoma* (*Pteranodon*) collected by C. H. Sternberg and continued to criticize the faults in Marsh's reconstructions. The following year, Williston (1897) published a "Restoration of *Ornithostoma* (*Pteranodon*)" and again indicated his belief that the American pterodactyls probably should be included in the European genus since there were no discernible differences between the two groups.

Following the discovery of an unusually complete specimen of *Nyctodactylus* (*Nyctosaurus*) by H. T. Martin, Williston (1902a) provided a further description of this smaller pterosaur. The general measurements he gives for the specimen (FMNH 25026) are of interest because they provide a more accurate look at these strange animals. Williston noted that

> while the wings gave a spread of very nearly eight feet, the body proper was less than four inches in diameter and not more than six inches in length, exclusive of the small tail. . . . One wonders where sufficient surface was presented for attachment of the strong muscles necessary for control of the wings. When it is remembered, however, that even the largest bones of the skeleton had walls less than a millimeter in thickness, and that many of the smaller ones were almost like cylinders of writing paper, he will perceive that, notwithstanding the extraordinary development of the anterior extremities and head, the creature, when alive, must have weighed but little. I very much doubt whether the living animal attained a weight of five pounds. (Ibid.:300)

In conclusion, Williston stated again, "I still believe that the genus *Pteranodon* is identical to *Ornithostoma*, and that the former term must

10.8. Ventral (top) and dorsal views of the skull of *Nyctosaurus gracilis* (FMNH 25026) from Williston (1902b). According to Williston, the skull is 31 cm (12 in) in length.

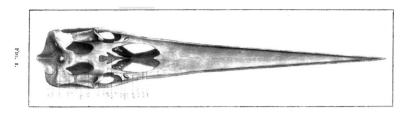

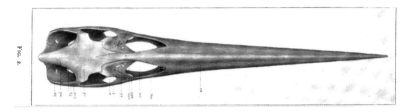

be abandoned" (1902a:305). In making a point that will be discussed later in this chapter, Williston (ibid.) noted at the end of the paper that when the skull of *Nyctodactylus* was fully prepared, it had no occipital crest. This statement is of current interest in a more modern controversy regarding the crest of *Nyctosaurus*. A follow-up paper on the same specimen (Williston, 1902b) provided a detailed description of the skull of *Nyctodactylus* and included one of the first known photographs of a pterosaur skull. The picture (ibid.:pl. I) shows a ventral view of the posterior of the skull. Two drawings (ibid.:pl. II, figs. 1 and 2; Fig. 10.8), showing the first ventral and dorsal views of the skull of *Nyctodactylus* (*Nyctosaurus*), are also included. Williston (1902c) also wrote "Winged Reptiles" for *Popular Science Monthly* magazine, in which he discussed the *Pteranodon* in more general terms. It is interesting to note at this point that he had largely given up on the idea of changing the genus back to *Ornithostoma*, although he still included the older term in parentheses.

As mentioned earlier, George Eaton (1903) wrote a short note on a newly discovered crest of *Pteranodon longiceps* in which he criticized Williston's work. Eaton was a graduate student at Yale at the time and was in the process of completing his doctorate. Eaton's critical comments did not go unnoticed by Williston, who replied to them in his 1903 paper on the osteology of *Nyctosaurus* and American pterosaurs. In 1902, Williston had accepted a new job as professor of paleontology at the University of Chicago and had been able to secure a complete *Nyctosaurus* specimen (FMNH 25026) that had been collected by H. T. Martin for the Field Museum. Williston also noted, with some venom in his comments, that "had Mr. Eaton done me the honor to have read more attentively the article which he quotes; or had he examined the extended article on the skull of *Nyctodactylus* with plates, published in the Journal of Geology for August, 1902; or even had he examined the figure of the skull in Zittel's text-book of Paleontology, published last autumn, of all which he seems strangely ignorant, he would have learned that *Nyctodactylus has no crest whatsoever*, not even the vestige of one" (1903:162). Ouch!

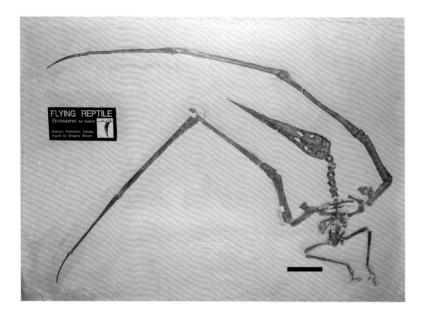

10.9. A nearly complete skeleton of *Nyctosaurus gracilis* (UNSM 93000) collected by Gregory Brown from the Smoky Hill Chalk in Gove County in 1974. Scale bar = 10 cm (4 in).

Eaton (1904) followed up with a second article in which he again improperly criticized Williston on the sclerotic ring in *Pteranodon* (see above) and in the count of vertebrae in the neck. So far as I am aware, Williston did not bother to reply in publication to these unfounded criticisms.

As complete as his description was of *Nyctosaurus*, Williston missed a couple of things. When you look at his detailed skeletal figures of *Nyctosaurus* (1902a:299; 1925:figs. 190, 191), you might notice that his reconstruction shows three wing fingers with claws and four phalanges. Williston did not find any of the clawed wing fingers (1903:145), and noted that he had recovered only the proximal portion of wing phalange III in the specimen that he was describing (ibid.:147). He assumed that the distal portion of wing phalange III and phalange IV had not been preserved.

In fact, phalange IV was not there to begin with. In 1974 Greg Brown collected a nearly complete and articulated specimen (UNSM 93000; Fig. 10.9; see also Bennett, 1994b:fig. 3) for the University of Nebraska State Museum. Brown (1986a, 1986b) noted that *Nyctosaurus* has three phalanges supporting the wing, not four as seen in *Pteranodon*. Both Williston (1903) and Brown (1986a, 1986b) reported that *Nyctosaurus* had the slender bones representing metacarpals I–III, but no wing carpals (finger bones) or claws. Did *Nyctosaurus* have wing fingers, or were they missing like phalange IV? Examination of four other relatively complete Kansas specimens—LACM 51130 collected by Marion Bonner from Logan County in 1965, the two strangely crested *Nyctosaurus* specimens collected by Kenneth Jenkins from Trego County (see Bennett, 2003a), and a privately owned specimen collected by Glenn Rockers from Logan County—as well as a new nyctosaur genus from Mexico (*Muzquizopteryx*; Frey et al., 2006), showed no evidence of wing fingers or claws. In regard to the crested specimens from Trego County, Bennett

(2003a) states that neither the KJ1 or KJ2 specimens preserve any trace of phalange IV or the metacarpals and phalanges of manual digits I–III. The lack of phalange IV in *Nyctosaurus* leads me to suspect that the wing fingers were no longer present, or at least vestigial, in this much smaller, lighter species.

Williston's 1903 paper also included considerably more detail in the descriptions of other specimens. One item that was briefly mentioned (ibid.:160) was the discovery of the upper end of a femur from the Kiowa Shale (Early Cretaceous) of Clark County that Williston believed was from a pterosaur. He named it *Apatomerus mirus* (KUVP 1198; Fig. 10.10). Years earlier, the same bone fragment had been figured by Williston (1894:pl. I; 1898:fig. 3) when he believed it might be part of a crocodile. If it was in fact a pterosaur femur, then it would have been an early record for a North American species that would have been substantially larger than *Pteranodon*.

Brown (1943:108) accepted Williston's later identification of the bone as a pterosaur but did not indicate that he had actually examined the material. Schultze et al. (1985:58), however, disagreed that it belonged to a pterosaur, but he was unable to identify it. Bennett (pers. comm., 2004) also indicated that he did not believe it was from a pterosaur. In May 2004, when I examined the KUVP 1198 specimen with Larry Martin, it was apparent that the limb bone was much too heavily constructed to be from a *Pteranodon*. In July 2004, I located another specimen (KUVP 16216) of the upper end of a "femur" in the KU collection that was similar to the fragment described by Williston. The second specimen was collected in 1969 by Orville Bonner in Clark County, Kansas, in association with two plesiosaur vertebrae. The shaft of both specimens is hollow, but the bone is much too thick for a pterosaur limb. Both specimens were collected from the Kiowa Shale, and I have concluded that both represent the weathered upper end of a plesiosaur limb (propodial). The hollow shaft is apparently an artifact of preservation, similar to a number of hollowed plesiosaur vertebral centra from the Kiowa Shale that Larry Martin pointed out to me.

Williston (1904) published a short note in which he discussed the clawed fingers and elongated wing finger of *Pteranodon*. In his final paper on pterosaurs, Williston (1911) described his restoration of *Nyctosaurus*, including a discussion of the hand and wing finger as they relate to their reptilian origins. Williston makes the point that the wing finger is the fourth finger in the hand (ring finger) and not the fifth or little finger as is sometimes assumed. Both *Pteranodon* and *Nyctosaurus* lack a 'thumb,' but do have a sharply pointed bone, called the pteroid, attached at the wrist and supporting the wing membrane in front of the radius/ulna. Whether or not it represents the thumb, and its function, is the subject of some debate (Bennett, 2007).

In 1910, Eaton published his doctoral dissertation on the osteology of *Pteranodon*. Besides updating the description of *Pteranodon*, the paper provides excellent photographs and line drawings of the catalogued

10.10. The upper end of a femur (KUVP 1198) that was described Williston as the type specimen of a huge Early Cretaceous pterosaur called *Apatomerus mirus*. Originally, it was suggested to be the femur of a crocodile, but now appears to be most likely part of a plesiosaur. Scale bar = 10 cm (4 in).

specimens in the Yale Peabody Museum collection. In the text, Eaton was not as apologetic for Marsh's mistakes as he had been in his 1903 and 1904 criticisms of Williston's work. He wrote pointedly that "to the critical reviewer of Professor Marsh's early work on this subject, it is evident that the two skulls referred by him to *P. ingens* [YPM 2594] and *P. occidentalis* were specifically identified by their size alone" (Eaton, 1910:2). Because he believed that many of the specimens were so incomplete as to have little value for describing the differences between species, Eaton limited his discussion to the skeleton of *Pteranodon*. In doing so, he quoted repeatedly from Williston's earlier work and credited his descriptions. It appears that at least some of what Williston had to say in his 1903 rebuke had made a point with Eaton.

On one issue, however, Eaton followed Williston too closely. On the subject of whether or not *Pteranodon* had a fibula (the small bone paired with the tibia in the lower leg of most animals, including *Pteranodon*), Eaton quotes from Williston (1903): "There is no trace of any fibula, in either of the preserved remains, or of any tibial articulation" (Eaton, 1910:35). Eaton (1910) then goes on to say, "This is also true of the *Pteranodon* material contained in the Marsh Collection."

That same year, however, in a short article regarding a recently mounted specimen of *Platecarpus*, Williston uses the opportunity to state "incidentally I may mention that in *Pteranodon* among pterodactyls the fibula is supposed to be absolutely wanting, yet in a specimen in our collection I find distinct remains of it fused with the tibia" (1910:540). Wiman (1920:9–10) noted the exchange between Eaton and Williston in his description of specimens purchased from C. H. Sternberg for the Palaeontological Museum in Uppsala, Sweden. One of the specimens he figured (ibid.:fig. 1a) was a *Pteranodon* tibia that included the remnant of the fibula (see also Mateer, 1975:fig. 12). Bennett (2001:105, fig. 108; pers. comm., 2004) indicated that "the fibula is readily apparent in well preserved specimens" of *Pteranodon*.

In a brief review published in the *Journal of Geology*, Williston took the opportunity to comment on Eaton's work, noting that "the writer

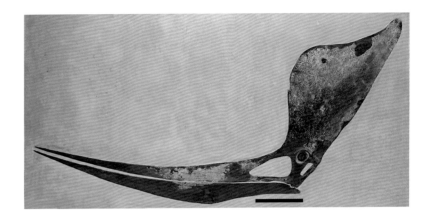

10.11. The type specimen of *Pteranodon sternbergi* (FHSM VP-339) in the Sternberg Museum of Natural History. The anterior portion of the skull and lower jaw has been reconstructed. Scale bar = 20 cm (8 in).

[Williston], whose acquaintance with vertebrate paleontology began with the collection of a specimen of *Pteranodon*, takes especial pleasure in the expression of his appreciation for the present memoir by Dr. Eaton. The rich material of this genus in the Yale collections is unsurpassed, and it has been well utilized in the present paper, with its large number of excellent illustrations" (1912:288). In noting that the occipital crest published by Marsh was well documented by the photograph of the specimen itself, Williston admits his 1892 criticism of the reconstruction was probably unjustified. Ending on a note suggesting that more could have been done, Williston (ibid.) wrote that "one could wish that Dr. Eaton had entered more fully into some of the disputed points about the relationships and characters of the genus, but the omissions are immaterial in comparison with what he has given" (ibid.:288).

In his review of the fossil vertebrates of Kansas, Lane (1946) had relatively little to say about the *Pteranodon* collection at the University of Kansas. This reflected the fact that few additional remains had been collected since Williston had left Kansas in 1902 and that no one had been actively working on *Pteranodon* since Eaton's major publication.

In 1952, however, a large and unusual *Pteranodon* skull was collected by George F. Sternberg in the low chalk of Graham County. Instead of the usually long, tapering crest that projected behind the skull, the new specimen had a short, wide crest that sat more or less on top of the skull. The strange-looking skull (FHSM VP-339) was briefly described as a new species by Harksen (1966) and named *Pteranodon sternbergi* (Fig. 10.11). Although the upper and lower jaws were fragmentary and incomplete, Harksen's (ibid.:76) illustration of the skull was impressive. He estimated the length of the lower jaw to be 121 cm (47.6 in), more than twice that of the type specimen of *P. longiceps* (58 cm; 22.8 in), and suggested that it was a truly huge flying reptile. Harksen (1966:77) estimated that the wingspread of *P. sternbergi* would have been about 9 m (30 ft). Bennett, however, suggested that "the mandible was probably not as long as reconstructed" (1994a:33) in the exhibited specimen by Sternberg. In a handwritten note that accompanies a black and white photo of the right side of the skull in

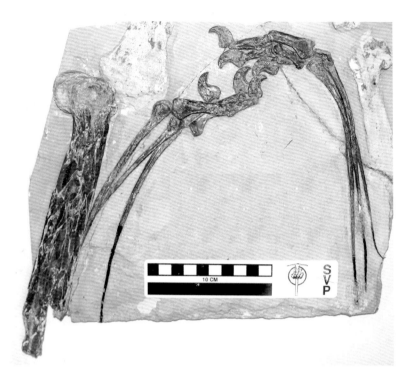

10.12. As collected, the distal end of metacarpal IV and wing fingers from the FHSM VP-2082 *Pteranodon*, including both sets of metacarpals I, II, and III, carpals, phalanges, and claw cores (ungals). Scale bar = 10 cm (4 in).

the Sternberg Museum archives, Sternberg wrote, "Length of the skull is 39 in [100 cm]. . . . The bone is poorly preserved."

Following his discovery of the type specimen of *Pteranodon stern-bergi*, Sternberg discovered a smaller, headless specimen in September 1956 near WaKeeney in Trego County in the lower chalk (Sternberg and Walker, 1958). The remains (FHSM VP-2062) included both wings with the wing fingers and claws (Fig. 10.12), the pelvis and both legs and feet, along with the sternum and fragments of vertebrae. The wings and legs were still articulated. From the measurements given (ibid.:84–85), the wingspread would have been about 3.8 m (12.5 ft). This specimen is currently on display at the Sternberg Museum of Natural History. In 2003, a complete and very large (1.8 m; 6 ft) skull of *Pteranodon sternbergi* was discovered in Trego County by a private collector, but nothing else is known about it.

In 1962, George Sternberg collected a nearly complete specimen of *Nyctosaurus* (FHSM VP-2148) near Elkader in Logan County (Fig. 10.13). The specimen was small enough (skull length 30 cm; 11.8 in) to be contained within a single slab of chalk roughly 14.5 x 27.5 inches (Bonner, 1964:11). The wingspread (ibid.:table 6) was a little over 2.6 m (8 ft). By comparing the measurements of it and other known specimens, Bonner determined that the specimen was a new species and named it *Nyctosaurus sternbergi*. However, the species name was preoccupied, and Miller (1971a:13) changed the name to *N. bonneri* in honor of Orville Bonner's work.

Miller (1971a) reviewed the literature regarding *Pteranodon*, provided brief descriptions of major specimens, including those in the collection

Nyctosaurus bonneri
TYPE SPECIMEN
Logan County, Kansas

Nyctosaurus, like its larger cousin
Pteranodon, lacked teeth. These
pterosaurs probably ate fish
whole, much like modern pelicans.

10.13. The type specimen of *Nyctosaurus bonneri* (FHSM VP- 2148). Originally described as new species (*N. sternbergi*) by Orville Bonner in 1964, it was renamed by Miller (1971a). Bennett (1994a) considered *N. bonneri* to be a junior synonym of *N. gracilis*. The skull is 29 cm (11.5 in) long. Scale bar = 10 cm (4 in).

of the Sternberg Museum, and suggested several changes in the systematics of the genus. Miller's proposed revision was faulty, however, and has been disregarded by subsequent workers (Bennett, 1994a:8). In the same journal, Miller (1971b) also described a new specimen of *Pteranodon longiceps* (FHSM VP-2183) that also had been collected near Elkader in Logan County by George Sternberg. Unlike many specimens, it included a skull and postcranial material (ibid.:21).

In the first hundred years after the discovery of *Pteranodon*, little work had been done to understand how such strange-looking animals were able to fly. Early reconstructions of pteranodons that showed their wing membranes attached to their legs as bats have (Williston, 1902c; Lucas, 1901; Eaton, 1910) were probably influenced by the drawings made of Jurassic pterodactyls by earlier European workers, including Seeley (1901). Padian notes that there "is no evidence that the hind limb was attached to the wing in any pterosaur" (1983:220). Essentially, this suggests that pteranodon wings were long and narrow, like those of modern marine soaring species such as albatrosses and frigate birds, and this factor is important in determining how they flew.

However, the actual shape and point of attachment of the pteranodon wing is still a point of contention among researchers. Was the back of the wing membrane attached at the back of the ribs, at the hips, or at the feet? This is important in terms of whether you consider a pteranodon to be adapted for long-distance soaring like an albatross or frigate bird (long, narrow wings, attached at the hip), or something more maneuverable with broad, triangular wings attached to the feet. The wings of *Pteranodon* have been reconstructed in a variety of ways by artists since their discovery. Elgin, Hone, and Frey (2011) reviewed soft tissue (wing membrane) preservation in the pterosaur linage, including extremely well preserved specimens from the Jurassic, and confirmed their view that the wing was attached at the ankle in several cases. They also noted that there was no soft-tissue evidence for the idea that the wing was not attached along the lower limbs. Much of this, of course, is based on lack of preservation of soft tissue impressions in later-occurring pterosaurs, including *Pteranodon*. So at this point, the subject is still open. I favor Padian's view, but have seen nothing in the Kansas fossil record that contradicts Elgin, Hone, and Frey. (ibid.) and a leg attachment for the *Pteranodon* wing.

Lucas (1901) provided some of the first analysis of the weight and surface area of the wings of modern birds (e.g., wing loading) compared to *Pteranodon*. And then at a time when humans were just beginning to learn about heavier-than-air flight, Hankin and Watson provided the first aerodynamic analysis of pterosaur flying capabilities, suggesting that it was "highly probable that their habitual mode of flight was soaring rather than flapping" (1914:11). In a later study, Bramwell and Whitfield (1970) calculated that an adult male *Pteranodon* with a 6.7 m (22 ft) wingspread would have had a minimal flight speed of between 24 and 27 km/h (15 and 17 mph), based on an estimated weight of 18–25 kg (40–55 lb).

The dynamics of flight in *Pteranodon ingens* (*longiceps*) were reexamined by Stein (1975) by conducting wind tunnel tests using models of pterosaur wings. Stein concluded that *Pteranodon* "was a highly evolved, maneuverable, and unstable flyer" (ibid.:547). In this case, 'unstable' is a good thing and means that when the animal's flight is disturbed, it has to be able to make changes to restore itself to its course. This requires a highly evolved sensory and nervous system. In contrast, outside forces acting on the wings of a stable flyer will tend to restore stability automatically. The critical issue here is that "high stability and high maneuverability are mutually incompatible" (ibid.: 547). His analysis of the data indicated that *Pteranodon* was a slow-speed, long-distance flyer and suggested (ibid.:541) that 4.5 m/s (about 10 miles/hr) would have been about the minimum flight speed of a large (15 kg; 33 lb) *Pteranodon*.

More recently, Henderson (2010) used computer modeling to estimate the weight (mass) of a variety pterosaurs. The model of *Pteranodon longiceps* that he used was almost identical in size to the juvenile male specimen that I collected in 1996 with a 5.3 m (17.5 ft) wingspread. The model would have weighed about 18.6 kg (41 lbs) with a wing area of about 1.8 sq m (19.4 sq ft). The weight of the pterosaur divided by the estimated wing area gives a ratio called wing loading. In the case of the medium-size *Pteranodon* that Henderson modeled, the wing loading would be similar to that of a modern albatross, well known for long flights over the ocean.

Stein (1975) also discussed the means by which *Pteranodon* would have been able to take off, maneuver, feed, and land, all of which he considered to be more important adaptations than pure flying. He also concluded that *Pteranodon* would have had to land on its back feet (as opposed to landing on all four feet) because in order to get the front feet on the ground, the wings would have to be collapsed and would no longer provide lift. It seems more likely that the back feet touched down first, followed quickly by the folding of the wings as the body leaned forward and allowed the front limbs to touch down.

Getting a large animal like *Pteranodon* into the air is another issue. Were they able to just flap their wings and take off, or did they need a running start like some large modern seabirds? Habib (2008) proposed an unusual quadrupedal launch posture in which the folded wings were used to literally spring the animal forward and into the air. As the pteranodon vaulted forward on its front feet and left the ground, the wings then snapped open laterally and began to flap, lifting it farther into the air. While I see this scenario as being possible, it seems to require overly complicated movement and very rapid extension of the wings. However, Bramwell and Whitfield (1970) didn't consider getting a *Pteranodon* into the air would be much of a problem, given their light weight and large wingspread. They believed that simply spreading their wings in a moderate wind would be sufficient.

Using the flight characteristics of a modern hang glider, Brower (1983) modeled the performance of *Pteranodon* and *Nyctosaurus* and noted that both were probably limited to slow horizontal and vertical speeds. While he believed that both were accomplished at long-duration soaring, Brower (ibid.) suggested that unlike *Nyctosaurus*, *Pteranodon* was probably not capable of continuous flapping flight. This certainly raises the question of why so many *Pteranodon* specimens (and a smaller number of *Nyctosaurus* remains) have been collected hundreds of miles from the nearest shoreline. It is certainly possible that the remains we find were those of animals lost from long-distance migrations, but judging from the relative abundance of the remains we find, it seems more likely that they maintained a relatively constant presence in the Western Interior Sea. To do so, in my opinion, means they had to have been strong long-distance flyers.

One feature that has been used to distinguish *Pteranodon* and *Nyctosaurus* from many other pterosaurs is that they are considered to be tailless. That said, they do have a short tail. Eaton noted that "the number of caudal vertebrae in *Pteranodon* is not definitely known, but the opinion of Professor Marsh and Professor Williston expressed in various papers, that the tail was short, is well supported by the material preserved in the Marsh collection" (1910:23). However, a study of three specimens in the Marsh collection in the Yale Peabody Museum (YPM 2489, 2546, and 2462) and another in the American Museum of Natural History (AMNH 6158) by Bennett (1987) indicated that *Pteranodon* had a somewhat longer tail that was complete in form and may have functioned to support a rearward extension of the wing membrane. As illustrated (ibid.:fig. 2), the tail consisted of about 10 small vertebrae and a terminal caudal rod. Bennett (ibid.:23) suggested that the caudal vertebrae and membrane would have formed a control surface, which would have provided pitch control when moved up or down. The caudal vertebrae would have been at least 19 cm (7.5 in) long on a *Pteranodon* with a 7.5 m (24.6 ft) wingspread (ibid.). Since that time, however, Bennett (pers. comm., 2004) has concluded the there was no connection between the wing membrane and the caudal vertebrae.

One of the problems in the study of pteranodons from the very start has been the determination of exactly where they occurred geographically, and more importantly their stratigraphic occurrence. Locality information, especially on the early discoveries, is generally lacking or inaccurate. During the 1870s, western Kansas was largely unsettled, and localities were often described as being "near the Smoky Hill River." Maps from the 1870s show an elongated Wallace County that included present-day Logan County. However, Logan County was not incorporated until 1885 (Bennett, 2000a), and localities recorded as Logan County for specimens collected before 1885 are therefore suspect. Later specimens had better locality information, but their stratigraphic occurrence was often limited to 'upper' or 'lower' chalk. Hattin (1982) provided detailed descriptions of 23 'marker units' within the Smoky Hill Chalk and enabled workers to make accurate determinations of the stratigraphic occurrence of specimens for the first time. Stewart (1990) based his six biostratigraphic zones on Hattin's work and noted that *Pteranodon* and *Nyctosaurus* made their first appearances near the middle of the chalk (Santonian). In that regard, I have collected *Pteranodon* remains as low as Hattin's marker unit 4 (late Coniacian) in Stewart's (1990) zone of *Protosphyraena perniciosa*. The distal end of a possible *Pteranodon* femur reported from the basal Lincoln Limestone Member of the Greenhorn Formation (Upper Cenomanian) of Russell County (Liggett et al., 1997; 2005) suggests that pterosaurs were at least occasional inhabitants of the Western Interior Sea before the deposition of the Niobrara Chalk. The bottom line, however, is that *Pteranodon* is known positively only from the early Coniacian and into the early Campanian, a time span of about 8 million years. *Pteranodon* remains are collected most often in the Kansas chalk, with only a few specimens known from the Pierre Shale, and from South Dakota and Wyoming.

In 2010 Bruce Schumacher collected an uncrushed fragment of a pterosaur wing bone from the basal (Upper Cenomanian) Lincoln Limestone (FHSM VP-17966; Fig. 10.14). The diameter of the bone (2.5 cm; 1 in) indicates a large pterosaur, most likely *Pteranodon*. The partial wing (left humerus, distal metacarpal IV, and proximal 1st phalanx IV) of a *Pteranodon* was reported by Myers (2010) from the Austin Chalk of Texas. The early Coniacian age of this specimen suggests that is the oldest known confirmed *Pteranodon*.

Bennett (2000a) inferred the stratigraphic occurrence of several specimens from various sources of historical and geographical information, noting that the type specimen of *Pteranodon sternbergi* (FHSM VP-339) came from the low chalk (near Hattin's marker unit 4). Bennett (ibid.) also suggested that the type specimen of *Pteranodon longiceps* (YPM 1177) came from the upper chalk between marker units 15 and 16. The stratigraphic separation of these two species was an important discovery that was not readily evident until the right stratigraphic tools became

10.14. An uncrushed fragment of a pterosaur wing bone (FHSM VP-17966) from the Upper Cenomanian Lincoln Limestone Member of the Greenhorn Formation in Russell County. Scale bar = 2 cm.

available. Bennett noted that "the post-cranial skeleton [of *Pteranodon*] is of no taxonomic value at the species level" (ibid.:2). This means that without certain portions of the skull, there is no way to determine the species of *Pteranodon* remains except by their stratigraphic occurrence.

After a study involving the measurement of more than 400 sets of *Pteranodon* remains from the Smoky Hill Chalk, Bennett (1992) discovered that there were two 'size-classes' represented among the specimens, with the average large size-class measurements being about 50 percent larger than those of the average small size-class specimen. In addition, it was the larger size-class that had the large crests, while the pelvis of the smaller size-class was proportionately larger. Bennett (ibid.) interpreted these findings to indicate that individuals in the larger size-class were males while those in the smaller size-class were females. He also determined that the number of specimens of females outnumbered the number of males by about two to one.

A recent discovery (Wang and Zhou, 2004) in China of an Early Cretaceous (Aptian, about 121 Ma) pterosaur embryo preserved inside an egg appears to have answered the question of whether or not they laid eggs. The 53 mm x 41 mm (2.1 x 1.6 in) fossilized egg contained a nearly-ready-to-hatch embryo that would have had a wingspread of about 27 cm (11 in.). Lü et al. (2010) described a specimen of a small Chinese pterosaur that died with an egg in close association with the body. It appears that the smaller pterosaurs were egg layers and that they most likely buried their eggs in a nest. Apparently their tiny hatchlings (flaplings) were able to fly upon hatching and were not tended by the parents. While the question of egg laying in the much larger *Pteranodon* is still open to discussion, it seems unlikely that they would have reproduced in a similar fashion, especially given the fact that we have never discovered hatchlings or early juveniles. I would even suggest that *Pteranodon* gave live birth to a single baby and that one or both parents cared for the young pterosaur for several months until it was large enough to fly. This would be similar to modern seabirds, such as the frigate bird, who take six months or more to fledge a single chick. That said, without locating a *Pteranodon* nesting site, I seriously doubt that the question will be answered anytime soon.

In a discussion of possible functions of the crest, Bennett (1992:430) concluded that it was most likely used as a display since the larger crests were associated only with the larger size-class individuals that he considers to be males. Regarding the stratigraphic occurrence of the different crests, and noting that *Pteranodon sternbergi* occurs in the lower chalk and *P. longiceps* in the upper chalk, Bennett suggested that that both *P. sternbergi* and *P. longiceps* are part of "a chronospecies forming parts of a single anagenetic lineage" (ibid.:422). In other words, it is likely that only a single species of *Pteranodon* occurs throughout the chalk, rather than two, and that that species would have to be called *P. longiceps* because it was the first one to be described. The shape of the crest and other small differences observed in the skull would be evolutionary changes that occurred gradually over the 5-million-year deposition of the chalk. I would even go so far as proposing that the strange crest on the type specimen of *P. sternbergi* is the result of pathology, from either an injury or disease. Until another example is discovered, it's difficult to understand the unusual shape compared to all of the other slender crests associated with *P. longiceps*. To me, it is odd that the crest would get smaller over time, rather than larger.

The taxonomy and systematics of *Pteranodon* were revised by Bennett, who noted again that "*Pteranodon sternbergi* seems to be ancestral to *P. longiceps*" (1994a:1) and that *Nyctosaurus* "*bonneri* is a junior synonym of *N. gracilis*" (ibid.:18). Since then, most researchers agreed that *P. sternbergi* and *P. longiceps* are closely related, if not the same species evolving over time. That is, until Kellner (2010) published an extensive article in a Brazilian journal suggesting that there were actually four species of pteranodon representing three genera flying over the Western Interior Sea at various times. Kellner (ibid.:1067) brings back *Geosternbergia sternbergi* (originally from Miller, 1976) in place of *P. sternbergi*, on the basis of several generic level differences that he observed between it and the skull of *P. longiceps*. Then he added a new species, *G. maiseyi*, based on the partial (and poorly preserved) skull of a *P. longiceps* (KUVP 27821; Fig. 10.15) from the Pierre Shale of South Dakota (Lower Campanian; see also Carpenter, 2006:fig. 14B). Finally, he created a new genus and species from a nearly complete specimen of a female *P. sternbergi* (*Dawndraco*

10.15. The partial skull of *Pteranodon longiceps* (KUVP 27821) in right lateral view from the Lower Campanian Pierre Shale of South Dakota. Scale bar = 10 cm (4 in).

kanzai, UALVP 24238) collected by the University of Alberta from the Lower Santonian Smoky Hill Chalk of northern Lane County, Kansas. Note that the locality where UALVP 24238 was collected was within a mile or so of where my wife discovered the skull and wing bones of the *Pteranodon sternbergi* (CMC VP7203) in 1996, and most likely near the same stratigraphic level (Hattin's [1982] marker unit 7).

So far as I can determine, the paper did not receive favorable reviews from anyone working with pterosaurs. Chris Bennett (pers. comm, 2016) has examined all of the specimens involved, including our CMC VP7203, and is in the process of rebutting the claims made by Kellner. Witton noted that "Bennett's (1994a) approach sees morphology trump stratigraphy in that the ranges of his species are dictated wholly by specimen anatomy" (pers. comm., 2016), while Kellner (2010) argued "that *Pteranodon* skulls found several levels [stratigraphically] away from each other were not contemporaries and thus cannot be reliably assessed for intraspecific variation. . . . The implication here is that there is a stratigraphic limit to when similar-looking animals might be considered conspecific, and that morphological similarity is eventually overruled by provenance" (Witton, pers. comm., 2016). While recognizing the implications of Kellner's (2010) analysis, Witton sides with Bennett 's more generally recognized concept of a *Pteranodon* being a chronospecies. In any case,

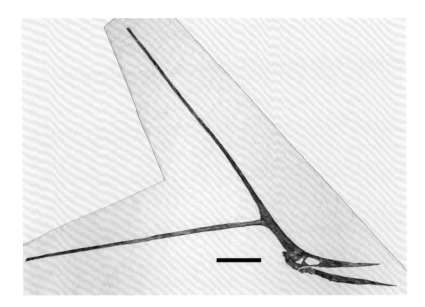

I am certainly against changing the names and creating new genera and species on the basis of single specimens.

At the Society of Vertebrate Paleontology annual meeting in Mexico City, Bennett (2000b) reported the discovery of three specimens of *Nyctosaurus* from western Kansas that were significantly different from previous known remains. One of the new specimens had a wingspan of 4.5 m (14.8 ft) and was thus the largest *Nyctosaurus* known. The other two specimens had been collected near WaKeeney (Trego County) in the middle chalk (Santonian), and both had crests that were large and rather bizarre for such small individuals (Fig. 10.16). It is likely that these specimens represent mature males. All other known specimens of *Nyctosaurus* are crestless.

Since *Nyctosaurus* had been considered to be 'crestless' almost since Williston's original description of the skull, this discovery obviously and abruptly changed the status quo of this genus. Bennett (2003a) provided a complete description, measurements, and detailed photographs of the specimens, and he suggested that the hyper-attenuated crest occurred only in mature males. According to Bennett (ibid.:62), the tall, slender, and branched crest of the smaller but more complete specimen is roughly three times the length of the skull (71.7 cm vs. 24.5 cm; 28.2 vs. 9.6 in). The skull of the larger specimen is about 1/3 longer (31.6 cm; 12.4 in), but the crest is incomplete. Viewed another way, the crest is almost as long as each of the wings (43 percent of the total wingspan; ibid.:73). There is no evidence of a membrane between the branches of either crest (ibid.). It is important to note, however, that Bennett did not see any reason to indicate that these animals represented a new species of *Nyctosaurus*. While Bennett concluded that "the aerodynamic effect of the crest could have been rather minor [since] the base and rami of the crest are streamlined and would have caused only a small amount of drag" (ibid.). I find it difficult to believe that a small, winged reptile (wingspread 1.7 m; 5.6 ft) could fly hundreds of miles from the nearest land with what was essentially a long pole sticking out of its skull. However, I have no other explanation for these bizarre specimens.

It is worth noting here that Frey et al. (2006) described a new genus and species of nyctosaur, *Muzquizopteryx coahuilensis*, from a nearly complete specimen collected from the early Coniacian Austin limestone of northeastern Mexico. The remains include the back half of the skull (crestless) and almost all of the body, including both of the 'hatchet-shaped' humeri that are typical of nyctosaurs. Although the specimen is older than *Nyctosaurus* remains in Kansas, it is not surprising that this smaller genus has not yet been identified from that part of the Western Interior Sea, given the few and incomplete pterosaur specimens that have been collected from that time.

Pteranodons were often portrayed in the old drawings of life in the oceans of Kansas as hanging by their wing claws from tall cliffs. This raises a couple of questions for me. The first is, where are these cliffs? I don't have a good answer for that because most of the shorelines, especially on the eastern side of the interior sea, were eroded away millions of years ago. One thing that can be said, however, is that the cliffs were never in western Kansas as was suggested by many of the early art works. While there were certainly cliffs (or large trees) someplace along the east and west coasts of the Western Interior Sea, the closest land of any sort that we can be reasonably sure about was probably in the southwestern corner of present-day Arkansas, more than 200 miles away from the chalk of western Kansas. The other coasts at the time were much farther away. Did they fly that great distance just to feed in the middle of the ocean? I don't think so. During the season when the young were born or hatched, at least one of the parents would have foraged close to where they nested because hungry, fast-growing young usually require frequent feedings. In addition, the shallower coastal areas, swamps, and estuaries provided a much richer bounty of prey in the small sizes that pteranodons probably fed upon. It is possible, however, that adult pteranodons were solitary animals and lived a lifestyle much like a modern albatross or frigate bird, flying for long periods and staying at sea most of their lives. We simply do not know at this point.

Second, wherever they nested, I don't believe they hung from anything as a general practice. Their wing claws may have been useful for climbing or clinging, but it is more likely that they nested near the coast in large, reasonably level rookeries like most modern seabirds. They apparently were able to walk fairly well (after folding up those huge wings, of course). Whether they were bipedal or quadrupedal is still the subject of some debate, although the analysis of pterosaur skeletal structure by Padian (1983) appears to indicate that they were primarily bipedal. There are pterosaur tracks along a shore in Jurassic rocks in Wyoming that preserve marks that are apparently from the front limb touching the substrate, but it is not clear that the front limbs necessarily were used to support the animal's walking. We are also relatively certain that some pterosaurs in other places and times, such as *Pterodaustro*, fed while standing in shallow water, much like modern flamingos. While they may not have been fast runners, they were certainly able to walk around while on the ground.

The question of why we find their remains hundreds of miles from the nearest land is still unanswered. The story at the beginning of this chapter has a flock of *Pteranodon* males flying for hours to find a school of fishes to feed upon because that seems to be the generally accepted idea. That being said, however, I don't believe they were flying clear out to the middle of the Western Interior Sea just to feed and then fly 10 or 15 hours back to their nests. One direction or the other would have been against the wind and, as lightly as *Pteranodons* were built, I don't see them as being able to actively fly against a headwind for long distances. If they were feeding far out to sea, they would be using more energy getting back and forth than they gained from what they were able to catch. At this point, I believe that what we are seeing in the many fossil remains from the Smoky Hill Chalk are the *Pteranodons* that died while migrating across the seaway. These would be the sick, old, and infirm flyers that dropped out of massive seasonal migrations—or, quite possibly, the results of migratory flights that ran into bad weather. A kite in a thunderstorm is probably a fairly good representation of a pteranodon's chances in bad weather. In regard to the lack of healed injuries in *Pteranodon* skeletons, Bennett notes that their relative absence "suggests that they were either uncommon or catastrophic and fatal" (2003b:185). In any case, while I do not believe that these flying reptiles were routinely feeding in mid-ocean, I don't have a better explanation for why we find so many of their remains in the Smoky Hill Chalk.

Most *Pteranodon* specimens from the Smoky Hill Chalk are fragmentary. Generally the specimens are fragments of the wings, which is understandable because the wing bones were the largest and most robust bones in the *Pteranodon* skeleton. Even so, they are almost always crushed to a thickness of 3 mm (0.12 in) or less. There are very few skulls known, and only a few of those are relatively complete (note earlier arguments between Williston and Marsh/Eaton over the size of the crest). The remains of the bones from the lower body (behind the wings) are also rare, most likely since it was the most consumable part of a dead *Pteranodon*. This suggests that the head and body were eaten by scavengers and the less nutritional membranous wings were left to sink to the bottom (Hargrave, 2007).

While the abundance of fragmentary remains suggests predation or scavenging on *Pteranodon*, not a lot of actual physical evidence for that occurs in the fossil record. In 1876, Williston collected a *Pteranodon* cervical vertebra (YPM 2458) from the chalk in Wallace County that was in association with a *Squalicorax* sp. tooth (YPM 56586). In 1965, Marion Bonner collected the articulated remains of a *Pteranodon longiceps* in

Logan County that had the tooth of a large *Cretoxyrhina mantelli* shark lying next to the third cervical vertebra. That specimen (LACMNH 50926) is currently on display in the Los Angeles County Museum of Natural History (R. Konuki, 2008, fig. 56; G. Takeuchi (LACMNH), pers. comm., 2008).

Given the history of fossil collecting in Kansas, it is fitting that *Pteranodon longiceps* and *Tylosaurus proriger*, discovered first and occurring mostly in Kansas, were named as our official State Fossils in 2014. I was able to take part in the legislative process, testifying in support of House Bill HB2595 before the House and Senate committees, and was pleased with the outcome.

Some current *Pteranodon* research opens up new possibilities for finding answers. Laura Wilson at the Sternberg Museum of Natural History in Hays is taking the study of *Pteranodon* to the microscopic level by looking at the histology (tissue structure or organization) of their bones as a means of determining the maturity of individual specimens. Under a microscope, the rapidly growing bone of young (juvenile or subadult) individuals looks very different from the cellular structure of older, mature individuals. Although her research is still in the preliminary stages, Wilson reported that "analysis of additional *Pteranodon* and *Nyctosaurus* specimens should shed light on intraspecific variation, sexual dimorphism, and taxonomic uncertainties" (2015:239). Anything that clears up the many uncertainties related to *Pteranodon* is a good thing!

Feathers and Teeth

11

On a bright, cloudless day, a flock of tern-size white birds flew in wide, lazy circles over the nearly calm surface of the Western Interior Sea. They were searching for their favored prey, a school of small fishes feeding near the surface. From their vantage point high in the air, such schools appeared as dark, ever-changing shapes within the lighter blue background of the water. More often, however, the flock was attracted to the feeding activities of other, larger predators. From experience, they had learned that watching for the skin-winged pterosaurs skimming the surface, or for marine reptiles or large fishes as they swarmed around the shifting masses of tiny fishes, was often a more successful strategy. It was also more dangerous.

Today they located a group of sharks swimming around a school of tiny fishes. Occasionally, the sharks would dart through the mass of trapped prey with their mouths open, scooping in large numbers of the much smaller silver fishes. Almost as one, the birds wheeled and flew as quickly as they could toward the feeding activity. As the birds approached, the circle of sharks tightened and the surface of the water erupted in a spray of water and fishes with the prey trying in vain to escape upward, the only direction left to them. Their escape was just momentary.

When the birds reached the surface, they folded their wings at the last moment and dove in headfirst, creating a barrage of small impacts. Once underwater, they opened their strong wings again and used them to propel themselves in pursuit of the little fishes, much as some modern seabirds do. They didn't have to swim very far to catch the panicked prey in their long jaws filled with sharp teeth. Soon, the birds popped back to the surface, each with a small wiggling fish in its mouth. With a jerk of the head, birds flipped each fish up into the air, caught it headfirst, and then swallowed. After bobbing their heads up and down several times to help move the prey into their gullets, the birds again plunged into the roiling mass of small fishes. This frenzied feeding activity continued for several minutes.

Then, without warning, a huge, 6 m (nearly 20 ft) long ginsu shark flashed into the tightly packed school of fishes. As he passed through the melee, taking hundreds of the small fishes into his open mouth, he also unknowingly swallowed two of the white toothed birds. The momentum of his sudden attack scattered the ring of smaller sharks and allowed the school of fishes to briefly disperse. Startled by his unexpected appearance, the remaining birds took wing and regrouped in the air, to begin their search for another feeding opportunity.

11.1. This broken and incomplete hesperornithid tarsometatarsus bone (FHSM VP-6318) from the Upper Cenomanian Lincoln Limestone of Russell County is one of the oldest known bird bones in the United States. Fragments of ichthyornithid bird bones and jaw fragments with teeth (Fig. 11.16) of similar age occur in the same formation. Scale bar in mm.

Birds

The story above, of course, is mostly fiction. We don't know very much about the daily activities of *Ichthyornis* ('fish bird'), or the other toothed (flying or swimming) birds of the Late Cretaceous. Most paleontologists who work with these fossil birds believe they were most like modern terns or gulls, although Clarke (2004) has recently provided a more accurate reconstruction based on her study of *Ichthyornis* specimens in the collection at the Yale Peabody Museum. Given the variety of ways that modern seabirds dive, paddle, or even 'fly' underwater, however, I tend to believe that *Ichthyornis* may have pursued their prey in the water, feeding directly on the bounty of small fishes that was available to them near the surface. This might help explain the relative 'wealth' of bird remains in the middle of the Western Interior Sea. More specimens and further study are certainly needed in this area. Like modern migratory birds, *Ichthyornis* and other species of toothed flying birds that occur in the chalk were apparently capable of remaining in the air for long periods of time and flying hundreds of miles from nesting or rookery areas on shore.

The much larger *Hesperornis* ('western bird') and related genera might be considered the only 'marine dinosaurs.' Having theropod ancestors, they are certainly the only feathered 'dinosaurs' for which we have a fossil record in the Western Interior Sea. These birds apparently returned to the oceans of the Late Cretaceous about the same time as the demise of the ichthyosaurs, and successfully competed for prey with plesiosaurs, mosasaurs, and large fishes in the more northern (cooler?) oceans. While hesperornithids were much smaller than the adults of most marine reptiles, it is important to remember that most of the marine reptile population at any one time consisted of subadult or juvenile animals that would have been feeding on the same size prey as the adult birds did.

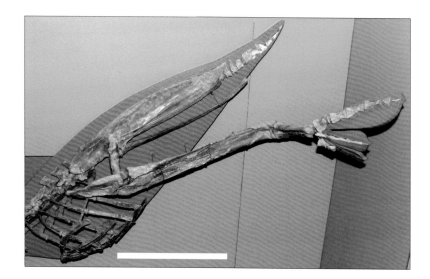

11.2. A reconstruction of *Hesperornis regalis* (FHSM VP-2069), showing the specialized arrangement of the leg bones. *Hesperornis* and related genera would have been strong swimmers but poorly adapted to walking on land. Scale bar = 20 cm (8 in).

From the fossil record, we believe that *Hesperornis* and related genera arrived in what is now Kansas during the early Campanian. The 1979 discovery of a much earlier hesperornithiform tarsometatarsal (lower leg) bone in the upper Cenomanian Lincoln Limestone (Everhart and Bell, 2009; Fig. 11.1), however, indicates that these birds were here much earlier, if only in small numbers. It is one of the earliest bird specimens from North America. Based on radiometric dating of the underlying 'X-bentonite,' the specimen is approximately the same age (95 million years) as the surprisingly abundant hesperornithid material collected from the Carrot River Bonebed in east central Saskatchewan, Canada (Cumbaa et al., 2006). However, none of these marine birds were ever as common in the Smoky Hill Chalk as in Late Cretaceous marine deposits further north.

The remains of *Hesperornis* in the Smoky Hill Chalk indicate that it was very much a bird, with feathers, but it still had teeth and solid (not hollow) bones. Its wings were reduced to a single, splint-like bone (the humerus) and atrophied to the point of being almost useless. Its powerful lower limbs were so specialized for efficient swimming that *Hesperornis* probably could not walk very well on land, if at all (Fig. 11.2). Reynaud (2006) studied the mode of aquatic locomotion in modern foot-propelled birds and compared them to the skeletal structure of *Hesperornis*. She suggested that hesperornids (*H. regalis*, *H. gracilis*, and *Parahesperornis alexi*) were most like the common loon, *Gavia immer*, in terms of hind limb bone-length ratios, pelvis shape, and position of the acetabulum (hip socket) on the pelvis, and she used *Gavia* as a model for her reconstruction of hesperornid hind limb musculature and locomotion. Her data supported the generally accepted idea that hesperornids swam with both feet thrusting backward simultaneously, and that they avoided land except for nesting.

Because of their poor ability to move on land, it is likely that *Hesperornis* built their nests close to the water's edge and, outside of nesting and raising their chicks, spent very little time on shore. In that regard, they may well be considered as an analog to modern penguins. *Hesperornis* and its kin were apparently very much at home in mid-ocean, since all of the remains collected from the Kansas chalk would have been more than 200 miles from the nearest coast. They apparently fed on fishes and other small prey, and in turn were occasionally eaten by mosasaurs (Martin and Bjork, 1987) and other large predators. A recent paper (Martin, Rothschild, and Burnham, 2016) suggests that a young hesperornithid survived an unsuccessful attack by a polycotylid plesiosaur, based on conical bite marks preserved on a lower leg bone (tibiotarsus). While they make a strong case for the identity of the attacker, I think it is just as likely that the predator was a mosasaur or a crocodile, or even that the failed attack could have been made on shore by a small dinosaur or another terrestrial predator.

Wilson et al. (2011) described the second specimen of another hesperornithiform bird (*Canadaga arctica*) from Devon Island in the Canadian High Arctic. The specimen represents the northernmost known occurrence of hesperornids and also indicates that this species from the far north appears to be significantly larger than those occurring in the portion of the Western Interior Sea over Kansas. A new specimen of a subadult *Hesperornis* sp. was described by Wilson, Chin, and Cumbaa (2016). The specimen included the first teeth ever associated with *Hesperornis* remains from the Arctic.

In an effort to see how cold weather might affect the formation of bones in these birds, Wilson and Chin (2014) compared the bone growth of *Hesperornis regalis* to that of modern penguins. More specifically, since some penguins are known to migrate long distances and *Hesperornis* has been assumed to have migrated from cooler northern waters to Kansas, the authors were looking for evidence in the histology of the bones from these two groups. However, there was no evidence of stress from long-distance swimming preserved in the bones of the penguins that were examined, and none associated with the fossilized bones of *Hesperornis*. At this point, we have no way to prove that *Hesperornis* migrated into the ocean over Kansas or was a full-time resident of that part of the Western Interior Sea. I'm fairly certain that research will continue in this area.

Bell, Irwin, and Davis (2015) reported two new specimens of *Hesperornis* from the Ozan and Marlbrook formations of Arkansas. A third specimen, the distal portion of a *Hesperornis* tarsometatarsus, had been reported earlier by Davis and Harris (1997). These discoveries represent more recent occurrences (Middle to Upper Campanian) of the genus than those from the Smoky Hill Chalk (Lower Campanian), and are also the first known examples from the Mississippi Embayment on the Gulf Coast (also the southernmost occurrences). Although the report was based on the isolated occurrences of fragments of two lower limb bones (a tibiotarsus and a tarsometatarsus) and a third phalanx (toe bone) of the foot, it appears that these three individuals were somewhat larger than those occurring earlier in the waters over Kansas. The third phalanx specimen (FHSM VP-17988) was discovered by coauthor Kelly Irwin and donated to the Sternberg Museum of Natural History.

Recent work by Aotsuka and Sato (2016) documented six species of hesperornithiform birds from the Campanian Pierre Shale in southern Manitoba, Canada, including a new species they called *Hesperornis lumgairi* from the Pembina Member. Naming a new species from a single damaged bone is of some concern to me, since it is a reminder of what Cope and Marsh were doing more than 100 years ago. The authors (ibid.) also noted that *H. regalis* was not present in the uppermost Millwood Member of the Pierre Shale (Middle Campanian). Unfortunately, rocks of that age have been eroded away in Kansas and we have no way of determining what species were present here during the final years of the Late Cretaceous.

Benjamin F. Mudge (1866a, 1866b), then a professor at the Kansas State Agricultural College (KSAC, now Kansas State University, Manhattan, Kansas), first reported the discovery of four three-toed 'Ornithichnites,' or footprints, in a piece of what was probably Dakota Sandstone from a bluff above the Republican River, several miles east of Concordia (Cloud County), Kansas. From differences observed in their size, Mudge believed that the tracks were made by at least two different individuals, both "long-legged waders" (1866b:11). Although he went to great pains to measure and otherwise document the tracks, the slab containing them was lost when his horse-drawn wagon was overturned the next day while fording a flooded stream. About 20 years later, Snow (1887; see also Everhart, 2015) reported a single bird track recovered by Judge E. P. West from the spoils of a well dug into the Dakota Sandstone of Ellsworth County. That specimen is curated in the collection of the University of Kansas. Williston (1898c) reprinted Snow's note along with a photograph of the footprint. Everhart (2015) published a recent photograph of the bird track specimen (KUVP 65696; Fig. 11.3) in the University of Kansas collection. The track represents the world's first documented discovery of a Mesozoic bird footprint.

While it was O. C. Marsh who collected the first remains of a bird (*Hesperornis regalis*) in the Smoky Hill Chalk of Trego County on July 25, 1871 (YPM 1200), the specimen was without a skull, and Marsh had no way of knowing the full significance of his discovery. Marsh wrote:

> One of the treasures secured during our explorations this year [1871] was the greater portion of the skeleton of a large fossil bird, at least five feet in height, which I was fortunate enough to discover in the Upper Cretaceous of Western Kansas. This interesting specimen, although a true bird—as is clearly shown by the vertebrae and some other parts of the skeleton—differs widely from any known recent or extinct forms of that class, and affords a fine example of a comprehensive type. The bones are all well preserved. The femur is very short, but the other portions of the legs are quite elongated. The metatarsal bones appear to have been

separated. On my return, I shall fully describe this unique fossil under the name *Hesperornis regalis.* (1872a:56–57)

Note that the locality given in the Yale Peabody Museum records ("Trego County?") is in error for several reasons. First, there is little or no Campanian-age chalk in that county, and no other remains of *Hesperornis* have since been collected that far east. Secondly, handwritten records were kept with all fossils collected by the Yale College scientific expeditions, and it is relatively easy to track the progress of the Marsh party as they moved eastward from Fort Wallace. I visited the Yale Peabody Museum in 2010 and examined many of the Kansas fossils in the collection. From those notes and the online information on the museum website, I was able to piece together a timeline of when and where many of the specimens were collected.

The type specimen of *Hesperornis regalis* (YPM 1200) was collected on July 25, 1871, and Marsh noted that it was discovered "on the south side of the Smoky Hill River about 20 miles east of Fort Wallace" (1880:195). The only locality that fits this description is an exposure of chalk referred to as Goblin Hollow, well inside Logan County, about a mile southwest of the town of Russell Springs, Kansas (Everhart, 2011). The first *H. regalis* specimen (YPM 1206) with a skull was discovered by Thomas H. Russell, a Yale graduate, in the same locality the following year. Marsh appeared to take some credit when he wrote that "a nearly perfect skeleton was obtained in Western Kansas by Mr. T. H. Russell and the

writer in November, 1872" (1875:404). Perfect as it was, a reconstruction (drawing) of the skeleton wasn't published until Marsh, 1880(pl. XX). A smaller version from Marsh, 1883 (fig. 18) is shown in Fig. 11.4.

Even so, the reconstruction of the skeleton, and the skull by Marsh (1880:pl. I) and subsequently by Heilmann (1926:fig. 28; Fig. 11. 7) and others, were missing a small but very important bone. It took a hundred years for the predentary bone to be recognized in *Hesperornis*, and then published by Larry Martin (1983; 1987:fig. 1; see also Martin and Naples, 2008:fig. 2). The specimen (KUVP 71012), which included a nearly complete, but disarticulated, skull and a few limb bones, was collected by Charles Bonner (pers. comm., 2016) from the upper chalk, southeast of Russell Springs, in Logan County around 1982. The predentary is a small, triangular bone that connects the two dentaries at the front via a ligament, forming a flexible joint that allows the lower jaws to move apart in the process of swallowing prey. Martin and Naples (2008) report, however, that Marsh (1880:pl. II, fig. 12) had actually located the predentary bone

in the YPM 1206 specimen and published a figure of it, but had misidentified it as the basihyal (part of the hyoid arch). Martin also discovered the
predentary bone in another marine bird, the holotype of *Parahesperornis
alexi* (KUVP 2287; discussed in a following section). See Zhou and Martin
(2010:figs. 1, 2) for photographs of the KUVP 71012 and KUVP 2287 pre-
dentaries, and additional discussion of the occurrence and function of the
predentary bone in *Hesperornis* and other Mesozoic ornithurine birds.

Actually, Marsh had collected a single *Hesperornis* foot bone (tarsal;
YPM 1205) during the first expedition to Kansas, but didn't realize it.
Williston noted that

> late in the season of 1870, Professor Marsh, with an escort of United States
> soldiers, spent a short time on the upper part of the Smoky Hill River
> collecting vertebrate fossils. The material then collected served for the
> description of a number of interesting types by Marsh. It included the first
> known specimen of 'Odontornithes,' a foot bone brought in with other
> material, but which was not discovered in the material until after other
> specimens had been obtained later. In June of the following year [1871]
> Marsh again visited the same region, with a larger party and a stronger
> escort of United States troops, and was rewarded by the discovery of the
> skeleton which forms the type of *Hesperornis regalis* Marsh, together with
> other material. (1898a:30)

The first recognized remains of a bird with teeth from the Smoky
Hill Chalk were collected the following year by B. F. Mudge. Williston
noted "that Professor Mudge found the remarkable specimen of *Ichthyornis*, from the North Fork of the Solomon, which furnished to the
world the discovery of the then startling fact of birds with genuine teeth"
(1898a:30–31). It was a small, very delicate specimen, contained in a slab
of chalk, and, although it eventually created considerable excitement, it
also led to a bit of confusion.

As was the custom at the time, Mudge had been sending his better
specimens to well-known paleontologists on the East Coast for examination and identification. In previous years, he had sent most of his material to E. D. Cope. In an 1870 letter to Mudge (Williston, 1898a:29–30),
Cope complimented him on the scientific value of the material he had
sent and named a new species of mosasaur (*Liodon mudgei*) after him. In
fact, many of the new species of fishes and mosasaurs described by Cope
between 1869 and 1872 were from specimens sent to him by Mudge (Everhart, 2002). Cope (1872) also visited Mudge in late 1871 and examined his
collection in Manhattan, Kansas. However, it was a letter from Marsh a
year later that initiated a chain of events that would deny a spectacular
specimen to Cope and give his rival, O. C. Marsh, the honor of describing the first toothed bird from the chalk.

According to Williston, on September 2, 1872, Marsh wrote to Mudge
"inquiring about his summer collections in the Cretaceous, with the offer
to 'determine any reptilian or bird remains without expense,' and stating
that he would give him 'full credit' for their discovery" (1898a:31). At the
time, Mudge had a shipment of specimens already prepared to send to

Cope. Apparently, he changed his mind at the last moment and instead sent the box containing the bird bones to O. C. Marsh for examination. Williston related the following story of what had occurred:

> An incident related to me by Professor Mudge in connection with this specimen is of interest. He had been sending his vertebrate fossils previously to Professor Cope for determination. Learning through Professor Dana that Professor Marsh, who as a boy had been an acquaintance of Professor Mudge, was interested in these fossils, he changed the address upon the box containing the bird specimen after he had made it ready to send to Professor Cope, and sent it instead to Professor Marsh. Had Professor Cope received the box, he would have been the first to make known to the world the discovery of "Birds with Teeth." (1898b:44)

Upon seeing the unusual specimen, Marsh quickly realized that he was looking at the skeleton of a small bird that was very different from his newly discovered *Hesperornis*. The skull, however, was not visible, and Marsh recognized what he thought were the lower jaws of a small reptile in the same block of chalk. Williston noted that Marsh wrote to Mudge again on September 25, 1872, acknowledging the receipt of a box of fossils, and stating that the "hollow bones are part of a bird, and the two jaws belong to a small saurian. The latter is peculiar, and I wish I had some of the vertebrae for comparison with other Kansas species" (1898a:31). Having no additional remains available at the time (or at least not recognizing them), Marsh concluded that Mudge's specimen actually represented the remains of two new species, and he briefly described the new bird in the *American Journal of Science*:

> One of the most interesting of recent discoveries in Paleontology is the skeleton of a fossil bird, found, during the past summer, in the upper Cretaceous shale of Kansas, by Prof. B. F. Mudge, who has kindly sent the specimen to me for examination. The remains indicate an aquatic bird, about as large as a pigeon, and differing widely from all known birds in having *biconcave vertebrae* [italics by Marsh]. The cervical, dorsal, and caudal vertebrae preserved all show this character, the ends of the centra resembling those in *Plesiosaurus*. The rest of the skeleton presents no marked deviation from the ordinary avian type. The wings were large in proportion to the posterior extremities. The humerus is 58.6 mm. in length, and has the radial crest strongly developed. The femur is small, and has the proximal end compressed transversely. The tibia is slender, and 44.5 mm. long. Its distal end is incurved, as in swimming birds, has no supratendinal ridge. This species may be called *Ichthyornis dispar*. A complete description will appear in an early number of this journal. (1872b:334)

While the records of the Yale Peabody collection indicate that the remains (YPM 1450) came from the Niobrara Formation of Scott County (which didn't even exist as an incorporated county in 1872—the boundaries were not defined until 1873), it is more likely that Mudge collected them much farther to the east (Williston, 1898a; Peterson, 1987). In

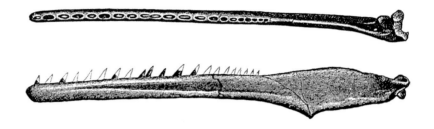

another paper, Williston noted that the specimen "was found by him [Mudge] near Sugar Bowl Mound, in northwestern Kansas" (1898b:44). Williston made a slight error in the name and was actually referring to Sugarloaf Mound (see Mudge, 1876:212), a well-known geologic feature near Bow Creek in northwestern Rooks County. Walker researched the occurrence of toothed birds in Kansas and noted that the slab containing the bird remains came from an exposure "on Bow Creek in northwestern Rooks County" (1967:61), a tributary of the Solomon River. This locality is about 75 miles from the closest chalk in what is now northeast Scott County. Later in the same year, Marsh followed with another brief note describing and naming a new species of marine reptile:

> An interesting addition to the reptilian fauna of the Cretaceous shale of Kansas is a very small Saurian, which differs widely from any hereto discovered. The only remains at present known are two lower jaws, nearly perfect, and with many of the teeth in good preservation [Fig. 11.5]. The jaws resemble in general form those of the Mosasauroid reptiles, but, aside from their very diminutive size, present several features which no species of that group has been observed to possess. The teeth are implanted in distinct sockets, and are directed obliquely backward. There were apparently twenty teeth in each jaw, all compressed, and with very acute summits. The rami were united in front only by cartilage. There is no distinct groove on their inner surface, as in all known Mosasauroids. The dentigerous portion of the jaw is 41 mm in length, its depth below the last tooth is 5 mm and below the first tooth in front 3 mm. The specimen clearly indicates a new genus which may be called *Colonosaurus*, and the species may be named *Colonosaurus Mudgei* for the discover Prof. B. F. Mudge, who found the remains in the upper Cretaceous shale of Western Kansas. [Yale College, Oct. 7th, 1872]. (1872c:406)

By the following January, however, Marsh had apparently recognized his mistake and noted that the specimen proved "on further investigation to possess some additional characters, which separate them still more widely from all known recent and fossil forms. The type species of this group, *Ichthyornis dispar* Marsh, has well developed *teeth in both jaws* [italics by Marsh]" (1873a:161). In February of that year, Marsh redescribed *Ichthyornis* as "a bird, about as large as a Pigeon, and differing from all known birds in having *teeth* and *biconcave vertebrae* [italics by Marsh]. The known remains were found in the Upper Cretaceous shale of Kansas, and are preserved in the collection of Yale College" (1873b:230). *Hesperornis* was also mentioned in the same note but without any indication that Marsh was then aware that it also had teeth. *Colonosaurus*, the new

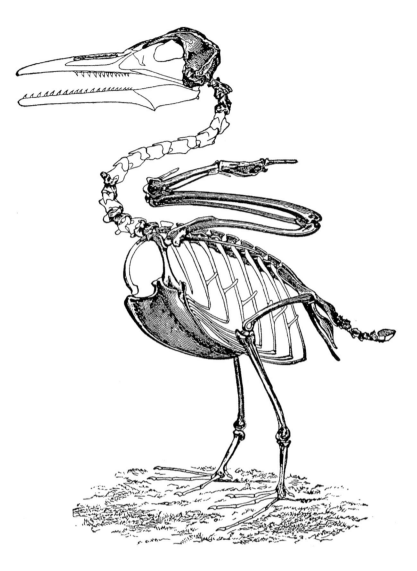

genus based on the jaws of the little reptile that wasn't, was left to die a quiet death. Williston (1898a) recounted the story briefly in volume 4 of the *Geological Survey of Kansas*.

In 1875, Marsh published a major article, "*On the Odontornithes, or Birds with Teeth*," in which he established a new subclass and orders for the recently discovered toothed birds. In the article, the history of the discovery of birds in western Kansas appears to have been rewritten slightly when Marsh noted that "the first species in which teeth were detected was *Ichthyornis dispar* Marsh, described in 1872. Fortunately the type specimen of this remarkable species was in excellent condition, and the more important portions both of the skull and skeleton were secured" (Marsh, 1875:403). No further mention was made of the discoverer (Mudge) or the resulting confusion over naming the lower jaws as a new species of reptile. Years later, Marsh (1883:fig. 27) provided the first drawing of the skeleton of *Ichthyornis dispar* (Fig. 11.6).

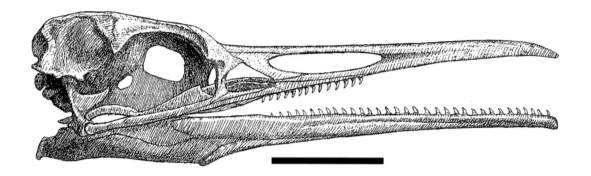

11.7. The skull of *Hesperornis regalis*, based on YPM 1206, in right lateral view exhibiting teeth on the maxilla and dentary. Adapted from Heilmann (1926:fig. 28). Scale bar = 5 cm.

Marsh noted that the "most interesting bird with teeth yet discovered is perhaps *Hesperornis regalis*, a gigantic diver, also from the Cretaceous of Kansas, and discovered by the writer in 1870" (1875:404). Marsh neglected to mention that the original discovery (the small foot bone mentioned by Williston, 1898a) did not include a skeleton or teeth, but adds that "a nearly perfect skeleton was obtained in Western Kansas by Mr. T. H. Russell and the writer in November, 1872" (ibid.). The reader is left to assume that the 1870 fossil was the first specimen of *Hesperornis* (YPM 1206) discovered with jaws containing teeth (Figure 11.7). Although this was not an outright lie, Marsh was certainly guilty of clouding the history of the discovery of *Hesperornis*.

As a side note here, I would like to mention that my own research into the locality where those first *Hesperornis* fossils had been collected by Marsh and Russell benefited greatly from an inquiry from Jill Hunting, the great-granddaughter of Dr. Thomas H. Russell, who wanted to visit the place where her ancestor had discovered this important specimen. In 2010, nearly 140 years later, I was able to show her and her daughter the locality where their ancestor had collected fossils with Professor Marsh in 1872 (Everhart, 2011).

In a brief report on Kansas geology, Mudge noted, with more than a bit of humor intended, that "it may not be amiss to add that birds with teeth have been found only in the United States, and one genus, the *Ichthyornis*, only in Kansas. Thus, our fossil pterodactyls have no teeth, and our birds have, in direct contrast with those found in other parts of the world" (1877:5).

Marsh published *Odontornithes: A Monograph on the Extinct Toothed Birds of North America* in 1880. In the appendix entitled "Synopsis of American Cretaceous Birds" (1880:191–201) he listed and briefly described nine genera and twenty species, most of which were in the Yale College collection. Shortly before Marsh's death in 1899, Williston (1898b:44) published an article on 'birds' in volume IV of the *University Geological Survey of Kansas*, crediting Mudge with the "most important" discovery (*Ichthyornis*), giving a brief history of the specimens known at the time, and describing additional specimens in the University of Kansas collection. In a short paragraph, Williston also demolished Marsh's Odontornithes: "The systematic position of the toothed birds from Kansas is

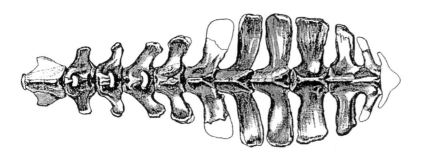

11.8. A dorsal view of the caudal vertebrae of *Hesperornis regalis* (Marsh, 1883:fig. 17). These vertebrae supported a broad, flat tail could be moved up or down and was probably used by *Hesperornis* for directional control while swimming underwater. No scale.

by no means yet settled. All ornithologists are, however, agreed that they do not form a separate group, and the name Odontornithes is in consequence generally abandoned. The value of the teeth is subordinate; they do not in themselves justify a separate subclass" (ibid.:47). So, like Cope's Pythonomorpha and Streptosauria years before, Marsh's Odontornithes vanished from the language of paleontology.

Marsh reported that the total number of bird specimens in the Yale museum, made up mostly of those collected from the Smoky Hill Chalk, represented "about one hundred and fifty different individuals" (1883:51). Clarke (2004) reported that 78 of these specimens were *Ichthyornis*. An earlier tally by Russell (1993) estimated that a total of 225 Late Cretaceous bird specimens (or about 3 percent of the total) were located in museum collections from the Smoky Hill Chalk, similar to the number of turtle specimens (210), and only about one-quarter of the number of *Pteranodon* specimens (878).

Lane (1946:393) noted that *Hesperornis* had 14 teeth in each maxilla and 33 teeth in each side of the lower jaw. There are no teeth in the premaxilla. Lane repeated (ibid.:293) Marsh's explanation of the mode by which the teeth were replaced, noting that the new tooth bud forms on the lingual (inner) side of the older tooth, and dissolves the root of the older tooth as it grows. Eventually the older tooth falls out and the new tooth fills the vacated space. Lane (ibid.:394) also indicated that the pelvis of *Hesperornis* is more reptilian in form than that of any modern bird. The tail of *Hesperornis* is also unusual because it is composed of 12 vertebrae (more than any modern bird), some of which have expanded transverse processes (Fig. 11.8) "and clearly indicate that the tail was moved mostly up-and-down, evidently an adaptation to diving" (ibid.:395).

Until recently (Clarke, 2004), little new information had been added to our knowledge of the toothed birds from Kansas. Williston's (1898b) summary of Late Cretaceous bird specimens in the Museum of Natural History at the University of Kansas was the previous major work on the subject. At that time, Williston reported that a few new specimens had been collected, but nowhere near the number discovered when Marsh made them a priority for his field workers. Williston (ibid.) noted that there were six species of *Ichthyornis* known at the time, all of which were named by Marsh between 1872 and 1880: *I. dispar, I. agilis, I. anceps, I. tener, I. validus*, and *I. victor*. Clarke's (2004) study of the same specimens

in the Yale Peabody collection indicated that only one species of *Ichthyornis* was valid—*I. dispar*, with '*I. tener*' becoming *Guildavis tener* and the other names becoming junior synonyms of *dispar*. Clarke also named a new ichthyornithid species (*Iaceornis marshi*) from one of the other specimens in the collection.

Partially because of their small size and fragility, bird fossils remain rare and elusive. Putting things into a personal perspective, I've collected fossils in the Smoky Hill Chalk since 1969 and the only *Ichthyornis* remains I have ever collected is an 11 mm (0.5 in) fragment of a shoulder bone (proximal end of the coracoid; FHSM VP-15574) from Gove County in 2002. Worse yet, the only reason that I saw it was that I was down on my hands and knees looking for shark teeth. A friend of mine, George Klein was on his first field trip to the chalk in Logan County in 2007 when he picked up a complete *Ichthyornis* right coracoid without realizing what it was. When I identified it for him, he donated to the Sternberg Museum (FHSM VP-17317). Such are the vagaries of collecting fossils in the chalk.

In her review of the Yale Peabody collection, Clarke (2004; pers. comm., 2003) further humbled my little discovery when she noted that the bones of the wing and shoulder, including the humerus and coracoid,

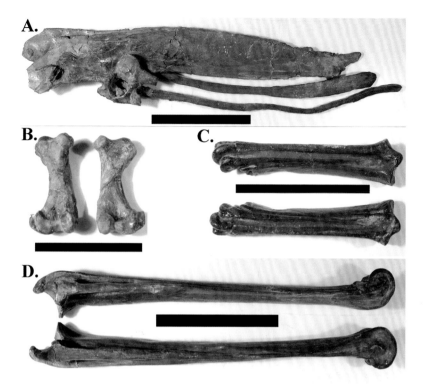

11.10. The pelvis and limb bones of a *Hesperornis regalis* specimen (UNSM 1212) collected by George Sternberg from the chalk in Logan County: A, the pelvis in left lateral view; B, both femora; C, both tarsometatarsi; D, both tibiotarsi. Scale bars = 10 cm (4 in).

are the most commonly collected bones of *Ichthyornis*. Her research indicated that limb bones (wings and legs) were among the first pieces to drop off a floating bird carcass, and she concluded that many of the specimens in museum collections were derived from the decomposing bodies of dead birds that may have floated some distance from the location where they died to where they were buried. While I have trouble understanding how the body of a small bird could float on the surface for any period of time without attracting scavengers, it was a big ocean, and we do know that several much larger dinosaur carcasses floated hundreds of miles into the Western Interior Sea during the same time period before settling to the bottom (Chapter 12).

There are some bright spots in the research on these marine birds since Williston's 1898 report. A 'headless' but otherwise nearly perfect articulated specimen of *Hesperornis* (AMNH FR 5100) was acquired by the American Museum of Natural History from Charles H. Sternberg in 1907. It had been discovered in Logan County, Kansas, probably very close to Elkader, where the Sternberg family made many of their discoveries. Sternberg (1909) credits his second son, Charles M. Sternberg, with the discovery. The specimen (ibid.:fig. 41; Fig. 11.9) is interesting in that the remains were discovered with the hind limbs positioned on either side of the body as they were in life.

Martin and Tate (1966) noted that the exhibit specimen of *Hesperornis* in the University of Nebraska State Museum (UNSM 1212; Fig. 11.10) was a composite of two specimens collected from the Smoky Hill Chalk in the 1930s in eastern Logan County by George F. Sternberg.

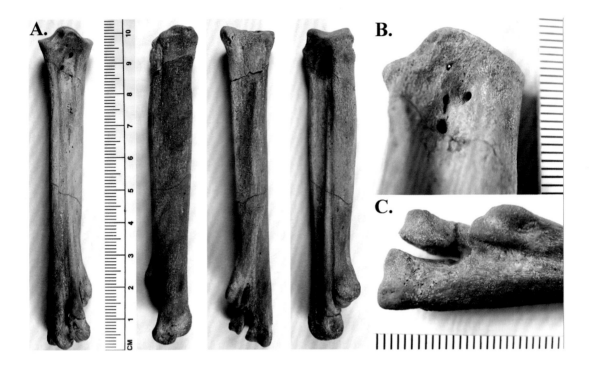

11.11. *Parahesperornis* tarsometatarsus (FHSM VP-17312) collected by Pete Bussen from the upper Smoky Hill Chalk in Wallace County: A, tarsometatarsus in (L–R) cranial, lateral, medial, caudal views; B, nutrient foramina in the proximal end of the bone; C, trochlea at the distal end of the metatarsal. Scale in mm.

One of the hesperornithid fossils (KUVP 2287) in the University of Kansas collection described by Williston (1896, 1898b) was a nearly complete specimen collected by H. T. Martin in 1894 from Graham County, Kansas. Williston referred the specimen to a new species, *H. gracilis* Marsh, and noted that this species is smaller than *H. regalis*, then added that it "has never been adequately described" (1898b:46). He also noted that the "specimen now in the University museum . . . is remarkable in showing the scuta of the tarso-metatarsal region, together with feathers. . . . Indications of feathers are also seen on back portion of the head, and everywhere they appear to be more plumulaceous than the ordinary type of feather" (ibid.:48). Williston (ibid.:pl. VIII) also figured the scuta and feathers, but the published image is not very informative.

Possibly as an indication of the animosity that may have existed between Williston and his employer at Yale, Marsh (1897) mentioned the discovery in *Nature*, as an 'I told you so,' but didn't mention Williston by name in the text of the article. Marsh's comment that the discovery of feathers proved his earlier contention that *Hesperornis* was "a carnivorous, swimming ostrich" (1880:114) drew several negative responses at the time, most notably by Schufeldt (1897).

Martin (1984) reexamined the KU specimen and concluded that it represented both a new genus and new species of an even more primitive swimming bird from the Late Cretaceous that he redescribed and named *Parahesperornis alexi* (*para* means 'near'). Curiously, Martin only mentioned the feathers and scutes in his description, and I can find no recent publication that describes them further. When I asked David Burnham (pers. comm., 2016) at the University of Kansas about this specimen, he

11.12. Colonites from the UNSM 20030 *Baptornis* (Now *Fumicollis hoffmani*) specimen. Note the fish jaw (cf. *Enchodus*) at upper left. Scale in mm.

indicated that a paper was being prepared to describe the feather and scute impressions. Alyssa Bell (pers. comm., 2016) is also working on a revised description of *Parahesperornis*.

In 2008, my friend Pete Bussen donated a bird bone (FHSM VP-17312; Fig. 11.11) that he had collected in 1995 from the upper chalk in Logan County. The bone was identified as the right tarsometatarsus of a *Parahesperornis alexi* (Bell and Everhart, 2009). There is only one other known specimen of this rare species (KUVP 24090).

Additional specimens of another genus of smaller swimming bird, *Baptornis advenus* (Marsh, 1877), were described by Larry Martin and Orville Bonner (1977). In an earlier paper, Martin and Tate (1976) had noted the presence of several coprolites in association with a specimen that they identified as *Baptornis advenus* (UNSM 20030) and were able to identify the jaw of a small fish (*Enchodus* sp.) in one of them. When I visited the University of Nebraska State Museum in 2008, I was able to examine the specimen and photograph the coprolites. In my opinion, rather than being coprolites (Fig. 11.12), the six small masses containing fragments of fish bones are most likely colonites, that is, wastes that were inside the bird's colon at the time of death. In their recent review of this specimen, however, Wilson, Chin, and Cumbaa (2016) noted that the association is questionable. While I understand their concern, in my experience, the likelihood of having another species' coprolites mixed with the skeletal remains is pretty remote.

The UNSM 20030 specimen was collected by G. F. Sternberg in 1936, and the same coprolites/colonites that I photographed are shown clearly in his original photograph of the remains (Sternberg Collection, archives of the Forsyth Library, Fort Hays State University). A note signed by Sternberg curated with the specimen at the University of Nebraska reads: "There are 7 or 8 coprolites . . . 2 show small fish bones. These

11.13. A partial skeleton of *Hesperornis regalis* (FHSM VP- 2069) on exhibit at the Sternberg Museum of Natural History. This specimen was collected by M. C. Bonner in 1958 from the early Campanian chalk of Logan County. No scale; however the tibiotarus leg bone is 31 cm (12 in) in length.

are small compared to the other coprolites I have seen and were found mingled with the bones" (G. Corner, pers. comm., 2007).

The UNSM 20030 specimen was subsequently examined by Bell and Chiappe (2015) and redescribed as a new genus and species, *Fumicollis hoffmani*. The genus name is from the Latin, 'fumi' meaning 'smoke' and 'collis' meaning 'hill,' a reference to the Smoky Hill River near the Logan County locality where it was discovered by Harold Shepherd and collected by George Sternberg in 1936 (GFS 101–36). The locality provided by G. F. Sternberg was in northwest Gove County, "about five miles northeast of Elkader" (G. Corner, pers. comm., 2007).

In 1991, the partial skeleton of a *Baptornis* was discovered by Dan Varner (the paleo-artist who created the paintings reproduced as the plates and the cover for this book) in the Sharon Springs member of the Pierre Shale in southwestern South Dakota (Martin and Varner, 1992). At the time, the specimen (SDSM 68430) was regarded by James Martin as the latest known occurrence of *Baptornis* since all other known specimens had been collected from the underlying Smoky Hill Chalk. Subsequent examination of the remains resulted in the determination that the specimen represented a new species (*Baptornis varneri*; Martin and Cordes-Person, 2007), named in honor of Dan Varner (1949–2012). The species has since been moved into a new family and genus of hesperornithiforms created by Larry Martin and others as *Brodavis varneri* (Martin, Kurochkin, and Tokaryk, 2012).

11.14. A reconstruction of *Ichthyornis dispar* (FHSM VP- 2503) at the Sternberg Museum of Natural History. This specimen was collected by J. D. Stewart in 1970 in the lower chalk of Graham County.

In September 1958, M. C. Bonner collected the fairly complete, but headless, remains of a *Hesperornis regalis* from middle chalk southeast of Russell Springs in Logan County. The specimen (FHSM VP-2069) is currently on exhibit in the Sternberg Museum of Natural History (Fig. 11.13).

An excellent *Ichthyornis* specimen (Fig. 11.14), including portions of the skull and lower jaws, was collected by J. D. Stewart in eastern Graham County in 1970 (Martin and Stewart, 1977). The locality is probably within a few miles of the original Rooks County site where Mudge collected the first specimen in 1872. Stewart's discovery (FHSM VP-2503) is currently on exhibit in the Sternberg Museum of Natural History.

Another *Ichthyornis* specimen (KUVP 119673) in the University of Kansas collection was discovered in 1992 by Greg Winkler and Pete Bussen in northern Lane County. They recognized that the delicate bones of the small bird were beyond their ability to prepare and generously donated the specimen to the University of Kansas. Like Mudge's first specimen, KUVP 119673 was collected in a slab of chalk and was mostly complete, except for the anterior part of the skull. Years later, Larry Martin showed me an x-ray photograph of the slab and pointed out many of the bones that it contained.

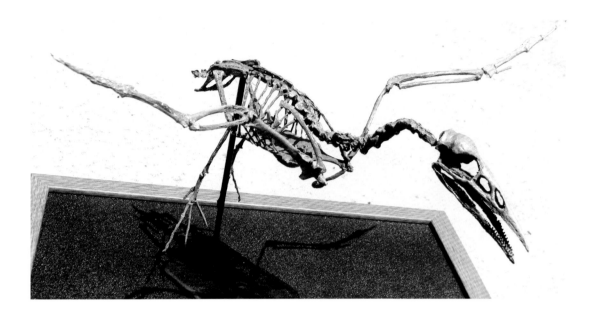

11.15. A reconstruction of *Ichthyornis dispar* in a flight pose based on the KUVP 119673 specimen collected by Greg Winkler and Pete Bussen in 1992. Cast provided by Triebold Paleontology. No scale.

The complicated and delicate preparation of the specimen, however, was delayed for several years until Martin made a deal for funding for the work with a museum in Japan that was interested in purchasing a mounted cast. At that point, David Burnham was able to use an innovative process to remove and cast the tiny bones, including the very delicate ribs, and then to assemble them into a three-dimensional mount (Burnham, 2005). Then Triebold Paleontology was allowed to make a copy of Burnham's original cast and make the specimen available for sale to other museums. Triebold Paleontology generously provided a copy of the cast to Greg and Pete, and Anthony Maltese assembled it into a dramatic taking-flight pose for them (Figure 11.15). The *Ichthyornis* cast is now a permanent exhibit at the Fort Wallace Museum, Wallace, Kansas, where Pete Bussen spent many years as a volunteer.

Ichthyornis is also known from somewhat older rocks in Kansas, Texas, and Canada. Walker (1967; see also Shimada and Fernandes, 2006) reported the proximal end of a tarsometatarsus (FHSM VP-2139) from the lower Fairport Chalk or upper Greenhorn Limestone (middle Turonian) of Ellis County. The known geographical range of *Ichthyornis* was extended some 2,300 km (1,400 mi) to the northwest of Kansas when a left humerus was discovered in the early Turonian of Alberta, Canada (Fox, 1984). More recently, a diverse bird fauna was reported by Tokaryk, Cumbaa, and Storer (1997) from the Upper Cenomanian of Saskatchewan, Canada.

Marsh (1877) described a new genus and species (*Graculavus lentus*, YPM 1796) on the basis of a partial tarsometatarsus from the Cretaceous Austin Chalk of Texas. At one point it was renamed *Ichthyornis lentus*, but then Clarke (2004) noted many differences between it and the type specimen, and placed it into a new provisional genus: *Austinornis*. Four *Ichthyornis* specimens were reported by Parris and Echols (1992) from the

11.16. Two teeth from an ichthyornithid bird (FHSM VP-17461) from the base of the Lincoln Limestone (Upper Cenomanian) in Russell County, Kansas. Scale bar = 5 mm (0.2 in).

Cretaceous of northern Texas, ranging in age from Coniacian to Campanian. Olson (1975) reported fragments of *Ichthyornis* from the Selma Chalk of Alabama. The most southern occurrence of *Ichthyornis* comes from the Austin Chalk in Coahuila State, northern Mexico (Porras-Múzquiz, Chatterjee, and Lehman, 2014). Together, the Texas, Alabama, and Mexican specimens represent the known southern distribution of *Ichthyornis* (see also Shimada and Wilson, 2016).

In 2004, Keith Ewell and I collected scattered fragments of small bird bones and teeth from the basal Lincoln Limestone Member of the Greenhorn Formation (Middle Cenomanian) of Russell County, Kansas. The bones included the upper portion of a femur and part of a coracoid (shoulder blade). The two bird teeth and attached jaw fragments (FHSM VP-17461; Fig. 11.16) were so small (about 2 mm; 0.1 in) that I actually discovered them while searching for microfossils in the matrix sample with a binocular microscope. Several years later, the specimens were published as remains of some of the oldest ornithurine birds in the United States (Bell and Everhart, 2011). Although the remains were too fragmentary to be certain of their genus, they most likely represent early ichthyornithid birds. Recently the humerus of a small bird identified as *Ichthyornis* sp. was reported from this same stratigraphic level by Shimada and Wilson (2016).

The most recent specimen of *Ichthyornis* (FHSM VP-18702) to be collected in Kansas was discovered by a Fort Hays State University student, Kris Super, near Castle Rock in southeastern Gove County. The remains are remarkably complete, considering that it is a bird preserved in chalk deposited in the middle of an ocean, and include the wing bones and sternum (Fig. 11.17). The specimen was prepared out of the chalk by David Burnham at the University of Kansas, and a description will follow.

Earlier in this chapter, I briefly mentioned the fragments of a bird tarsometatarsus (lower leg bone) from Russell County (FHSM VP-6318;

11.17. The most recent (2015) specimen (FHSM VP-18702) of an *Ichthyornis dispar* bird, from the middle of the Smoky Hill Chalk in southeastern Gove County. The large bone at upper left is the sternum in left lateral view. Scale bar = 10 cm (4 in).

Fig. 11.1). It was stored in the Sternberg Museum *Hesperornis* collection, and, not being particularly interested in fossil birds at the time, I just assumed that is what it was. Then in early 2008, Alyssa Bell (then a graduate student studying *Hesperornis*, now Dr. Bell) was examining our specimens for her dissertation. When she picked up VP-6318, she looked puzzled for a bit and then handed to me, saying, "This is not *Hesperornis*." I looked at the information on the curation card for the specimen and it suddenly dawned on me that the rocks in Russell County were much older than any known *Hesperornis regalis* specimen. Then I recognized the locality and stratigraphy, and realized that what I was holding was probably the oldest bird bone ever discovered in Kansas. The bone had been discovered by a student (D. M. Shumaker) on a field trip in 1979, given to the trip leader (Dr. Zakrzewski), who properly identified it as a bird tarsometatarsus, and then was curated into the collection by J. D. Stewart in 1984. Numerous paleontologists had looked at it over the previous 25 years but had not connected the 'age' to the 'bird.' The combination of Alyssa's revised identification and my knowledge of the locality and age of the specimen led us to publishing what was then the oldest bird remains in Kansas (Everhart and Bell, 2009). Although it was not *Hesperornis* exactly, it did represent a much older and related genus of a currently unknown hesperornithiform bird from the Middle Cenomanian (about 95 Ma) Lincoln Limestone Member of the Greenhorn Formation.

In the more than 100 years since Williston's (1898b) note on Kansas toothed birds, it has become apparent that Kansas was near the southern range of *Hesperornis*. Remains of these birds are extremely rare to the south of Kansas and have only recently been located in the Gulf region (David Schwimmer, pers. comm., 2003; Bell, Irwin, and Davis, 2015). As you go farther north, however, *Hesperornis* becomes much more common. According to Russell (1967), almost one-third of the early Campanian vertebrate remains collected from the Anderson River (Northwest Territories, Canada) in 1965 were identified as *Hesperornis*. Nicholls (1988:33) reported that bird remains are "one of the dominant components" of collections made from the Pembina Shale of that region (172 of 476 specimens counted, or 36 percent). *Hesperornis* may have been a Cretaceous analog to modern penguins, preferring cooler waters or feeding on prey that was more abundant in cooler waters. During the Cretaceous, however, these birds were limited to the Northern Hemisphere, just the opposite of where penguins are living today. As a point of interest, Alyssa Bell (pers. comm., 2016) noted that the Smoky Hill Chalk in Kansas is the only place that reasonably complete remains of hesperornithiform birds have ever been recovered. Almost all specimens from outside Kansas are limited to isolated elements of the skeleton.

As Marsh noted, "the remains of birds are among the rarest of fossils" (1883:49). Complete bird skeletons are extremely rare in the Western Interior Sea. Because of their smaller size and relatively light construction (generally hollow), most of the material in museum collections consists of a few parts or just a single bone. A survey of the collection at the Sternberg Museum in 2002 indicated only five partial specimens of *Hesperornis*, including the exhibit specimen (FHSM VP-2069), and eight specimens of *Ichthyornis*, including the exhibit specimen (FHSM VP-2503). In most cases, the remains consisted of elements of the lower limbs. It is difficult from fragmentary remains such as these to figure out what these birds were like and how they lived. Since it was without wings and had strong, well-developed legs, it is fairly obvious that *Hesperornis* was a swimming bird. *Ichthyornis*, on the other hand, had strong wings and a well-developed keel (sternum or breastbone) for anchoring flight muscles. Chinsamy, Martin, and Dodson noted that it was "generally accepted that *Ichthyornis* used its long jaws and recurved teeth for scooping fish from surface waters (as gulls and terns scoop fish from surface waters today)" (1998:226). While that is certainly a strong possibility, another method of feeding that would be well suited for jaws with teeth would be diving and pursuing prey underwater, as is done by modern seabirds such as shearwaters, guillemots, and puffins. Such a scenario might also better explain the occurrence of the remains of these relatively small birds in the chalk.

Sharing the ocean with other large predators sometimes made life short for the unwary. The remains of a *Hesperornis* were discovered in 1979 as gut contents of a large mosasaur, *Tylosaurus proriger* (SDSM 10439), along with another mosasaur, a large fish, and possibly a shark (Martin and Bjork, 1987). Hanks and Shimada (2002) reported the serrated bite marks of a shark (*Squalicorax*) on a hesperornithiform bird bone from the Turonian of South Dakota. The fragmentary remains that make up most of the collections of the Yale Museum and other institutions are indicative of the 'leftovers' when these birds were partially consumed by other predators. Kear, Boles, and Smith (2003) reported a bird bone and the remains of small turtles as gut contents of an Early Cretaceous ichthyosaur from Australia.

Both ichthyornithids and hesperornithids are now regarded as unsuccessful side branches in avian evolution. In spite of their superficial similarities to extant birds, they have no modern relatives. While the number of specimens collected since the first remains were collected by Marsh and Mudge increases but slowly, research into these fascinating toothed birds continues.

Dinosaurs?

12

The small herd of plant-eating dinosaurs moved noisily through thick underbrush as they fed on the lush vegetation growing along the bank of the flood-swollen river. It was late in the rainy season, and the new growth near ground level was easy picking for the bladelike teeth of the squat, heavily built nodosaurs. In order to get the nourishment they needed to feed their large bodies, these animals fed almost continuously on low-growing woody shrubs. The floor of the forest was still muddy from the last rain, and the slow-moving animals were churning it into a sticky muck that clung to their stubby legs.

One of the nodosaurs was a 4-year-old female. She was one of only seven surviving nodosaurs of her age group in the herd. At just over 4 m (15 ft) in length, she was not yet full-grown. However, she was old enough and large enough to feed on the outer edge of the herd with the adults and to help provide a ring of protection for the lightly armored juveniles and the much smaller hatchlings in the center. While the older animals were heavily armored and built low to the ground for protection against attack by predators, the younger ones were at risk of becoming prey if left unattended for long. The large carnivores that roamed the same scrub forest were always looking for an opportunity to make a meal out of a young nodosaur.

The young female worked her way slowly through the thick brush, her sharp teeth clipping efficiently at the tender new growth on the plants around her. Although she couldn't see it through the undergrowth, she was on the side of the herd nearest the river. The roaring sound of the river at flood stage seemed to be getting louder, drowning out the feeding noises made by the herd. Then, as she moved forward across a small clearing, the ground began to tremble under her. Suddenly, the edge of the riverbank shifted and gave way beneath her feet. Startled, she raised her head and saw a crack open in the muddy soil between her and the rest of the herd. Before she could move, the earth dropped from under her with a loud splash as it crashed into the rain-swollen river below. As she fell, her heavy body and all the vegetation around her were immediately immersed by muddy water. She was pulled under the surface by the current and tumbled along with the flow. Unable to raise her head out of the water, she soon drowned. Her body was carried downstream along with many uprooted trees and other flood debris.

After few miles, the river reached the edge of a shallow ocean, and the current slowed. The body of the dead nodosaur was too heavy for the diminished current to carry it further, and it soon lodged under the edge of a tangle of brush and tree limbs in a brackish estuary along the coast. Microorganisms living in the gut of the dead nodosaur continued to break down the plants the animal had eaten that day but also began working to decompose her internal organs. In doing so, they generated gases that bloated the nodosaur's body and made it steadily more buoyant. After a few days in the warm water, the bloated carcass tugged free of the last of the branches that held it and drifted slowly up to the surface. The weight of the heavy dermal scutes that made up the protective armor of the dinosaur turned the carcass over so that it now floated upside down. Once the current from the diminishing floodwaters caught the body, it began to move seaward. Slowly but relentlessly, the swollen body of the nodosaur moved away from the land and into the reaches of the vast shallow sea.

After drifting at sea for many days, the bloated carcass was rotting from the inside out and beginning to come apart. The thick skin of the dinosaur still held the body together, and the bony armor protected most of it from scavenging by all but the largest sharks, but it was only a matter of time before the abdominal cavity ruptured. Eventually, the layer of skin and muscle covering the belly ripped open, releasing the gases trapped inside, and the dinosaur began to sink toward the muddy floor of the chalk sea. Still largely intact after its long journey, the nodosaur's large bones and dermal scutes made it heavy enough to drop rapidly toward the bottom some 400 feet below. As the carcass landed hard on its right side amid a sparsely populated community of inoceramid clams, some of the remains were driven several inches into the soft mud. At last at rest in the pitch-black darkness of the sea bottom, the carcass would be slowly buried without much further damage by scavengers. (See Reisdorf et al., 2012, for a recent and more detailed discussion of the decomposition of carcasses of air-breathing vertebrates in the ocean.)

Dinosaurs

The preceding story is fiction, except for the fact that the reasonably complete remains of a nodosaur now called *Niobrarasaurus coleii* were actually discovered in the chalk of Gove County in 1930. We will never know for certain how or where this nodosaur died and was carried out to sea. Even more of a mystery is how it traveled hundreds of miles from the nearest shore to the middle of the Western Interior Sea, where the carcass sank and was eventually covered by millions of years of accumulated marine sediments. Although it is a rare occurrence, this specimen is not the only dinosaur to be discovered in the Smoky Hill Chalk. The five sets of dinosaur remains that had been discovered since 1871 were reported by Carpenter, Dilkes, and Weishampel (1995), and another, more recent one by Hamm and Everhart (2001: see also Everhart and Hamm, 2005). Since then, at least four new specimens have been collected (Liggett, 2005; Everhart and Ewell, 2006; Buskuskie, 2015). Of the total number, two are hadrosaurs and the others are nodosaurs. As I will explain below, the first two nodosaur specimens may actually be from the same dinosaur.

In the summer of 1871, Professor O. C. Marsh and his scientific expedition of students from Yale College were the first to collect the remains of a dinosaur in the Niobrara chalk. In Marsh's words, the specimen "was the greater part of a skeleton of a small *Hadrosaurus*, discovered by the writer in the blue Cretaceous shale near the Smoky Hill River, in Western Kansas" (1872:301). The specimen was nearly complete except for the skull. Although the exact locality is unknown, the dinosaur remains (YPM 1190) were probably collected 5 or 6 miles east of Russell Springs in Logan County according to Carpenter, Dilkes, and Weishampel (1995:289). Records of the Yale Peabody Museum indicate that the specimen was collected on July 30, 1871, on the north side of the Smoky Hill River, 26 miles east of Fort Wallace. Marsh initially called the new dinosaur *Hadrosaurus agilis* and indicated that it "would be fully described in this Journal [*American Journal of Science*] at an early day" (1872:301).

The specimen was mounted as an exhibit in the Yale Peabody Museum, but Marsh never made good on his promise to fully describe the remains. Some years later, Marsh (1890) did reassign the remains to a new genus, renaming it *Claosaurus agilis*. According to Carpenter, Dilkes, and Weishampel, however, "*C. agilis* has never been adequately described and figured" (1995:290). That said, the wall mount of the first dinosaur discovered in Kansas has remained on exhibit at the Yale Peabody Museum since its discovery. It was amazing to see it on the wall when I visited Yale University in 2010.

It was more than 40 years before the next dinosaur remains were discovered in the Smoky Hill Chalk. In 1905, Charles Sternberg (1909) discovered the remains of what he thought was "a large new sea-tortoise with an ossified carapace." By then Sternberg had collected a number of large marine turtles from the Smoky Hill Chalk (see Chapter 6) and should have been an expert on them. The locality information he provided ("five miles south of Castle Rock, and 3 miles south of Hackberry Creek": Wieland, 1909:250) is unclear (Hackberry Creek is actually north of Castle Rock), but it is likely the remains were collected from the low chalk (late Coniacian in age) of southeastern Gove County. At the time, Sternberg thought the bones were too weathered to bother collecting, but his son, George F., later discovered the same specimen and collected it anyway.

At some point, Charles Sternberg sent two of the "peculiar" pieces he believed to be part of the shell (neurals) of the 'new turtle' to the Yale Museum for study (Fig. 12.1). There, Dr. G. R. Wieland, who was working with fossil turtles at the time, saw the unusual remains included in Sternberg's turtle material. Wieland told Sternberg they were dermal scutes of an armored dinosaur. The Sternbergs went back to the site, collected the rest of the remains and sent at least some of them to Yale, where they were later described by Wieland (1909:250) as *Hierosaurus Sternbergii* (now *H. sternbergi* per Carpenter, Dilkes, and Weishampel, 1995, and a nomen dubium). Although Sternberg (1909) suggested that they had collected a large portion of the skeleton and dermal armor, it is unclear what

12.1. Dermal scutes (YPM 1847 and 55419) from the first nodosaur remains discovered in the Smoky Hill Chalk (Wieland, 1909, figs. 1–7). The specimen was discovered by Charles H. Sternberg in 1905 (probably in Gove or Trego counties) and named *Hierosaurus sternbergi* in his honor by Wieland.

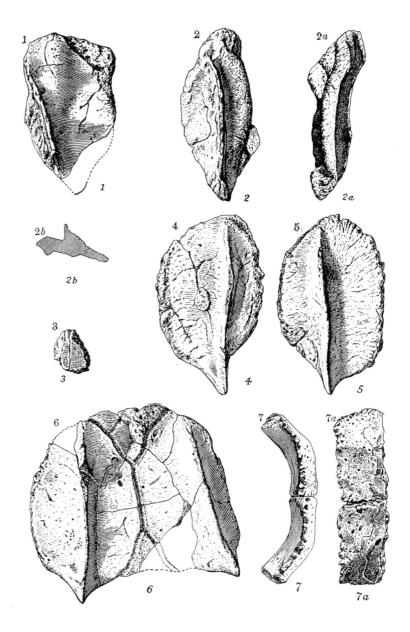

may have happened to a substantial portion of the specimen. According to Carpenter, Dilkes, and Weishampel, about "34 scutes, skull and rib fragments" (1995:276) are all that remain of the skeleton (YPM 1847). To make things even more difficult, we are unsure if Sternberg collected one dinosaur from the chalk or two. Wieland (1909:250) figured two dermal elements from the original material provided by Charles Sternberg in 1907, then acknowledged the receipt of "six dermal scutes" from the "last season [1908]". The Yale Peabody Museum collection has two specimens donated by Charles Sternberg. YPM 1847 is the original specimen, and YPM 55419, curated later, apparently contains the dermal scutes donated the following year. Wieland later reported that "Mr. Sternberg was able to make the needed reexamination of the locality before the frosts of another winter had set in, securing and sending every last remaining fragment" (1911:112).

It is unclear if they represent the same set of remains (Carpenter, Dilkes, and Weishampel, 1995:275). Given the relative rarity of dinosaurs in the Smoky Hill Chalk, however, I think they are most likely from the same individual.

Years passed, and in February 1930, Virgil B. Cole, a geologist doing surface mapping in western Kansas for Gulf Oil, discovered the remains of what he initially thought was a 'baby dinosaur' in southeastern Gove County (roughly 20 miles south of Quinter). A handwritten letter by Cole to Dr. M. G. Mehl at the University of Missouri at Columbia described the discovery and said that he was shipping it (300 pounds of bones and plaster) to Missouri. Cole had been a student of Mehl and received his master's degree from the University of Missouri at Columbia in 1923. The letter and some of Cole's field notes were retained with the specimen. The notes were transcribed and published posthumously as Cole, 2007.

From his letter, it is apparent that Cole initially thought the remains were those of a plesiosaur. However, after excavating the fairly complete articulated right hind limb, he realized that it was a dinosaur. From his letter, he seemed disappointed at first that it wasn't a plesiosaur, but was justifiably proud of his discovery for the rest of his life (Walters, 1986).

After receiving the specimen from Cole and removing it from the plaster that had been poured directly on the bones, Mehl briefly reported on the discovery in an abstract for the Geological Society of America (GSA) meeting in 1931. He then followed with a more complete description (Mehl, 1936), naming the new dinosaur *Hierosaurus coleii* in honor of Virgil Cole. In the title of his paper, Mehl (1936) indicated his belief that *Hierosaurus* was an aquatic dinosaur, possibly to explain why it had been discovered in the middle of the Western Interior Sea, so far from the nearest land.

After being described and named, the specimen more or less sat unnoticed in the collection of the University of Missouri at Columbia until Carpenter, Dilkes, and Weishampel (1995) reexamined the type material and published a revised description. Because of the differences the authors noted between it and the type specimen of *Hierosaurus* at the Yale Peabody Museum, they renamed it as '*Niobrarasaurus coleii*,' a separate genus of nodosaur from *Hierosaurus* (now a nomen dubium, ibid.).

In 1973, J. D. Stewart, who was then a student at the University of Kansas, discovered the badly weathered remains (rib and limb fragments and dermal scutes) of still another Niobrara nodosaur in the middle chalk (Santonian) from Rooks County. This specimen (KUVP 25150) is currently in the Museum of Natural History at the University of Kansas. It was also described by Carpenter, Dilkes, and Weishampel (1995) but is so damaged and incomplete that it could be identified only as Nodosauridae *incertae sedis*. Stewart was only the fourth person to find dinosaur remains in the chalk, and it would be almost 30 years before it happened again.

12.2. The right radius (top) and ulna of a juvenile nodosaur (FHSM VP-13985) collected from the Smoky Hill Chalk in Lane County in 2000 by Shawn Hamm of Wichita, Kansas. There are two parallel bite marks on the distal end of the radius (top right, under number). These bones appear to have been partially digested by a large shark, probably *Cretoxyrhina mantelli*. Scale bar = 10 cm (4 in).

In October 2000, Shawn Hamm, a geology student at Wichita State University in Wichita, Kansas, became the fifth person known to have discovered dinosaur remains in the Smoky Hill Chalk. He was collecting fossils in northeast Lane County when he came across two unusual bones lying on the surface of the chalk near Hattin's marker unit 8 (early Santonian). The bones were quite dense and had been already bleached gray-white by exposure to the sun. Unable to identify them, he showed them to me. I was also puzzled because they were like nothing I had ever seen from the chalk. My first impression was that they were limb bones of a large marine turtle, like *Protostega*, but they didn't look like the figures in the references that I had available. The larger one also appeared to have been crushed, suggesting that it had originally been hollow, something that is not characteristic of turtle bone. I told him that was my best guess, but that I would continue to try to get them properly identified.

A few days later, when I was going through Carpenter, Dilkes, and Weishampel (1995) while looking for information on another subject, I realized that Hamm's strange bones were from a dinosaur. The two bones (Fig. 12.2) were the right radius and ulna from a young nodosaur (ibid.:figs. 10, 12), apparently closely related to *Niobrarasaurus*. When I called Shawn and told him the good news, we decided the discovery was worth reporting at the upcoming Society of Vertebrate Paleontology (SVP) meeting.

In early 2001, after Shawn Hamm and I submitted our abstract (Hamm and Everhart, 2001) for the SVP annual meeting in Bozeman, Montana, I decided we needed to compare the two bones Shawn had discovered (FHSM VP-13985) with those of the type specimen (MU 650 VP). So, after tracking down the type material at the University of Missouri at Columbia, I contacted the geology department in March 2001

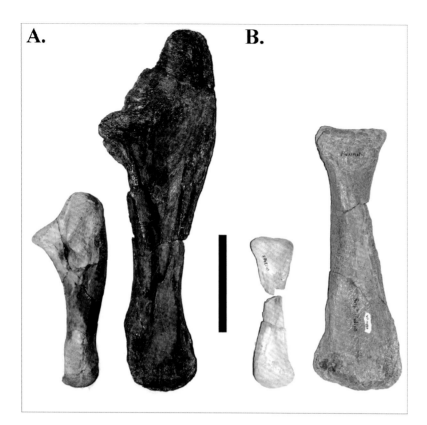

12.3. A comparison between the front limb bones of the juvenile *Niobrarasaurus coleii* (FHSM VP-13985) and the type specimen (FHSM VP-14855): A, ulna; B, radius. Scale bar = 10 cm (4 in).

and requested the right radius and ulna be sent on loan to me in care of the Sternberg Museum. My request was granted, and the right lower limb of *Niobrarasaurus coleii* arrived for study within a month.

Even though VP-13985 represents a much younger individual (about half the length of the comparable elements in the type specimen), their overall characteristics appeared to be very similar to those of the holotype of *Niobrarasaurus* (Fig. 12.3). From the bite marks on the radius and the corroded appearance of both bones, it appeared that the lower portion of the front limb had been ripped off a floating carcass, swallowed, and partially digested by a large shark. In August 2002, when I returned the type material to the University of Missouri, I took a chance and asked if it might be possible that the entire specimen could be transferred to the Sternberg Museum for use in a 'Kansas Dinosaur' exhibit. The answer was an enthusiastic 'yes,' and in December 2002, my wife and I stopped by the University of Missouri at Columbia and picked up the entire specimen. The transfer between museums was duly recorded by Everhart (2004).

While I was packing the specimen for shipment, I discovered an unpublished note from Virgil Cole, written on a scrap of an old cardboard box, that carried an additional bit of information about Cole's discovery (later published as Cole, 2007), some of which was confusing: "*Hierosaurus coleii* more pieces picked up after the rains June 8, 1935, Trego Co. Kan. Horizon 122′ up in Niobrara Chalk V. B. Cole."

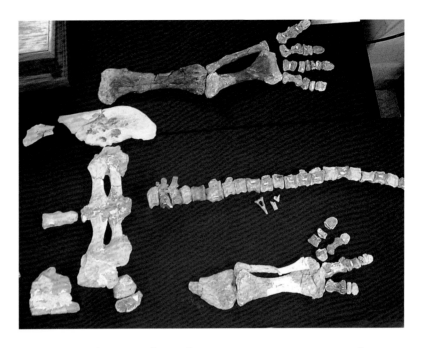

This bit of stratigraphic information was important to me because Cole's 1930 letter to Mehl had originally placed the specimen some 190 feet above the base of the chalk. Cole had also provided information on the section, range, and township where the remains had been collected, and I was very familiar with the area, having collected there for more than 10 years. However, his original determination of the horizon of the specimen had me worried because the whole area was much closer to the base of the chalk than he indicated. This could mean his locality data was wrong, or his estimate of the horizon was wrong, or both. His second determination of the horizon was much closer to the actual distance above the Fort Hays Limestone in his designated section. I was also a little concerned about the "Trego Co." notation since the original locality given by Cole is in Gove County. However, the given locality was also just a few miles to the east of the county line between Gove and Trego Counties. I hoped his change in the name of the county was just an honest mistake.

On May 1, 2003, a media conference was arranged at the Sternberg Museum to announce the acquisition of the type specimen of *Niobrarasaurus coleii* (now officially FHSM VP-14855). I spent the morning arranging a display of the bones of *Niobrarasaurus* on a piece of dark blue cloth and talking to reporters and photographers who had arrived early. It was possibly the first time the specimen had been laid out since it was first prepared (Fig. 12.4), and certainly the first time it had been photographed so extensively.

Then, on May 3, after most of the news stories had played out, I took a group of amateur paleontologists into the field. By coincidence, we were collecting fossils in the same map section (square mile) that had supposedly produced the bones of *Niobrarasaurus* in 1930. The locality also happens to be one of my favorite areas of the chalk and one for which we had access from the property owner. And there begins the rest of this strange and wonderful story . . .

The morning promised a nice day for fossil hunting. It was cool and cloudy, with just a bit too much wind at times. I was leading a group of my friends from the New Jersey Paleontological Society on a field trip in the chalk. When we arrived at the site in southeastern Gove County, just south of the Smoky Hill River, the five of us scattered quickly across the shallow valley. We were all carrying radios, but, surprisingly, there was little of the chatter we usually had when we were having good luck finding fossils. My day started pretty well, finding a nice, though partially digested, mosasaur axis vertebra (shark food), and it ended with the discovery of my first *Martinichthys* rostrum (a rare plethodid fish—see Chapter 5) since 1996. About 5:00 p.m., just as I began to walk back to the van to leave for the day, Tom Caggiano called me on the radio and told me he had picked up some unusual bones.

After making a joke on the radio about the 'Bovinasaurus' (cow bones) I had seen in the pasture I could see him walking through, I asked him what they were. Tom said he didn't know. When I met up with him, he showed me four bones (three metacarpals and a terminal phalanx), bleached gray-white by exposure to the Kansas sun. I recognized them immediately as dinosaur bones, just like the ones I had been arranging for the *Niobrarasaurus* news conference at the Sternberg two days earlier.

"Great!" I thought to myself. Tom had discovered another Smoky Hill Chalk dinosaur—only the sixth person to ever do so (after O. C. Marsh, Charles Sternberg, Virgil Cole, J. D. Stewart, and Shawn Hamm). We were both excited about the discovery as we walked back to the site where he had discovered the bones.

Tom had seen the terminal phalanx (end of the toe) lying on the side wall of a 7-foot-deep gully that he was walking in, at about eye level. When he had climbed up to see where it came from, he saw the three metacarpals lying close together near the edge. Then, as I examined the bones again, I noticed that all three of the metacarpals still had bits of white plaster clinging to them. I suddenly realized that this wasn't a "new" dinosaur discovery at all, but rather the lingering remains of Cole's original *Niobrarasaurus* specimen. By sheer chance, Tom had located the site where Cole discovered the holotype material. To me, this was even more important than a new dinosaur because it provided accurate stratigraphic information on the original specimen. When I looked around, I determined that the remains occurred below Hattin's (1982) marker unit 3. That meant it was located about 85 feet above the contact with the underlying Fort Hays Limestone, not far from Cole's estimate, and was definitely late Coniacian in age. Tom's discovery also verified what I had concluded from their localities about how old (Lower Coniacian) most of these nodosaurs were.

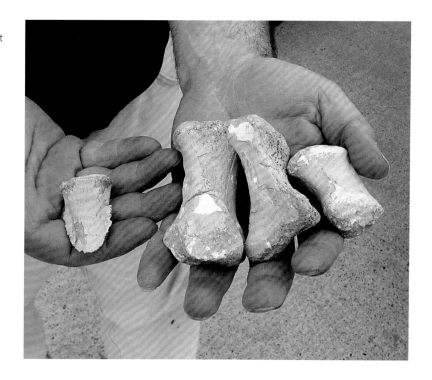

Of course, it also added four of the missing bones (Fig. 12.5) of the right front foot noted by Cole in his letter to Dr. Mehl: "Some of the joints of the toes are missing but I am sure they are present" (Cole 2007:132). As best I could determine from the description of the dig in his letter, Cole had poured plaster on everything (literally, directly on the bones) and then probably covered the right front foot with rock and dirt as he dug around to remove the larger bones of the right front limb. This pile of chalk and the plaster cap protected the bones for many years before it finally weathered away, and it enabled them to survive in place until they were discovered. It was simply a matter of having a lot of good luck, having good eyes, and being in the right place at the right time on the part of Tom Caggiano.

We got the rest of the group together and did a thorough search of the area. Although we saw bits of plaster still clinging to some inoceramid shell fragments, we were unable to locate any additional remains of the *Niobrarasaurus*. If anything else had been left on the site by Cole, it had probably washed down the nearby gully long ago. Another heavy rain and the four little bones discovered that day by Tom would have suffered the same fate.

And there you have the rest of the story of *Niobrarasaurus coleii*, newly arrived back in Kansas, where it had spent most of the last 86 million years and more recently was reunited with part of its right front foot.

Although the type specimen of *Niobrarasaurus coleii* does not include a complete skull, it does include elements of the skull, including a jaw fragment and a tooth (Figure 12.6). Using the fragments of the skull, Ken Carpenter was able to construct a reasonable facsimile of

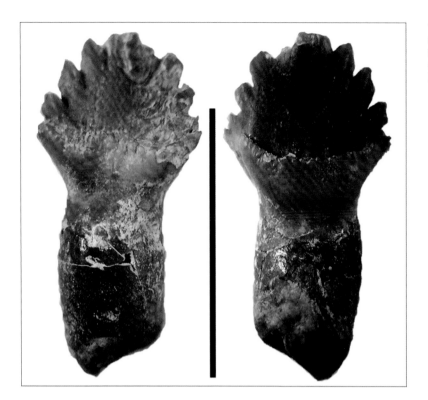

the original skull based on other specimens. We reported this along with the recovered bones of the foot and the new stratigraphic information in Carpenter and Everhart (2007).

In June 2005, just as the first edition of this book was being published, I was spending a day in the chalk at my favorite locality with Keith Ewell. We weren't finding much and had decided to move to another site. As Keith came up to the van, he was carrying a large Ziploc bag containing vertebrae. I asked him what he had collected; he said he wasn't sure but thought it was part of a *Xiphactinus*. When I looked at the vertebrae, I knew they didn't come from a fish. Yes, they were cupped on both ends like *Xiphactinus* vertebrae, but they were much heavier, and included attached dorsal processes. My first thought was that they were the cervical vertebrae from an elasmosaur, but I was wrong, too. When I sent photos of them to Ken Carpenter the next day, he came back right away with the correct ID: hadrosaur caudal vertebrae. Based on size, they had come from near the end of the tail. Keith had discovered what turned out to be only the second set of remains of a hadrosaur ever collected from the Smoky Hill Chalk.

The new dinosaur specimen (FHSM VP-15824) came from the lower Smoky Hill Chalk in southeastern Gove County and consists of nine articulated caudal (tail) vertebrae from an adult hadrosaur (credit again for the ID goes to Ken Carpenter). The vertebrae were lying on their right sides. The two most anterior vertebrae of the series had partially eroded

Another Hadrosaur Specimen

12.7. Nine caudal vertebrae from a hadrosaur (FHSM VP-15824) that had been scavenged by sharks. The specimen was collected from the lower chalk in Gove County. Scale bar = 10 cm (4 in).

out and were whitened from exposure on the surface of the chalk. The other seven vertebrae were still completely enclosed in the matrix. The specimen is 55 cm (22 in) in length, and is from the distal part of the tail (Fig. 12.7). Scaled up, it represents an adult animal that was about 10 m (33 ft) in length. There are deep, nonserrated bite marks across both sides of the last vertebra, suggesting that a big shark had used a head-shaking bite to shear off the end of the tail. After that, another shark literally had swum up the tail and bitten off the piece containing the nine vertebrae. The bone surfaces at both ends of the articulated series are severely eroded and appear to have been partially digested. Consistent with many other specimens of large vertebrates (mosasaurs, plesiosaurs, turtles, etc.) from this time period during the deposition of the Smoky Hill Chalk, these remains appear to have been consumed by a large shark, partially digested, and then regurgitated while they were still covered with connective tissue. The thought that stays with me in regard to this specimen is that the tail did not get out there in the middle of the chalk by itself. Is the rest of that hadrosaur buried somewhere not too far away? It may have eroded out 10,000 years ago or still be buried in the chalk and covered with prairie sod.

Around the same time, two other isolated nodosaur caudal vertebrae were discovered in the lower chalk, bits of dinosaur carcasses that had been eaten by sharks and regurgitated. The latest specimen from the chalk, the partial sacrum and dorsal vertebrae of another nodosaur (FHSM VP-18524; Buskuskie, 2015), was collected by Mike Urban from the low chalk of Trego County. The bones appear to be from a smaller individual than is represented by the type specimen. There is no evidence of scavenging visible on these remains. They appear to be from a floating carcass that just fell apart and scattered across the sea floor.

One other dinosaur specimen deserves an honorable mention because it actually 'lived' in Kansas at the edge of the sea about 10 million years before the deposition of the Smoky Hill Chalk. During Cenomanian time, the Western Interior Sea was still spreading across

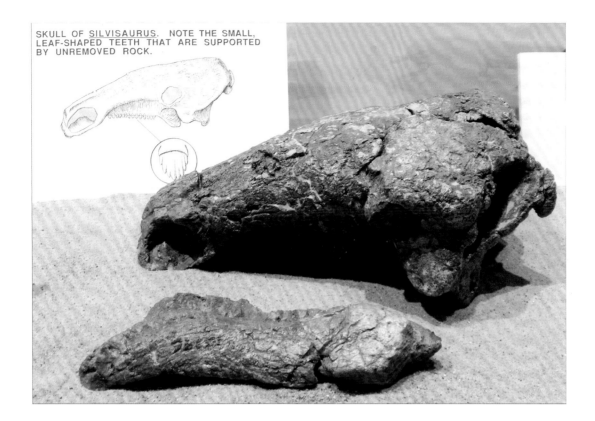

SKULL OF SILVISAURUS. NOTE THE SMALL, LEAF-SHAPED TEETH THAT ARE SUPPORTED BY UNREMOVED ROCK.

12.8. Skull and lower jaw of the type specimen of *Silvasaurus condrayi* (KUVP 10296) discovered in the Dakota Formation of Ottawa County, Kansas. The skull is about 33 cm (13 in) in length.

the middle of North America at the same time as a large river flowing from the northeast was depositing a huge delta of sand westward across the middle of Kansas. This river delta would eventually become the Dakota Formation, a huge deposit of sand and clay that stretches across (and under) the western half of Kansas. The north central portion of the state was still just above sea level and hosted a heavily forested tangle of islands and smaller braided streams where the river met the ocean. The Dakota Formation is well known for the thousands of leaf imprints that have been collected from the sandstone (Lesquereux, 1891; Sternberg, 1909; Everhart, 2015). However, the Dakota is not well known for vertebrate remains.

In 1955, Warren Condray, a landowner in Ottawa County, discovered bones eroding from a stream channel on his property. As the story goes (Eaton, 1960), Condray contacted his U.S. senator, who called the Museum of Natural History at the University of Kansas. The specimen was collected from the Terra Cotta Clay Member of the Dakota by the then curator of vertebrate paleontology, Russell Camp, and students, and subsequently prepared out of the hard ironstone. The remains consisted of the skull and lower jaws (Fig. 12.8), vertebrae, ribs, sacrum, and dermal scutes of an unknown nodosaurian dinosaur. The limbs were missing except for the right femur. The specimen (KUVP 10296) was described by Eaton (1960) and named *Silvasaurus* (forest lizard) *condrayi* in honor of the landowner.

12.9. Ventral view of a natural mold of the sacrum of a nodasaurian dinosaur (FHSM VP-10041) from the Dakota Formation in Russell County on exhibit at the Sternberg Museum of Natural History, Hays, Kansas. No scale, about .3 m (12 in) in length.

The ironstone mold of a probable *Silvasaurus* sacrum (FHSM VP-10041; Fig. 12.9) was also collected from the Dakota Formation at Wilson Lake in Russell County (Liggett, 2005). A similar specimen was reported from the Dakota of Cloud County, Kansas, by Parmenter (1899) as the internal cast of a small turtle. Although the specimen has been lost, after reviewing the published photographs, Liggett (2005:fig. 3) suggested that it was more likely to be the sacrum of an anklyosaurid dinosaur. Note here that the Dakota Formation in Lincoln County also preserves hadrosaur tracks (KUVP 5914 and FHSM VP-138; Lane, 1946; Liggett, 2005).

At least 10 specimens of dinosaur remains have been collected in the Smoky Hill Chalk over the past 130 years, and they are coming more rapidly now that they are being recognized. If Sternberg's *Hierosaurus* remains are counted as one dinosaur, this figures out roughly to be a new discovery about every 15 years. While it is likely that additional remains of dinosaurs will be discovered where they were buried in the middle of the Western Interior Sea, they will never be more than a rare occurrence. They are, however, valuable additions to our knowledge of dinosaurs during the Late Cretaceous. As noted by Carpenter, Dilkes, and Weishampel (1995:275), they represent "the best known assemblage presently available" from this relatively undocumented time interval in the terrestrial fauna of Late Cretaceous of North America.

The Big Picture

13

When the mosasaur died, its body settled to the darkness of the sea bottom, where it lay undisturbed for many months until decomposition of the flesh left nothing but the bones. Over a period of years, the bones were covered with a thin layer of chalky sediment. Chemicals in the watery mud began to enter the open spaces within the bones, but mostly the physical makeup of the bones and teeth remained unchanged. Many years passed and more chalky mud piled up, slowly becoming heavier and pressing against the bones. As the sediment deepened, the weight began to crush the bones, flattening the skeleton slightly and forcing out much of the surrounding water. The mud began to compact, becoming denser as the increasing mass of the overlying sediment crushed it. Eventually there would be several hundred feet of additional chalk over the bones and then several thousand feet of shale on top of that, layers of rock that faithfully recorded the last days of a vast inland sea.

The world continued to change. Immense forces from far below the marine sediments lifted these layers above the surface of the dying ocean, and slowly heaved them upward hundreds and then thousands of feet above sea level. The waters slowly receded to the north and south, and the surface dried out for the first time in millions of years. Even as this was happening, the wind and rain began tearing the exposed rocks down again, eroding the surface with rivulets, gullies, ravines, channels, and rivers, carrying tons of dislodged sediment, and the fossils it contained, downward and back toward the sea, now hundreds of miles away. At the same time as they were being eroded and washed away, the layers of the ancient sea floor continued to be uplifted as a range of jagged mountains rose to the west. Slowly, over the course of millions of years, layers of shale and chalk weathered away. As the chalk layers thinned over the mosasaur remains, the soft rock was subjected to freezing and thawing, which opened tiny crevices and allowed plant roots to probe deep for water and nutrients. The combined effects of the weather and the growth of vegetation caused the chalk to erode even faster; within a few years, the bones of the mosasaur were exposed to sunlight for the first time in 85 million years.

The Western Interior Sea during the Late Cretaceous extended across the middle of North America from the present-day Gulf of Mexico to the Arctic Ocean. Its width varied considerably through time, depending on changes in sea level and the movement of continents (plate tectonics), as

The Big Picture

well as other planetwide events we don't yet fully understand. My rough estimate of the size of the sea during its maximum expansion near the end of the Turonian (90–89 Ma—million years ago) would be a length of about 5,000 km (3,000 mi) from north to south, and a maximum width of about 1,600 km (1,000 mi) from east to west, across the center of North America. The sea narrowed at the north (Arctic) and south (Gulf) ends, but even so would have covered something in the neighborhood of 1–2 million square miles of what is mostly dry land today. As a comparison, this area would be equivalent to roughly half that of the present day United States. Although relatively small and much shallower in comparison to the Atlantic or Pacific oceans, the Western Interior Sea was still a big place.

In previous chapters, I have described many of the animals whose remains have been discovered in the Smoky Hill Chalk and, to a lesser extent, elsewhere in the Cretaceous marine deposits in Kansas. Most of these animals lived in the Western Interior Sea during a relatively brief (in geologic terms) 5-million-year span of time (87–82 Ma) during the latter half of the Late Cretaceous. Five million years, however, is a long period of time by human standards, and is roughly the same amount of time that humans in some form have existed on this planet. Another way of looking at this time span is to think of it in terms of days—approximately 1.8 BILLION of them. That is a lot of sunrises and sunsets, and is enough time for many changes to occur in the climate and the environment that affected life in the oceans of Earth.

The deposition of the Smoky Hill Chalk occurred some 20 million years before the end the Cretaceous, when sea levels were high and the Western Interior Sea was slowly decreasing in size from what had been its greatest expansion across the middle of North America at the end of the Turonian. For life in the sea, times were relatively good. Some major marine groups were already extinct, including the ichthyosaurs and the giant pliosaurids (most recently, *Megacephalosaurus eulerti* and *Brachauchenius lucasi* that were living during the Turonian over Kansas), while others, including the bony fishes, turtles, mosasaurs, pterosaurs, and birds, were evolving rapidly. We know relatively little about what was going on during the period when the Fort Hays Limestone (early Coniacian) was deposited prior to the Smoky Hill Chalk because only a few fossils have been discovered in this unit so far (Shimada, 1996; Shimada and Everhart, 2003; Everhart, 2005a; Shimada et al., 2009). However, based on what I have seen in the last 30 years, it is likely that the fauna was not too different from that of the chalk. The marine animals living near the middle of the Western Interior Sea were hundreds of miles from the nearest land and relatively isolated from terrestrial influences. The birds, pterosaurs, and egg-laying marine reptiles (turtles) still were connected to the land, but what I consider to be the major players in this ecosystem lived their entire lives at sea. On the following pages, I will describe the changes in the fauna that occurred as this epicontinental sea narrowed and became shallower.

As discussed briefly in Chapter 3, the productivity of the Western Interior Sea, as that of modern oceans and seas, was based on the abundance of microscopic single-celled plants (algae) that used photosynthesis to capture the energy in sunlight and convert it into biomass for growth and reproduction. In the process, these cells (called coccolithophores) also released oxygen and fixed carbon dioxide as calcium carbonate from the seawater to construct tiny wheel-like structures (coccoliths) that they used to cover themselves. Besides being the major food source for microscopic and macroscopic consumers at the base of the food chain, vast quantities of their coccoliths settled to the sea floor over that same 5-million-year period to form the calcareous ooze that eventually become the Smoky Hill Chalk. It was this chalk that faithfully preserved the wealth of vertebrate remains that we find today.

By the early part of the Campanian (about 82 Ma), the Western Interior Sea had decreased in size (width and depth) to the point that what is now western Kansas was much closer to land, and deposition of the Pierre Shale began as mud from the debris that was being eroded from nearby land, especially from the Rocky Mountains rising to the west. However, life continued to flourish in the sea at the time of this transition, and relatively little change has been noted in the fauna (Carpenter, 1990, 2003).

The 5-million-year deposition of the Smoky Hill Chalk may be broken up into shorter periods of time that are defined here as the end of the late Coniacian (a period of about 1.2 million years), all of the Santonian (a period of about 2.3 million years), and the beginning of the Campanian (a period of about 1.5 million years). The fauna living in the sea during that time changed slowly as new species evolved and others became extinct or retreated to other places because of changing environmental conditions.

Cope and Marsh, who initially described most of the vertebrate fauna from the chalk, were mainly interested in discovering new species, and bigger and better specimens, and apparently paid little attention to the stratigraphy. To some degree, this was because of a general lack of knowledge regarding the geology of Kansas, but it is also apparent that knowing exactly where a specimen came from within a geologic formation wasn't considered to be important at the time. In his fictional book, *Buffalo Land*, Webb (1872) provided one of the first illustrations of a reconstructed fauna from the Smoky Hill Chalk. The print (Fig. 13.1) is attributed to a popular Western artist of the time (Henry Worrall), but it is readily apparent that his work was influenced by Webb's contacts with E. D. Cope (Davidson, 2003; Everhart, 2016). It is also apparent from the association of species illustrated in the fanciful scene that it is representative of fossils collected at the time from the upper chalk (early Campanian). *Elasmosaurus*, *Liodon* (*Tylosaurus*) *proriger*, *Polycotylus*, *Protostega*, and the two species of *Pteranodon* shown were all first described and named by Cope. While *Elasmosaurus* is actually from the Pierre Shale, the formation was not recognized as being distinct from the Niobrara Chalk at the time *Buffalo Land* was published.

THE SEA WHICH ONCE COVERED THE PLAINS.

Elasmosaurus platyurus.   2. Liodon proriger.   3, 4, 5. Ornithochirus umbrosus.   6. Ornithochirus harpyia.
                7. Protostega jigas.   8. Polycotylus latipinnis.

13.1. One of the first illustrations depicting several species of marine reptiles and pterosaurs from the Western Interior Sea (by Henry Worrall, from *Buffalo Land*, W. E. Webb, 1872). The caption reads: "The sea which once covered the plains. 1. *Elasmosaurus platyurus* 2. *Liodon* [*Tylosaurus*] *proriger* 3, 4, 5. *Ornithochirus umbrosus* 6. *Ornithochirus harpyia* 7. *Protostega jigas* [*gigas*] 8. *Polycotylus latipinnis*." Note that the pterosaur (4) is shown with teeth (see Chapter 10).

As noted in Chapter 2, Logan (1897) was the first to refer to the chalk as the 'Pteranodon Beds.' Williston (1897, 1898b) then informally divided the Pteranodon Beds into the lower Rudistes Beds and upper Hesperornis Beds and was among the first to report that some species occurred at different stratigraphic levels within the formation. Russell (1988) listed marine vertebrates from the Cretaceous of North America, including the fauna of the Kansas Niobrara, by their geologic and stratigraphic occurrence. Stewart (1990a) provided a comprehensive listing of species collected from the Smoky Hill Chalk by their occurrence in six biostratigraphic zones based on Hattin's (1982) earlier work, and implored others to provide accurate stratigraphic information with newly discovered specimens. Bennett (2000) inferred the stratigraphic occurrence of *Pteranodon* specimens from the localities where they had been collected, and Everhart (2001) further refined the ranges of eight mosasaur species from new material. Other authors have documented recent discoveries of various species in the chalk (see previous chapters) and improved our knowledge of the life in the oceans over Kansas during the Late Cretaceous.

Table 13.1 provides a summary of most of the species that have become known from the Smoky Hill Chalk of Kansas since 1868. It is based on my collecting experience as well as the previous work by Hattin (1982), Russell (1988), Stewart (1990a), Carpenter (1990, 2003), Shimada and Fielitz (2006), and many others. I have reduced the six biostratigraphic zones of Stewart (1990a) to the three time periods mentioned above: late

Late Coniacian	Santonian	Early Campanian
87–85.8 Ma (1.2 m.y.)	85.8–83.5 Ma (2.3 m.y.)	83.5–82 Ma (1.5 m.y.)
Invertebrates	Invertebrates	Invertebrates
Volviceramus grandis		
	Cladoceramus undulatoplicatus	
Platyceramus platinus	*Platyceramus platinus*	*Platyceramus platinus*
Pseudoperna congesta	*Pseudoperna congesta*	*Pseudoperna congesta*
Durania maxima		*Durania maxima*
Tusoteuthis longa	*Tusoteuthis longa*	*Tusoteuthis longa*
	Uintacrinus socialis	
	Clioscaphites vermiformis	
	Clioscaphites choteauensis	
	Heteromorph ammonite	
Spinaptychus (2 new species)	*Spinaptychus sternbergi*	*Rugaptychus* sp.
Unidentified ammonite (cast)	*Baculites* sp. (molds)	*Baculites* sp. (molds)
Actinocamax sp.	Belemnites	Belemnites
Sharks / Rays	Sharks / Rays	Sharks / Rays
Ptychodus rugosus	*Ptychodus mortoni*	
Ptychodus mortoni		
Ptychodus martini		
Ptychodus latissimus		
Ptychodus marginalis		
Rhinobatos incertus	*Rhinobatos incertus*	*Rhinobatos incertus*
Scapanorhynchus raphiodon		
Johnlongia sp.		
Pseudocorax laevis	*Pseudocorax laevis*	
Squalicorax microserratodon		
Squalicorax falcatus	*Squalicorax falcatus*	
	Squalicorax kaupi	*Squalicorax kaupi*
		Squalicorax pristodontus
Cretalamna appendiculata	*Cretalamna appendiculata*	*Cretalamna appendiculata*
Cretalamna cwelli		*Cretalamna hattini*
Cretoxyrhina mantelli	*Cretoxyrhina mantelli*	*Cretoxyrhina mantelli*
Bony Fish	Bony Fish	Bony Fish
Asarotus arcanus		
Micropycnodon kansasensis	*Micropycnodon kansasensis*	*Hadrodus marshi*
	Unidentified pycnodont	*Anomoeodus* sp.
Lepisosteus sp.		
Protosphyraena nitida		
Protosphyraena perniciosa		*Protosphyraena perniciosa* (?)
	Protosphyraena tenuis	*Protosphyraena tenuis*
Bonnerichthys gladius	*Bonnerichthys gladius*	*Bonnerichthys gladius*

Table 13.1 A partial listing of invertebrate and vertebrate species from the Smoky Hill Chalk by stage. Adapted from lists by Hattin (1982); Russell (1988); Stewart (1990, pers. comm., 1992); Carpenter (1990, 2003); Shimada and Fielitz, 2006; other individual papers; and my personal observations.

Table 13.1 Continued . . .

Paraliodesmus guadagnii	*Paraliodesmus guadagnii*	
Xiphactinus audax	*Xiphactinus audax*	*Xiphactinus audax*
Ichthyodectes ctenodon	*Ichthyodectes ctenodon*	*Ichthyodectes ctenodon*
Gillicus arcuatus	*Gillicus arcuatus*	*Gillicus arcuatus*
Saurodon leanus	*Saurodon leanus*	
	Prosaurodon pygmaeus	*Saurocephalus lanciformis*
	Urenchelys abditus (eel)	*Leptecodon* sp.
	Apateodus sp.	*Apateodus busseni*
Apsopelix anglicus	*Apsopelix anglicus*	*Apsopelix anglicus*
Pachyrhizodus minimus	*Pachyrhizodus minimus*	*Pachyrhizodus minimus*
Pachyrhizodus caninus	*Pachyrhizodus caninus*	*Pachyrhizodus caninus*
Pachyrhizodus leptopsis	*Pachyrhizodus leptopsis*	
Unidentified albulid	Unidentified albulid	
Cimolichthys nepaholica	*Cimolichthys nepaholica*	*Cimolichthys nepaholica*
Enchodus shumardi	*Enchodus shumardi*	*Enchodus shumardi*
Enchodus gladiolus	*Enchodus gladiolus*	*Enchodus gladiolus*
Enchodus petrosus	*Enchodus petrosus*	*Enchodus petrosus*
Enchodus dirus	*Enchodus dirus*	
Stratodus apicalis	*Stratodus apicalis*	*Stratodus apicalis*
Thryptodus zitteli		
Martinichthys ziphioides		
Martinichthys brevis		
'Other plethodids'	*Niobrara encarsia*	*Pentanogmius evolutus*
	Zanclites xenurus	
	Caproberyx sp.	
	Trachichthyoides sp.	
Various holocentrids	Various holocentrids	*Kansius sternbergi*
Unidentified coelacanth	*Megalocoelacanthus dobiei*	*Megalocoelacanthus dobiei*
		Aethocephalichthys hyainarhinos
Turtles	Turtles	Turtles
Toxochelys latiremis	*Toxochelys latiremis*	*Toxochelys latiremis*
Porthochelys laticeps	*Ctenochelys stenopora*	*Bothremys barberi*
Chelosphargis advena	*Protostega gigas*	*Protostega gigas*
Plesiosaurs	Plesiosaurs	Plesiosaurs
Unidentified *polycotylid*	Unidentified *polycotylid*	*Polycotylus latipinnis*
		Dolichorhynchops osborni
	Styxosaurus snowii (?)	*Styxosaurus snowii*
		'Elasmosaurus' sternbergi
		Unidentified elasmosaur
Mosasaurs	Mosasaurs	Mosasaurs
Clidastes liodontus	*Clidastes liodontus*	*Clidastes propython*
Plesioplatecarpus planifrons	*Plesioplatecarpus planifrons*	
	Platecarpus tympaniticus	*Platecarpus tympaniticus*
Tylosaurus kansasensis	*Selmasaurus johnsoni*	

Table 13.1 Continued . . .

Tylosaurus nepaeolicus	Tylosaurus proriger	Tylosaurus proriger
	Tylosaurus nepaeolicus	
	Ectenosaurus clidastoides	Eonatator sternbergi
Pterosaurs	Pterosaurs	Pterosaurs
Pteranodon sternbergi	Pteranodon longiceps	Pteranodon longiceps
	Nyctosaurus gracilis	Nyctosaurus gracilis
Birds	Birds	Birds
Ichthyornis dispar	Ichthyornis dispar	Ichthyornis dispar
		Guildavis tener
		Apatornis celer
		Iaceornis marshi
		Baptornis advenus
		Parahesperornis alexi
		Hesperornis regalis
		Fumicollis hoffmani
Dinosaurs	Dinosaurs	Dinosaurs
Niobrarasaurus coleii	Niobrarasaurus coleii	Claosaurus agilis
'Hierosaurus sternbergi'	Nodosauridae incertae sedis	
	Hadrosauridae incertae sedis	

Coniacian, Santonian, and early Campanian. This simplifies the data and is reasonably accurate for most species. I will go into more detail on some of the issues regarding the occurrence of the various species in the discussion that follows.

Late Coniacian

During the late Coniacian (87.0–85.8 Ma), the middle portion of the Western Interior Sea was probably at its deepest, around 200 m (650 ft) (Hattin, 1982). This is not very deep compared to what we know about modern oceans, but it was certainly a much different habitat than present at the time closer to the eastern and western shores or along the Gulf of Mexico. It was deep enough, however, that very little, if any, light reached the bottom. Hattin (ibid.) also noted that the water just above the sea bottom was generally poorly circulated and may have been anoxic (with low or no oxygen) at times. In any case, the sea floor was not a very hospitable environment for most invertebrates and fishes, and would certainly not have supported the number of species present in shallower, better-circulated, and well-oxygenated waters. Without much competition for resources and with few predators, the inoceramid clams that could live there grew to become very large and abundant, feeding on the continuous rain of organic debris from overhead. There were relatively few other vertebrates living or feeding on the sea bottom, and we know little about most of them.

Volviceramus grandis was the largest of the clams in the lower chalk, reaching the size and general shape of a bathroom basin. The shells of

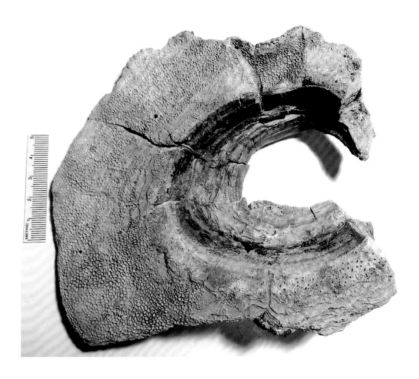

a single species of rudist clam (*Durania maxima*) usually occurred as solitary individuals (Fig. 13.2), but they sometimes occur in groups that have grown together in a small reef-like mass (see Fig. 3.2). The outer surfaces of most shells (and any other solid surface) were usually covered with oysters (*Pseudoperna congesta*), with new generations of these oysters building on top of the older shells. Even the bones of dead animals were sometimes briefly colonized after they were defleshed and before they were covered by the chalky ooze. It was not a good place to attach, however, because the bones were covered by sediments within the space of a couple of years and any oysters that were living there were smothered while they were relatively young. Suitable sites for attachment were at a premium, and only the inoceramids and the rudists evolved a way to keep at least part of their shells above the soft and oxygen-poor mud bottom.

Ammonites were probably fairly numerous in the water column, but because their aragonite shells were usually dissolved before they could be preserved as fossils, we know little about them. The only evidence of their presence is a few partial external casts of their shells and the delicate aptychi (a paired jaw structure; e.g., *Spinaptychus* n. sp.; Fig. 3.9), which were not dissolved because they were composed of calcite. A single species of giant squid (*Tusoteuthis longa*) probably competed with the ammonites for prey and may have grown as large as 9 m (30 ft) in length. The 2002 discovery of a 1.9 m (6 ft) gladius from a *Tusoteuthis longa* in the Pierre Shale of North Dakota certainly supports the existence of a giant squid (Hoganson, 2009). It is important to note that unlike many modern giant squids (e.g., *Architeuthis*), the Late

13.3. Close-up ventral view of the head and teeth of a *Cretoxyrhina mantelli* 'mummy' in the collection of the Sternberg Museum of Natural History. The specimen (FHSM VP-2187) is about 4.6 m (15 ft) in length, and preserves the calcified cartilage of the cranium and jaws, including a complete dentition, as well as most of the calcified centra in the vertebral column.

Cretaceous *Tusoteuthis* comes from the order Vampyromorphida and most likely did not have extremely long arms and the body was a larger portion of its total length. However, since those soft body parts are not preserved in the chalk, we are can only estimate the total length. While it is interesting to think of these giant squid attacking mosasaurs and visa versa, no evidence exists in either case. More than likely, smaller squid were an abundant food source for mosasaurs and many other predators (e.g., *Cimolichthys*, Chapter 5).

During the early deposition of the chalk (late Coniacian) we see the greatest variety of shark species, including several species of shell crushers (ptychodontids), and it appears that the giant ginsu shark, *Cretoxyrhina mantelli*, had reached its maximum size and abundance, based on the number and size of the shed teeth that we find. Almost all of the *Cretoxyrhina* 'shark mummies' that I am aware of come from the lower chalk (Fig. 13.3). In these rare cases (Chapter 4), a major portion of the body of the shark, including the calcified cartilage of the braincase, jaws, fins, and vertebrae, complete sets of teeth, and dermal denticles (shagreen) are preserved. *Cretalamna appendiculata* and *Scapanorhynchus raphiodon* teeth are collected relatively rarely in the low chalk, but *Squalicorax falcatus* seems to be the most abundant, represented by shed teeth (an estimated two-thirds of all teeth in collections from the Smoky Hill Chalk), bite marks on the bones of many vertebrate specimens, and occasional well-preserved 'mummies.'

The ptychodontid sharks (*Ptychodus mortoni*, etc.) are relatively abundant during this period, at the same time that the bowl-shaped inoceramid *Volviceramus grandis* is the most common bivalve. Given their crushing dentitions and the wear on their teeth, it seems apparent

13.4. This relatively large (0.5 m; 20 in) and complete specimen of *Apsopelix anglicus* is in the fossil fishes exhibit at the Sternberg Museum of Natural History. Many specimens of this species are less than 0.3 m (12 in) in length. Scale bar = 5 cm (2 in).

Apsopelix
Logan County, Kansas

that the ptychodontids were feeding on hard-shelled prey on the bottom of the sea, although preying on ammonites and other cephalopods closer to the surface cannot be ruled out. Patches of ground-up small inoceramid prisms that occur in the chalk could be interpreted as coprolites from something feeding on young inoceramids, but the producer of these fragmented shell masses is unknown.

Two genera of bony fishes (*Martinichthys* and *Thryptodus*) that were possibly feeding on bivalves, or the oysters growing on them, were also fairly common at this time, and then disappeared as *Volviceramus grandis* became extinct. Little is known about these odd fishes with their 'battering-ram' noses (e.g., Figs. 5.19, 5.21, and 5.22), but we do have hundreds of small coprolites composed of ground-up oyster shells (Fig. 5.20) at the same stratigraphic interval where most of the *Martinichthys* specimens occur.

Relatively few complete specimens of small fishes are preserved from the late Coniacian chalk, although tiny fish bones, teeth, and vertebrae occur occasionally in coprolites, as well as in acid-washed samples of chalk. The tiny, distinctive jaws of *Enchodus shumardi* with their long front fangs are seen occasionally when layers of chalk are split apart. They represent minnow-size fishes that would have been less than 15 cm (6 in) in length. *Apsopelix anglicus* (Fig. 13.4) and *Pachyrhizodus minimus* (Fig. 5.8) are the smallest fishes for which we have complete specimens from the chalk (excluding the tiny fishes that are preserved inside clam shells), ranging from about 0.3 m (12 in) to 1 m (40 in) in length. For unknown reasons, these small fishes seem to preserve far better than the larger individuals of other species.

Occasionally the fragmentary remains of other small fishes have been discovered in the lower chalk. The distinctive nipping teeth and crushing jaw plates of a small pycnodont (*Micropycnodon kansasensis*; Fig. 13.5) occur occasionally and are collected only in the lower chalk and underlying Fort Hays Limestone. It is likely that their feeding habits were similar to those of modern parrot fishes. The discovery of a late Coniacian coprolite containing the bones of *M. kansasensis* and prisms

13.5. The second specimen of *Micropycnodon kansasensis* (KUVP 7030) from Trego County, Kansas. Anterior is to the left where the mouth is open and several of the jaw plates are visible. Scale bar = 10 cm (4 in).

of small inoceramid shells raises the question of what these little fishes were feeding on (Everhart, 2007a). Pycnodont remains are usually associated with shallow-water deposits, from nearer to shore. While the existing specimens suggest that *Micropycnodon* was feeding on the epifauna that was attached to the large inoceramids on the sea bottom, Earl Manning (pers. comm., 2004) suggested it is more likely that these remains were from shallow-water fishes that strayed into the middle of the deep sea and died there.

Another fish, a small coelacanth (LACM 131958; Chapter 5) that may have made a 'wrong turn' and ended up in deep water, where it died, was discovered by my wife in 1990 near marker unit 4 in Gove County. At the time of collection, it was the only coelacanth known from the Smoky Hill Chalk. Since then, two specimens of the giant coelacanth, *Megalocoelacanthus dobiei* (AMNH FF20267 and FHSM VP-18758), have been collected from the Smoky Hill Chalk (Dutel et al., 2012; Everhart and Maltese, in press). The AMNH specimen was collected from the early Santonian while the FHSM specimen comes from the early Campanian. It now seems clear that coelacanths were a normal part of the Smoky Hill Chalk fauna, if relatively rare.

Among the medium-size bony fish, *Enchodus petrosus, Cimolichthys nepaholica, Saurodon leanus, Gillicus arcuatus*, and several species of *Protosphyraena* appear to be the most common lower chalk varieties. *Enchodus*, with its oversized palatine and anterior dentary fangs, occurs from the beginning of the Late Cretaceous and, unlike most Late Cretaceous fishes, apparently survived for a time after the end of the Cretaceous. Its close relative, *Cimolichthys*, lived through the deposition of the chalk and into the Pierre Shale (middle Campanian) and has been

13.6. Partially prepared skull of *Saurodon leanus* (FHSM VP-18690) from the Upper Coniacian chalk of Gove County, collected in 2013 by Matthew Everhart and the author. Scale bar = 10 cm (4 in).

collected occasionally with the preserved remains of its last meal, including *Enchodus*, or in one case a squid (*Tusoteuthis longa*).

Saurocephalids were long, eel-like fishes that were unusual for the swordlike beak (predentary) attached to the front of the lower jaw. They were represented by three genera in the Smoky Hill Chalk: *Saurodon leanus* in the lower chalk, *Saurocephalus lanciformis* in the upper chalk, and *Prosaurodon pygmaeus* more or less in between (Stewart, 1999). Note that S. lanciformis was the first fossil collected from the Smoky Hill Chalk along the Missouri River by the Lewis and Clark Corps of Discovery (Chapter 5). Since then many specimens have been collected from the chalk, indicating that it was fairly common genus of Late *Cretaceous* fishes. Most recently a complete, articulated skull of S. *leanus* (FHSM VP-18690) was discovered in Gove County in 2013 by my son, Matthew (Everhart and Everhart, 2014; Fig. 13.6).

Protosphyraena, a primitive swordfish, also possessed a billfish-like snout, although in its case, it was attached as would be expected to the upper jaw. In addition, *Protosphyraena* had large, forward-pointing, bladelike teeth, and two species (*P. perniciosa* and *P. tenuis*) had long, sawtoothed fins. Whether this assortment of dangerous-looking attachments were offensive or defensive weapons, or were weapons at all, we do not know. Charles Sternberg wrote a fanciful description of this fish and its predatory habits:

> Notice the head is prolonged in front into a long round bony snout, or ram. On account of this I called it a snout fish when I first discovered their bones in the Kansas chalk. The ram ends, you notice, in a sharp point eight or ten inches long. Then at the end of the mouth are four lance like teeth projecting forward, and outward. The object was for these to cut the breach his ram had made in the quivering flesh of a mosasaur

wider, so he could force his head into the bleeding flesh to the eye rims. But his most terrible weapons are his pectoral fins. See, they are four feet long. Serrated on the cutting or outer edge, enameled and sharp as a knife. (1917:178)

However, it is highly unlikely that *Protosphyraena* attacked mosasaurs, sharks, or other large predators. Although we've never collected a *Protosphyraena* specimen with gut contents, it's probable that this early version of a swordfish fed on small fishes.

Gillicus, on the other hand, was a fairly large fish (up to 2 m or more; up to 6 ft or more) with tiny teeth and no other obvious armament. While its main claim to fame was ending up as preserved stomach contents in a giant fish called *Xiphactinus* (George Sternberg's famous fish-within-a-fish specimen; Fig. 5.1), I have no doubt that it was a serious menace to smaller fishes, and a successful predator in its own right. *Ichthyodectes* was about the same size, a long slender fish with a mouth full of medium-size teeth. We know that *Ichthyodectes* fed on *Enchodus* from a fish-within-a-fish specimen collected by Charlie Sternberg (Everhart, Hageman, and Hoffman, 2010).

The largest of the predatory bony fishes during the late Coniacian were *Xiphactinus audax* and *Pachyrhizodus caninus*. Both of these species had massive jaws and large teeth for capturing prey. A nearly 3 m (9 ft) long *Pachyrhizodus* specimen (FHSM VP-2189) at the Sternberg Museum has a skull and teeth that are nearly as large as those of any *Xiphactinus* that I have ever seen. The body of *Pachyrhizodus*, however, is much shorter and more compact than that of *Xiphactinus*. Both would have been fearsome threats to squid, smaller fishes, and just maybe the occasional baby marine reptile. We know from the number of *Xiphactinus* specimens preserved with large prey inside (Chapter 5) that sometimes things went wrong, and the predator died before the meal could be digested. I am not aware of any specimens of *Pachyrhizodus minimus* that have been discovered with preserved stomach contents, but with a large mouth and sharp, recurved teeth, they probably fed on medium-size fishes like the larger species of *Enchodus*.

The late Coniacian represents a sort of 'coming-out' party for the mosasaurs (Chapter 9). Although the first mosasaur remains in Kansas occur in the lower Middle Turonian Fairport Chalk (Martin and Stewart, 1977; Polcyn et al., 2008; Schumacher, 2011; Everhart and Pearson, 2014), they are certainly few and far between. Within a few million years, however, by the beginning of deposition of the chalk, mosasaurs are the second-most-common vertebrate fossil collected (fish being the most common) in the chalk. Four species occur in the lower chalk: *Clidastes liodontus*, *Plesioplatecarpus planifrons* (see Konishi and Caldwell, 2007, 2011), *Tylosaurus nepaeolicus*, and my recently described *T. kansasensis* (Everhart, 2005b).

Tylosaurus kansasensis is the most abundant mosasaur in terms of the number of specimens occurring in the lower chalk (pers. obs.) and is the smallest of the three species. Based on the identification of a badly scavenged series of mosasaur vertebrae (Everhart, 2004a), *Ectenosaurus* may also have been a rare inhabitant of the Western Interior Sea as early as the end of the Coniacian. Contrary to earlier reports by Williston (1898b) and others, mosasaurs of all ages occur in the lower chalk (Everhart, 2002), ranging from the very young (2 m in length or less; 6.6 ft or less) to full-grown adults (7 m or more; 23 ft or more). As noted in Chapter 9, it appears likely that mosasaurs gave birth to their young in mid-ocean, very much like modern whales.

Actually, the most common mosasaur specimens that are collected in the lower chalk are the pieces (skulls, vertebrae, and limbs) that have been severed by large sharks, swallowed, and partially digested (Chapter 4). We are fairly certain of this scenario because we sometimes find the broken tips of shark teeth embedded in these bones (Fig. 4.1; Shimada, 1997) and know that the only predator capable of doing that kind of bone-severing damage to a mosasaur carcass at the time was the giant ginsu shark, *Cretoxyrhina mantelli* (Fig. 13.3). Although it may have been a bad time for the new guys on the block (mosasaurs) as they spread across the Western Interior Sea and encountered the larger ginsu sharks, other mosasaurs were probably also feeding on baby sharks. While there may not be a cause and effect relationship, within a period of about 5 million years (late Coniacian, 87 Ma, to early Campanian, 82 Ma), mosasaurs were much larger and more numerous, and the giant ginsu shark had become extinct.

Sharks and mosasaurs may have made life difficult not just for each other, but also for the few species of plesiosaurs living in the Western Interior Sea (Chapter 7). Until recently (Everhart, 2003), no remains of any plesiosaurs had been documented from the late Coniacian or Santonian of the Smoky Hill Chalk. The fragmentary, partially digested remains that have been collected in the last 25 years indicate that few plesiosaurs ventured into the middle of the sea, and those that did were likely to have been preyed on by sharks. It is equally possible, however, that the very limited remains of plesiosaurs (25+ specimens—probably polycotylid) that have been discovered in this stratigraphic interval are pieces scavenged from carcasses that had floated in from other places. It is worth noting here that the remains of dinosaurs, terrestrial animals that did not live in the ocean, are almost as common as those of plesiosaurs in the lower chalk.

Elasmosaurids were certainly still living around the edges of the Western Interior Sea, closer to shore and a possibly safer habitat, but have not yet been discovered in the chalk deposited during the late Coniacian. The remains of a large shark (KUVP 68979; probably *Cretoxyrhina*) collected by H. T. Martin from the lower chalk near Hackberry Creek in Trego County does provide indirect evidence of the limited presence of elasmosaurids during that time. Moodie (1912) briefly mentioned the incomplete remains of a large shark that contained "many hundreds of greatly abraded, very smooth and polished stones." Based on the size of the 41 calcified vertebral centra (10 cm; 4 in) that were preserved, I estimate that the shark was at least 6 m (20 ft) in length. It apparently died with more than 120 black chert gastroliths as gut contents (Everhart, 2000). The larger stones are about 7 cm (2.6 in) long. Since sharks do not normally have gastroliths and elasmosaurids usually have large numbers of them (ibid.), it appears likely that the shark fed on an elasmosaur sometime before it died (Shimada, 1997). I would not be surprised if the gastroliths were somehow involved in the death of the shark.

Although the remains of an unidentified pterosaur reported from the Cenomanian Greenhorn Limestone represent the earliest occurrence of this group in Kansas (Liggett et al., 1997), the oldest identifiable *Pteranodon* remains occur in the chalk near the end of the Coniacian. The remains are few and fragmentary, and as discussed in Chapter 10, may represent individuals that died during migrations across the sea rather than those who were actively feeding in it. It is also likely that large storms or hurricanes occurred over the Western Interior Sea just as they do in our modern oceans. We do know that the bodies of dinosaurs living along the shores of the Western Interior Sea during this time could be transported hundreds of miles out to sea to the spot where they sank and were buried in the Smoky Hill Chalk. The same process could have happened to the pteranodons and the plesiosaurs mentioned above. The fossil record shows that the remains of pterosaurs appear to occur much more frequently in the upper chalk, when the nearest land would have been much closer.

The only remains of birds discovered in the lower chalk are those of the pigeon-size *Ichthyornis dispar*, and they are generally limited to small fragments that appear to have fallen off floating carcasses as they decomposed or to have been removed by scavengers. In terms of total numbers, such specimens are very rare. Clarke (2004) recently reported that the majority of the *Ichthyornis* bones (mostly from the upper chalk) she examined in the Marsh collection at the Yale Peabody Museum were from the limbs (wings and legs), and interred that many of those bones fell off floating carcasses. Regardless of how the remains ended up in the chalk, it seems unlikely to me that there were large flocks of birds flying over or feeding in the middle of the Western Interior Sea during the late Coniacian.

Santonian

The middle portion of the Smoky Hill Chalk was deposited during the Santonian stage, roughly 85.8–83.5 Ma. From the fossil record, it appears that there was a major species turnover during the early part of the Santonian. Unfortunately, the record for vertebrates is not as complete for the middle of the chalk as it is for the beginning and the end. In part, this may be because of a thick layer of resistant chalk near the middle that effectively tends to cap the underlying layers and results in relatively tall vertical exposures that are more difficult to search and to collect. In any case, we know a bit less about the Santonian vertebrates in the chalk than about those of the late Coniacian or early Campanian

By the beginning of the Santonian, the large bowl-shaped inoceramid, *Volviceramus grandis*, had been replaced by another even larger species (*Cladoceramus undulatoplicatus*). The new species was relatively thin-shelled and had a 'rippled' appearance, not unlike corrugated sheet metal. It shared the sea floor with *Platyceramus platinus* clams that were also somewhat larger than during the late Coniacian. The few remaining rudist clams were gone, but would reappear briefly during the early Campanian. *C. undulatoplicatus*, however, became extinct within the early Santonian, and P. *platinus* continued to get larger. While inoceramids were present almost to the top of the chalk P. *platinus* became less abundant after about the middle of the Santonian.

Following the extinction of *C. undulatoplicatus* sometime before the middle of the Santonian, an unusual crinoid, *Uintacrinus socialis* (Fig. 3.15), made a brief appearance in the Western Interior Sea and around the world. From the remains of this crinoid that have been collected in Kansas, it appears that they almost always lived together in a group. Most of the fossil crinoids (sea lilies) that we are familiar with from the Paleozoic were anchored to the sea bottom by a long stem and holdfasts and were solitary. It appears that *Uintacrinus* was a free-floating or swimming species, much like the few remaining modern crinoids that are mobile. All that is usually visible in the preserved specimens is the central calyx and a mass of tangled arms. It appears likely that they somehow floated at or near the surface with their long arms extending downward to feed. How they died is unknown, but when they died the colonies remained relatively intact on the way to the sea bottom, where dozens of individuals were covered and preserved en masse as a thin layer of limestone.

Two small, coiled scaphitid ammonites occurred in large numbers near the middle of the Santonian: *Clioscaphites vermiformis* and *C. choteauensis* (Fig. 3.10). They may have occurred earlier, but bottom conditions may have not been favorable for their preservation. As with their larger cousins, the ammonites that are represented in the late Coniacian only by their delicate aptychi, the aragonite shells of these smaller cephalopods were dissolved before they could be preserved. However, the process apparently occurred at a slow enough pace that the shape of the outer surface of the shell was retained as a finely detailed mold that is discovered occasionally when layers of chalk are carefully split. Although these scaphitids were probably predators on smaller prey or scavengers, they were also a likely source of food for many of the larger predators, including mosasaurs. A small heteromorph ammonite, not yet named, was also present during the Santonian (Fig. 3.11).

In 2006 I received three unusual bones (Fig. 4.22) that I could not identify from an amateur collector. When I sent pictures to several of my friends in paleontology, the bones were quickly identified (first by Ken Carpenter) as the tooth plates of a chimaeroid fish (FHSM VP-16685), the first known specimen from the Smoky Hill Chalk. Chimaeroids (ratfish) are cartilaginous fishes, but are not sharks. Still living in our modern oceans, they also have a fairly complete fossil record at least back to Permian time. After comparing the tooth plates with other specimens from the East Coast, we were able to identify them as remains of a chimaeroid called *Edaphodon laqueatus* Leidy (Cicimurri, Parris and Everhart, 2008). In the process, we were also able to locate a chimaeroid dorsal fin spine (BMNH P 10343) that had been collected from Gove County by H. T. Martin about 1900 and sent to the British Museum of Natural History. Reported and figured by Stahl (1999), it was otherwise unknown until then to those of us working in the fauna of the Western Interior Sea.

Most of the ptychodontid sharks were already extinct in the Western Interior Sea by the early Santonian. One species, *Ptychodus mortoni*, survived for a while longer in small numbers. We don't know whether this extinction was due to climate change, loss of their favored prey, or other factors. In 2007, I was able to photograph a beautiful specimen of *Ptychodus mortoni* in a private collection that included the upper and lower jaw plates with teeth that were articulated when collected (Fig. 4.20). In 2010, Steve Mense contacted me about a larger specimen of *P. mortoni* that he had discovered in the middle chalk of Logan County. Together we collected more than 550 teeth, calcified cartilage of the jaws, a few vertebrae, and literally thousands of dermal denticles from the shark's skin. Steve generously donated the specimen to the Sternberg Museum (FHSM VP-17606). Both of these specimens represent some of the last ptychodontid sharks occurring in the Western Interior Sea.

Cretoxyrhina and *Cretalamna* were present, although in somewhat lower numbers as measured by the number of their teeth collected from Santonian-age chalk. *Squalicorax falcatus* appears to have been the most common shark at the time, again measured in terms of shed teeth, and appears to have become slightly larger. Its teeth and serrated bite marks continue to be associated with the remains of other vertebrates, including other sharks (Bourdon and Everhart, 2011), and it seems likely that these sharks were active scavengers. Another species, *S. kaupi*, first appeared in the chalk in small numbers during the early Santonian (pers. obs.) and replaced *S. falcatus* by the early Campanian. The other, smaller varieties of sharks (*Pseudocorax, Johnlongia, Scapanorhynchus*) that have been collected in the late Coniacian do not seem to occur in the Santonian.

Shimada (pers. comm., 2004; Beeson and Shimada, 2004) reported the discovery of the teeth of the guitarfish, *Rhinobatos incertus*, and several other unreported species from an unusual bone bed sample (FHSM VP-644) collected from near Monument Rocks in the middle of the chalk. The discovery was formally reported by Shimada and Fielitz (2006), but like ships passing in the night, I was unaware of their publication when I was researching the occurrence of the genus in Kansas in 2006. I had collected *Rhinobatos* teeth from Late Coniacian (MU 2) and early Campanian (MU 24) chalk samples, and from most of the other Cretaceous rock units in Kansas, but I was unable to recover their teeth from the middle chalk (Everhart, 2007b).

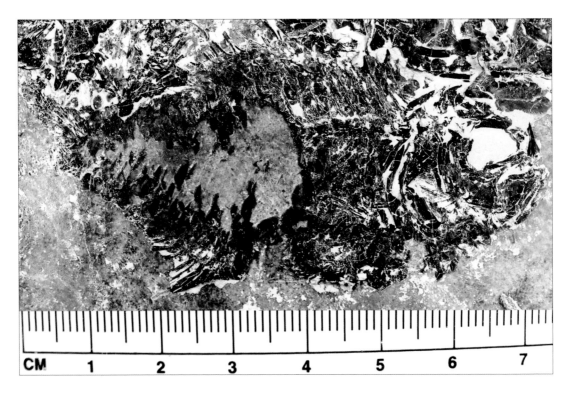

| CM | 1 | 2 | 3 | 4 | 5 | 6 | 7 |

13.7. Close-up in right lateral view (head to right) shows the partial remains of a small (less than 7.6 cm; less than 3 in) holocentrid fish (cf. *Kansius sternbergi*)preserved inside a *Platyceramus platinus* shell from the Smoky Hill Chalk (early Santonian). Scale in mm.

Most of the larger bony fishes occurring in the late Coniacian continued into the Santonian. *Protosphyraena perniciosa* and *P. nitida*, however, appear to have become extinct, or at least extirpated, at the end of the Coniacian, along with the plethodids, *Martinichthys* and *Thryptodus*. The remains of the three ichthyodectid species (*Xiphactinus*, *Ichthyodectes*, and *Gillicus*) are common, although most specimens of *Gillicus* we have discovered were represented by only a detached caudal fin (pers. obs.—the remains of feeding by an unknown predator). The 4 m (13 ft) fish-within-a-fish specimen at the Sternberg Museum comes from the early to middle Santonian, along with a larger (5.2 m; 17 ft) *Xiphactinus* with a partially digested *Gillicus* inside that I discovered in 1996. Specimens of *Pachyrhizodus*, *Cimolichthys*, and *Enchodus* are also relatively common in the middle chalk.

Stewart (1990b; 1990c) reported on the discovery of numerous small (less than 12 cm or 5 in) holocentrid bony fishes (e.g., squirrelfish) that lived and died inside the giant inoceramid clams (*Platyceramus platinus*). Such preservation is apparently not unusual in a relatively narrow interval of chalk near Hattin's (1982) marker unit 8. The bottom conditions must have been favorable to support large schools of these small fishes at the same time as there were many of the giant clams present to shelter them. Many of the fish species (Fig. 13.7) have not yet been formally described, but represent sizes and species not commonly seen anywhere else in the Smoky Hill Chalk. It seems likely that small schools (up to about a hundred individuals of a single species discovered in one specimen—KUVP 49403) of these fishes were living inside the clams and were trapped there when the clams died (Stewart, 1990b). After the death of the clam, the remains of the fishes were effectively sealed inside the closed shell, in a anaerobic environment, and protected from scavengers. Stewart (ibid.) also notes that other small species, including *Kansius sternbergi*, the small amioid *Paraliodesmus guadagnii*, the eel-like dercetid, *Leptecodon rectus*, and a true eel, *Urenchelys abditus*, were discovered preserved inside the shells of *Platyceramus platinus* from the late Coniacian and early Santonian. In 2010, high school student Kris Super discovered a clam shell in Gove County that included a second specimen of *Urenchelys abditus* (FHSM VP-17591; Fig. 5.25) along with several specimens of *Kansius sternbergi*.

Toxochelys latiremis was the most common sea turtle occurring in the chalk through the Santonian and into the overlying early Campanian Pierre Shale. Several other species of small sea turtles, including *Chelospargis advena*, *Prionochelys galeotergum*, *Ctenochelys stenopora* and *Porthochelys laticeps*, lived in the Western Interior Sea (Chapter 6), but their remains are too rare and too fragmentary to say much about their occurrence. The scattered partial remains of smaller sea turtles in the chalk provide evidence, however, that very few, if any, of these turtles died of old age. They appear to have been a favored prey of both *Cretoxyrhina* and *Squalicorax* sharks. It also seems likely that mosasaurs would have eaten them. The 2011 discovery of a *Protostega gigas* turtle in the early Santonian chalk that exhibits more than 90 bite marks from a large mosasaur clearly indicates attempted predation on even larger individuals (Everhart, 2013; Fig. 6.13).

Plesiosaurs appear to be largely absent during most of the Santonian in Kansas. Stewart (1990a) notes the occurrence of a possible polycotylid in the middle Santonian, but I am unaware of other specimens.

Relatively few remains of mosasaurs have been collected from the middle chalk. The earliest known example of *Tylosaurus proriger* (FFHM 1997–10) was reported by Everhart (2001) from the middle Santonian of Gove County (Fig. 13.8). A new species called *Selmasaurus johnsoni* (FHSM VP-13910), a relative of *Ectenosaurus*, was described from the early Santonian of Gove County by Polcyn and Everhart (2008). *Platecarpus tympaniticus* (Konishi et al., 2010) appears to be the most common mosasaur in the upper Santonian; *Tylosaurus proriger* became larger and approached 9 m (30 ft) in length. The much smaller *Clidastes* was rare but still present in small numbers in the relatively deep water of the Western Interior Sea. They were much more common in the near shore environment of the Gulf Coast at that time. The nearly complete specimen of *Ectenosaurus clidastoides* (FHSM VP-401; Fig. 9.5) in the Sternberg Museum, discovered in northwest Trego County in 1963, most likely came from chalk deposited near the end of the Santonian.

The type specimen of the strangely crested *Pteranodon sternbergi* is also from the Santonian chalk. The size of the skull indicates that it came from a large male, with a wingspan of at least 7 m (24 ft). The smaller pterosaur, *Nyctosaurus gracilis*, also appears for the first time in the Santonian. Bennett (2003) reported on the discovery of two unusual

13.8. Dorso-left lateral view of the skull of the earliest known example of *Tylosaurus proriger* (FFHM 1997–10), in the collection of the Fick Fossil and History Museum in Oakley, Kansas. The remains were collected from the early to middle Santonian chalk of Gove County. The articulated skull is 1.2 m (4 ft) in length and represents a mosasaur that would have been about 8.6 m (28 ft) long. The bony sclerotic ring around the left eye and the tympanic membrane on the left quadrate were both preserved intact in this specimen.

Nyctosaurus specimens with very large crests in the Santonian chalk of Trego County (Fig. 13.9). The occurrence of pterosaur specimens becomes much more frequent as you go higher in the chalk as the Western Interior Sea becomes shallower and land becomes closer. While this may mean that there were larger populations of pterosaurs flying over the Western Interior Sea at that time, I think the most likely reason is that land was much closer to where these remains occurred. The question of whether these specimens represent animals that died during routine feeding at sea, or instead constitute losses during migration or storms, is likely to remain unanswered for now. As the sea became narrower, it is also possible that more floating carcasses of pterosaurs could be carried into the middle, although one would also expect to see the remains of more terrestrial animals (dinosaurs) if that was the case. To the contrary, most dinosaur remains occur in the late Coniacian chalk.

Ichthyornis is the only bird represented in collections from the Santonian chalk. Several of these are fairly complete specimens, including the type specimen of *Ichthyornis dispar* (YPM 1450; see Clarke, 2004) and another recently collected set of remains in the Sternberg collection (FHSM VP-2503, Fig. 11.14). Both of these specimens are from the middle chalk of Rooks County and appear to represent remains that dropped to the sea floor relatively soon after death, before their wings and legs had a chance to fall off or be detached by scavengers. A nearly complete specimen of *Ichthyornis* (KUVP 119673) was collected by Greg Winkler and Pete Bussen in the early 1990s. It has since been prepared out of the chalk by David Burnham at the University of Kansas, and reconstructed in 3-D as an impressive skeletal mount (Fig. 11.15). The latest specimen of *Ichthyornis* (FHSM VP-18702) was collected from the middle Santonian chalk near Castle Rock in Gove County by Kris Super.

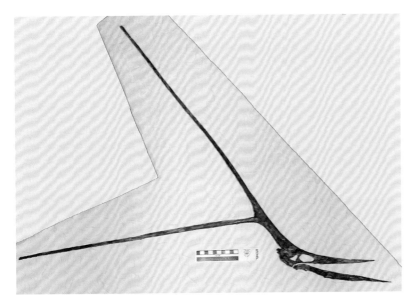

13.9. Cast of the skull in right lateral view of the smaller of two crested *Nyctosaurus gracilis* specimens collected by Kenneth Jenkins in the Santonian chalk of Trego County. The skull measures 24.5 cm (9.6 in), while the crest is almost three times longer (71.7 cm; 28 in). Cast by Kenneth Jenkins (Sternberg Museum of Natural History).

By the early Campanian (83.5–82 Ma), there were very few large invertebrates (notably inoceramid clams) living on the sea bottom where the Smoky Hill Chalk was being deposited. Although less common, *Platyceramus platinus* still served as a substrate for attachment of oysters. *Pseudoperna congesta* appears to be on the decline during the early Campanian and was replaced by other species of oysters. Environmental conditions on the sea floor may have changed to the point that it was no longer suitable even for the giant bivalves. The continued presence of ammonites is indicated by the preservation of their aptychi (*Rugaptychus* sp.), along with external casts of *Baculites* sp. Squid (*Tusoteuthis longa*) remains occur throughout the chalk and into the overlying Pierre Shale. Vertebrate remains (Fig. 13.10) are unusually well preserved (uncrushed and relatively undisturbed) in the early Campanian chalk as compared to preservation in the lower and middle chalk. The incidence of scavenging by sharks on vertebrate remains appears to be much less, based on fewer observed bite marks.

Cretoxyrhina mantelli becomes extinct in the Western Interior Sea sometime before the end of deposition of the chalk. My wife and I have collected their teeth as high in the chalk as Hattin's (1982) marker unit 21. The species has not been discovered in the overlying Pierre Shale, but a smaller lamniform shark, *Cretalamna appendiculata*, continues past the chalk/shale contact. Shimada (2007) described the most complete known specimen of *C. appendiculata* (LACM 128126) from the Lower Campanian chalk of Logan County. This specimen has since been redescribed by Siverson et al. (2015) as the type of a new species, *Cretalamna hattini*.

Squalicorax kaupi replaced *S. falcatus* as the most common shark, and in turn was replaced by *S. pristodontus* later in the Campanian. The only two *S. pristodontus* teeth (FHSM VP-15009 and VP-15010)

13.10. Two dorsal vertebrae from a medium-sized mosasaur (probably *Platecarpus tympaniticus*) that I discovered in the upper chalk (early Campanian) of Logan County. The uncrushed condition is typical of remains that occur in the upper chalk and was the reason that many of the early collectors favored working there rather than lower in the formation. Compare these vertebrae with examples of flattened vertebrae from the late Coniacian (Fig. 13.5). This photograph also shows the characteristic rounded condyle and cupped cotyle of mosasaur vertebrae. Scale in cm.

that I know of from the Smoky Hill Chalk were collected by my friend Pete Bussen from Logan County right below the contact with the Pierre Shale. This species becomes more common and larger in the middle of the Campanian. While S. *pristodontus* may have replaced *Cretoxyrhina mantelli* as the dominant large shark in the Western Interior Sea, it never approached the size or achieved the bone-severing bite of the ginsu shark.

Although many of the more common fish species that occurred in the lower chalk continued through to the overlying Pierre Shale (Carpenter 1990, 2003), their remains are not preserved as frequently in the upper levels of the chalk. A rare species of pycnodont fish, *Anomoeodus* cf. *A. barberi* Hussakof, was described from near the top of the chalk (Shimada and Everhart, 2009) based on a specimen collected by Pete Bussen from western Logan County.

Until recently, *Protostega gigas*, a giant turtle with a body about the size of a small car, seemed to have appeared out of nowhere during the early Campanian. A chance discovery of a much smaller (younger?) specimen of *P. gigas* in the middle Santonian Chalk in 2011 suggests that this turtle was at least a visitor to the middle portion of the Western Interior Sea a million or so years earlier. It is likely that this species migrated from the Gulf coast with changing environmental conditions and decided to stay. My wife discovered the complete left side of the plastron of a subadult *Protostega* (FHSM VP-13448) in the upper chalk near the Santonian–Campanian boundary (MU18) in 1994 (Fig. 6.3). Although there were no bite marks visible on the delicate bones of the plastron, the fact that there were no other remains of the turtle in the immediate area suggests that the turtle's body had been dismembered by scavengers. Both *Cretoxyrhina* and *Squalicorax* are known to have scavenged the remains of *Protostega* (Chapter 6).

A related but slightly larger species of marine turtle (*Archelon ischyros*) appears later in the Campanian and occurs in the Pierre Shale of South Dakota and Wyoming. Although *Archelon* probably swam over what is now Kansas during that same period, the upper levels of the Pierre Shale that would have preserved their remains were eroded away long ago.

Conditions for plesiosaurs appear to have improved in the Campanian. At least two species of polycotylids were present in the early Campanian: *Polycotylus latipinnis* Cope and *Dolichorhynchops osborni* Williston. Although they are never common, their remains do occur for the first time in the chalk as reasonably complete skeletons (Chapter 8). A juvenile was discovered by Charlie Sternberg as stomach contents of a large *Tylosaurus proriger* (Everhart, 2004b), and the remains of a mother *Polycotylus* and baby were collected by Charles Bonner (O'Keefe and Chiappe, 2011).

Elasmosaurids reappear for the first time in the upper chalk and are represented by at least ten fragmentary specimens (Table 7.1). All except one (USNM 11910) of these specimens were collected before 1900. I find it amazing that no additional elasmosaur remains have been collected from the chalk since 1927. Two of the more interesting ones are *Styxosaurus snowii* (KUVP 1301) and the giant '*Elasmosaurus*' *sternbergi* (KUVP 1312). Neither specimen is anywhere near complete. The type specimen of *Styxosaurus snowii* (Williston, 1890a) includes the only skull of an elasmosaurid known from the chalk, or from Kansas for that matter. The skull is 48 cm (19 in) in length and is attached to the anterior portion of the neck containing 28 cervical vertebrae (Williston, 1890b; Everhart, 2006).

According to Charles H. Sternberg (1917), the remains of '*E.*' *sternbergi* (Fig. 13.11) were originally much more complete before being nearly destroyed by the landowner during the construction of a hillside corral for livestock. Although he was unable to save more than three vertebrae and a few ribs, he was pleased when Williston named the species in honor of him.

Mosasaurs were much larger and more diverse in the upper chalk, although most of those changes are better exhibited in specimens from just above the contact of the Smoky Hill Chalk and the Pierre Shale. The largest documented *Tylosaurus proriger* specimen from the early Campanian chalk is about 10 m (33 ft) in length, although I have seen bones from something significantly larger. The slightly younger 'Bunker' *Tylosaurus proriger* specimen (KUVP 5033) from near the base of the Sharon Springs Member of the Pierre Shale is about 12 m (40 ft) in length. There are also several other incomplete and unreported specimens of these very large tylosaurs in the University of Kansas and the Sternberg Museum collections.

Platecarpus tympaniticus remains the most common mosasaur in the upper chalk, and *Clidastes propython* is present in larger numbers. Neither species continued very far into the Campanian. *Ectenosaurus clidastoides* and *Halisaurus sternbergi* also occur in the upper chalk

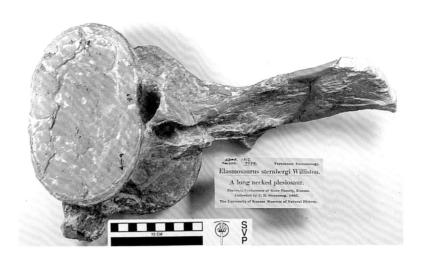

13.11. The centrum of this large dorsal vertebra of *"Elasmosaurus sternbergi"* (KUVP 1312) from the early Campanian chalk of Logan County measures about 16 cm (6.5 in) across the widest part and is at least (33%) wider than the largest dorsal vertebra of *E. platyurus.* Note that the picture shows the anterior (front) of the vertebra with the dorsal process to the right. The vertebra has been crushed from front to back and was originally somewhat thicker. Scale in cm.

(early Campanian), and *Globidens* (FHSM VP-13828; Everhart, 2008) occurs in Kansas for the first time in the overlying early Campanian Sharon Springs Member of the Pierre Shale along with *Plioplatecarpus* (pers. obs.). Recent discoveries of two new *Mosasaurus* specimens in the Weskan Member of the Pierre Shale (Wallace County; Everhart et al., 2015; Fig. 9.13) document the first occurrence of this species in Kansas. Actually the first *Mosasaurus* specimen was collected around 1930 and briefly mentioned by Elias (1931) in his Kansas Geological Survey report on the geology of Wallace County. Although the specimen was collected, it was never prepared or reported. With the help of David Burnham at the University of Kansas, I was able to locate the material and get it officially curated into their collection (KVUP 155838; Fig. 9.14).

It is likely that *Prognathodon* also occurs at this level in the Pierre Shale, but it is yet to be documented there. However, the number and size of mosasaurs discovered in other early Campanian deposits indicate that even as it narrowed, the Western Interior Sea was still teeming with abundant life and was capable of supporting large numbers of very large predators.

Pteranodon longiceps appears to have replaced P. *sternbergi* by the latter part of the Santonian in the Smoky Hill Chalk and continues upward into the Pierre Shale. Full-grown, presumed males of the species have a wingspread of at least 7 m (24 ft), although the average wingspread of the male specimens (Bennett, 1992) that have been collected is about 5.6 m (18 ft). A few specimens preserve stomach contents (Chapter 11) and indicate that small fishes were the usual prey of pteranodons. The remains of a smaller pterosaur (*Nyctosaurus*) also occur more frequently in the upper chalk, and this may indicate a significant reduction in the flying distance from the nearest land.

More bird species occur in the early Campanian chalk than during the Santonian or late Coniacian. Besides *Ichthyornis dispar*, the list of flying birds includes *Guildavis tener*, *Apatornis celer*, and a recently named species, *Iaceornis marshi* (Clarke, 2004). The first examples of the large and flightless *Hesperornis regalis* also occur in the upper chalk. Martin (1984) described a new species of more primitive hesperornithid from the upper chalk (*Parahesperornis alexi* – KUVP 2287) that was about 30 percent smaller than the type specimen of *Hesperornis regalis*, but larger than *Baptornis advenus*. The new species had originally been referred to *Hesperornis gracilis* by Williston (1898a). Another specimen of a swimming bird that had been originally identified as *Baptornis advenus* (UNSM 20030) was redescribed as a new species, *Fumicollis hoffmani* by Bell and Chiappe (2015).

The fact that more numerous remains of *Hesperornis* are discovered farther north in Canadian deposits clearly indicates a connection of some sort with cooler waters. The arrival of these hesperornithids in more southern waters during the early Campanian may indicate a cooler climate, closer shores, adaptations to changes in prey, or other factors. The presence of these diving birds certainly seems to distinguish the upper chalk fauna from that of the Coniacian and Santonian. While hesperornithids fed primarily on small fishes and cephalopods, it is likely that they provided yet another food source for hungry mosasaurs (Chapter 9).

The preceding 'big picture' demonstrates that the fauna of the Western Interior Sea (and the sea itself) was undergoing fairly constant changes during the 5 million years when the Smoky Hill Chalk was deposited. While we know a lot about the marine creatures of this 'ocean of Kansas,' clearly much more work remains to be done in the chalk to discover and describe what species were living there during the Late Cretaceous.

E

What happened to the oceans of Kansas? Where did all that water go? The answer is simple and complex at the same time. The simple answer is that near the end of the Cretaceous, the land beneath the Western Interior Sea rose, and as a result the ocean slowly receded over millions of years to its present shorelines.

Long before the end of the Cretaceous, however, the collision of two tectonic plates along the western edge of North America began the final process of closing the shallow sea covering the Midwest for the last time. Beginning during the Jurassic, the Farallon Plate in the Pacific Ocean subducted beneath (slid under) the western edge of the North American Plate as it was being pushed westward by the spreading of the Atlantic Ocean. Over a period of millions of years, the collision eventually created the mountain chains along the western shore of North America, and eventually the Rocky Mountains, and slowly uplifted much of the Midwest. This process would eventually create what we now call the Great Plains by raising what had been sea bottom thousands of feet above sea level. During the Cretaceous, a nearly continuous series of volcanic eruptions caused by the heating generated by the collision of tectonic plates dumped many cubic miles of volcanic ash over the expanse of the shallow sea covering the midsection of North America. Some of these volcanic eruptions were much larger than any humans have ever observed. Marine life (and dinosaurs) persevered through the hundreds of major eruptions and resulting ash falls that periodically blanketed much of the western half of North America and the Western Interior Sea. Although these ash falls are accurately recorded as bentonite layers in the Smoky Hill Chalk, so far as I am aware there are few, if any, records of a mass mortality of marine life associated with these catastrophic events.

In the middle of the Campanian (about 75 Ma), a stony asteroid about 1.5 miles in diameter impacted along the eastern edge of the Western Interior Sea in what is now northwestern Iowa near the small town of Manson. There is no evidence of the crater visible now on the surface. It was reduced by millions of years of erosion following the impact, and then ground down further by the glaciers that covered Iowa during the ice ages. The crater was so well hidden beneath relatively flat farm fields that it remained unrecognized until the 1990s.

Since then, studies have confirmed that it is a crater left by the impact of a large meteorite during the latter stages of the Cretaceous. It is still uncertain if the Manson impact point was in shallow water or on the nearby land, but the resulting crater was nearly 22 miles across. The environmental effects of this impact are still largely unstudied, but it is likely they were catastrophic for all forms of terrestrial and marine life within hundreds of miles of 'ground zero.' Based on studies involving the shape of the buried crater, it appears that the meteor came from the southeast at a low angle. Ejecta from the impact has been noted to the northwest in South Dakota in the shallow marine Crow Creek Member of the Pierre Shale, and there are indications of a major tsunami that would have inundated most of the coastal areas along the Western Interior Sea. While the environmental effects were certainly devastating, they were much less so than those of the impact in the Yucatan that would follow in about 10 million years at the end of the Cretaceous. The rich fossil record in marine deposits to the northwest in the Pierre Shale, however, does indicate that fishes, birds, and marine reptiles returned relatively quickly to the seaway. In 2003, I was able to participate in the dig of a large elasmosaur in the Mobridge Chalk Member (middle Maastrichtian — dated at 68 Ma) of the Pierre Shale in northeast Nebraska, not far from the Manson Crater. From the size and condition of the elasmosaur and the large number of other species that were collected (mosasaurs, fishes, invertebrates), it certainly appeared to me that the life in the sea was back to normal by that time.

By the middle Maastrichtian (about 68 Ma), however, in response to the uplift of the Rocky Mountains, the Western Interior Sea had been roughly divided through the middle into 'Gulf' and 'Arctic' remnants, with the land across northern Oklahoma and western Kansas probably among the first areas to dry out. As the land continued to rise, the new shorelines retreated farther to the south and north. At the same time, formations of shale, sandstone, chalk, and limestone that had been underwater for millions of years were exposed to weathering and began to erode away. Over time, literally thousands of feet of marine sediments would disappear completely, leaving little or no evidence of the Cretaceous sea that had once covered most of the state. Enough remained, however, to give us a glimpse of the richness of life that existed in the oceans of Kansas.

So what happened at the very end of the Cretaceous? Did the asteroid/comet impact in the Yucatan peninsula of Mexico alone cause the extinction of the dinosaurs, pteranodons, mosasaurs, plesiosaurs, and many other species? That question has yet to be resolved. As this edition is being written, new cores are being drilled in the Chicxulub Crater in Mexico to better understand the effects of that impact. My opinion is that it was a worldwide, long-term (millions of years) combination of events, including continental drift, that resulted in changes in oceanic currents (El Niño on a planetary scale), climate (especially temperature decreases), and increased volcanic activity around the world that significantly affected the environment on the land and in the ocean. At some point, large portions of Earth's ecology collapsed, and those species that were unable to adapt became extinct. While the Yucatan impact may have been the final straw, my 'gut feeling' is that the extinction that we presume occurred at or near what we call the Cretaceous–Paleogene (K–Pg) boundary would have happened anyway, even if we cannot yet understand how it happened.

As a final note in that regard, I leave you with a bit of mathematical juggling that puts the Yucatan impact into better perspective (at least for me). Reams of data, and many papers and books, have been produced about this event, popularizing the phrase 'nuclear winter' and generally comparing it to the effect of thousands of the largest nuclear weapons going off at the same moment and place. Although it was certainly the 'end of the world' for anything living within hundreds or thousands of miles of the point of impact, or possibly all of North America, my opinion is that compared to the overwhelming size and mass of Earth, it was much less devastating than those descriptions portray. I readily admit, however, that I am not an astrophysicist and have no experience in evaluating the effects of a large piece of rock hitting Earth at cosmic speeds.

In an attempt to put things into perspective for myself, I downsized Earth to something that can be more readily visualized on a human scale. Imagine a nearly smooth globe of rock, roughly 66 inches in diameter. This is Earth reduced to a scale of 1 inch for every 120 miles. At this scale, Earth's 24,900-mile circumference at the equator would be reduced to slightly more than 17 feet. The atmosphere around our micro-planet is contained almost entirely within an inch of the surface. The average depth of the ocean at this scale is only 1/50 of an inch. Within the upper layer of the ocean and the base of the atmosphere, the whole life-zone containing the planet's biosphere is but a thin film.

Coming out of space at very high speed, a tiny rock, about the size of a BB (1/10 in) impacts the surface of our micro-planet at the edge of the ocean. The impact generates enough heat to vaporize the rock and some of the surface of the little planet. The resulting crater is a circle about an inch wide and 1/120 of an inch deep. The diameter of the crater is 1/207 of the circumference of the globe and the area (less than one square inch) is about 0.00005 percent of the micro-planet's surface. Seen from even a short distance away (several feet), the impact hardly leaves a blemish on the surface of our little model. The real-life event at the end of the Cretaceous would be of similar proportion to the real Earth, huge by human standards but rather small compared to the size of our planet. Another comparison to the size of the Yucatan crater (111 miles in diameter) would be the 109-mile-wide Grimaldi crater on the west edge of the moon, not visible to the naked eye.

In real life, was the Yucatan impact devastating to at least some portion of the life on Earth? Certainly. Was it the sole cause of the major extinction at the end of the Cretaceous? I don't think so. Whatever killed off the marine reptiles and the dinosaurs of the Late Cretaceous was a far more complex mix of environmental problems than just a big rock falling out of the sky.

What happened at the end of the Cretaceous is considered to be the fifth major extinction event affecting life on our planet, and it was hardly the worst. A much more severe extinction happened about 180 million years earlier at the end of the Permian. This fourth extinction killed off an estimated 96 percent of all life on the planet and allowed the rise of the dinosaurs, and eventually the sea monsters whose remains we collect in the Kansas chalk. Some geologists now believe that we have entered a new geological age, something called the Anthropocene, a time when humans have a major effect on Earth's environment. Products made by humans—plastic, glass, chemicals, and other materials—are now entering the geologic record in huge amounts wherever sedimentation is occurring.

More importantly, we are also observing what many scientists believe to be the sixth major extinction event, and we humans are, to some degree, the cause of it. Overpopulation, depletion of resources, deforestation, desertification, air and water pollution, and climate change are humankind's contribution to the current and rapid extinction of many species. If the fossils of extinct creatures mean anything, they are a dramatic indication that life on this planet is fragile and subject to change without prior notice.

References

Buchanan, R. C. (ed.). 1984. Kansas Geology: An Introduction to Landscapes, Rocks, Minerals and Fossils. Reprint, University Press of Kansas, Lawrence, Kansas, 208 pp.

Buchanan, R. C., and J. R. McCauley. 1987. Roadside Kansas: A Traveler's Guide to Its Geology and Landmarks. Reprint, University Press of Kansas, Lawrence, Kansas, 365 pp.

Carpenter, K., and M. J. Everhart. 2007. Skull of the ankylosaur *Niobrarasaurus coleii* (Ankylosauria: Nodosauridae) from the Smoky Hill Chalk (Coniacian) of western Kansas. Transactions of the Kansas Academy of Science 110(1–2):1–9.

Carpenter, K., D. Dilkes, and D. B. Weishampel. 1995. The dinosaurs of the Niobrara Chalk Formation (Upper Cretaceous, Kansas). Journal of Vertebrate Paleontology 15(2):275–297.

Cope, E. D. 1872. [Sketch of an expedition in the valley of the Smoky Hill River in Kansas]. Proceedings of the American Philosophical Society 12(87):174–176.

Eaton, T. H., Jr. 1960. A new armored dinosaur from the Cretaceous of Kansas. University of Kansas Paleontology Contributions 8:1–14.

Everhart, M. J. 2004. Notice of the transfer of the holotype specimen of *Niobrarasaurus coleii* (Ankylosauria; Nodosauridae) to the Sternberg Museum of Natural History. Transactions of the Kansas Academy of Science 107(3–4):173–174.

Everhart, M. J. 2015. Elias Putnam West (1820–1892)—lawyer, attorney general, militia commander, judge, postmaster, archaeologist, and paleontologist. Transactions of the Kansas Academy of Science 118(3–4):285–294.

Everhart, M. J., and K. Ewell. 2006. Shark bitten dinosaur (Hadrosauridae) caudal vertebrae from the Niobrara Chalk (Upper Coniacian) of western Kansas. Transactions of the Kansas Academy of Science 109(1–2):27–35.

Everhart, M. J., and S. A. Hamm. 2005. A new nodosaur specimen (Dinosauria: Nodosauridae) from the Smoky Hill Chalk (Upper Cretaceous) of western Kansas. Transactions of the Kansas Academy of Science 108(1–2):15–21.

Everhart, M. J., P. A. Everhart, and K. Ewell. 2004. A marine ichthyofauna from the Upper Dakota Sandstone (Late Cretaceous). Abstracts of the oral presentations and posters, Joint Annual Meeting of the Kansas and Missouri Academies of Science, p. 48.

Hattin, D. E. 1982. Stratigraphy and Depositional Environment of the Smoky Hill Chalk Member, Niobrara Chalk (Upper Cretaceous) of the Type Area, Western Kansas. Kansas Geological Survey Bulletin No. 225. University of Kansas, Lawrence, Kansas, 108 pp.

Hattin, D. E., and C. T. Siemers. 1978. Upper Cretaceous Stratigraphy and Depositional Environments of Western Kansas. Guidebook Series 3, Kansas Geological Survey, Lawrence, Kansas, 55 pp.

Leidy, J. 1870. [Remarks on *Elasmosaurus platyurus*]. Proceedings of the Academy of Natural Sciences of Philadelphia 22:9–10.

Lesquereux, L. 1868. On some Cretaceous fossil plants from Nebraska. American Journal of Science, ser. 2, 46(136):91–105.

Liggett, G. A. 2005. A review of the dinosaurs from Kansas. Transactions of the Kansas Academy of Science 108(1–2):1–14.

Marsh, O. C. 1872. Notice of a new species of *Hadrosaurus*. American Journal of Science 16:301.

Mehl, M. G. 1941. *Dakotasuchus kingi*, a crocodile from the Dakota of Kansas. Denison University Bulletin. Journal of the Scientific Laboratory 36:47–66.

Merriam, D. F. 1963. The Geologic History of Kansas. State Geological Survey of Kansas Bulletin No. 162, University of Kansas, Lawrence, Kansas, 317 pp.

Polcyn, M. J., and M. J. Everhart. 2008. Description and phylogenetic analysis of a new species of *Selmasaurus* (Mosasauridae: Plioplatecarpinae) from the Niobrara Chalk of western Kansas.

1. Introduction

Proceedings of the Second Mosasaur Meeting, Fort Hays Studies Special Issue 3, Fort Hays State University, Hays, Kansas, 13–28.

Scott, R. W. 1970. Paleoecology and Paleontology of the Lower Cretaceous Kiowa Formation, Kansas. University of Kansas Paleontological Contributions, Article 52. University of Kansas Paleontological Institute, Lawrence, Kansas, 94 pp.

Vaughn, P. P. 1956. A second specimen of the Cretaceous crocodile *Dakotasuchus* from Kansas. Transactions of the Kansas Academy of Science 59(3):379–381.

Wieland, G. R. 1909. An armored saurian. American Journal of Science 27:250–252.

2. Our Discovery of the Western Interior Sea

Almy, K. J. 1987. Thof's dragon and the letters of Capt. Theophilus Turner, M.D., U.S. Army. Kansas History Magazine 10(3):170–200.

Baur, G. 1892. On the morphology of the skull of the Mosasauridae. Journal of Morphology 7(1):1–22.

Buchanan, R. C. 1989. To Bring Together, Correlate, and Preserve—A History of the Kansas Geological Survey, 1864–1989. Kansas Geological Survey Bulletin No. 227. University of Kansas, Lawrence, Kansas, 96 pp.

Cope, E. D. 1868a. Remarks on a new enaliosaurian, *Elasmosaurus platyurus*. Proceedings of the Academy of Natural Sciences of Philadelphia 20:92–93.

Cope, E. D. 1868b. Note on the fossil reptiles near Fort Wallace; p. 68 in J. L. LeConte (ed.), Notes on the Geology of the Survey for the Extension of the Union Pacific Railway, E. D., from the Smoky Hill River, Kansas, to the Rio Grande. Review Printing House, Philadelphia, Pennsylvania.

Cope, E. D. 1869a. [Remarks on *Holops brevispinus*, O*rnithotarsus immanis*, and *Macrosaurus proriger*]. Proceedings of the Academy of Natural Sciences of Philadelphia 11(81):123.

Cope, E. D. 1869b. Synopsis of the Extinct Batrachia and Reptilia of North America, pt. 1. Transactions of the American Philosophical Society, n.s., 14:1–235.

Cope, E. D. 1869c. [A resolution thanking Dr. Theophilus Turner for his donation of the skeleton of *Elasmosaurus platyurus*]. Proceedings of the Academy of Natural Sciences of Philadelphia 20:314.

Cope, E. D. 1870. Synopsis of the extinct Batrachia and Reptilia of North America. Transactions of the American Philosophical Society, n.s., 14:1–252.

Cope, E. D. 1871. On the fossil reptiles and fishes of the Cretaceous rocks of Kansas; pp. 385–424 in F. V. Hayden (ed.), [Fourth Annual] Preliminary Report of the United States Geological Survey of Wyoming and Portions of Contiguous Territories (Being a Second Annual Report of Progress). Government Printing Office, Washington, D.C.

Cope, E. D. 1872a. Note on some Cretaceous Vertebrata in the State Agricultural College of Kansas. Proceedings of the American Philosophical Society 12(87):168–170.

Cope, E. D. 1872b. [Sketch of an expedition in the valley of the Smoky Hill River in Kansas]. Proceedings of the American Philosophical Society 12(87):174–176.

Cope, E. D. 1872c. On the geology and paleontology of the Cretaceous strata of Kansas; pp. 318–349 in F. V. Hayden (ed.), [Fifth Annual] Preliminary Report of the United States Geological Survey of Montana and Portions of the Adjacent Territories, Being a Fifth Annual Report of Progress, pt. 3: Paleontology. Government Printing Office, Washington, D.C.

Cope, E. D. 1872d. Catalogue of the Pythonomorpha found in the Cretaceous strata of Kansas. Proceedings of the Academy of Natural Sciences of Philadelphia 12(87):264 and 12(88):265–287.

Cope, E. D., 1872e. [On a species of *Clidastes* and *Plesiosaurus gulo* Cope]. Proceedings of the Academy of Natural Sciences of Philadelphia 29:127–124.

Cope, E. D. 1875. The Vertebrata of the Cretaceous Formations of the West; in F. V. Hayden (ed.), Report of the U.S. Geological Survey of the Territories, Volume 2. Government Printing Office, Washington, D.C., 302 pp.

Cope, E. D. 1894. Observations on the geology of adjacent parts of Oklahoma and North West Texas. Proceedings of the Academy of Natural Sciences of Philadelphia 46:63–68.

Davidson, J. P. 2016. The last time: Edward Drinker Cope's last trip to Kansas. Kansas Academy of Science, Transactions 119(3–4):363–374.

Everhart, M. J. 2001. Revisions to the biostratigraphy of the Mosasauridae (Squamata) in the Smoky Hill Chalk Member of the Niobrara Chalk (Late Cretaceous) of Kansas. Transactions of the Kansas Academy of Science 104(1–2):56–75.

Everhart, M. J. 2016. William E. Webb—Civil War correspondent, railroad land baron, town founder, Kansas legislator, adventurer, fossil collector, author. Transactions of the Kansas Academy of Science 119(2):179–192.

Goldfuss, A. 1845. Der Schädelbau des *Mosasaurus*, durch Beschreibung einer neuen Art dieser Gattung erläutert. Nova Acta Academa Caesar Leopoldino-Carolinae Germanicae Natura Curiosorum 21:1–28.

Goldfuss, A. 2013. The skull structure of the *Mosasaurus*, explained by means of a description of a new species of this genus. English translation edited by M. J. Everhart. Transactions of the Kansas Academy of Science 116(1–2):27–46.

Harlan, R. 1824. On a new fossil genus of the order Enalio Sauri (of Conybeare). Proceedings of the Academy of Natural Sciences of Philadelphia, ser. 1, 3, pt. 2:331–337.

Harlan, R. 1834. Notice of the discovery of the remains of the *Ichthyosaurus* in Missouri, N. A. Transactions of the American Philosophical Society 4:405–409.

Hattin, D. E. 1982. Stratigraphy and Depositional Environment of the Smoky Hill Chalk Member, Niobrara Chalk (Upper Cretaceous) of the Type Area, Western Kansas. Kansas Geological Survey Bulletin No. 225. University of Kansas, Lawrence, Kansas, 108 pp.

Hayden, F. V. 1872. Final Report of the United States Geological Survey of Nebraska and Portions of the Adjacent Territories. Government Printing Office, Washington, D.C., 264 pp.

Hays, I. 1830. Description of a fragment of the head of a new fossil animal, discovered in a marl pit near Moorestown New Jersey. Transactions of the American Philosophical Society, ser. 2, 3(18):471–477.

LeConte, J. L. 1868. Notes on the Geology of the Survey for the Extension of the Union Pacific Railway, E. D., from the Smoky Hill River, Kansas, to the Rio Grande. Review Printing House, Philadelphia, Pennsylvania, 76 pp.

Leidy, J. 1856. Remarks on certain extinct species of fishes. Proceedings of the Academy of Natural Sciences of Philadelphia 8:301–302.

Leidy, J. 1868. [Photographs of fossil bones]. Proceedings of the Academy of Natural Sciences of Philadelphia 20:316.

Leidy, J. 1870. [Remarks on ichthyodorulites and on certain fossil Mammalia]. Proceedings of the Academy of Natural Sciences of Philadelphia 22:12–13.

Leidy, J. 1873. Contributions to the Extinct Vertebrate Fauna of the Western Territories; in F. V. Hayden (ed.), Report of the U.S. Geological Survey of the Territories, Volume 1. Government Printing Office, Washington, D.C., 358 pp.

Lesquereux, L. 1868. On some Cretaceous fossil plants from Nebraska. American Journal of Science, ser. 2, 46(136):91–105.

Lesquereux, L. 1873. Fossil flora; pp. 283–318 in Part III, Paleontology, in F. V. Hayden (ed.), Sixth Annual Report of the United States Geological Survey of the Territories, Embracing Portions of Montana, Idaho, Wyoming, and Utah, Being a Report of Progress of the Explorations for the Year 1872. Government Printing Office, Washington, D.C.

Liggett, G. A. 2001. Dinosaurs to Dung Beetles: Expeditions through Time—A Guide to the Sternberg Museum of Natural History. Sternberg Museum of Natural History, Hays, Kansas, 127 pp.

Logan, W. N. 1897. The Upper Cretaceous of Kansas: with an introduction by Erasmus Haworth. University Geological Survey of Kansas 2:194–234.

Meek, F. B., and F. V. Hayden. 1861. Descriptions of New Lower Silurian (Primordial), Jurassic, Cretaceous, and Tertiary fossils, collected in Nebraska. Proceedings of the Academy of Natural Sciences of Philadelphia 13:415–447.

Mitchell, S. L. 1818. Observations on the geology of North America, illustrated by the description of various organic remains found in that part of the world; pp. 319–431 in G. Cuvier (ed.), Essay on the Theory of the Earth. Kirk and Mercein, New York, New York.

Moore, R. C., and W. P. Haynes. 1917. Oil and Gas Resources of Kansas. Kansas Geological Survey Bulletin No. 3. University of Kansas, Lawrence, Kansas, 391 pp.

Moulton, G. E. (ed.). 1983–1997. The Journals of the Lewis and Clark Expedition. Vols. 1–11. University of Nebraska Press, Lincoln, Nebraska.

Mudge, B. F. 1866a. Discovery of fossil footmarks in the Liassic (?) Formation in Kansas. American Journal of Science, ser. 2, 41(122):174–176.

Mudge, B. F. 1866b. First Annual Report on the Geology of Kansas. State Printer, Lawrence, Kansas, 57 pp.

Osborn, H. F. 1931. Cope: Master Naturalist. Princeton University Press, Princeton, New Jersey, 740 pp.

Peterson, J. M. 1987. Science in Kansas: the early years, 1804–1875. Kansas History Magazine 10(3):201–240.

Rogers, K. 1991. The Sternberg Fossil Hunters: A Dinosaur Dynasty. Mountain Press Publishing Company, Missoula, Montana, 288 pp.

Russell, D. A. 1967. Systematics and Morphology of American Mosasaurs. Peabody Museum of Natural History, Yale University, Bulletin No. 23, New Haven, Connecticut, 241 pp.

Russell, D. A. 1970. The vertebrate fauna of the Selma Formation of Alabama, pt. 7: the mosasaurs. Fieldiana Geology Memoirs 3(7):369–380.

Schumacher, B. A. 1993. Biostratigraphy of Mosasauridae (Squamata, Varanoidea) from the Smoky Hill Chalk Member, Niobrara Chalk (Upper Cretaceous) of western Kansas. M.S. thesis, Fort Hays State University, Hays, Kansas, 68 pp.

Sheldon, M. A. 1996. Stratigraphic distribution of mosasaurs in the Niobrara Formation of Kansas. Paludicola 1:21–31.

Spamer, E. E., R. M. McCourt, R. Middleton, E. Gilmore, and S. B. Duran. 2000. A national treasure: accounting for the natural history specimens from the Lewis and Clark Expedition (western North America, 1803–1806) in the Academy of Natural Sciences of Philadelphia. Proceedings of the Academy of Natural Sciences of Philadelphia 150:47–58.

Stewart, J. D. 1988. The stratigraphic distribution of Late Cretaceous *Protosphyraena* in Kansas and Alabama: geology, paleontology and biostratigraphy of western Kansas; pp. 80–94 in M. E. Nelson (ed.), Geology, Paleontology and Biostratigraphy of Western Kansas: Articles in Honor of Myrl V. Walker. Fort Hays Studies, Third Series, No. 10, Science, Fort Hays State University, Hays, Kansas.

Stewart, J. D. 1990. Niobrara Formation vertebrate stratigraphy; pp. 19–30 in S. C. Bennett (ed.), 1990 Niobrara Chalk Excursion Guidebook. University of Kansas Museum of Natural History and Kansas Geological Survey, Lawrence, Kansas.

Taft, R. 1953. Artists and Illustrators of the Old West, 1850–1900. Scribner, New York, New York, 492 pp.

Webb, W. E. 1872. Buffalo Land: An Authentic Account of the Discoveries, Adventures, and Mishaps of a Scientific and Sporting Party in the Wild West. Hubbard Bros., Philadelphia, Pennsylvania, 503 pp.

Wetzel, C. R. 1960. Monument Station, Gove County. The Kansas Historical Quarterly 26:250–254.

Williston, S. W. 1895. New or little-known extinct vertebrates. Kansas University Quarterly 3(3):165–176.

Williston, S. W. 1897. The Kansas Niobrara Cretaceous. University Geological Survey of Kansas 2:235–246.

Williston, S. W. 1898a. Addenda to Part I. University Geological Survey of Kansas 4:28–32.

Williston, S. W. 1898b. Mosasaurs. University Geological Survey of Kansas 4(5):81–221.

Zakrzewski, R. J. 1996. Geologic studies in western Kansas in the 19th Century. Transactions of the Kansas Academy of Science 99(3–4):124–133.

Blaire, S. A., and D. K. Watkins. 2009. High-resolution calcareous nannofossil biostratigraphy for the Coniacian/Santonian Stage boundary, western Interior Basin. Cretaceous Research 30:367–384.

Brown, R. W. 1940. Fossil pearls from the Colorado Group of western Kansas. Washington Academy of Science 30(9):365–374.

Buckland, W. 1829. On the discovery of a new species of *Pterodactyle*; and also of the faeces of the *Ichthyosaurus*; and of a black substance resembling sepia, or India ink, in the Lias at Lyme Regis. Proceedings of the Geological Society of London 1:96–98.

Carpenter, K. 1996. Sharon Springs Member, Pierre Shale (Lower Campanian) depositional environment and origin of its vertebrate fauna, with a review of North American Plesiosaurs. Ph.D. Dissertation, University of Colorado, Boulder, Colorado, 251 pp.

Elias, M. K. 1933. Cephalopods of the Pierre Formation of Wallace County, Kansas, and adjacent area. University of Kansas Science Bulletin 21(9):289–263.

Everhart, M. J. 1999. Evidence of feeding on mosasaurs by the Late Cretaceous lamniform shark, *Cretoxyrhina mantelli*. Journal of Vertebrate Paleontology 17(3, Supplement):43A-44A.

Everhart, M. J. 2001. Revisions to the biostratigraphy of the Mosasauridae (Squamata) in the Smoky Hill Chalk Member of the Niobrara Chalk (Late Cretaceous) of Kansas. Transactions of the Kansas Academy of Science 104(1–2):56–75.

Everhart, M. J. 2003. First records of plesiosaur remains in the Lower Smoky Hill Chalk Member (Upper Coniacian) of the Niobrara Formation in western Kansas. Transactions of the Kansas Academy of Science 106(3–4):139–148.

Everhart, M. J. 2004a. Plesiosaurs as the food of mosasaurs: new data on the stomach contents of a *Tylosaurus proriger* (Squamata; Mosasauridae) from the Niobrara Formation of western Kansas. Mosasaur 7:41–46.

3. Invertebrates, Plants, and Trace Fossils

Everhart, M. J. 2007. Remains of a pycnodont fish (Actinopterygii: Pycnodontiformes) in a coprolite; an upper record of *Micropycnodon kansasensis* in the Smoky Hill Chalk, western Kansas. Transactions of the Kansas Academy of Science 110(1–2):35–43.

Everhart, M. J. 2015. Elias Putnam West (1820–1892)—lawyer, attorney general, militia commander, judge, postmaster, archaeologist, and paleontologist. Transactions of the Kansas Academy of Science 118(3–4):285–294.

Everhart, M. J., and P. A. Everhart. 1992. Oyster-shell concentrations: a stratigraphic marker in the Smoky Hill Chalk (Upper Cretaceous) of western Kansas. Transactions of the Kansas Academy of Science 11:12.

Everhart, M. J., and A. Maltese. 2010. First report of a heteromorph ammonite, cf. *Glyptoxoceras*, from the Smoky Hill Chalk (Santonian) of western Kansas, and a brief review of Niobrara cephalopods. Transactions of the Kansas Academy of Science 113(1–2):64–70.

Gill, J. R., W. A. Cobban, and L. G. Schultz. 1972. Stratigraphy and composition of the Sharon Springs Member of the Pierre Shale in western Kansas. Geological Survey Professional Paper 728, U.S. Government Printing Office, Washington, D.C., 50 pp.

Green, R. G. 1977. *Niobrarateuthis walkeri*, a new species of teuthid from the Upper Cretaceous Niobrara Formation of Kansas. Journal of Paleontology 51(5):992–995.

Grinnell, G. B. 1876. On a new crinoid from the Cretaceous Formation of the West. American Journal of Science, ser. 3, 12(3):81–83.

Gudger, E. W. 1949. Natural history notes on tiger sharks, *Galeocerdo tigrinis*, caught in Key West, Florida with emphasis on food and feeding habits. Copeia 1949:39–47.

Hattin, D. E. 1982. Stratigraphy and Depositional Environment of the Smoky Hill Chalk Member, Niobrara Chalk (Upper Cretaceous) of the Type Area, Western Kansas. Kansas Geological Survey Bulletin No. 225. University of Kansas, Lawrence, Kansas, 108 pp.

Hattin, D. E. 1988. Rudists as historians: Smoky Hill Member of Niobrara Chalk (Upper Cretaceous) of Kansas. Fort Hays Studies, 3rd ser., 10:4–22.

Hattin, D. E. 1996. Fossilized regurgitate from Smoky Hill Chalk Member of Niobrara Chalk (Upper Cretaceous) of Kansas, USA. Cretaceous Research 17:443–450.

Hawkins, T. 1834. Memoirs of Ichthyosauri and Plesiosauri, Extinct Monsters of the Ancient Earth. Relfe and Fletcher, London, England, 58 pp.

Hawn, F. 1858. The Trias of Kansas. Transactions of the Academy of Science of St. Louis 1(2):171–172.

Hayden, F. V. 1873. Sixth Annual Report of the United States Geological Survey of the Territories, Embracing Portions of Montana, Idaho, Wyoming, and Utah, Being a Report of Progress of the Explorations for the Year 1872. Washington, D.C., Government Printing Office, 844 pp.

Hess, H. 1999. *Uintacrinus* beds of the Upper Cretaceous Niobrara Formation, Kansas, USA; pp. 225–232 in H. Hess, W. I. Ausich, C. E. Brett, and M. J. Simms (eds.), Fossil Crinoids. Cambridge University Press, Cambridge, England, xv–275 pp.

Jeletzky, J. A. 1955. *Belemnitella praecursor*, probably from the Niobrara of Kansas, and some stratigraphic implications. Journal of Paleontology 29(5):876–885.

Jeletzky, J. A. 1961. *Actinocamax* from the Upper Cretaceous Benton and Niobrara formations of Kansas. Journal of Paleontology 35(3):505–531.

Kauffman, E. G. 1990. Giant fossil inoceramid bivalve pearls; pp. 66–68 in A. J. Boucot (ed.), Evolutionary Paleobiology of Behavior and Coevolution. Elsevier, Amsterdam, the Netherlands, 750 pp.

Kauffman, E. G., P. J. Harries, C. Meyer, T. Villamil, C. Arango, and G. Jaecks. 2007. Paleoecology of giant Inoceramidae (*Platyceramus*) on a Santonian seafloor in Colorado. Journal of Paleontology 81(1):64–81.

LeConte, J. L. 1868. Notes on the Geology of the Survey for the Extension of the Union Pacific Railway, E. D., from the Smoky Hill River, Kansas, to the Rio Grande. Review Printing House, Philadelphia, 76 pp.

Lehman, U. 1979. The jaws and radula of the Jurassic ammonite *Dactylioceras*. Paleontology 22(1):265–271.

Lesquereux, L. 1868. On some Cretaceous fossil plants from Nebraska. American Journal of Science, ser. 2, 46(136):91–105.

Logan, W. N. 1897. The Upper Cretaceous of Kansas: with an Introduction by Erasmus Haworth. University Geological Survey of Kansas 2:195–234.

Logan, W. N. 1898. The Invertebrates of the Benton, Niobrara and Fort Pierre Groups. University Geological Survey of Kansas 4:432–518.

Logan, W. N. 1899. Some additions to the Cretaceous invertebrates of Kansas. Kansas University Quarterly 8(2):87–98.

Marsh, O. C. 1871a. Scientific expedition to the Rocky Mountains. American Journal of Science, ser. 3, 1(2):142–143.

Marsh, O. C. 1871b. On the geology of the Eastern Uintah Mountains. American Journal of Science, ser. 3, 1(3):191–198.

Martin, J. E., and P. R. Bjork. 1987. Gastric residues associated with a mosasaur from the Late Cretaceous (Campanian) Pierre Shale in South Dakota. Dakoterra 3:68–72.

McAllister, J. A. 1985. Reevaluation of the formation of spiral coprolites. University of Kansas Paleontology Contributions, Paper 114, 12 pp.

Meek, F. B., and F. V. Hayden. 1859. On the so-called Triassic rocks of Kansas and Nebraska. American Journal of Science, ser. 2, 27(79):31–35.

Miller, H. W., Jr. 1957a. *Niobrarateuthis bonneri*, a new genus and species of squid from the Niobrara Formation of Kansas. Journal of Paleontology 31(5):809–811.

Miller, H. W., Jr. 1957b. Intestinal casts in *Pachyrhizodus*, an elopid fish, from the Niobrara Formation of Kansas. Transactions of the Kansas Academy of Science 60(4):399–401.

Miller, H. W., Jr. 1968a. Invertebrate Fauna and Environment of Deposition of the Niobrara (Cretaceous) of Kansas. Fort Hays Studies, Science Series, no. 8. Fort Hays Kansas State College, Fort Hays, Kansas, 90 pp.

Miller, H. W., Jr. 1968b. *Enchoteuthis melanae* and *Kansasteuthis lindneri*, new genera and species of teuthids, and a septid from the Niobrara Formation of Kansas. Transactions of the Kansas Academy of Science 71(2):176–183.

Miller, H. W., Jr. 1969. Additions to the fauna of the Niobrara Formation of Kansas. Transactions of the Kansas Academy of Science 72(4):533–546.

Miller, H. W., Jr., G. F. Sternberg, and M. V. Walker. 1957. *Uintacrinus* localities in the Niobrara Formation of Kansas. Transactions of the Kansas Academy of Science 60(2):163–166.

Morton, N. 1981. Aptychi: The myth of the ammonite operculum. Lethaia 14(1):57–61.

Mudge, B. F. 1866. First Annual Report on the Geology of Kansas. State Printer, Lawrence, Kansas, 57 pp.

Mudge, B. F. 1876. Notes on the Tertiary and Cretaceous periods of Kansas. US Geological and Geographical Survey of the Territories Bulletin 2(3):211–221.

Mudge, B. F. 1877. Notes on the Tertiary and Cretaceous periods of Kansas; pp. 277–294 in F. V. Hayden (ed.), Ninth Annual Report of the United States Geological and Geographical Survey of the Territories, Embracing Colorado and Parts of Adjacent Territories; Being a Report of Progress of the Exploration for the year 1875, pt. 1: Geology. Government Printing Office, Washington, D.C., 827 pp.

Nicholls, E. L., and H. Isaak. 1987. Stratigraphic and taxonomic significance of *Tusoteuthis longa* Logan (Coleoidea, Teuthida) from the Pembina Member, Pierre Shale (Campanian), of Manitoba. Journal of Paleontology 61(4):727–737.

Nicollet, J. N. 1843. Report Intended to Illustrate a Map of the Hydrographical Basin of the Upper Mississippi River. 26th Congress, 2nd Session, Senate Document 237, serial 380. 170 pp.

Shimada, K. 1997. Shark-tooth-bearing coprolite from the Carlile Shale (Upper Cretaceous), Ellis County, Kansas. Transactions of the Kansas Academy of Science 100(3–4):133–138.

Shimada, K., M. J. Everhart, R. Decker and P. D. Decker. 2009. A new skeletal remain of the durophagous shark, *Ptychodus mortoni*, from the Upper Cretaceous of North America: an indication of gigantic body size. Cretaceous Research 31(2):249–254.

Springer, F. 1901. *Uintacrinus*: Its structure and relations. Memoirs of the Museum of Comparative Zoology 25(1):1–89.

Stenzel, H. B. 1971. Oysters; pp. 953–1224 in R. C. Moore, Treatise on Invertebrate Paleontology, pt. N, vol. 3: Mollusca 6: Bivalvia. Geological Society of America and University Press of Kansas, Lawrence, Kansas.

Sternberg, C. H. 1909. The Life of a Fossil Hunter. Henry Holt and Company, 286 pp.

Sternberg, C. H. 1917. Hunting Dinosaurs in the Badlands of the Red Deer River, Alberta, Canada. Published by the author, San Diego, California, 232 pp.

Sternberg, C. H. 1922. Explorations of the Permian of Texas and the chalk of Kansas, 1918. Transactions of the Kansas Academy of Science 30(1):119–120.

Sternberg, M. 1920. George Miller Sternberg: A Biography. Chicago: American Medical Association. 331 pp.

Stewart, J. D. 1976. Teuthids of the North American Late Cretaceous. Transactions of the Kansas Academy of Science 79(3–4):74.

Stewart, J. D. 1978. Enterospirae (Fossil Intestines) from the Upper Cretaceous Niobrara Formation of western Kansas.

University of Kansas Paleontological Contributions, Article 89. University of Kansas Paleontological Institute, Lawrence, Kansas, 8 pp.

Stewart, J. D. 1990a. Niobrara Formation Vertebrate Stratigraphy; pp. 19–30 in S. C. Bennett (ed.), Niobrara Chalk Excursion Guidebook. University of Kansas Museum of Natural History and Kansas Geological Survey, Lawrence, Kansas.

Stewart, J. D. 1990b. Niobrara Formation symbiotic fish in inoceramid bivalves; pp. 31–41 in S. C. Bennett (ed.), 1990 Niobrara Chalk Excursion Guidebook. University of Kansas Museum of Natural History and Kansas Geological Survey, Lawrence, Kansas.

Stewart, J. D. 1990c. Preliminary account of holecostome-inoceramid commensalism in the Upper Cretaceous of Kansas; pp. 51–57 in A. J. Boucot (ed.), Evolutionary Paleobiology and Coevolution. Elsevier, Amsterdam, the Netherlands.

Stewart, J. D., and K. Carpenter. 1990. Examples of vertebrate predation on cephalopods in the Late Cretaceous of the Western Interior; pp. 203–208 in A. J. Boucot (ed.), Evolutionary Paleobiology of Behavior and Coevolution. Elsevier, Amsterdam, the Netherlands.

Turner, M. M., K. Weed, C. D. Burke, and W. Yang. 2001. Disentangling conflicting stratigraphic subdivisions through ammonoid biostratigraphy, the Sharon Springs Member of the Pierre Shale Formation (Upper Cretaceous), western Kansas. Abstracts, GSA 2001 Rocky Mountain and South Central Sections, GSA, Joint Annual Meeting, 33(5):13.

Wang, H. 2002. Diversity of angiosperm leaf megafossils from the Dakota Formation (Cenomanian, Cretaceous), north western interior, USA. Ph.D. Dissertation, University of Florida, Gainesville, Florida, 395 pp.

Williston, S. W. 1897. The Kansas Niobrara Cretaceous. University Geological Survey of Kansas 2:235–246.

Williston, S. W. 1898. Mosasaurs. University Geological Survey of Kansas 4(5):81–347.

Williston, S. W. 1902. On certain homoplastic characters in aquatic air-breathing vertebrates. Kansas University Science Bulletin 1(9):259–266.

Agassiz, J. L. R. 1833–1844. Recherches sur les Poissons Fossiles. [5 Volumes]. Neuchâtel, Switzerland, Imprimerie de Petitpierre, 1420 pp.

Beamon, J. C. 1999. Depositional environment and fossil biota of a thin clastic unit of the Kiowa Formation, Lower Cretaceous (Albian), McPherson County, Kansas. M.S. thesis, Fort Hays State University, Hays, Kansas, 97 pp.

Bice, K. N., and K. Shimada. 2016. Fossil marine vertebrates from the Codell Sandstone Member (middle Turonian) of the Upper Cretaceous Carlile Shale in Jewell County, Kansas, USA. Cretaceous Research 65:172–198.

Bourdon, J., and M. J. Everhart. 2011. Analysis of an associated *Cretoxyrhina mantelli* dentition from the Late Cretaceous (Smoky Hill Chalk, Late Coniacian) of western Kansas. Transactions of the Kansas Academy of Science 114(1–2):15–32.

Cappetta, H. 1973. Selachians from the Carlile Shale (Turonian) of South Dakota. Journal of Paleontology 47:504–514.

Carpenter, K., D. Dilkes, and D. B. Weishampel. 1995. The dinosaurs of the Niobrara Chalk Formation (Upper Cretaceous, Kansas). Journal of Vertebrate Paleontology 15(2):275–297.

Cicimurri, D. J. 2004. Late Cretaceous Chondrichthyans from the Carlile Shale (Middle Turonian to Early Coniacian) of the Black Hills Region, South Dakota and Wyoming. The Mountain Geologist 41(1):1–16.

Cicimurri, D. J., D. C. Parris, and M. J. Everhart. 2008. Partial dentition of a chimaeroid fish (Chondrichthyes, Holocephali) from the Upper Cretaceous Niobrara Chalk of Kansas, USA. Journal of Vertebrate Paleontology 28(1):34–40.

Cope, E. D. 1872a. On the geology and paleontology of the Cretaceous strata of Kansas; pp. 318–349 in F. V. Hayden (ed.), [Fifth Annual] Preliminary Report of the United States Geological Survey of Montana and Portions of the Adjacent Territories, Being a Fifth Annual Report of Progress, pt. 3: Paleontology. Government Printing Office, Washington, D.C.

Cope, E. D. 1872b. On the families of fishes of the Cretaceous Formation in Kansas. Proceedings of the American Philosophical Society 12(88):327–357.

Cope, E. D. 1874. Review of the Vertebrata of the Cretaceous Period found west of the Mississippi River. Bulletin of the U.S. Geological and Geographical Survey of the Territories 1(2):3–48.

Cope, E. D. 1875. The Vertebrata of the Cretaceous Formations of the West; in F. V. Hayden (ed.), Report of the U.S. Geological Survey of the Territories, Volume 2. Government Printing Office, Washington, D.C., 302 pp.

4. Sharks

Corrado, C. A., D. A. Wilhelm, K. Shimada, and M. J. Everhart. 2003. A new skeleton of the Late Cretaceous lamniform shark, *Cretoxyrhina mantelli*, from western Kansas. Journal of Vertebrate Paleontology 23(3, Supplement):43A.

Dickerson, A. A., K. Shimada, B. Reilly, and C. K. Rigsby. 2012. New data on the Late Cretaceous lamniform shark, *Cardabiodon* sp., based on an associated specimen from Kansas. Transactions of the Kansas Academy of Science 115(3–4):125–133.

Dixon, F. 1850. The Geology and Fossils of the Tertiary and Cretaceous Formations of Sussex. London, 422 pp.

Druckenmiller, P. S., A. J. Daun, J. L. Skulan, and J. C. Pladziewicz. 1993. Stomach contents in the Upper Cretaceous shark *Squalicorax falcatus*. Journal of Vertebrate Paleontology 13(3, Supplement):33A.

Eastman, C. R. 1895. Bietrage Kenntniss Gattung *Oxyrhina* mit besonderer Berucksichtigung von *Oxyrhina mantelli* Agassiz. Palaeontographica 41:149–192.

Everhart, M. J. 1999. Evidence of feeding on mosasaurs by the Late Cretaceous lamniform shark, *Cretoxyrhina mantelli*. Journal of Vertebrate Paleontology 17(3, Supplement):43A-44A.

Everhart, M. J. 2002. New data on cranial measurements and body length of the mosasaur, *Tylosaurus nepaeolicus* (Squamata; Mosasauridae), from the Niobrara Formation of western Kansas. Transactions of the Kansas Academy of Science 105(1–2):33–43.

Everhart, M. J. 2004a. First record of the hybodont shark genus, "*Polyacrodus*" sp. (Chondrichthyes; Polyacrodontidae) from the Kiowa Formation (Lower Cretaceous) of McPherson County, Kansas. Transactions of the Kansas Academy of Science 107(1–2):39–43.

Everhart, M. J. 2004b. Plesiosaurs as the food of mosasaurs: new data on the stomach contents of a *Tylosaurus proriger* (Squamata; Mosasauridae) from the Niobrara Formation of western Kansas. Mosasaur 7:41–46.

Everhart, M. J. 2004c. Late Cretaceous interaction between predators and prey: evidence of feeding by two species of shark on a mosasaur. PalArch Journal of Vertebrate Paleontology 1(1):1–7.

Everhart, M. J. 2005. Bite marks on an elasmosaur (Sauropterygia; Plesiosauria) paddle from the Niobrara Chalk (Upper Cretaceous) as probable evidence of feeding by the lamniform shark, *Cretoxyrhina mantelli*. PalArch Journal of Vertebrate Palaeontology 2(2):14–24.

Everhart, M. J. 2007. New stratigraphic records (Albian-Campanian) of the guitarfish, *Rhinobatos sp.* (Chondrichthyes; Rajiformes), from the Cretaceous of Kansas. Transactions of the Kansas Academy of Science 110(3–4):225–235.

Everhart, M. J. 2009. Probable plesiosaur remains from the Blue Hill Shale (Carlile Formation; Middle Turonian) of north central Kansas. Transactions of the Kansas Academy of Science 112(3–4):215–221.

Everhart, M. J. 2012. Exceptional preservation of shark integument in the Greenhorn Formation (Turonian; Late Cretaceous), Republic County, Kansas. Transactions of the Kansas Academy of Science 115(1–2):56–57.

Everhart, M. J. 2013. The palate bones of a fish?—the first specimen of *Ptychodus mortoni* (Chondrichthyes; Elasmobranchii) from Alabama. Bulletin of the Alabama Museum of Natural History 31(1):98–104

Everhart, M. J., and T. Caggiano. 2004. An associated dentition of the Late Cretaceous elasmobranch, *Ptychodus anonymous* Williston 1900. Paludicola 4(4):125–136.

Everhart, M. J., and M. K. Darnell. 2004. Note on the occurrence of *Ptychodus mammillaris* (Elasmobranchii) in the Fairport Chalk Member of the Carlile Shale (Upper Cretaceous) of Ellis County, Kansas. Transactions of the Kansas Academy of Science 107(3–4):126–130.

Everhart, M. J., and P. A. Everhart. 2003. First report of the Paleozoic shark *Ctenacanthus amblyxiphias* Cope 1891 from the Lower Permian of Morris County, Kansas. Transactions of the Kansas Academy of Science 22:13.

Everhart, M. J., and K. Ewell. 2006. Shark-bitten dinosaur (Hadrosauridae) vertebrae from the Niobrara Chalk (Upper Coniacian) of western Kansas. Transactions of the Kansas Academy of Science 109(1–2):27–35.

Everhart, M. J., P. A. Everhart, and K. Ewell. 2004. A marine ichthyofauna from the Upper Dakota Sandstone (Late Cretaceous). Abstracts of the oral presentations and posters, Joint Annual Meeting of the Kansas and Missouri Academies of Science, p. 48.

Everhart, M. J., P. Everhart, E. M. Manning, and D. E. Hattin. 2003. A Middle Turonian marine fish fauna from the Upper Blue Hill Shale Member, Carlile Shale, of North Central Kansas. Journal of Vertebrate Paleontology 23(3, Supplement):49A.

Ewell, K., and M. J. Everhart. 2004. A Paleozoic shark fauna from the Council

Grove Group (Lower Permian). Abstracts of the oral presentations and posters, Joint Annual Meeting of the Kansas and Missouri Academies of Science, pp. 48–49.

Hamm, S. A. 2010a. The Late Cretaceous shark, *Ptychodus rugosus*, (Ptychodontidae) in the Western Interior Sea. Transactions of the Kansas Academy of Science 113(1–2):44–55.

Hamm, S. A. 2010b. The Late Cretaceous shark *Ptychodus marginalis* in the Western Interior Seaway, USA. Journal of Paleontology 84(3):538–548.

Hamm, S. A. 2013. The first associated tooth set of *Ptychodus mammillaris* in North America. Transactions of the Kansas Academy of Science 1167(1–2):76.

Hamm, S. A., and M. J. Everhart. 1999. The occurrence of a rare ptychodontid shark from the Smoky Hill Chalk (Upper Cretaceous) of western Kansas. Transactions of the Kansas Academy of Science 18:34.

Hamm, S. A., and M. J. Everhart. 2001. Notes on the occurrence of nodosaurs (Ankylosauridae) in the Smoky Hill Chalk (Upper Cretaceous) of western Kansas. Journal of Vertebrate Paleontology 21(3, Supplement):58A.

Hamm, S. A., and K. Shimada. 2002. Associated tooth set of the Late Cretaceous lamniform shark, *Scapanorhynchus raphiodon* (Mitsukurinidae), from the Niobrara Chalk of western Kansas. Transactions of the Kansas Academy of Science 105(1–2):18–26.

Hattin, D. E. 1962. Stratigraphy of the Carlile Shale (Upper Cretaceous) in Kansas. State Geological Survey of Kansas, Bulletin no. 156. University of Kansas, Lawrence, Kansas, 155 pp.

Hattin, D. E. 1982. Stratigraphy and Depositional Environment of the Smoky Hill Chalk Member, Niobrara Chalk (Upper Cretaceous) of the Type Area, Western Kansas. Kansas Geological Survey Bulletin No. 225. University of Kansas, Lawrence, Kansas, 108 pp.

Hattin, D. E. 1996. Fossilized Regurgitate from Smoky Hill Chalk Member of Niobrara Chalk (Upper Cretaceous) of Kansas, USA. Cretaceous Research 17:443–450.

Herman, J. 1977. Les sélaciens des terrains néocrétacés et paléocenes de Belgique et des contrées limitrophes. Eléments d'une biostratigraphique intercontinentale. Mémoires pour sérvir a l'explication des Cartes géologiques et miniéres de la Belgique. Service Géologique de Belgique, Mémoire 15, 401 pp.

Hoffman, B. L., G. D. Claycomb, and S. A. Hageman. 2015. Triple-layered enameloid in teeth of the Late Cretaceous durophagous shark *Ptychodus rhombodus*.

Transactions of the Kansas Academy of Science 118(1–2):143–144.

Hoffman, B. L., S. A. Hageman, and G. D. Claycomb. 2016. Scanning electron microscope examination of the dental enameloid of the Cretaceous durophagous shark *Ptychodus* supports neoselachian classification. Journal of Paleontology 90(4):741–762.

Hoganson, J. W., and J. M. Erickson. 2005. A new species of *Ischyodus* (Chondrichthyes: Holocephali: Callorhynchidae) from the upper Maastrichtian shallow marine facies of the Fox Hills and Hell Creek Formations, Williston Basin, North Dakota, USA. Palaeontology 48(4):709–721.

Hoganson, J. W., J. M. Erickson, and M. J. Everhart. 2015. *Ischyodus rayhaasi* (Chimaeroidei; Callorhynchidae) from the Campanian-Maastrichtian Fox Hills of northeastern Colorado, U.S.A. Transactions of the Kansas Academy of Science 1118(1–2):27–40.

Hussakof, L. 1908. Catalogue of the type and figured vertebrates in the American Museum of Natural History. Bulletin of the American Museum of Natural History XXV, 103 pp.

Ikejiri, T., and M. J. Everhart. 2015. Notes on the authorship and the holotype of the Late Cretaceous durophagous shark *Ptychodus mortoni* (Chondrichthyes, Ptychodontidae); pp. 69–73 in R. M. Sullivan and S. G. Lucas (eds.), Fossil Record 4. New Mexico Museum of Natural History and Science Bulletin 67.

Kauffman, E. G. 1972. *Ptychodus* predation upon a Cretaceous *Inoceramus*. Palaeontology 15(3):439–444.

Lane, H. H. 1944. A Survey of the fossil vertebrates of Kansas, pt 1: the fishes. Transactions of the Kansas Academy of Science 47(2):129–176.

LeConte, J. L. 1868. Notes on the Geology of the Survey for the Extension of the Union Pacific Railway, E. D., from the Smoky Hill River, Kansas, to the Rio Grande. Philadelphia: Review Printing House. 76 pp.

Leidy, J. 1859. [*Xystracanthus*, *Cladodus*, and *Petalodus* from the Carboniferous of Kansas]. Proceedings of the Academy of Natural Sciences of Philadelphia 11:3.

Leidy, J. 1868. Notice of American species of *Ptychodus*. Proceedings of the Academy of Natural Sciences of Philadelphia 20:205–208.

Leidy, J. 1873. Contributions to the Extinct Vertebrate Fauna of the Western Territories; in F. V. Hayden (ed.), Report of the U.S. Geological Survey of the Territories, Volume 1. Government Printing Office, Washington, D.C., 358 pp.

Liggett, G. A., K. Shimada, S. C. Bennett, and B. Schumacher. 2005. Cenomanian (Late Cretaceous) reptiles from northwestern Russell County, Kansas. PaleoBios 25(2):9–17.

Mantell, G. A. 1836. A Descriptive Catalogue of the Objects of Geology, Natural History, and Antiquity (chiefly discovered in Sussex,) in the Museum attached to the Sussex Scientific and Literary Institution at Brighton. Relfe and Fletcher, London, England, 41 pp.

Marsh, O. C. 1871. Note on a new and gigantic species of pterodactyle. American Journal of Science, ser. 3, 1(6):472.

Marsh, O. C. 1872. Notice of a new species of *Hadrosaurus*. American Journal of Science 16:301.

Martin, J. E., and G. L. Bell Jr. 1995. Abnormal caudal vertebrae of Mosasauridae from Late Cretaceous marine deposits of South Dakota. Proceedings of the South Dakota Academy of Sciences 7:23–27.

Martin, L. D., and B. M. Rothschild. 1989. Paleopathology and diving mosasaurs. American Scientist 77:460–467.

Meyer, R. L. 1974. Late Cretaceous elasmobranchs from the Mississippi and east Texas embayments of the Gulf Coastal Plain. Ph.D. dissertation, Southern Methodist University, Dallas, Texas, 400 pp.

Morton, S. G. 1834. Synopsis of the organic remains of the Cretaceous group of the United States. Key and Biddle, Philadelphia, 88 pp.

Mudge, B. F. 1876. Notes on the Tertiary and Cretaceous periods of Kansas. US Geological and Geographical Survey of the Territories Bulletin 2(3):211–221.

Mudge, B. F. 1877a. Annual Report of the Committee on Geology, for the Year Ending November 1, 1876. Transactions of the Kansas Academy of Science, Ninth Annual Meeting 4–5.

Mudge, B. F. 1877b. Notes on the Tertiary and Cretaceous periods of Kansas; pp. 277–294 in F. V. Hayden (ed.), Ninth Annual Report of the United States Geological and Geographical Survey of the Territories, Embracing Colorado and Parts of Adjacent Territories; Being a Report of Progress of the Exploration for the year 1875, pt. 1: Geology. Government Printing Office, Washington, D.C., 827 pp.

Rothschild, B., and M. J. Everhart. 2015. Co-ossification of adjacent vertebrae in mosasaurs (Squamata, Mosasauridae); evidence of habitat interactions and susceptibility to disease. Transactions of the Kansas Academy of Science 118(3–4):265–275.

Schultze, H.-P., J. D. Stewart, A. M. Neuner, and R. W. Coldiron. 1982. Type and Figured Specimens of Fossil Vertebrates in the Collection of the University of Kansas Museum of Natural History, pt. 1: Fishes. University of Kansas Museum of Natural History, Lawrence, Kansas, 53 pp.

Schumacher, B. A., and M. J. Everhart. 2005. A stratigraphic and taxonomic review of plesiosaurs from the old Fort Benton Group of central Kansas: a new assessment of old records. Paludicola 5(2):33–54.

Schwimmer, D. R., J. D. Stewart, and G. D. Williams. 1997. Scavenging by sharks of the genus *Squalicorax* in the Late Cretaceous of North America. Palaios 12:71–83.

Scott, R. W. 1970. Paleoecology and Paleontology of the Lower Cretaceous Kiowa Formation, Kansas. University of Kansas Paleontological Contributions, Article 52. University of Kansas Paleontological Institute, Lawrence, Kansas, 94 pp.

Shimada, K. 1996. Selachians from the Fort Hays Limestone Member of the Niobrara Chalk (Upper Cretaceous), Ellis County, Kansas. Transactions of the Kansas Academy of Science 99(1–2):1–15.

Shimada, K. 1997a. Stratigraphic record of the Late Cretaceous lamniform shark, *Cretoxyrhina mantelli* (Agassiz), in Kansas. Transactions of the Kansas Academy of Science 100(3–4):139–149.

Shimada, K. 1997b. Dentition of the Late Cretaceous lamniform shark, *Cretoxyrhina mantelli*, from the Niobrara Chalk of Kansas. Journal of Vertebrate Paleontology 17(2):269–279.

Shimada, K. 1997c. Paleoecological relationships of the Late Cretaceous lamniform shark, *Cretoxyrhina mantelli* (Agassiz). Journal of Paleontology 71(5):926–933.

Shimada, K. 1997d. Gigantic lamnoid shark vertebra from the Lower Cretaceous Kiowa Shale of Kansas. Journal of Paleontology 71(3):522–524.

Shimada, K. 2007. Skeletal and dental anatomy of lamniform shark, *Cretalamna appendiculata*, from upper Cretaceous Niobrara Chalk of Kansas. Journal of Vertebrate Paleontology 27(3):584–602.

Shimada, K. 2008. New anacoracid shark from Upper Cretaceous Niobrara Chalk of western Kansas, U.S.A. Journal of Vertebrate Paleontology 28(4):1189–1194.

Shimada, K. 2012. Dentition of Late Cretaceous shark, *Ptychodus mortoni* (Elasmobranchii, Ptychodontidae).

Journal of Vertebrate Paleontology 32:6:1271–1284.

Shimada, K., and C. Fielitz. 2006. Annotated checklist of fossil fishes from the Smoky Hill Chalk of the Niobrara Chalk (Upper Cretaceous) in Kansas; pp. 193–213 in S. G. Lucas and R. M. Sullivan (eds.), Late Cretaceous vertebrates from the Western Interior. New Mexico Museum of Natural History and Science Bulletin 35.

Shimada, K., and G. E. Hooks, III. 2004. Shark-bitten protostegid turtles from the Upper Cretaceous Mooreville Chalk, Alabama. Journal of Paleontology 78(1):205–210.

Shimada, K., and D. J. Martin. 2008. Fossil fishes from the basal Greenhorn Limestone (Upper Cretaceous, Late Cenomanian) in Russell County, Kansas; pp. 89–103 in G. H. Farley and J. R. Choate (eds.), Unlocking the Unknown; Papers Honoring Dr. Richard Zakrzewski, Fort Hays Studies, Special Issue No. 2, Fort Hays State University, Hays, Kansas, 153 pp.

Shimada, K., K. Ewell, and M. J. Everhart. 2004. The first record of the lamniform shark Genus, Johnlongia, from the Niobrara Chalk (Upper Cretaceous), western Kansas. Transactions of the Kansas Academy of Science 107(3–4):131–135.

Shimada, K., C. K. Rigsby, and S. H. Kim. 2009. Partial skull of Late Cretaceous durophagous shark, Ptychodus occidentalis (Elasmobranchii: Ptychodontidae), from Nebraska, U.S.A. Journal of Vertebrate Paleontology 29(2):336–349.

Shimada, K., M. J. Everhart, R. Decker, and P. D. Decker. 2010. A new skeletal remain of the durophagous shark, Ptychodus mortoni, from the Upper Cretaceous of North America: an indication of gigantic body size. Cretaceous Research 31(2):249–254.

Shimada, K., M. J. Everhart, B. Reilly, and C. Rigsby. 2011. First associated specimen of the Late Cretaceous shark, Cretodus (Elasmobranchii: Lamniformes). Abstracts of the 71st Meeting, Journal of Vertebrate Paleontology, 31:194.

Siverson, M. 1996. Lamniform sharks of the Mid Cretaceous Alinga Formation and Beedagong Claystone, Western Australia. Palaeontology 39:813–849.

Siverson, M. 1999. A new large lamniform shark from the uppermost Gearle Siltstone (Cenomanian, Late Cretaceous) of Western Australia. Transactions of the Royal Society of Edinburgh: Earth Sciences 90:49–66.

Siverson, M., and J. Lindgren. 2005. Late Cretaceous sharks Cretoxyrhina and Cardabiodon from Montana, USA. Acta Palaeontology Polonica 50(2):301–314.

Siverson, M., J. Lindgren, M. G. Newbrey, P. Cederstrom, and T. D. Cook. 2015. Cenomanian-Campanian (Late Cretaceous) mid-plaeolatitude sharks of Cretalamna appendiculata type. Acta Palaeontologica Polonica 60(2):339–384.

Spamer, E. E., E. Daeschler, and L. G. Vostreys-Shapiro. 1995. A study of fossil vertebrate types in the Academy of Natural Sciences of Philadelphia; taxonomic, systematic, and historical perspectives. Academy of Natural Sciences of Philadelphia, Special Publication 16, 434 pp.

Stahl, B. J. 1999. Chondrichthyes III: Holocephali; in H.-P. Schultze (ed.), Handbook of Paleoichthyology, Volume 4. Verlag Dr. Friedrich Pfeil, Munich, Germany, 164 pp.

Sternberg, C. H. 1907. Some animals discovered in the fossil beds of Kansas. Transactions of the Kansas Academy of Science 20:122–124.

Sternberg, C. H. 1909. The Life of a Fossil Hunter. Henry Holt and Company, New York, New York, 286 pp.

Sternberg, C. H. 1911. In the Niobrara and Laramie Cretaceous. Transactions of the Kansas Academy of Science 23:70–74.

Sternberg, C. H. 1917. Hunting Dinosaurs in the Badlands of the Red Deer River, Alberta, Canada. Published by the author, San Diego, California, 232 pp.

Sternberg, M. 1920. George Miller Sternberg: A Biography. Chicago: American Medical Association. 331 pp.

Stewart, J. D. 1980. Reevaluation of the phylogenetic position of the Ptychodontidae. Transactions of the Kansas Academy of Science 83(3):154.

Stewart, J. D. 1990. Niobrara Formation Vertebrate Stratigraphy; pp. 19–30 in S. C. Bennett (ed.), Niobrara Chalk Excursion Guidebook. University of Kansas Museum of Natural History and Kansas Geological Survey, Lawrence, Kansas.

Varricchio, D. J. 2001. Gut contents from a Cretaceous tyrannosaurid; implications for theropod dinosaur digestive tracts. Journal of Paleontology 75(2):401–406.

Welton, B. J., and R. F. Farish. 1993. The Collector's Guide to Fossil Sharks and Rays from the Cretaceous of Texas. Horton Printing Company, Dallas, Texas, 204 pp.

Williston, S. W. 1898. Mosasaurs. University Geological Survey of Kansas 4:81–347.

Williston, S. W. 1900. Cretaceous fishes: selachians and pycnodonts. University Geological Survey of Kansas 6(2):237–256.

Woodward, A. S. 1887. On the definition and affinities of the selachian genus *Ptychodus* Agassiz. Quarterly Journal of the Geological Society 13:121–131.

5. Fishes

Agassiz, L. 1833–1843. Recherches sur les Poissons Fossiles. [5 Volumes]. Neuchâtel, Switzerland, Imprimerie de Petitpierre, 1420 pp.

Arbour, V. M., and P. J. Currie. 2011. An istiodactylid pterosaur from the Upper Cretaceous Nanaimo Group, Hornby Island, British Columbia, Canada. Canadian Journal of Earth Sciences 48:63–69.

Bardack, D. 1965. Anatomy and Evolution of Chirocentrid Fishes. University of Kansas Paleontological Contributions, Article 10. University of Kansas Paleontological Institute, Lawrence, Kansas, 88 pp.

Bardack, D. 1976. Paracanthopterygian and acanthopterygian fishes from the Upper Cretaceous of Kansas. Fieldiana Geology 33(20):355–374.

Beamon, J. C. 1999. Depositional environment and fossil biota of a thin clastic unit of the Kiowa Formation, Lower Cretaceous (Albian), McPherson County, Kansas. M.A. thesis, Fort Hays State University, Hays, Kansas, 97 pp.

Carpenter, K. 1996. Sharon Springs member, Pierre Shale (Lower Campanian) depositional environment and origin of its vertebrate fauna, with a review of North American plesiosaurs. Ph.D. dissertation, University of Colorado, Boulder, Colorado, 251 pp.

Carrillo-Briceño, J., J. Alvarado-Ortega, and C. Torres. 2012. Primer regristro de *Xiphactinus* Leidy, 1870 (Teleostei, Ichthyodectiformes) en el Cretácico de América del Sur (Formación La Luna, Venezuela). Revista Brazileira de Paleontologia 15(3):327–335.

Cicimurri, D. J., and M. J. Everhart. 2001. An elasmosaur with stomach contents and gastroliths from the Pierre Shale (Late Cretaceous) of Kansas. Transactions of the Kansas Academy of Science 104(3–4):129–143.

Cope, E. D. 1868. On a new large enaliosaur. American Journal of Science, Series 2 46(137):263–264.

Cope, E. D. 1871. On the fossil reptiles and fishes of the Cretaceous rocks of Kansas; pp. 385–424 in F. V. Hayden (ed.), [Fourth Annual] Preliminary Report of the United States Geological Survey of Wyoming and Portions of Contiguous Territories (Being a Second Annual Report of Progress). Government Printing Office, Washington:, D.C.

Cope, E. D. 1872a. Note on some Cretaceous vertebrata in the State Agricultural College of Kansas. Proceedings of the American Philosophical Society 12(87):168–170.

Cope, E. D. 1872b. On the families of fishes of the Cretaceous Formation in Kansas. Proceedings of the American Philosophical Society, 12(88):327–357.

Cope, E. D. 1872c. On the geology and paleontology of the Cretaceous strata of Kansas; pp. 318–349 in F. V. Hayden (ed.), [Fifth Annual] Preliminary Report of the United States Geological Survey of Montana and Portions of Adjacent Territories; Being a Fifth Annual Report of Progress. Government Printing Office, Washington, D.C.

Cope, E. D. 1872d. [On an extinct genus of saurodont fishes]. Proceedings of the Academy of Natural Sciences of Philadelphia 24:280–281.

Cope, E. D. 1873. On two new species of Saurodontidae. Proceedings of the Academy of Natural Sciences of Philadelphia 25:337–339.

Cope, E. D. 1874. Review of the vertebrata of the Cretaceous Period found west of the Mississippi River. Bulletin of the U.S. Geological and Geographical Survey of the Territories 1(2):3–48.

Cope, E. D. 1875. The Vertebrata of the Cretaceous Formations of the West; in F. V. Hayden (ed.), Report of the U.S. Geological Survey of the Territories, Volume 2. Government Printing Office, Washington, D.C., 302 pp.

Cope, E. D. 1877a. On some new or little known reptiles and fishes of the Cretaceous no. 3 of Kansas. Proceedings of the American Philosophical Society 17(100):176–181.

Cope, E. D. 1877b. On the genus *Erisichthe*. Bulletin of the U.S. Geological and Geographical Survey, Article 34, 3(4):821–823.

Cope, E. D. 1886. Note on *Erisichthe*. Geology Magazine 3:239.

Druckenmiller, P. S., A. J. Daun, J. L. Skulan, and J. C. Pladziewicz. 1993. Stomach Contents in the Upper Cretaceous Shark *Squalicorax falcatus*. Journal of Vertebrate Paleontology 13(3, Supplement):33A.

Dunkle, D. H. 1969. A new amioid fish from the Upper Cretaceous of Kansas. Kirtlandia, Cleveland Museum of Natural History 7:1–6.

Dunkle, D. H., and C. W. Hibbard. 1946. Some comments upon the structure of a pycnodontid fish from the Upper Cretaceous of Kansas. University of Kansas Science Bulletin 31(8):161–181.

Dutel, H., J. Maisey, D. Schwimmer, P. Janvier, and G. Clément. 2011. Giant coelacanth *Megalocoelacanthus dobiei* from the Upper Cretaceous of North America and its bearings on the phylogeny of Mesozoic coelacanths. Journal of Vertebrate Paleontology 31(Supplement to 3):103.

Dutel, H, J. G. Maisey, D. R. Schwimmer, P. Janvier, M. Herbin, and G. Clément. 2012. The giant Cretaceous coelacanth (Actinistia, Sacropterygii) *Megalocoelacanthus dobiei* Schwimmer, Stewart and Williams, 1994, and its bearing on Latimerioidei interrelationships. PLoS ONE 7(11):1–27.

Everhart, M. J. 2002. New data on cranial measurements and body length of the mosasaur *Tylosaurus nepaeolicus* (Squamata; Mosasauridae), from the Niobrara of Western Kansas. Transactions of the Kansas Academy of Science 105(1–2):33–43.

Everhart, M. J. 2004. First record of the hybodont shark genus, "*Polyacrodus*" sp., (Chondrichthyes; Polyacrodontidae) from the Kiowa Formation (Lower Cretaceous) of McPherson County, Kansas. Transactions of the Kansas Academy of Science 107(1–2):39–43.

Everhart, M. J. 2007. Remains of a pycnodont fish (Actinopterygii: Pycnodontiformes) in a coprolite; an upper record of *Micropycnodon kansasensis* in the Smoky Hill Chalk, western Kansas. Transactions of the Kansas Academy of Science 110(1–2):35–43.

Everhart, M. J. 2009. First occurrence of marine vertebrates in the Early Cretaceous of Kansas: champion shell bed, basal Kiowa Formation. Transactions of the Kansas Academy of Science 112(3–4):201–210.

Everhart, M. J., and P. A. Everhart. 1992. Oyster-shell concentrations: a stratigraphic marker in the Smoky Hill Chalk (Upper Cretaceous) of western Kansas. Transactions of the Kansas Academy of Science 11:12.

Everhart, M. J., and P. A. Everhart. 1993. Notes on the biostratigraphy of the Plethodid *Martinichthys* in the Smoky Hill Chalk (Upper Cretaceous) of western Kansas. Transactions of the Kansas Academy of Science 12:36.

Everhart, M. J., P. A. Everhart, and K. Ewell. 2004. A marine ichthyofauna from the Upper Dakota Sandstone (Late Cretaceous). Abstracts of the oral presentations and posters, Joint Annual Meeting of the Kansas and Missouri Academies of Science, p. 48.

Everhart, M. J., S. A. Hageman, and B. L. Hoffman. 2010. Another Sternberg "fish-within-a-fish" discovery: first report of *Ichthyodectes ctenodon* (Teleostei; Ichthyodectiformes) with stomach contents. Transactions of the Kansas Academy of Science 113(3–4):197–205.

Everhart, M. J., P. Everhart, E. M. Manning, and D. E. Hattin. 2003. A Middle Turonian marine fish fauna from the Upper Blue Hill Shale Member, Carlile Shale, of north central Kansas. Journal of Vertebrate Paleontology 23(3, Supplement):49A.

Fielitz, C. 1999. Phylogenetic analysis of the family Enchodontidae and its relationship to recent members of the order Aulopiformes. Ph.D. dissertation, University of Kansas, Lawrence, Kansas, 86 pp.

Fielitz, C., and K. Shimada. 1999. A new species of *Bananogmius* (Teleostei; Tselfatiiformes) from the Upper Cretaceous Carlile Shale of western Kansas. Journal of Paleontology 73(3):504–511.

Fielitz, C., and K. Shimada. 2009. A new species of *Apateodus* (Teleostei: Aulopiformes) from the Upper Cretaceous Niobrara Chalk of western Kansas, U.S.A. Journal of Vertebrate Paleontology 28(3):650–658.

Fielitz, C., J. D. Stewart, and J. Wiffen. 1999. *Aethocephalichthys hyainarhinos* gen. et sp. nov., a New and Enigmatic Late Cretaceous Actinopterygian from North America and New Zealand; pp. 95–106, in G. Arrantia and H.-P. Shultze (eds.), Mesozoic Fishes 2: Systematics and Fossil Record. Proceedings of the International Meeting. Munich: F. Pfeil.

Friedman, M., K. Shimada, and A. Maltese. 2007. New insights on the Upper Cretaceous pachycormid '*Protosphyraena*' *gladius* (Actinopterygii: Teleostei) from North America. Abstracts, 55th Symposium of Vertebrate Paleontology and Comparative Anatomy and 16th Symposium of Palaeontological Preparation and Conservation, University of Glasgow, Scotland, p. 14.

Friedman, M., K. Shimada, M. J. Everhart, K. J. Irwin, B. S. Grandstaff, and J. D. Stewart. 2013. Geographic and stratigraphic distribution of the Late Cretaceous suspension-feeding bony fish *Bonnerichthys gladius* (Teleostei, Pachycormiformes). Journal of Vertebrate Paleontology 33:35–47.

Friedman, M., K. Shimada, L. Martin, M. J. Everhart, J. Liston, A. Maltese, and M. Triebold. 2010 100-million-year dynasty of giant planktivorous bony fishes in the Mesozoic seas. Science 327:990–993.

Goody, P. C. 1970. The Cretaceous teleostean fish *Cimolichthys* from the Niobrara Formation of Kansas and the Pierre Shale of Wyoming. American Museum Noviates 2434:1–29.

Goody, P. C. 1976. *Enchodus* (Teleostei: Enchodontidae) from the Upper Cretaceous Pierre Shale of Wyoming and South Dakota with an evaluation of the North American enchodontid species. Palaeontographica Abteilung A 152:91–112.

Gregory, J. T. 1950. A large pycnodont from the Niobrara Chalk. Postilla 5:1–10.

Harlan, R. 1824. On a new fossil genus of the order Enalio sauri, (of Conybeare). Journal of the Academy of Natural Sciences of Philadelphia, ser. 1, 3(pt. 2):331–337.

Hattin, D. E. 1962. Stratigraphy of the Carlile Shale (Upper Cretaceous) in Kansas. State Geological Survey of Kansas, Bulletin no. 156. University of Kansas, Lawrence, Kansas. 155 pp.

Hattin, D. E. 1982. Stratigraphy and Depositional Environment of the Smoky Hill Chalk Member, Niobrara Chalk (Upper Cretaceous) of the Type Area, Western Kansas. Kansas Geological Survey Bulletin No. 225. University of Kansas, Lawrence, Kansas, 108 pp.

Hay, O. P. 1898. Observations on the genus of Cretaceous fishes called by Professor Cope *Portheus*. Science 7(175):646.

Hay, O. P. 1903. On certain genera and species of North American Cretaceous actinopterous fishes. Bulletin of the American Museum of Natural History 19:1–95.

Hays, I. 1830. Description of a fragment of the head of a new fossil animal, discovered in a marl pit, near Moorestown, New Jersey. Transactions of the American Philosophical Society, ser. 2, 3(18):471–477.

Hibbard, C. W. 1939. A new pycnodont fish from the Upper Cretaceous of Russell County, Kansas. University of Kansas Science Bulletin 26(9):373–375.

Hibbard, C. W., and A. Graffham. 1941. A new pycnodont fish from the Upper Cretaceous of Rooks County, Kansas. University of Kansas Science Bulletin 27(5):71–77.

Hussakof, L. 1929. A new teleostean fish from the Niobrara of Kansas. American Museum Noviates 357:1–4.

Jordan, D. S. 1924. A collection of fossil fishes in the University of Kansas from the Niobrara Formation of the Cretaceous. University of Kansas Science Bulletin 15(2):219–245.

Kauffman, E. G. 1990. Cretaceous fish predation on a large squid; pp. 195–196 in J. Boucot (ed.), Evolutionary Paleobiology of Behavior and Coevolution. Elsevier, Amsterdam, the Netherlands, 750 pp.

Leidy, J. 1856. Notes on the fishes in the collection of the Academy of Natural Sciences of Philadelphia. Proceedings of the Academy of Natural Sciences of Philadelphia 8:221.

Leidy, J. 1857. Remarks on Saurocephalus and its allies. Transactions of the American Philosophical Society 11:91–95.

Leidy, J. 1859. [*Xystracanthus*, *Cladodus*, and *Petalodus* from the Carboniferous of Kansas]. Proceedings of the Academy of Natural Sciences of Philadelphia 11:3.

Leidy, J. 1865. Cretaceous reptiles of the United States. Smithsonian Contributions to Knowledge 14(6):1–135.

Leidy, J. 1870. [Remarks on ichthyodorulites and on certain fossil Mammalia]. Proceedings of the Academy of Natural Sciences of Philadelphia 22:12–13.

Liggett, G. A. 2001. Dinosaurs to Dung Beetles: Expeditions through Time. Guide to the Sternberg Museum of Natural History. Sternberg Museum of Natural History, Hays, Kansas, 127 pp.

Loomis, F. B. 1900. Die anatomie und die verwandtschaft der Ganoidund Knochen-fische aus der Kreide-Formation von Kansas, U.S.A. Palaeontographica 46:213–283.

Mantell, G. 1822. The fossils of the South Downs; or, Illustrations of the Geology of Sussex. Lupton Relfe, London, xiv + 327 pp.

Mantell, G. 1833. The Geology of the South-East of England. Longman, Rees, Orme, Brown and Longman, London, 415 pp.

Marsh, O. C. 1871. Scientific expedition to the Rocky Mountains. American Journal of Science, ser. 3, 1(2):142–143.

Martin, H. T. 1920. *Anguillavus hackberryensis*. A new species and a new genus of fish from the Niobrara Cretaceous of Kansas. University of Kansas Science Bulletin 13:95–98.

McClung, C. E. 1908. Ichthyological notes on the Kansas Cretaceous, I. University of Kansas Science Bulletin, 4:235–246.

McClung, C. E. 1926. *Martinichthys*, a new genus of Cretaceous fish from Kansas, with descriptions of six new species. Proceedings of American Philosophical Society 65(Supplement 5):20–26.

Miller, H. W. 1957. Intestinal casts in *Pachyrhizodus*, an elopid fish, from the Niobrara Formation of Kansas.

Transactions of the Kansas Academy of Science 60(4):399–401.

Mkhitaryan, T. G., and A. O. Averianov. 2011. New material and phylogenetic position of *Aidachar paludalis* Nesov, 1981 (Actinopterygii, Ichthyodectiformes) from the Late Cretaceous of Uzbekistan. Proceedings of the Zoological Institute RAS 315(2):181–192.

Mudge, B. F. 1866. Discovery of fossil footmarks in the Liassic(?) Formation in Kansas. American Journal of Science, ser. 2, 41(122):174–176.

Mudge, B. F. 1874. Rare forms of fish in Kansas. Transactions of the Kansas Academy of Science 3:121–122.

Mudge, B. F. 1876. Notes on the Tertiary and Cretaceous periods of Kansas. Bulletin of the U.S. Geological and Geographical Survey of the Territories (Hayden) 2(3):211–221.

Newton, E. T. 1878. Remarks on *Saurocephalus*, and on the species which have been referred to this genus. Quarterly Journal of the Geological Society of London 34:786–796.

Osborn, H. F. 1904. The great Cretaceous fish *Portheus molossus* Cope. Bulletin of the American Museum of Natural of Natural History 20(31):377–381.

Rogers, K. 1991. The Sternberg Fossil Hunters: A Dinosaur Dynasty. Mountain Press Publishing Company, Missoula, Montana, 288 pp.

Romer, A. S. 1966. Vertebrate Paleontology. 3rd ed. University of Chicago Press, Chicago, Illinois, 468 pp.

Russell, D. A. 1988. A check list of North American marine Cretaceous vertebrates including fresh water fishes. Occasional Paper of the Tyrrell Museum of Palaeontology, no. 4. Tyrrell Museum of Palaeontology, Drumheller, Alberta, Canada, 57 pp.

Schumacher, B. A., K. Shimada, J. Liston, and A. Maltese. 2016. Highly specialized suspension-feeding bony fish *Rhinconichthys* (Actinopterygii: Pachycormiformes) from the mid-Cretaceous of the United States, England, and Japan. Cretaceous Research 61:71–85.

Schwimmer, D. R., J. D. Stewart, and G. D. Williams. 1994. Giant fossil coelacanths of the Late Cretaceous in the eastern United States. Geology 22:503–506.

Schwimmer, D. R., J. D. Stewart, and G. D. Williams. 1997. *Xiphactinus vetus* and the distribution of *Xiphactinus* species in the eastern United States. Journal of Vertebrate Paleontology 17(3):610–615.

Shimada, K. 1997. Paleoecological relationships of the Late Cretaceous lamniform shark *Cretoxyrhina mantelli* (Agassiz). Journal of Paleontology 71(5):926–933.

Shimada, K. 2006. Marine vertebrates from the Blue Hill Shale Member of the Carlile Shale (Upper Cretaceous: Middle Turonian) in Kansas; pp. 165–175 in S. G. Lucas and R. M. Sullivan (eds.), Late Cretaceous vertebrates from the Western Interior. New Mexico Museum of Natural History and Science Bulletin 35.

Shimada, K., and M. J. Everhart. 2004. Shark bitten *Xiphactinus audax* (Teleostei: Ichthyodectiformes) from the Niobrara Chalk (Upper Cretaceous) of Kansas. Mosasaur 7:35–39.

Shimada, K., and M. J. Everhart. 2009. First record of *Anomoeodus* (Osteichthyes: Pycnodontiformes) from the Upper Cretaceous Niobrara Chalk of western Kansas. Kansas Academy of Science, Transactions 112(1–2):98–102.

Shimada, K., and C. Fielitz. 2006. Annotated checklist of fossil fishes from the Smoky Hill Chalk of the Niobrara Chalk (Upper Cretaceous) in Kansas. Bulletin of the New Mexico Museum of Natural History 35:193–213.

Shimada, K., and D. J. Martin. 2008. Fossil fishes from the basal Greenhorn Limestone (Upper Cretaceous, Late Cenomanian) in Russell County, Kansas; pp. 89–103 in G. H. Farley and J. R. Choate (eds.), Unlocking the Unknown; Papers Honoring Dr. Richard Zakrzewski. Fort Hays Studies, Special Issue No. 2, 153 pp., Fort Hays State University, Hays, Kansas.

Shimada, K., and B. A. Schumacher. 2003. The earliest record of the Late Cretaceous Fish *Thryptodus* (Teleostei: Tselfatiiformes) from central Kansas. Transactions of the Kansas Academy of Science 106(1–2):54–58.

Shor, E. N. 1971. Fossils and Flies: The Life of a Compleat Scientist—Samuel Wendell Williston, 1851–1918. University of Oklahoma Press, Norman, Oklahoma, 285 pp.

Spamer, E. E., R. M. McCourt, R. Middleton, E. Gilmore, and S. B. Duran. 2000. A national treasure: accounting for the natural history specimens from the Lewis and Clark Expedition (Western North America, 1803–1806) in the Academy of Natural Sciences of Philadelphia. Proceedings of the Academy of Natural Sciences of Philadelphia 150:47–58.

Stenzel, H. B. 1945. Decapod crustaceans from the Cretaceous of Texas. University of Texas Publication 4401:401–476, pl. 34–45.

Sternberg, C. H. 1917. Hunting Dinosaurs in the Badlands of the Red Deer River, Alberta, Canada. Published by the author, San Diego, California, 232 pp.

Stewart, A. 1898. A contribution to the knowledge of the ichthyic fauna of the Kansas Cretaceous. Kansas University Quarterly 7(1):22–29, pl. 1.

Stewart, A. 1899. *Pachyrhizodus minimus*, a new species of fish from the Cretaceous of Kansas. Kansas University Quarterly 8(1):37–38.

Stewart, A. 1900. Teleosts of the Upper Cretaceous. The University Geological Survey of Kansas 6:257–403.

Stewart, J. D. 1979. Biostratigraphic distribution of species of *Protosphyraena* (Osteichthyes: Actinopterygii) in the Niobrara and Pierre Formations of Kansas. Proceedings of the Nebraska Academy of Sciences and Affiliated Societies 89:51–52.

Stewart, J. D. 1988. The Stratigraphic Distribution of Late Cretaceous *Protosphyraena* in Kansas and Alabama, Geology; pp. 80–94 in M. E. Nelson (ed.), Geology, Paleontology and Biostratigraphy of Western Kansas: Articles in Honor of Myrl V. Walker. Fort Hays Studies, Third Series, no. 10, Science, Fort Hays State University, Hays, Kansas.

Stewart, J. D. 1990a. Niobrara Formation Vertebrate Stratigraphy; pp. 19–30 in S. C. Bennett (ed.), Niobrara Chalk Excursion Guidebook. University of Kansas Museum of Natural History and Kansas Geological Survey, Lawrence, Kansas.

Stewart, J. D. 1990b. Preliminary account of Holecostome-Inoceramid Commensalism in the Upper Cretaceous of Kansas; pp. 51–59 in A. J. Boucot (ed.), Evolutionary Paleobiology of Behavior and Coevolution. Elsevier, Amsterdam, the Netherlands, 750 pp.

Stewart, J. D. 1996. Cretaceous acanthomorphs of North America; pp. 283–294 in G. Arratia and G. Viohl (eds.)—Mesozoic Fishes: Systematics and Paleoecology. F. Pfeil, Munich, Germany.

Stewart, J. D. 1999. A new genus of Saurodontidae (Teleostei: Ichthyodectiformes) from Upper Cretaceous Rocks of the Western Interior of North America; pp. 335–360 in G. Arrantia, and H.-P. Schultze (eds.), Mesozoic Fishes 2: Systematics and Fossil Record. F. Pfeil, Munich, Germany.

Stewart, J. D., and G. L. Bell Jr. 1994. North America's oldest mosasaurs are teleosts. Natural History Museum of Los Angeles County Contributions to Science 441:1–9.

Stewart, J. D., and K. Carpenter. 1990. Examples of vertebrate predation on cephalopods in the Late Cretaceous of the Western Interior; pp. 203–208 in A. J. Boucot (ed.), Evolutionary Paleobiology of Behavior and Coevolution. Elsevier, Amsterdam, the Netherlands, 750 pp.

Stewart, J. D., and V. Friedman. 2001. Oldest North American Record of Saurodontidae (Teleostei: Ichthyodectiformes). Journal of Vertebrate Paleontology 21(3, Supplement):104A.

Stewart, J. D., P. A. Everhart, and M. J. Everhart. 1991. Small coelacanths from Upper Cretaceous rocks of Kansas. Journal of Vertebrate Paleontology 11(3, Supplement):56A.

Taverne, L. 1999. Revision de *Zanclites xenurus*, Teleosteen (Pisces, Tselfatiiformes) marin du Santonian (Crétacé supérieur) du Kansas (États-Unis). Belgian Journal of Zoology 129(2):421–438.

Taverne, L. 2000a. Révision du genre *Martinichthys*, poisson marin (Teleostei, Tselfatiiformes) du Crétacé supérieur du Kansas (États-Unis). Geobios 33(2):211–222.

Taverne, L. 2000b. Osteology et position systematique du genre *Plethodus*, et des nouveaux genres *Dixonangomius* et *Pentanogmius*, ossils du Crétacé (Telostei, Tselfatiiformes). Biologisch Jaarboek Dodonaea 67(1):94–123.

Taverne, L. 2001a. Révision du genre *Bananogmius* (Teleostei, Tselfatiiformes), poisson marin du Crétacé supérieur d'Amérique du Nord et d'Europe. Geodiversitas 23(1):17–40.

Taverne, L. 2001b. Révision de *Niobrara encarsia* téléostéen (Osteichthyes, Tselfatiiformes) du Crétacé supérieur marin du Kansas (États-Unis). Belgian Journal of Zoology 131(1):3–16.

Taverne, L. 2001c. Révision de *Syntegmodus altus* (Teleostei, Tselfathformes), poisson marin du Crétacé supérieur du Kansas (États-Unis). Cybium 25(3):251–260.

Taverne, L. 2002a. Révision de *Luxilites striolatus* poisson marin (Teleostei, Tselfatiiformes) du Crétacé supérieur du Kansas (États-Unis). Belgian Journal of Zoology 132(1):25–34.

Taverne, L. 2002b. Étude de *Pseudanogmius maiseyi* gen. et. sp. nov., poisson marin (Teleostei, Tselfatiiformes) du Crétacé supérieur du Kansas (États-Unis). Geobios 35:605–614.

Taverne, L. 2003. Redescription critique des genres *Thryptodus*, P*seudothryptodus* et *Paranogmius*, ossils marin (Teleostei, Tselfatiiformes) du Crétacé supérieur des États-Unis, d'Egypte et de Libye. Belgian Journal of Zoology 133(2):163–173.

Taverne, L. 2004. Ostéologie de *Pentanogmius evolutus* (Cope, 1877) n. comb. (Teleostei, Tselfatiiformes) du Crétacé supérieur marin des États-Unis. Remarques sur la systématique du genre *Pentanogmius* Taverne 2000. Geodiversitas 26(1):89–113.

Thurmond, J. T. 1969. Notes on mosasaurs from Texas. Texas Journal of Science 21(1):69–79.

Vavrek, M. J., A. M. Murray, and P. R. Bell. 2016. *Xiphactinus audax* Leidy 1870 from the Puskwaskau Formation (Santonian to Campanian) of northwestern Alberta, Canada, and the distribution of *Xiphactinus* in North America. Vertebrate Anatomy Morphology Palaeontology 1(1):89–100.

Vullo, R., E. Buffetaut, and M. J. Everhart. 2012. Reappraisal of *Gwawinapterus beardi* from the Late Cretaceous of Canada: a saurodontid fish, not a pterosaur. Journal of Vertebrate Paleontology 32(5):1198–1201.

Walker, M. V. 1982. The Impossible Fossil. University Forum, Fort Hays State University 26, 4pp.

Walker, M. V. (M. J. Everhart [ed.]). 2006. The impossible fossil—revisited. Transactions of the Kansas Academy of Science 109(1–2):87–96.

Wiley, E. O., and J. D. Stewart. 1977. A gar (*Lepisosteus* sp.) from the marine Cretaceous Niobrara Formation of western Kansas. Copeia 4:761–762.

Wiley, E. O., and J. D. Stewart. 1981. *Urenchelys abditus*, new species, the first undoubted eel (Teleostei: Anquilliformes) from the Cretaceous of North America. Journal of Vertebrate Paleontology 1(1):43–47.

Williston, S. W. 1894. On various vertebrate remains from the lowermost Cretaceous of Kansas. Kansas University Quarterly 3(1):1–4, pl. I.

Williston, S. W. 1898. Addenda to Part I. University Geological Survey of Kansas 4:28–32.

Williston, S. W. 1900. Cretaceous fishes: selachians and pycnodonts. University Geological Survey of Kansas 6:237–256.

Woodward, A. S. 1895. Catalogue of the Fossil Fishes in the British Museum: pt. 3. British Museum of Natural History, London, England, 544 pp.

Zielinski, S. L. 1994. First report of Pycnodontidae (Osteichthyes) from the Blue Hill Shale Member of the Carlile Shale (Upper Cretaceous; Middle Turonian), Ellis County, Kansas. Transactions of the Kansas Academy of Science 13:44.

Agassiz, Louis. 1849. [Remarks on crocodiles of the green sand of New Jersey and on *Atlantochelys*]. Proceedings of the Academy of Natural Sciences of Philadelphia 4:169.

Almy, K. J. 1987. Thof's dragon and the letters of Capt. Theophilus Turner, M.D., U.S. Army. Kansas History Magazine 10(3):170–200.

Beamon, J. C. 1999. Depositional environment and fossil biota of a thin clastic unit of the Kiowa Formation, Lower Cretaceous (Albian), McPherson County, Kansas. M.S. thesis, Fort Hays State University, Hays, Kansas, 97 pp.

Carpenter, K. 1996. Sharon Springs Member, Pierre Shale (Lower Campanian) depositional environment and origin of its vertebrate fauna, with a review of North American plesiosaurs. Ph.D. dissertation, University of Colorado, Boulder, Colorado, 251 pp.

Case, E. C. 1897. On the osteology and relationship of *Protostega*. Journal of Morphology 14:21–60.

Case, E. C. 1898. *Toxochelys*. University Geological Survey of Kansas 4:370–385.

Cope, E. D. 1870. [Observations on the Reptilia of the Triassic Formations of the Atlantic regions of the United States]. Proceedings of the American Philosophical Society 11(84):444–446.

Cope, E. D. 1872a. [Sketch of an expedition in the valley of the Smoky Hill River in Kansas]. Proceedings of the American Philosophical Society 12(87):174–176.

Cope, E. D. 1872b. On a new testudinate from the chalk of Kansas. Proceedings of the American Philosophical Society 12(88):308–310.

Cope, E. D. 1872c. A description of the genus *Protostega*, a form of extinct testudinata. Proceedings of the American Philosophical Society 12(88):422–433.

Cope, E. D. 1872d. On the geology and paleontology of the Cretaceous strata of Kansas; pp. 318–349 in F. V. Hayden (ed.), [Fifth Annual] Preliminary Report of the United States Geological Survey of Montana and Portions of Adjacent Territories; Being a Fifth Annual Report of Progress. Government Printing Office, Washington, D.C.

Cope, E. D. 1872e. [On a species of *Clidastes* and *Plesiosaurus gulo* Cope]. Proceedings of the Academy of Natural Sciences of Philadelphia 24:127–129.

6. Turtles

Cope, E. D. 1873. [On *Toxochelys latiremis*]. Proceedings of the Academy of Natural Sciences of Philadelphia 25:10.

Cope, E. D. 1875. The Vertebrata of the Cretaceous Formations of the West; in F. V. Hayden (ed.), Report of the U.S. Geological Survey of the Territories, Volume 2. Government Printing Office, Washington, D.C., 302 pp.

Cope, E. D. 1877. On some new or little known reptiles and fishes of the Cretaceous no. 3 of Kansas. Proceedings of the American Philosophical Society 17(100):176–181.

Densmore, M. and Brinkman, D. B. 2013. A new specimen of *Porthochelys* (Testudines: Chelonioidea) from the Late Cretaceous Niobrara Formation of Kansas. Kirtlandia 58:1–4.

Dollo, L. 1887. Le hainosaure et les nouveaux vertébrés ossils du Musée de Bruxelles. Revue des Questions Scientifiques 21:504–539 and 22:70–112.

Druckenmiller, P. S., A. J. Daun, J. L. Skulan, and J. C. Pladziewicz. 1993. Stomach contents in the Upper Cretaceous shark *Squalicorax falcatus*. Journal of Vertebrate Paleontology 13(3, Supplement):33A.

Elliot, D. K., G. V. Irby, and J. H. Hutchinson. 1997. *Desmatochelys lowi*, a marine turtle from the Upper Cretaceous; pp. 243–258 in J. M. Callaway and E. L. Nicholls (eds.) Ancient Marine Reptiles. Academic Press, San Diego, California.

Everhart, M. J. 2013. A new specimen of the marine turtle, *Protostega gigas* Cope (Cryptodira; Protostegidae), from the Late Cretaceous Smoky Hill Chalk of western Kansas. Transactions of the Kansas Academy of Science 116(1–2):73.

Everhart, M. J., and Pearson, G. 2009. First report on a marine turtle from the Fairport Chalk Member of the Carlile Shale of Mitchell County, Kansas. Transactions of the Kansas Academy of Science 112(1–2):138–139.

Gaffney, E. S., and R. Zangerl. 1968. A revision of the chelonian genus *Bothremys* (Pleurodira: Pelomedusidae). Fieldiana, Geology 16(7):193–239.

Hattin, D. E. 1982. Stratigraphy and Depositional Environment of the Smoky Hill Chalk Member, Niobrara Chalk (Upper Cretaceous) of the Type Area, Western Kansas. Kansas Geological Survey Bulletin No. 225. Kansas Geological Survey, University of Kansas, Lawrence, Kansas, 108 pp.

Hay, O. P. 1895. On certain portions of the skeleton of *Protostega gigas*. Field Columbian Museum Zoological Series 1(2):57–62.

Hay, O. P. 1896. On the skeleton of *Toxochelys latiremis*. Field Columbian Museum Zoological Series 1(5):101–106.

Hay, O. P. 1905. A revision of the species of the family of fossil turtles called Toxochelyidae, with descriptions of two new species of *Toxochelys* and a new species of *Porthochelys*. Bulletin of the American Museum of Natural History 21(10):177–185.

Hay, O. P. 1908. The Fossil Turtles of North America. Carnegie Institution of Washington, Publication No. 75. Carnegie Institution of Washington, Washington, D.C., 568 pp.

Hirayama, R. 1997. Distribution and diversity of Cretaceous chelonioids; pp. 225–241 in J. M. Callaway and E. L. Nicholls (eds.), Ancient Marine Reptiles. Academic Press, San Diego, California.

Hooks, G. E., III. 1998. Systematic revision of the Protostegidae, with a redescription of *Carcarichelys gemma* Zangerl, 1957. Journal of Vertebrate Paleontology 18(1):85–98.

Kear, B. P. 2006. First gut contents in a Cretaceous sea turtle. Biological Letters 2:113–115.

Lane, H. H. 1946. A survey of the fossil vertebrates of Kansas, pt. 3: the reptiles. Transactions of the Kansas Academy of Science 49(3):289–332.

Leidy, J. 1865. Cretaceous reptiles of the United States. Smithsonian Contributions to Knowledge 14(192):1–135.

Leidy, J. 1873. Contributions to the Extinct Vertebrate Fauna of the Western Territories; in F. V. Hayden (ed.), Report of the U.S. Geological Survey of the Territories, Volume 1. Government Printing Office, Washington, D.C., 358 pp.

Konuki, R. 2008. Biostratigraphy of sea turtles and possible bite marks on a *Toxochelys* (Testudine, Chelonioidea) from the Niobrara Formation (late Santonian), Logan County, Kansas and paleoecological implications for predator-prey relationships among large marine vertebrates. M.S. thesis, Fort Hays State University, Hays, Kansas, 141 pp.

Manning, E. M. 1994. Dr. William Spillman (1806–1886), pioneer paleontologist of Mississippi. Mississippi Geology 15(4):64–69.

Matzke, A. T. 2007. An almost complete juvenile specimen of the cheloniid turtle *Ctenochelys stenoporus* (Hay, 1905) from the Upper Cretaceous Niobrara Formation of Kansas, USA. Palaeontology 50(3):669–691.

Matzke, A. T. 2008. A juvenile *Toxochelys latiremis* (Testudines, Cheloniidae) from the Upper Cretaceous Niobrara Formation of Kansas, USA. Neues Jahrbuch für Geologie und Paläontologie. Abhandlungen 249(3):371–380.

Matzke, A. T. 2009. Osteology of the skull of *Toxochelys* (Testudines, Chelonioidea) (with 25 text-figures). Palaeontographica Abteilung A 288(4):93–150.

Nicholls, E. L. 1988. New material of *Toxochelys latiremis* Cope, and a revision of the genus *Toxochelys* (Testudines, Chelonioidea). Journal of Vertebrate Paleontology 8(2):181–187.

Nicholls, E. L. 1992. Note on the occurrence of the marine turtle *Desmatochelys* (Reptilia: Chelonioidea) from the Upper Cretaceous of Vancouver Island. Canadian Journal of Earth Sciences 29:377–380.

Nicholls, E. L. 1997. Introduction to Part III: Testudines; pp. 225–241 in J. M. Callaway and E. L. Nicholls (eds.), Ancient Marine Reptiles. Academic Press, San Diego, California.

Roth, D. D., and M. J. Everhart. 2012. Preparation of a protostegid turtle from the Fairport Chalk (Carlile Formation, Late Cretaceous). Transactions of the Kansas Academy of Science 115(1–2): 71–72.

Russell, D. A. 1993. Vertebrates in the Western Interior Sea; pp. 665–680 in W. G. E. Caldwell and E. G. Kauffman (eds.), Evolution of the Western Interior Basin. Geological Association of Canada, Special Paper 39. Geological Association of Canada, St. John's, Newfoundland.

Schultze, H.-P., L. Hunt, J. Chorn, and A. M. Neuner. 1985. Type and Figured Specimens of Fossil Vertebrates in the Collection of the University of Kansas Museum of Natural History, pt. 2: Fossil Amphibians and Reptiles. University of Kansas Museum of Natural History, Lawrence, Kansas, 66 pp.

Schwimmer, D. R., J. D. Stewart, and G. D. Williams. 1997. Scavenging by sharks of the genus *Squalicorax* in the Late Cretaceous of North America. Palaios 12:71–83.

Shimada, K., and G. E. Hooks III. 2004. Shark-bitten protostegid turtles from the Upper Cretaceous Mooreville Chalk, Alabama. Journal of Paleontology 78(1):205–210.

Shor, E. N. 1971. Fossils and Flies: The Life of a Compleat Scientist—Samuel Wendell Williston, 1851–1918. University of Oklahoma Press, Norman, Oklahoma, 285 pp.

Sternberg, C. H. 1884. Directions for collecting vertebrate fossils. Kansas City Review of Science and Industry 8(4):219–221.

Sternberg, C. H. 1905. *Protostega gigas* and other Cretaceous reptiles and fishes from the Kansas Chalk. Transactions of the Kansas Academy of Science 19:123–128.

Sternberg, C. H. 1909. The Life of a Fossil Hunter. Henry Holt and Company, New York, New York, 286 pp.

Stewart, J. D. 1978. Earliest record of the Toxochelyidae. Transactions of the Kansas Academy of Science 81(2):178.

Stewart, J. D. 1990. Niobrara Formation vertebrate stratigraphy; pp. 19–30 in S. C. Bennett (ed.), Niobrara Chalk Excursion Guidebook. University of Kansas Museum of Natural History and Kansas Geological Survey, Lawrence, Kansas.

Wieland, G. R. 1896. *Archelon ischyros*: a new gigantic cryptodire testudinate from the Fort Pierre Cretaceous of South Dakota. American Journal of Science, ser. 4, 2(12):399–412.

Wieland, G. R. 1898. The protostegan plastron. American Journal of Science, ser. 4, 5:15–20.

Wieland, G. R. 1900. Some observations on certain well-marked stages in the evolution of the testudinate humerus. American Journal of Science, ser. 4, 9(54):413–424.

Wieland, G. R. 1902. Notes on the Cretaceous turtles, *Toxochelys* and *Archelon*, with a classification of the marine Testudinata. American Journal of Science, ser. 4, 14:95–108.

Wieland, G. R. 1905. A New Niobrara *Toxochelys*. American Journal of Science, ser. 4, 20(119):325–343.

Wieland, G. R. 1906. The osteology of *Protostega*. Memoirs of the Carnegie Museum 2(7):279–305.

Wieland, G. R. 1909. Revision of the Protostegidae. American Journal of Science, ser. 4, 27(158):101–130.

Williston, S. W. 1894a. A new turtle from the Benton Cretaceous. Kansas University Quarterly 3(1):5–18.

Williston, S. W. 1894b. On various vertebrate remains from the lowermost Cretaceous of Kansas. Kansas University Quarterly 3(1):1–4.

Williston, S. W. 1898. Turtles. University Geological Survey of Kansas 4:349–369.

Williston, S. W. 1901. A new turtle from the Kansas Cretaceous. Transactions of the Kansas Academy of Science 17:195–199.

Williston, S. W. 1902. On the hind limb of *Protostega*. American Journal of Science, ser. 4, 13(76):276–278.

Williston, S. W. 1914. Water Reptiles of the Past and Present. University of Chicago Press, Chicago, Illinois, 251 pp.

Zangerl, R. 1953. The vertebrate fauna of the Selma Formation of Alabama, pt. 4: The turtles of the family Toxochelyidae. Fieldiana Geology Memoirs 3(4):137–277.

Zangerl, R., and R. E. Sloan. 1960. A new description of *Desmatochelys lowi* (Williston): a primitive cheloniid sea turtle from the Cretaceous of South Dakota. Fieldiana Geology Memoirs 14:7–40.

7. Where the Elasmosaurs Roamed

Adams, D. A. 1997. *Trinacromerum bonneri*, new species, last and fastest pliosaur of the Western Interior Seaway. Texas Journal of Science 49(3):179–198.

Alexander, R. M. 1989. Dynamics of Dinosaurs and Other Extinct Giants. Columbia University Press, New York, New York, 167 pp.

Almy, K. J. 1987. Thof's dragon and the letters of Capt. Theophilus Turner, M.D., U.S. Army. Kansas History Magazine 10(3):170–200.

Bakker, R. T. 1993. Plesiosaur extinction cycles: events that mark the beginning, middle and end of the Cretaceous; pp. 641–664 in W. G. E. Caldwell and E. G. Kauffman (eds.), Evolution of the Western Interior Basin. Geological Association of Canada Special Paper 39, St. John's, Newfoundland, Canada.

Bell, G. L., Jr., M. A. Sheldon, J. P. Lamb, and J. E. Martin. 1996. The first direct evidence of live birth in Mosasauridae (Squamata): exceptional preservation in Cretaceous Pierre Shale of South Dakota. Journal of Vertebrate Paleontology 16(3, Supplement):21A.

Bennett, S. C. 2000. Inferring stratigraphic position of fossil vertebrates from the Niobrara Chalk of western Kansas. Current Research in Earth Sciences. Kansas Geological Survey Bulletin No. 244. University of Kansas, Lawrence, Kansas, 26 pp.

Caldwell, M. W., and M. S. Y. Lee. 2001. Live birth in Cretaceous marine lizards (mosasauroids). Proceedings of the Royal Society of London: Biological Sciences 268(1484):2397–2401.

Carpenter, K. 1996. A review of short-necked plesiosaurs from the Cretaceous of the Western Interior, North America. Neues Jahrbuch für Geologie und Palaeontologie, Abhandlungen (Stuttgart) 201(2):259–287.

Carpenter, K. 1999. Revision of North American elasmosaurs from the Cretaceous of the western interior. Paludicola 2(2):148–173.

Carpenter K., F. Sanders, B. Reed, J. Reed, and P. Larson. 2010. Plesiosaur swimming as interpreted from skeletal analysis and experimental results. Transactions of the Kansas Academy of Science 113(1–2):1–34.

Chatterjee, S., and W. J. Zinsmeister. 1982. Late Cretaceous marine vertebrates from Seymour Island, Antarctic Peninsula. Antarctic Journal 17(5):66.

Cicimurri, D. J., and M. J. Everhart. 2001. An elasmosaur with stomach contents and gastroliths from the Pierre Shale (Late Cretaceous) of Kansas. Transactions of the Kansas Academy of Science 104(3–4):129–143.

Cope, E. D. 1868. Remarks on a new enaliosaurian, *Elasmosaurus platyurus*. Proceedings of the Academy of Natural Sciences of Philadelphia 20:92–93.

Cope, E. D. 1869. Synopsis of the extinct Batrachia and Reptilia of North America, pt. 1. Transactions of the American Philosophical Society, n.s., 14:1–235.

Cope, E. D. 1870. Synopsis of the extinct Batrachia and Reptilia of North America, pt. 1. Transactions of the American Philosophical Society, n.s., 14:1–235

Cope, E. D. 1872. On the geology and paleontology of the Cretaceous strata of Kansas; pp. 318–349 in F. V. Hayden (ed.), [Fifth Annual] Preliminary Report of the United States Geological Survey of Montana and Portions of the Adjacent Territories, Being a Fifth Annual Report of Progress, pt. 3: Paleontology. Government Printing Office, Washington, D.C.

Cope, E. D. 1875. The Vertebrata of the Cretaceous Formations of the West; in F. V. Hayden (ed.), Report of the U.S. Geological Survey of the Territories, Volume 2. Government Printing Office, Washington, D.C., 302 pp.

Cope, E. D. 1894. On the structure of the skull in the plesiosaurian Reptilia, and on two new species from the Late Cretaceous. Proceedings of the American Philosophical Society 33:109–113.

Cruikshank, A. R. I., P. G. Small, and M. A. Taylor. 1991. Dorsal nostrils and hydrodynamically driven underwater olfaction in plesiosaurs. Nature 352:62–64.

Davidson, J. P. 2002. Bonehead mistakes: the background in scientific literature and illustrations from Edward Drinker Cope's first restoration of *Elasmosaurus platyurus*. Proceedings of the Academy of Natural Sciences of Philadelphia 152:215–240.

De la Beche, H. T., and W. D. Conybeare. 1821. Notice of the discovery of a new animal, forming a link between the *Ichthyosaurus* and crocodile, together with general remarks on the osteology of Ichthyosaurus. Transactions of the Geological Society of London 5:559–594.

Druckenmiller, P. S., and A. P. Russell. 2006. A new elasmosaurid (Reptilia: Sauropterygia) from the Lower Cretaceous Clearwater Formation, northeastern Alberta, Canada. Paludicola 5:184–199.

Ellis, R. 2003. Sea Dragons: Predators of the Prehistoric Oceans. University Press of Kansas, Lawrence, Kansas, 313 pp.

Everhart, M. J. 2000. Gastroliths associated with plesiosaur remains in the Sharon Springs Member of the Pierre Shale (Late Cretaceous), western Kansas. Transactions of the Kansas Academy of Science 103(1–2):58–69.

Everhart, M. J. 2002. Remains of immature mosasaurs (Squamata; Mosasauridae) from the Niobrara Chalk (Late Cretaceous) argue against nearshore nurseries. Journal of Vertebrate Paleontology 22(3, Supplement):52A.

Everhart, M. J. 2003. First records of plesiosaur remains in the lower Smoky Hill Chalk Member (Upper Coniacian) of the Niobrara Formation in western Kansas. Transactions of the Kansas Academy of Science 106(3–4):139–148.

Everhart, M. J. 2004a. Plesiosaurs as the food of mosasaurs: new data on the stomach contents of a *Tylosaurus proriger* (Squamata; Mosasauridae) from the Niobrara Formation of western Kansas. Mosasaur 7:41–46.

Everhart, M. J. 2004b. Conchoidal fractures preserved on elasmosaur gastroliths are evidence of use in processing food. Journal of Vertebrate Paleontology 24(Supplement to 3):56A.

Everhart, M. J. 2005a. Elasmosaurid remains from the Pierre Shale (Upper Cretaceous) of western Kansas. Possible missing elements of the type specimen of *Elasmosaurus platyurus* Cope 1868? PalArch Journal of Vertebrate Palaeontology 4(3):19–32.

Everhart, M. J. 2005b. Probable plesiosaur gastroliths from the basal Kiowa Shale (Early Cretaceous) of Kiowa County, Kansas. Transactions of the Kansas Academy of Science 108(3–4):109–115.

Everhart, M. J. 2005c. Bite marks on an elasmosaur (Sauropterygia; Plesiosauria) paddle from the Niobrara Chalk (Upper Cretaceous) as probable evidence of feeding by the lamniform shark, *Cretoxyrhina mantelli*. PalArch Journal of Vertebrate Palaeontology 2(2):14–24.

Everhart, M. J. 2006. The occurrence of elasmosaurids (Reptilia: Plesiosauria) in the Niobrara Chalk of western Kansas. Paludicola 5(4):170–183.

Everhart, M. J. 2007. Use of archival photographs to rediscover the locality of the Holyrood elasmosaur (Ellsworth County, Kansas). Transactions of the Kansas Academy of Science 110(1–2):135–143.

Everhart, M. J. 2015. Elias Putnam West (1820–1892)—lawyer, attorney general, militia commander, judge, postmaster, archaeologist, and paleontologist. Transactions of the Kansas Academy of Science 118(3–4):285–294.

Everhart, M. J., and G. Pearson. 2014. An isolated squamate dorsal vertebra from the Late Cretaceous Greenhorn Formation of Mitchell County, Kansas. Transactions of the Kansas Academy of Science 117(3–4):261–269.

Gasparini, Z., N. Bardet, J. E. Martin, and M. Fernandez. 2003. The elasmosaurid plesiosaur *Aristonectes* Cabrera from the latest Cretaceous of South America and Antarctica. Journal of Vertebrate Paleontology 23(1):104–115.

Hector, J. 1874. On the fossil reptilia of New Zealand. Transactions and Proceedings of the New Zealand Institute 6:333–364.

Henderson, D. M. 2006. Floating point: a computational study of buoyancy, equilibrium, and gastroliths in plesiosaurs. Lethaia 39:227–244.

Kubo, T., M. T. Mitchell, and D. M. Henderson. 2012. *Albertonectes vanderveldei*, a new elasmosaur (Reptilia, Sauropterygia) from the Upper Cretaceous of Alberta. Journal of Vertebrate Paleontology 32:557–572.

LeConte, J. L. 1868. Notes on the Geology of the Survey for the Extension of the Union Pacific Railway, E. D., from the Smoky Hill River, Kansas, to the Rio Grande. Review Printing House, Philadelphia, Pennsylvania, 76 pp.

Leidy, J. 1870. [Remarks on *Elasmosaurus platyurus*]. Proceedings of the Academy of Natural Sciences of Philadelphia 22:9–10.

Lingham-Soliar, T. 2000. Plesiosaur locomotion: is the four-wing problem real or merely an atheoretical exercise? Neues Jahrbuch für Geologie und Paläeontologie, Abhandlungen (Stuttgart) 217:45–87.

Lingham-Soliar, T. 2003. Extinction of ichthyosaurs: a catastrophic or evolutionary paradigm? Neues Jahrbuch für Geologie und Palaeontologie Abhandlungen 228(3):421–452.

Liu, S., A. S. Smith, Y. Gu, C. K. Liu, and G. Turk. 2015. Computer simulations imply forelimb dominated underwater flight in plesiosaurs. PloS Computational Biology 11(12):18 pp.

Long, J. A. 1998. Dinosaurs of Australia and New Zealand and Other Animals of the Mesozoic Era. Harvard University Press, Cambridge, Massachusetts, 188 pp.

Mantell, G. A.1854. The medals of creation, or first lessons in geology, and the study of organic remains. Henry G. Bohn, London, England, vol. 2, xi + 447–930.

Marsh, O. C. 1873. Reply to Professor Cope's explanation. American Naturalist 7(6, Appendix):i–ix.

Martin, J. E., J. F. Sawyer, M. Reguero, and J. A. Case. 2007. Occurrence of a young elasmosaurid plesiosaur skeleton from the Late Cretaceous (Maastrichtian) of Antarctica; 4 pp. in A. K. Cooper and C. R. Raymond et al. (eds.), Antarctica. A Keystone in a Changing World—Online Proceedings of the 10th International Symposium on Antarctic Earth Sciences, USGS Open-File Report 2007–1047, Short Research Paper 066.

Martin, L. D., and J. D. Stewart. 1977. The oldest (Turonian) mosasaurs from Kansas. Journal of Paleontology 51(5):973–975.

Massare, J. A. 1987. Tooth morphology and prey preference of Mesozoic marine reptiles. Journal of Vertebrate Paleontology 7(2):121–137.

McGowan, C. 2001. The Dragon Seekers. Perseus Publishing, Cambridge, Massachusetts, 253 pp.

Mudge, B. F. 1876. Notes on the Tertiary and Cretaceous periods of Kansas. Bulletin of the U.S. Geological Survey of the Territories (Hayden) 2(3):211–221.

Mudge, B. F. 1877. Notes on the Tertiary and Cretaceous periods of Kansas; pp. 277–294 in F. V. Hayden (ed.), Ninth Annual Report of the United States Geological and Geographical Survey of the Territories, Embracing Colorado and Parts of Adjacent Territories; Being a Report of Progress of the Exploration for the year 1875, pt. 1: Geology. Government Printing Office, Washington, D.C., 827 pp.

Mulder, E. W. A., N. Bardet, P. Godefroit, and J. W. M. Jagt. 2003. Elasmosaur remains from the Maastrichtian type Area, and a review of the latest Cretaceous elasmosaurs (Reptilia, Plesiosauroidea); pp. 93–105 in E. W. A. Mulder (ed.), On Latest Cretaceous Tetrapods from the Maastrichtian Type Area. Publicaties van het Natuurhistorisch Genootschap in Limburg, Reeks 44, aflevering 1. Stichting Natuurpublicaties Limburg, Maastricht, the Netherlands, 188 pp.

Nakaya, H. 1989. Upper Cretaceous elasmosaurid (Reptilia, Plesiosauria) from Hobetsu, Hokkaido, northern Japan. Translated Proceedings of the Paleontology Society of Japan 154:96–116.

Nicholls, E. L. 1988. Marine vertebrates of the Pembina Member of the Pierre Shale (Campanian, Upper Cretaceous) of Manitoba and their significance to the biogeography. Ph.D. dissertation, University of Calgary, Calgary, Alberta, Canada, 317 pp.

O'Gorman, J.P., F. B. Olivero, S. Santillana, and M. J. Everhart. 2014. Gastroliths associated with an *Aristonectes* specimen (Plesiosauria, Elasmosauridae), López de Bertodano Formation (upper Maastrichtian) Seymour Island (*Is. Marambio*), Antarctic Peninsula. Cretaceous Research 50:228–237.

O'Keefe, F. R. 2008. Cranial anatomy and taxonomy of *Dolichorhynchops bonneri* new combination, a polycotylid plesiosaur from the Pierre Shale of Wyoming and South Dakota. Journal of Vertebrate Paleontology 28(3):664–676.

O'Keefe, F. R., and L. M. Chiappe. 2011. Viviparity and K-selected life history in a Mesozoic marine plesiosaur (Reptilia, Sauropterygia). Science 333(6044):870–873.

O'Neill, J. P. 1999. The Great New England Sea Serpent. Versa Press, East Peoria, Illinois, 256 pp.

Osborn, H. F. 1931. Cope: Master Naturalist. Princeton University Press, Princeton, New Jersey, 740 pp.

Reisdorf, A. G., R. Bux, D. Wyler, M. Benecke, C. Klug, M. W. Maisch, P. Fornaro, and A. Wetzel. 2012. Float, explode or sink: postmortem fate of lung-breathing marine vertebrates. Palaeobiodiversity and Palaeoenvironments 92:67–81.

Rothschild, B. M., and L. Martin. 1993. Paleopathology: Disease in the Fossil Record. CRC Press, Boca Raton, Florida, 386 pp.

Russell, D. A. 1967. Systematics and Morphology of American Mosasaurs. Bulletin no. 23, Peabody Museum of Natural History, Yale University, New Haven, Connecticut, 241 pp.

Russell, D. A. 1993. Vertebrates in the Western Interior Sea; pp. 665–680 in W. G. E. Caldwell and E. G. Kauffman (eds.), Evolution of the Western Interior Basin. Geological Association of Canada, Special Paper, St. John's, Newfoundland.

Sachs, S. 2005. Redescription of *Elasmosaurus platyurus* Cope 1868 (Plesiosauria: Elasmosauridae) from the Upper Cretaceous (lower Campanian) of Kansas, U.S.A. Paludicola 5(3):92–106.

Sachs, S., B. P. Kear, and M. J. Everhart. 2013. Revised vertebral count in the "longest-necked vertebrate" *Elasmosaurus platyurus* Cope 1868, and clarification of the cervical-dorsal transition in Plesiosauria. PloS ONE 8(8):6 pp.

Sato, T. 2003. *Terminonatator ponteixensis*, a new elasmosaur (Reptilia: Sauropterygia) from the Upper Cretaceous of Saskatchewan. Journal of Vertebrate Paleontology 23(1):89–103.

Schumacher, B. A. 2011. A '*woollgari*-zone mosasaur' (Squamata; Mosasauridae) from the Carlile Shale (Lower Middle Turonian) of central Kansas and the stratigraphic overlap of early mosasaurs and pliosaurid plesiosaurs. Transactions of the Kansas Academy of Science 114(1–2):1–14.

Schumacher, B. A., and M. J. Everhart. 2005. A stratigraphic and taxonomic review of plesiosaurs from the old Fort Benton Group of central Kansas: a new assessment of old records. Paludicola 5(2):33–54.

Shuler, E. W. 1950. A new elasmosaur from the Eagle Ford Shale of Texas, pt. II: the elasmosaur and its environment. Fondren Science Series 1(2): 1–32.

Sternberg, C. H. 1917 (1932). Hunting Dinosaurs in the Badlands of the Red Deer River, Alberta, Canada. Published by the author, San Diego, California, 261 pp.

Sternberg, C. H. 1922. Explorations of the Permian of Texas and the Chalk of Kansas, 1918. Transactions of the Kansas Academy of Science 30(1):119–120.

Storrs, G. W. 1984. *Elasmosaurus platyurus* and a page from the Cope-Marsh War. Discovery 17(2):25–27.

Storrs, G. W. 1993. Function and phylogeny in Sauropterygian (Diapsida) evolution. American Journal of Science 293A:63–90.

Storrs, G. W. 1999. An examination of Plesiosauria (Diapsida: Sauropterygia) from the Niobrara Chalk (Upper Cretaceous) of Central North America. University of Kansas Paleontological Contributions, n.s., Article 11. University of Kansas Paleontological Institute, Lawrence, Kansas, 15 pp.

Taylor, M. A. 1981. Plesiosaurs—rigging and ballasting. Nature 290:628–629.

Thompson, W., J. E. Martin, and M. Reguero. 2007. Comparison of gastroliths within plesiosaurs (Elasmosauridae) from the Late Cretaceous marine deposits of Vega Island, Antarctic Peninsula, and the Missouri River area, South Dakota. Geological Society of America, Special Paper 427, pp. 147–153.

Warren, L. 1998. Joseph Leidy: The Last Man Who Knew Everything. Yale University Press, New Haven, Connecticut, 320 pp.

Webb, W. E. 1872. Buffalo Land: An Authentic Account of the Discoveries, Adventures, and Mishaps of a Scientific and Sporting Party in the Wild West. Hubbard Brothers, Philadelphia, Pennsylvania, 504 pp.

Welles, S. P. 1943. Elasmosaurid plesiosaurs with a description of the new material from California and Colorado. University of California Memoirs 13:125–254.

Welles, S. P. 1952. A review of the North American Cretaceous elasmosaurs. University of California Publications in Geological Sciences, 29:47–143.

Welles, S. P. 1962. A new species of elasmosaur from the Aptian of Columbia and a review of the Cretaceous plesiosaurs. University of California Publications in Geological Sciences 44(1):96 pp.

Whittle, C. H., and M. J. Everhart. 2000. Apparent and implied evolutionary trends in lithophagic vertebrates from New Mexico and elsewhere; pp. 75–82 in S. G. Lucas and A. B. Heckert (eds.), Dinosaurs of New Mexico. New Mexico Museum of Natural History and Science Bulletin 17, Albuquerque, New Mexico.

Williston, S. W. 1890. A new plesiosaur from the Niobrara Cretaceous of Kansas. Transactions of the Kansas Academy of Science 7:174–178.

Williston, S. W. 1893. An interesting food habit of the plesiosaurs. Transactions of the Kansas Academy of Science 13:121–122.

Williston, S. W. 1903. North American plesiosaurs. Field Columbian Museum Geological Series 2(1):1–79.

Williston, S. W. 1906. North American plesiosaurs: *Elasmosaurus*, *Cimoliasaurus*, and *Polycotylus*. American Journal of Science, ser. 4, 21(123):221–234.

Williston, S. W. 1914. Water Reptiles of the Past and Present. University of Chicago Press, Chicago, Illinois, 251 pp.

Adams, D. A. 1997. *Trinacromerum bonneri*, new species, last and fastest pliosaur of the Western Interior Seaway. Texas Journal of Science 49(3):179–198.

Ballou, W. 1890. [O. C. Marsh's published letter about Cope.] New York Herald 19(508):11.

Bonner, O. W. 1964. An osteological study of *Nyctosaurus* and *Trinacromerum* with a description of a new species of *Nyctosaurus*. M.S. thesis, Fort Hays State University, Hays, Kansas, 63 pp.

Carpenter, K. 1996. A review of short-necked plesiosaurs from the Cretaceous of the western interior, North America. Neues Jahrbuch für Geologie und Palaeontologie Abhandlungen 201(2):259–287.

8. Pliosaurs and Polycotylids

Carpenter, K. 1997. Comparative cranial anatomy of two North American Cretaceous plesiosaurs; pp. 191–216 in J. M. Calloway and E. L. Nicholls (eds.), Ancient Marine Reptiles. Academic Press, San Diego, California.

Cicimurri, D. J., and M. J. Everhart. 2001. An elasmosaur with stomach contents and gastroliths from the Pierre Shale (Late Cretaceous) of Kansas. Transactions of the Kansas Academy of Science 104(3–4):129–143.

Cope, E. D. 1868. Remarks on a new enaliosaurian, Elasmosaurus platyurus. Proceedings of the Academy of Natural Sciences of Philadelphia 20:92–93.

Cope, E. D. 1869. [Remarks on fossil reptiles, Clidastes propython, Polycotylus latipinnis, Ornithotarsus immanis]. Proceedings of the American Philosophical Society 11:117.

Cope, E. D. 1871. Synopsis of the extinct batrachia and reptilia of North America. Transactions of the American Philosophical Society, n.s., 14:1–252.

Cragin, F. W. 1888. Preliminary description of a new or little known saurian from the Benton of Kansas. American Geologist 2:404–407.

Cragin, F. W. 1894. Vertebrata from the Neocomian of Kansas. Bulletin of the Washburn College Laboratory 2(9):69–73.

Davidson, J. P. 2003. Edward Drinker Cope, Professor Paleozoic and Buffalo Land. Transactions of the Kansas Academy of Science 106(3–4):177–191.

Ellis, R. 2003. Sea Dragons: Predators of the Prehistoric Oceans. University Press of Kansas, Lawrence, Kansas, 313 pp.

Everhart, M. J. 2003. First records of plesiosaur remains in the Lower Smoky Hill Chalk Member (Upper Coniacian) of the Niobrara Formation in western Kansas. Transactions of the Kansas Academy of Science 106(3–4):139–148.

Everhart, M. J. 2004a. New data regarding the skull of Dolichorhynchops osborni (Plesiosauroidea: Polycotylidae) from Rediscovered Photos of the Harvard Museum of Comparative Zoology Specimen. Paludicola 4(3):74–80.

Everhart, M. J. 2004b. Plesiosaurs as the food of mosasaurs: new data on the stomach contents of a Tylosaurus proriger (Squamata; Mosasauridae) from the Niobrara Formation of western Kansas. Mosasaur 7:41–46.

Everhart, M. J. 2004c. Late Cretaceous interaction between predators and prey. Evidence of feeding by two species of shark on a mosasaur. PalArch Journal of Vertebrate Palaeontology 1(1):1–7.

Everhart, M. J. 2007a. Historical note on the 1884 discovery of Brachauchenius lucasi (Plesiosauria; Pliosauridae) in Ottawa County, Kansas. Transactions of the Kansas Academy of Science 110(3–4):255–258.

Everhart, M. J. 2007b. Sea Monsters: Prehistoric Creatures of the Deep. National Geographic, Washington, D.C., 192 pp.

Everhart, M. J. 2009. Probable plesiosaur remains from the Blue Hill Shale (Carlile Formation; Middle Turonian) of north central Kansas. Transactions of the Kansas Academy of Science 112(3–4):215–221.

Everhart, M. J. 2016. William Edward Webb (1838–1906)—war correspondent, railroad land baron, town founder, Kansas legislator, adventurer, fossil collector, author, mining engineer. Transactions of the Kansas Academy of Science 119:179–192.

Everhart, M. J., R. Decker, and P. Decker. 2006. Earliest remains of Dolichorhynchops osborni (Plesiosauria: Polycotylidae) from the basal Fort Hays Limestone, Jewell County, Kansas. Transactions of the Kansas Academy of Science 109(3–4):251.

Gilmore, C. W. 1921. An extinct sea-lizard from western Kansas. Scientific American 124:273, 280.

Hattin, D. E. 1982. Stratigraphy and Depositional Environment of the Smoky Hill Chalk Member, Niobrara Chalk (Upper Cretaceous) of the Type Area, Western Kansas. Kansas Geological Survey Bulletin No. 225. University of Kansas, Lawrence, Kansas, 108 pp.

Liggett, G. A., S. C. Bennett, K. Shimada, and J. Huenergarde. 1997. A Late Cretaceous (Cenomanian) fauna in Russell County, Kansas. Transactions of the Kansas Academy of Science 16:26.

Lingham-Soliar, T. 2003. Extinction of ichthyosaurs: a catastrophic or evolutionary paradigm? Neues Jahrbuch für Geologie und Palaeontologie, Abhandlungen (Stuttgart) 228(3):421–452.

Liu, S., A.S. Smith, Y. Gu, C. K. Liu, and G. Turk. 2015. Computer simulations imply forelimb dominated underwater flight in plesiosaurs. PloS Computational Biology, 18 pp.,

Marsh, O. C. 1871. Scientific expedition to the Rocky Mountains. American Journal of Science, ser. 3, 1(2):142–143.

Moodie, R. L. 1908. Reptilian epiphyses. American Journal of Anatomy 7(4):443–467.

Moodie, R. L. 1911. An embryonic plesiosaurian propodial. Kansas Academy Science, Transactions 23:95–101.

O'Keefe, F. R. 2008. Cranial anatomy and taxonomy of *Dolichorhynchops bonneri* new combination, a polycotylid plesiosaur from the Pierre Shale of Wyoming and South Dakota. Journal of Vertebrate Paleontology 28(3):664–676.

O'Keefe, F. R., and C. J. Byrd. 2012. Ontogeny of the shoulder in *Polycotylus latipinnus* [*sic*] (Plesiosauria: Polycotylidae) and its bearing on plesiosaur viviparity. Paludicola 9(1):21–31.

O'Keefe, F. R., and L. M. Chiappe. 2011. Viviparity and K-selected life history in a Mesozoic marine plesiosaur (Reptilia, Sauropterygia). Science 333(6044):870–873.

Owen, R. 1842. Report on British fossil reptiles. Report of the Eleventh Meeting (at Plymouth, 1841), British Association for the Advancement of Science, Part 2:60–204.

Peterson, J. M. 1987. Science in Kansas: the early years, 1804–1875. Kansas History Magazine 10(3):201–240.

Riggs, E. S. 1944. A new polycotylid plesiosaur. University of Kansas Science Bulletin 30:77–87.

Rothschild, B. M., and L. D. Martin. 1993. Paleopathology: Disease in the Fossil Record. CRC Press, Boca Raton, Florida, 386 pp.

Schmeisser McKean, R. L. 2012. A new species of polycotylid plesiosaur (Reptilia: Sauropterygia) from the Lower Turonian of Utah: extending the stratigraphic range of *Dolichorhynchops*. Cretaceous Research 34:184–199.

Schmeisser, R. L., and D. D. Gillette. 2009. Unusual occurrence of gastroliths in a polycotylid plesiosaur from the Upper Cretaceous Tropic Shale, southern Utah. Palaios 2009(24):453–459.

Schultze, H.-P., L. Hunt, J. Chorn, and A. M. Neuner. 1985. Type and Figured Specimens of Fossil Vertebrates in the Collection of the University of Kansas Museum of Natural History, pt. 2: Fossil Amphibians and Reptiles. Miscellaneous Publications of the University Kansas Museum of Natural History 77, Lawrence, Kansas, 66 pp.

Schumacher, B. A. 2008. On the skull of a pliosaur (Plesiosauria; Pliosauridae) from the Upper Cretaceous (Early Turonian) of the North American western interior. Transactions of the Kansas Academy of Science 111(3–4):203–218.

Schumacher, B. A., and M. J. Everhart. 2004. A new assessment of plesiosaurs from the Old Fort Benton Group, Central Kansas. Abstracts of the oral presentations and posters, Joint Annual Meeting of the Kansas and Missouri Academies of Science, p. 50.

Schumacher, B. A., and M. J. Everhart. 2005. A stratigraphic and taxonomic review of plesiosaurs from the old "Fort Benton Group" of central Kansas: a new assessment of old records. Paludicola 5(2):33–54.

Schumacher, B. A., and J. E. Martin. 2016. *Polycotylus latipinnis* Cope (Plesiosauria, Polycotylidae), a nearly complete skeleton from the Niobrara Formation (Early Campanian) of southwestern South Dakota. Journal of Vertebrate Paleontology 36(1):15 pp.

Schumacher, B. A., K. Carpenter, and M. J. Everhart. 2013. A new Cretaceous pliosaurid (Reptilia, Plesiosauria) from the Carlile Shale (middle Turonian) of Russell County, Kansas. Journal of Vertebrate Paleontology 33(3):613–628.

Scott, R. W. 1970. Paleoecology and Paleontology of the Lower Cretaceous Kiowa Formation, Kansas. University of Kansas Paleontological Contributions, Article 52 (Cretaceous 1). University of Kansas Paleontological Institute, Lawrence, Kansas, 94 pp.

Sternberg, C. H. 1922. Explorations of the Permian of Texas and the chalk of Kansas, 1918. Transactions of the Kansas Academy of Science 30(1):119–120.

Sternberg, G. F., and M. V. Walker. 1957. Report on a plesiosaur skeleton from western Kansas. Transactions of the Kansas Academy of Science 60(1): 86–87.

Stewart, J. D. 1990. Niobrara Formation vertebrate stratigraphy; pp. 19–30 in S. C. Bennett (ed.), Niobrara Chalk Excursion Guidebook. University of Kansas Museum of Natural History and Kansas Geological Survey, Lawrence, Kansas.

Storrs, G. W. 1999. An examination of plesiosauria (Diapsida: Sauropterygia) from the Niobrara Chalk (Upper Cretaceous) of Central North America. University of Kansas Paleontological Contributions, n.s., Article 11. University of Kansas Paleontological Institute, Lawrence, Kansas, 15 pp.

Webb, W. E. 1872. Buffalo Land: An Authentic Account of the Discoveries, Adventures, and Mishaps of a Scientific and Sporting Party in the Wild West. Hubbard Brothers, Philadelphia, Pennsylvania, 503 pp.

Williston, S. W. 1893. An interesting food habit of the plesiosaurs. Transactions of the Kansas Academy Science 13:121–122.

Williston, S. W. 1902. Restoration of *Dolichorhynchops osborni*, a new Cretaceous plesiosaur. Kansas University Science Bulletin 1(9):241–244.

Williston, S. W. 1903. North American plesiosaurs. Field Columbian Museum Geological Series 2(1):1–79.

Williston, S. W. 1906. North American plesiosaurs: *Elasmosaurus*, *Cimoliasaurus*, and *Polycotylus*. American Journal of Science, ser. 4, 21(123):221–234.

Williston, S. W. 1907. The skull of *Brachauchenius*, with special observations on the relationships of the plesiosaurs. United States National Museum Proceedings 32:477–489.

Adams, D. A. 1997. *Trinacromerum bonneri*, new species, last and fastest pliosaur of the Western Interior Seaway. Texas Journal of Science 49(3):179–198.

Almy, K. J. 1987. Thof's dragon and the letters of Capt. Theophilus Turner, M.D., U.S. Army. Kansas History Magazine 10(3):170–200.

Bakker, R. T. 1993. Plesiosaur extinction cycles: events that mark the beginning, middle and end of the Cretaceous; pp. 641–664 in W. G. E. Caldwell and E. G. Kauffman (eds.), Evolution of the Western Interior Basin. Special Paper 39. Geological Association of Canada, St. John's, Newfoundland.

Bardet, N., and J. W. M. Jagt. 1996. *Mosasaurus hoffmanni*, le 'Grand Animal fossile des Carrieres de Maestricht': deux siècles d'histoire. Bulletin du Museum d'Histoire Naturelle, Paris, 4th ser., 18:569–593.

Bardet, N., and X. P. Suberbiola. 2001. The basal mosasaurid *Halisaurus sternbergii* from the Late Cretaceous of Kansas (North America): a review of the Uppsala type specimen. Comptes Rendus de l'Académie des Sciences, ser. IIA, Earth and Planetary Science 332:395–402.

Bardet, N., and C. Tunoglu. 2002. The first mosasaur (Squamata) from the Late Cretaceous of Turkey. Journal of Vertebrate Paleontology 22(3):712–715.

Bardet, N., J. W. M. Jagt, M. M. M. Kuypers, and R. W. Dortangs. 1998. Shark tooth marks on a vertebra of the mosasaur *Plioplatecarpus marshi* from the Late Maastrichtian of Belgium. Publicaties van het Natuurhistorisch Genootschap in Limburg 41(1):52–55.

Bardet N., X. P. Suberbiola, M. Iarochene, F. Bouyahyaoui, B. Bouya, and M. Amaghzaz. 2005. A new species of *Halisaurus* from the Late Cretaceous phosphates of Morocco, and the phylogenetical relationships of the Halisaurinae (Squamata: Mosasauridae). Zoological Journal of the Linnean Society 143:447–472.

Baur, G. 1892. On the morphology of the skull of the Mosasauridae. Journal of Morphology 7(1):1–22.

Beamon, J. C. 1999. Depositional environment and fossil biota of a thin clastic unit of the Kiowa Formation, Lower Cretaceous (Albian), McPherson County, Kansas. M.S. thesis, Fort Hays State University, Hays, Kansas, 97 pp.

Bell, G. L., Jr. 1993. A phylogenetic revision of Mosasauroidea. Ph.D Dissertation, University of Texas at Austin, Austin, Texas, 293 pp.

Bell, G. L., Jr. 1997a. Mosasauridae—introduction; pp. 281–292 in J. M. Callaway and E. L. Nicholls (eds.), Ancient Marine Reptiles. Academic Press, San Diego, California, 501 pp.

Bell, G. L., Jr. 1997b. A phylogenetic revision of North American and Adriatic Mosasauroidea; pp. 293–332 in J. M. Callaway and E. L Nicholls (eds.), Ancient Marine Reptiles. Academic Press, San Diego, California, 501 pp.

Bell, G. L., Jr., and K. R. Barnes. 2007. First records of stomach contents in *Tylosaurus nepaeolicus* and comments on predation among Mosasauridae. Abstracts, Second Mosasaur Meeting, Sternberg Museum of Natural History, Hays, Kansas, pp. 6–7.

Bell, G. L., Jr., and J. E. Martin. 1995. Direct evidence of aggressive intraspecific competition in *Mosasaurus conodon* (Mosasauridae: Squamata). Journal of Vertebrate Paleontology 15(Supplement to 3):18A.

Bell, G. L., Jr. and M. J. Polcyn. 2005. *Dallasaurus turneri*, a new primitive mosasauroid from the Middle Turonian of Texas and comments on the phylogeny of Mosasauridae (Squamata). Netherlands Journal of Geosciences / Geologie en Mijnbouw, 84(3):177–194.

Bell, G. L., Jr., and A. M. Sheldon. 2004. A gravid mosasaur (*Plioplatecarpus*) from South Dakota; p. 16 in A. S. Schulp and John W. M. Jagt (eds.), Abstract Book and Field Guide of the First Mosasaur Meeting. Natuurhistorisch Museum, Maastricht, the Netherlands, 107 pp.

Bell, G. L., Jr., and J. P. VonLoh. 1998. New records of Turonian mosasauroids from the western United States: fossil Vertebrates of the Niobrara Formation in South Dakota. Dakoterra 5:15–28.

Bell, G. L., Jr., K. R. Barnes, and M. J. Polcyn. 2013. Late Cretaceous mosasauroids (Reptilia, Squamata) of the Big Bend region in Texas, USA. Earth and Environmental Science Transactions of the Royal Society of Edinburgh 103:11 pp.

Bell, G. L., Jr., M. A. Sheldon, J. P. Lamb, and J. E. Martin. 1996. The first direct evidence of live birth in Mosasauridae (Squamata): exceptional preservation in Cretaceous Pierre Shale of South Dakota. Journal of Vertebrate Paleontology 16(3, Supplement):21A.

Bernard, A., C. Lécuyer, P. Vincent, R. Amiot, N. Bardet, E. Buffetaut, G. Cuny, F. Fourel, F. Martineau, J.-M. Mazin, and A. Prieur. 2010. Regulation of body temperatures by some Mesozoic marine reptiles. Science 328:1379–1382.

Betts, C. W. 1871. The Yale College Expedition of 1870. Harper's New Monthly Magazine 43(257):663–671.

Bice, K. N., and K. Shimada. 2016. Fossil marine vertebrates from the Codell Sandstone Member (middle Turonian) of the Upper Cretaceous Carlile Shale in Jewell County, Kansas, USA. Cretaceous Research 65:172–198.

Bullard, T. S., and M. W. Caldwell. 2010. Redescription and rediagnosis of the tylosaurine mosasaur Hainosaurus pembinensis Nicholls, 1988, as Tylosaurus pembinensis (Nicholls, 1988). Journal of Vertebrate Paleontology 30(2):416–426.

Burnham, D. A. 1991. A new mosasaur from the Upper Demopolis Formation of Sumter County, Alabama. M.S. thesis, University of New Orleans, New Orleans, Louisiana, 63 pp.

Caldwell, M. W. 1999. Squamate phylogeny and the relationships of snakes and mosasaurids. Zoological Journal of the Linnean Society 125:115–147.

Caldwell, M. W. 2012. A challenge to categories: "What, if anything, is a mosasaur?" Bardet, N. (ed.), Troisième congrès sur les Mosasaures, MNHN, Paris. Bulletin de la Société Géologique de France 183(1):7–34.

Caldwell, M. W., L. A. Budney, and D. O. Lamoureux. 2003. Histology of tooth attachment tissues in the Late Cretaceous mosasaurid Platecarpus. Journal of Vertebrate Paleontology 23(3):622–630.

Camp, C. L. 1942. California Mosasaurs. University of California Press, Berkeley, California, 67 pp.

Camper, A. G. 1800. Sur les ossemens fossiles de la montagne de St. Pierre, à Maëstricht (Lettre de A. G. Camper à G. Cuvier). Journal de Physique, de Chemie 51:278–291.

Carpenter, K. 1996. A review of short-necked plesiosaurs from the Cretaceous of the western interior, North America. Neues Jahrbuch für Geologie und Palaeontologie Abhandlungen 201(2):259–287.

Carroll, R. L., and M. Debraga. 1992. Aigialosaurs: Mid-Cretaceous varanid lizards. Journal of Vertebrate Paleontology 12(1):66–86.

Case, J. A., J. E. Martin, D. S. Chaney, M. Reguero, S. A. Marenssi, S. M. Santillana, and M. O. Woodburne. 2000. The first duck-billed dinosaur (family Hadrosauridae) from Antarctica. Journal of Vertebrate Paleontology 20(3):612–614.

Chatterjee, S., and W. J. Zinsmeister. 1982. Late Cretaceous marine vertebrates from Seymour Island, Antarctic Peninsula. Antarctic Journal 17(5):66.

Christiansen, P., and N. Bonde. 2002. A new species of gigantic mosasaur from the Late Cretaceous of Israel. Journal of Vertebrate Paleontology 22(3):629–644.

Cope, E. D. 1868a. Remarks on a new enaliosaurian, Elasmosaurus platyurus. Proceedings of the Academy of Natural Sciences of Philadelphia 20:92–93.

Cope, E. D. 1868b. On some Cretaceous Reptilia. Proceedings of the Academy of Natural Sciences of Philadelphia 20:233–242.

Cope, E. D. 1869a. [Remarks on Thoracosaurus brevispinus, Ornithotarsus immanis, and Macrosaurus proriger]. Proceedings of the Academy of Natural Sciences of Philadelphia 11(81):123.

Cope, E. D. 1869b. [Remarks on fossil reptiles, Clidastes propython, Polycotylus latipinnis, Ornithotarsus immanis]. Proceedings of the American Philosophical Society 11:117.

Cope, E. D. 1869c. [Remarks on Holops brevispinus, Ornithotarsus immanis, and Macrosaurus proriger.] Proceedings of the Academy of Natural Sciences of Philadelphia 1869:123.

Cope, E.D. 1869d. On the reptilian orders Pythonomorpha and Streptosauria. Proceedings of the Boston Society of Natural History 12:250–266.

Cope, E. D. 1870. Synopsis of the extinct Batrachia and Reptilia of North America. Transactions American Philosophical Society 14:1–252.

Cope, E. D. 1872a. Catalogue of the Pythonomorpha found in the Cretaceous strata of Kansas. Proceedings of the American Philosophical Society 12(87):264 and 12(88):265–287.

Cope, E. D. 1872b. [On the structure of the Pythonomorpha]. Proceedings of the Academy of Natural Sciences of Philadelphia 24:140–141.

Cope, E. D. 1872c. On the geology and paleontology of the Cretaceous strata of Kansas; pp. 318–349 in

F. V. Hayden (ed.), [Fifth Annual] Preliminary Report of the United States Geological Survey of Montana and Portions of the Adjacent Territories, Being a Fifth Annual Report of Progress, pt. 3: Paleontology. Government Printing Office, Washington, D.C.

Cope, E. D. 1874. Review of the vertebrata of the Cretaceous period found west of the Mississippi River. Bulletin of the U.S. Geological and Geographical Survey of the Territories 1(2):3–48.

Cope, E. D. 1875. The Vertebrata of the Cretaceous Formations of the West; in F. V. Hayden (ed.), Report of the U.S. Geological Survey of the Territories, Volume 2. Government Printing Office, Washington, D.C., 302 pp.

Cope, E. D. 1881. A new *Clidastes* from New Jersey. American Naturalist 15(7):587–588.

Cuthbertson, R. S., J. C. Mallon, N. E. Campione, and R. B. Holmes. 2007. A new species of mosasaur (Squamata: Mosasauridae) from the Pierre Shale (lower Campanian) of Manitoba. Canadian Journal of Science 44:593–606.

Dollo, L. 1882. Note sur l'ostéologie des Mosasauridae. Bulletin du Musée royale d'Histoire naturelle de Belgique 1:55–74.

Dollo, L. 1885. Premiere note le hainosaure. Bulletin du Musée d'Histoire naturelle de Belgique 4:25–35.

Dollo, L. 1887. Le hainosaure et les nouveaux vertébrés ossils du Musée de Bruxelles. Revue des Questions Scientifiques 21:504–539 and 22:70–112.

Dollo, L. 1889. Première note sur les mosasauriens de Mesvin. Mémoires de la Société Belge de Géologie de Paléontologie and D'Hydrologie 3:271–304.

Dortangs, R. W., A. S. Schulp, E. W. A. Mulder, J. W. M. Jagt, H. H. G. Peeters, and D. Th. De Graaf. 2002. A large new mosasaur from the Upper Cretaceous of the Netherlands. Netherlands Journal of Geosciences/Geologie en Mijnbouw 81(1):1–8.

Elias, M. K. 1931. The geology of Wallace County, Kansas. State Geological Survey Bulletin 18, 254 pp.

Ellis, R. 2003. Sea Dragons: Predators of the Prehistoric Oceans. University Press of Kansas, Lawrence, Kansas, 313 pp.

Emmons, E. 1858. Description of reptilian remains of the marl beds of North-Carolina–reptiles of the green sand; chapter 16 in Report of the North Carolina Geological Survey: Agriculture of the Eastern Counties; Together with

Descriptions of the Fossils from the Marl Beds. Raleigh, Henry D. Turner, 315 pp.

Everhart, M. J. 1999. Evidence of feeding on mosasaurs by the Late Cretaceous lamniform shark, *Cretoxyrhina mantelli*. Journal of Vertebrate Paleontology 17(3, Supplement):43A–44A.

Everhart, M. J. 2001. Revisions to the biostratigraphy of the Mosasauridae (Squamata) in the Smoky Hill Chalk Member of the Niobrara Chalk (Late Cretaceous) of Kansas. Transactions of the Kansas Academy of Science 104(1–2):56–75.

Everhart, M. J. 2002a. New data on cranial measurements and body length of the mosasaur, *Tylosaurus nepaeolicus* (Squamata; Mosasauridae), from the Niobrara Formation of western Kansas. Transactions of the Kansas Academy of Science 105(1–2):33–43.

Everhart, M. J. 2002b. Remains of immature mosasaurs (Squamata; Mosasauridae) from the Niobrara Chalk (Late Cretaceous) argue against nearshore nurseries. Journal of Vertebrate Paleontology 22(3, Supplement):52A.

Everhart, M. J. 2004a. Plesiosaurs as the food of mosasaurs: New data on the gut contents of a *Tylosaurus proriger* (Squamata; Mosasauridae) from the Niobrara Formation of western Kansas. Mosasaur 7:41–46.

Everhart, M. J. 2004b. *Tylosaurus novum* sp.: An update on an unnamed species of basal mosasaur; pp. 35–39 in A. S. Schulp and John W. M. Jagt (eds.), Abstract Book and Field Guide of the First Mosasaur Meeting. Natuurhistorisch Museum, Maastricht, the Netherlands, 107 pp.

Everhart, M. J. 2004c. Late Cretaceous interaction between predators and prey. Evidence of feeding by two species of shark on a mosasaur. PalArch Journal of Vertebrate Palaeontology 1(1): 1–7.

Everhart, M. J. 2005a. *Tylosaurus kansasensis*, a new species of tylosaurine (Squamata: Mosasauridae) from the Niobrara Chalk of western Kansas, U.S.A. Netherlands Journal of Geosciences / Geologie en Mijnbouw 84(3):231–240.

Everhart, M. J. 2005b. Earliest record of the genus *Tylosaurus* (Squamata; Mosasauridae) from the Fort Hays Limestone (Lower Coniacian) of western Kansas. Transactions of the Kansas Academy of Science 108(3–4):149–155.

Everhart, M. J. 2007. Remains of young mosasaurs from the Smoky Hill Chalk (Upper Coniacian–Lower Campanian) of western Kansas; p. 11 in A. S. Schulp and John W.M. Jagt (eds.), Abstract Book and Field Guide of the First Mosasaur Meeting. Natuurhistorisch Museum, 107 pp.

Everhart, M. J. 2008a. A bitten skull of *Tylosaurus kansasensis* (Squamata: Mosasauridae) and a review of mosasaur-on- mosasaur pathology in the fossil record. Transactions of the Kansas Academy of Science 111(3–4):251–262

Everhart, M. J. 2008b. Rare occurrence of a *Globidens* sp. (Reptilia; Mosasauridae) dentary in the Sharon Springs Member of the Pierre Shale (Middle Campanian) of Western Kansas; pp. 23–29 in G. H. Farley and J. R. Choate (eds.), Unlocking the Unknown; Papers Honoring Dr. Richard Zakrzewski. Fort Hays Studies, Special Issue No. 2, Fort Hays State University, Hays, Kansas, 153 pp.

Everhart, M. J. 2008c. The mosasaurs of George F. Sternberg, paleontologist and fossil photographer. Proceedings of the Second Mosasaur Meeting, Fort Hays Studies Special Issue 3, Fort Hays State University, Hays, Kansas 37–46.

Everhart, M. J. (ed.). 2008d. Proceedings of the Second Mosasaur Meeting. Fort Hays Studies Special Issue 3, Fort Hays State University, Hays, Kansas, 172 pp.

Everhart, M. J. 2013. A new specimen of the marine turtle, *Protostega gigas* Cope (Cryptodira; Protostegidae), from the Late Cretaceous Smoky Hill Chalk of western Kansas. Transactions of the Kansas Academy of Science 116(1–2):73.

Everhart, M. J. 2016a. William E. Webb— Civil War correspondent, railroad land baron, town founder, Kansas legislator, adventurer, fossil collector, author. Transactions of the Kansas Academy of Science 119(2):179–192.

Everhart, M. J. 2016b. Note on the rare occurrence of mosasaur remains in the Blue Hill Shale (Middle Turonian) of Mitchell County, Kansas. Transactions of the Kansas Academy of Science 119(3–4):375–380.

Everhart, M. J., and P. A. Everhart. 1996. First report of the shell crushing mosasaur, *Globidens* sp., from the Sharon Springs Member of the Pierre Shale (Upper Cretaceous) of western Kansas. Transactions of the Kansas Academy of Science 15:17.

Everhart, M. J., and P. A. Everhart. 1997. Earliest occurrence of the mosasaur *Tylosaurus proriger* (Mosasauridae: Squamata) in the Smoky Hill Chalk (Niobrara Formation, Upper Cretaceous) of western Kansas. Journal of Vertebrate Paleontology 17(3, Supplement):44a.

Everhart, M. J., and S. E. Johnson. 2001. The occurrence of the mosasaur, *Platecarpus planifrons*, in the Smoky Hill Chalk (Upper Cretaceous) of western Kansas. Journal of Vertebrate Paleontology 21(3, Supplement):48A.

Everhart, M. J., and G. Pearson. 2014. An isolated squamate dorsal vertebra from the Late Cretaceous Greenhorn Formation of Mitchell County, Kansas. Transactions of the Kansas Academy of Science 117(3–4):261–269.

Everhart, M. J., P. A. Everhart, and J. Bourdon. 1997. Earliest documented occurrence of the mosasaur, *Clidastes liodontus*, in the Smoky Hill Chalk (Upper Cretaceous) of western Kansas. Transactions of the Kansas Academy of Science 16:14.

Everhart, M. J., J. W. M. Jagt, E. W. A. Mulder, and A. S. Schulp. 2016. Mosasaurs—how large did they really get? Abstracts, 5th Triennial Mosasaur Meeting, Uppsala Universitet, Uppsala, Sweden, pp. 8–10.

Everhart, M. J., M. Triebold, A. Maltese, J. Jett, and B. Small. 2015. First occurrence of *Mosasaurus* (Squamata; Mosasauridae) from the Late Cretaceous (upper Campanian) of western Kansas. Transactions of the Kansas Academy of Science 118(1–2):139.

Field, D. J., A. LeBlanc, A. Gau, and A. D. Behlke. 2015. Pelagic neonatal fossils support viviparity and precocial life history of Cretaceous mosasaurs. Palaeontology (Rapid Communication), pp. 1–7.

Gilmore, C. W. 1912. A new mosasauroid reptile from the Cretaceous of Alabama. Proceedings of the U.S. National Museum 40(1870):489–484.

Gilmore, C. W. 1921. An extinct sea-lizard from western Kansas. Scientific American 124:273, 280.

Goldfuss, A. 1845. Der Schädelbau des *Mosasaurus*, durch beschreibung einer neuen art dieser gattung erläutert. Nova Acta Academa Caesar Leopoldino-Carolinae Germanicae Natura Curiosorum 21:1–28.

Goldfuss, A. 2013. The skull structure of the *Mosasaurus*, explained by means of a description of a new species of this genus. Edited by M. J. Everhart. Transactions of the Kansas Academy of Science 116(1–2):27–46.

Grigoriev, D. V. 2014. Giant *Mosasaurus hoffmanni* (Squamata, Mosasauridae) from the Late Cretaceous (Maastrichtian) of Penza, Russia. Proceedings of the Zoological Institute RAS 318(2):148–167.

Harlan, R. 1834a. Notice of fossil bones found in the Tertiary formation of the State of Louisiana. Transactions of the American Philosophical Society 4:397–403.

Harlan, R. 1834b. Notice of the discovery of the remains of the *Ichthyosaurus* in Missouri, N. A. Transactions of the American Philosophical Society 4:405–409.

Harrell, T. L., Jr, A. Pérez-Huerta, and C. A. Suarez. 2016. Endothermic mosasaurs? Possible thermoregulation of Late Cretaceous mosasaurs (Reptilia, Squamata) indicated by stable oxygen isotopes in fossil bioapatite in comparison with coeval marine fish and pelagic seabirds. Palaeontology, 13 pp.

Hattin, D. E. 1982. Stratigraphy and Depositional Environment of the Smoky Hill Chalk Member, Niobrara Chalk (Upper Cretaceous) of the Type Area, Western Kansas. Kansas Geological Survey Bulletin No. 225. University Press of Kansas, Lawrence, Kansas, 108 pp.

Holmes, R. B. 1996. *Plioplatecarpus primaevus* (Mosasauridae) from the Bearpaw Formation (Campanian, Upper Cretaceous) of the North American Western Interior Seaway. Journal of Vertebrate Paleontology 16(4):673–687.

Holmes, R., M. W. Caldwell, and S. L. Cumbaa. 1999. A new specimen of *Plioplatecarpus* (Mosasauridae) from the Lower Maastrichtian of Alberta: comments on allometry, functional morphology, and paleoecology. Canadian Journal of Earth Science 36:363–369.

Jacobs, L. L., M. J. Polcyn, L. H. Taylor, and K. Ferguson. 2005. Sea-surface temperatures and palaeoenvironments of dolichosaurs and early mosasaurs. Netherlands Journal of Geosciences / Geologie en Mijnbouw 84(3):269–282.

Jacobs, L. L., O. Mateus, M. J. Polcyn, A. S. Schulp, M. T. Antunes, L. M. Morais, and T. Silva-Tavares. 2006. The occurrence and geological setting of Cretaceous dinosaurs, mosasaurs, plesiosaurs, and turtles from Angola. Journal of the Paleontological Society of Korea 22(1):91–110.

Jiménez -Huidobro, P., and M. W. Caldwell. 2013. *Tylosaurus kansasensis*, *T. proriger*, and *T. nepaeolicus*: can they be differentially diagnosed? Program and abstracts, Society of Vertebrate Paleontology Annual Meeting, p. 139.

Jiménez-Huidobro, P., and M. W. Caldwell. 2016. Assessment and reassignment of the early Maastrichtian mosasaur *Hainosaurus bernardi* Dollo, 1885, to *Tylosaurus* Marsh, 1872. Journal of Vertebrate Paleontology, 13 pp. February 22, 2016.

Jiménez-Huidobro, P., T. R. Simões, and M. W. Caldwell. 2016. Re-characterization of *Tylosaurus nepaeolicus* (Cope, 1874) and *Tylosaurus kansasensis*

Everhart, 2005: ontogeny or sympatry? Cretaceous Research 65:68–81.

Kase, T., P. A. Johnston, A. Seilacher, and J. B. Boyce. 1998. Alleged mosasaur bite marks on Late Cretaceous ammonites are limpet (Patellogastropod) home scars. Geology 26(10):947–950.

Kass, M. S. 1999. *Prognathodon stadtmani* (Mosasauridae): a new Species from the Mancos Shale (Lower Campanian) of Western Colorado; pp. 275–294 in D. D. Gillette (ed.), Vertebrate Paleontology in Utah. Utah Geological Survey, Miscellaneous Publication, No. 99, Salt Lake City.

Kauffman, E. G., and R. V. Kesling. 1960. An Upper Cretaceous ammonite bitten by a mosasaur. University of Michigan Contributions from the Museum of Paleontology 15(9):193–248.

Konishi, T. 2008. A new specimen of *Selmasaurus* sp., cf. *S. russelli* (Mosasauridae: Plioplatecarpini) from Greene County, western Alabama, USA. Proceedings of the Second Mosasaur Meeting, Fort Hays Studies Special Issue 3, Fort Hays State University, Hays, Kansas, pp. 95–105.

Konishi, T., and M. W. Caldwell. 2007. New specimens of *Platecarpus planifrons* (Cope, 1874) (Squamata: Mosasauridae) and a revised taxonomy of the genus. Journal of Vertebrate Paleontology 27(1):59–72.

Konishi, T., and M. W. Caldwell. 2011. Two new plioplatecarpine (Squamata, Mosasauridae) genera from the Upper Cretaceous of North America, and a global phylogenetic analysis of plioplatecarpines. Journal of Vertebrate Paleontology 31(4):754–83.

Konishi, T., M. W. Caldwell, and G. L. Bell Jr. 2010. Redescription of the holotype of *Platecarpus tympaniticus* Cope 1869 (Mosasauridae: Plioplatecarpinae), and its implications for the alpha taxonomy of the genus. Journal of Vertebrate Paleontology 30(5):1410–1421.

Konishi, T., M. G. Newbrey, and M. W. Caldwell. 2014. A small, exquisitely preserved specimen of *Mosasaurus missouriensis* (Squamata, Mosasauridae) from the upper Campanian of the Bearpaw Formation, western Canada, and the first stomach contents for the genus. Journal of Vertebrate Paleontology 34(4): 802–819.

Konishi, T., J. Lindgren, M. W. Caldwell, and L. M. Chiappe. 2012. *Platecarpus tympaniticus* (Squamata, Mosasauridae): osteology of an exceptionally preserved specimen and its insights into the acquisition of a streamlined body shape in mosasaurs. Journal of Vertebrate Paleontology 32:1313–1327.

Konishi, T., M. W. Caldwell, T. Nishimura, K. Sakurai, and K. Tanoue. 2015. A new halisaurine mosasaur (Squamata: Halisaurinae) from Japan: the first record in the western Pacific realm and the first documented insights into binocular vision in mosasaurs. Journal of Systematic Paleontology, 31 pp.

Lee, M. S. Y. 2005. Molecular evidence and marine snake origins. The Royal Society, Biology Letters 1:227–230.

Lee, M. S., G. L. Bell Jr., and M. W. Caldwell. 1999. The origin of snake feeding. Nature 400:655–659.

Leidy, J. 1858. [Remarks on the teeth of *Mosasaurus*]. Proceedings of the Academy of Natural Sciences of Philadelphia 9:176.

Leidy, J. 1868. [Photographs of fossil bones]. Proceedings of the Academy of Natural Sciences of Philadelphia 20:316.

Leidy, J. 1873. Contributions to the Extinct Vertebrate Fauna of the Western Territories; in F. V. Hayden (ed.), Report of the U.S. Geological Survey of the Territories, Volume 1. Government Printing Office, Washington, D.C., 358 pp.

Lindgren, J., and M. Siverson. 2004. *Halisaurus sternbergi*, a Small Mosasaur with an Intercontinental Distribution. In J. Lindgren, Early Campanian Mosasaurs (Reptilia: Mosasauridae) from the Kristianstad Basin, Southern Sweden. Paper IV, Litholund Theses No. 4. Department of Geology, Lund University, Lund, Sweden, 11 pp.

Lindgren, J., and M. Siverson. 2005. *Halisaurus sternbergi*, a small mosasaur with intercontinental distribution. Journal of Paleontology 79(4):763–773.

Lindgren, J., M. W. Caldwell, and J. W. M. Jagt. 2008. New data on the postcranial anatomy of the California mosasaur *Plotosaurus bennisoni* (Camp, 1942) (Upper Cretaceous: Maastrichtian), and the taxonomic status of *P. tuckeri* (Camp, 1942). Journal of Vertebrate Paleontology 28:1043–1054.

Lindgren, J., M. J. Everhart, and M. W. Caldwell. 2011. Three-dimensionally preserved integument reveals hydrodynamic adaptations in the extinct marine lizard *Ectenosaurus* (Reptilia, Mosasauridae). PLoS ONE 6(11):5 pp.

Lindgren, J., J. W. M. Jagt, and M. W. Caldwell. 2007. A fishy mosasaur: the axial skeleton of *Plotosaurus* (Reptilia, Squamata) reassessed. Lethaia 40:153–160.

Lindgren, J., H. F. Kaddumi, and M. J. Polcyn. 2013. Soft tissue preservation in a fossil marine lizard with a bilobed tail fin. Nature Communications 4:2423, 8 pp.

Lindgren, J., M. J. Polcyn, and B. A. Young. 2011. Landlubbers to leviathans: evolution of swimming in mosasaurine mosasaurs. Paleobiology 37:445–469.

Lindgren, J., C. Alwmark, M. W. Caldwell, and A. R. Fiorillo. 2009. Skin of the Cretaceous mosasaur *Plotosaurus*: implications for aquatic adaptations in giant marine reptiles. Biology Letters 5:528–531.

Lindgren, J., M. W. Caldwell, T. Konishi, and L. M. Chiappe. 2010. Convergent evolution in aquatic tetrapods: insights from an exceptional fossil mosasaur. PLoS ONE 5(8):10 pp.

Lingham-Soliar, T. 1992. A new mode of locomotion in mosasaurs: subaqueous flying in *Plioplatecarpus marshii*. Journal of Vertebrate Paleontology 12(4):405–421.

Lingham-Soliar, T. 1994. Mosasaur '*Angolasaurus*' *bocagei* (Reptilia: Mosasauridae) from the Turonian of Angola re-interpreted as the earliest member of the genus *Platecarpus*. Paläeontology Z 68(1–2):267–282.

Lingham-Soliar, T. 1995. Anatomy and functional morphology of the largest marine reptile known, *Mosasaurus hoffmanni* (Mosasauridae, Reptilia) from the Upper Cretaceous, Upper Maastrichtian of the Netherlands. Philadelphia Transactions Royal Society of London B 347:155–180.

Lingham-Soliar, T. 1998. A new mosasaur *Pluridens walkeri* from the Upper Cretaceous Maastrichtian of the Lullemmeden Basin, Southwest Niger. Journal of Vertebrate Paleontology 18(4):709–717.

Lingham-Soliar, T. 1999a. A functional analysis of the skull of *Goronyosaurus nigeriensis* (Squamata: Mosasauridae) and its bearing on the predatory behavior and evolution of this enigmatic taxon. Neues Jahrbuch für Geologie und Palaeontologie, Abhandlungen (Stuttgart) 213(3):355–374.

Lingham-Soliar, T. 1999b. What happened 65 million years ago: the study of giant marine reptiles throws new light on the last major mass extinction. Science Spectra 17:20–29.

Lingham-Soliar, T. 2002. First occurrence of premaxillary caniniform teeth in the Varanoidea: presence in the extinct mosasaur *Goronyosaurus* (Squamata: Mosasauridae) and its functional and paleoecological considerations. Lethaia 35:187–190.

Lingham-Soliar, T. 2003. Extinction of ichthyosaurs: a catastrophic or evolutionary paradigm? Neues Jahrbuch für Geologie und Palaeontologie, Abhandlungen (Stuttgart) 228(3):421–452.

Lingham-Soliar, T. and D. Nolf. 1989. The mosasaur *Prognathodon* (Reptilia, Mosasauridae) from the Upper Cretaceous of Belgium. Bulletin de L'Institute Royal Des Sciences Naturelles de Belgique 59:137–190.

Makádi, L., M. W. Caldwell and A. Osi. 2012. The first freshwater mosasauroid (Upper Cretaceous, Hungary) and a new clade of basal mosasauroids. PLoS ONE 7(12):1–16.

Manning, E. M. 1994. Dr. William Spillman (1806–1886), pioneer paleontologist of Mississippi. Mississippi Geology 15(4):64–69.

Mantell, G. 1829. A tabular arrangement of the organic remains of the County of Sussex. Transactions of the Geological Society of London, 2nd ser., 3(1):201–216.

Marsh, O. C. 1869. Notice of some new mosasauroid reptiles from the Green-Sand of New Jersey. American Journal of Science 48(144):392–397.

Marsh, O. C. 1871a. Scientific expedition to the Rocky Mountains. American Journal of Science, ser. 3, 1(6):142–143.

Marsh, O. C. 1871b. Notice of some new fossil reptiles from the Cretaceous and Tertiary Formations. American Journal of Science, ser. 3, 1(6):447–459.

Marsh, O. C. 1872a. Discovery of the dermal scutes of mosasaurid reptiles. American Journal of Science, ser. 3, 3(16):290–292.

Marsh, O. C. 1872b. On the structure of the skull and limbs in mosasaurid reptiles, with descriptions of new genera and species. American Journal of Science, ser. 3, 3(18):448–464.

Marsh, O. C. 1872c. Note on *Rhinosaurus*. American Journal of Science 4(20):147.

Martin, J. E. 2007. A new species of the durophagous mosasaur, *Globidens* (Squamata: Mosasauridae) from the Late Cretaceous Pierre Shale Group of central South Dakota, USA; pp. 177–198 in J. E. Martin and D. C. Parris (eds.), The Geology and Paleontology of the Late Cretaceous Marine Deposits of the Dakotas. Geological Society of America, Special Paper 427. Boulder, Colorado, Geological Society of America.

Martin, J. E., and P. R. Bjork. 1987. Gastric residues associated with a mosasaur from the Late Cretaceous (Campanian) Pierre Shale in South Dakota. Dakoterra 3:68–72.

Martin, J. E., and M. Fernández. 2007. The synonymy of the Late Cretaceous mosasaur (Squamata) genus *Lakumasaurus* from Antarctica with *Taniwhasaurus* from New Zealand and its bearing upon faunal similarity within the Weddellian Province. Geological Journal 42(2):203–211.

Martin, J. E., and J. E. Fox. 2004. Molluscs in the stomach contents of *Globidens*, a shell-crushing mosasaur, from the Late Cretaceous Pierre Shale, Big Bend Area of the Missouri River, Central South Dakota. Abstracts with programs, Geological Society of America, 2004 Rocky Mountain and Cordilleran Regions Joint Meeting, 36(4):80.

Martin, J. E., and J. E. Fox. 2007. Stomach contents of *Globidens*, a shell-crushing mosasaur (Squamata), from the Late Cretaceous Pierre Shale Group, Big Bend area of the Missouri River, central South Dakota; pp. 167–176 in J. E. Martin and D. C. Parris (eds.), The Geology and Paleontology of the Late Cretaceous Marine Deposits of the Dakotas. Geological Society of America, Special Paper 427. Boulder, Colorado, Geological Society of America.

Martin, L. D., and B. M. Rothschild. 1989. Paleopathology and diving mosasaurs. American Scientist 77:460–467.

Martin, L. D., and J. D. Stewart. 1977. The oldest (Turonian) mosasaurs from Kansas. Journal of Paleontology 51(5):973–975.

Massare, J. A. 1987. Tooth morphology and prey preference of Mesozoic marine reptiles. Journal of Vertebrate Paleontology 7(2):121–137.

Mehl, M. G. 1941. *Dakotasuchus kingi*, a crocodile from the Dakota of Kansas. Denison University Bulletin, Journal of the Scientific Laboratories 35:47–65.

Meijer, A. W. F. 1971. Natuurhistorisch Museum Maastricht [Hageman *Mosasaurus* dentary]. Natuurhistorisch Maandblad 60e Jaargang 1971:3.

Merriam, J. C. 1894. Über die Pythonomorphen der Kansas-Kreide. Palaeontographica 41. 39 pp.

Mitchell, S. L. 1818. Observations on the geology of North America, illustrated by the description of various organic remains found in that part of the world; pp. 319–431, in G. Cuvier, Essay on the Theory of the Earth. Kirk and Mercein, New York, New York.

Motani, R. 2010. Warm-blooded 'Sea Dragons'? Science 328:1361–1362.

Moulton, G. E. (ed.). 1983–1997. The Journals of the Lewis and Clark Expedition, vols. 1–11. University of Nebraska Press, Lincoln, Nebraska.

Mudge, B. F. 1876. Notes on the Tertiary and Cretaceous periods of Kansas.

Bulletin of the U.S. Geological Survey of the Territories (Hayden) 2(3):211–221.

Mulder, E. W. A. 2003. On the Latest Cretaceous Tetrapods from the Maastrichtian Type Area. Publicaties van het Natuurhistorisch Genootschap in Limburg, Reeks XLIV, aflevering 1. Maastricht: Stichting Natuurpublicaties Limburg, Maastricht, the Netherlands, 188 pp.

Nicholls, E. L., 1988. The first record of the mosasaur *Hainosaurus* (Reptilia: Lacertilia) from North America. Canadian Journal of Earth Science 25:1564–1570.

Nicholls, E. L., and S. J. Godfrey. 1994. Subaqueous flight in mosasaurs: a discussion. Journal of Vertebrate Paleontology 14(3):450–452.

Novas, F. E., M. Fernández, Z. B. Gasparini, J. M. Lirio, H. J. Nuñez, and P. Puerta. 2002. *Lakumasaurus antarcticus*, n. gen. et sp., a new mosasaur (Reptilia, Squamata) from the Upper Cretaceous of Antarctica. Ameghiniana 39(2):245–249.

O'Keefe, F. R. 2008. Cranial anatomy and taxonomy of *Dolichorhynchops bonneri* new combination, a polycotylid plesiosaur from the Pierre Shale of Wyoming and South Dakota. Journal of Vertebrate Paleontology 28(3):664–676.

Osborn, H. F. 1899. A complete mosasaur skeleton, osseous and cartilaginous. Bulletin of the Peabody Museum of Natural History 1(4):167–188.

Ott, C. J., A. D. B. Behlke, and D. K. Kelly. 2002. An unusually large specimen of *Clidastes* (Mosasauroidea) from the Niobrara Chalk of western Kansas. Transactions of the Kansas Academy of Science 21:32.

Parkinson, J. 1822. An Introduction to the Study of Fossil Organic Remains; Especially Those Found in the British Strata: Intended to Aid the Student in His Enquiries Respecting the Nature of Fossils and Their Connection with the Formation of the Earth. Sherwood, Neely, and Jones, London, 346 pp.

Persson, P. O. 1963. Studies on Mesozoic marine reptile faunas with particular regard to the Plesiosauria. Publications from the Institutes of Mineralogy, Paleontology, and Quaternary Geology, University of Lund, Sweden 118: 1–15.

Polcyn, M. J., and G. L. Bell Jr. 2005. The rare mosasaur genus *Globidens* from north central Texas (Mosasaurinae: Globidensini). Journal of Vertebrate Paleontology 25(Supplement to 3):101A.

Polcyn, M. J., and G. L. Bell Jr. 2016. A reassessment of two problematic Kansas mosasaurs. Abstracts and Program, 5th Triennial Mosasaur: A Global Perspective

on Mesozoic Marine Amniotes. Uppsala University, Sweden, pp. 30–32.

Polcyn, M. J., and M. J. Everhart. 2008. Description and phylogenetic analysis of a new species of *Selmasaurus* (Mosasauridae: Plioplatecarpinae) from the Niobrara Chalk of western Kansas; pp. 13–28 in Proceedings of the Second Mosasaur Meeting, Fort Hays Studies Special Issue 3, Fort Hays State University, Hays, Kansas.

Polcyn, M. J., G. L. Bell Jr., K. Shimada, and M. J. Everhart. 2008. The oldest North American mosasaurs (Squamata: Mosasauridae) from the Turonian (Upper Cretaceous) of Kansas and Texas with comments on the radiation of major mosasaur clades; pp. 137–155 in Proceedings of the Second Mosasaur Meeting, Fort Hays Studies Special Issue 3, Fort Hays State University, Hays, Kansas.

Pyron, R., and F. T. Burbrink. 2014. Early origin of viviparity and multiple reversions to oviparity in squamate reptiles. Ecology Letters 17:13–21.

Rothschild, B., and M. J. Everhart. 2015. Co-ossification of adjacent vertebrae in mosasaurs (Squamata, Mosasauridae); evidence of habitat interactions and susceptibility to disease. Transactions of the Kansas Academy of Science 118(3–4):265–275.

Russell, D. A. 1967. Systematics and Morphology of American Mosasaurs. Peabody Museum of Natural History, Yale University, Bulletin No. 23. Peabody Museum of Natural History, Yale University, New Haven, Connecticut, 241 pp.

Russell, D. A. 1970. The vertebrate fauna of the Selma Formation of Alabama, pt. 7: the mosasaurs. Fieldiana Geology Memoirs 3(7):369–380.

Russell, D. A. 1975. A new species of *Globidens* from South Dakota. Fieldiana Geology 33(13):235–256.

Russell, D. A. 1988. A checklist of North American marine Cretaceous vertebrates including fresh water fishes. Occasional Paper of the Tyrrell Museum of Paleontology, no. 4, 57 pp.

Russell, D. A. 1993. Vertebrates in the Western Interior Sea; pp. 665–680 in W. G. E. Caldwell and E. G. Kauffman (eds.), Evolution of the Western Interior Basin. Geological Association of Canada Special Paper 39, St. John's, Newfoundland, Canada.

Saint-Fond, B. F. 1799. Historie Naturelle de la Montage de Saint-Pierre de Maestricht. Chez H, J. Jansen, Imprimeur-Libraire, Paris, France, 263 pp.

Suzuki, S. 1985. A new species of *Mosasaurus* (Reptilia, Squamata) from the Upper Cretaceous Hakobuchi Group in Central Hokkaido, Japan. Monograph of the Association for Geological Collaboration in Japan 30:45–66.

Schulp, A. S. 2005. Feeding the mechanical mosasaur: what did *Carinodens* eat? Netherlands Journal of Geosciences / Geologie en Mijnbouw 84(3):345–358.

Schumacher, B. A. 1993. Biostratigraphy of Mosasauridae (Squamata, Varanoidea) from the Smoky Hill Chalk Member, Niobrara Chalk (Upper Cretaceous) of western Kansas. M.S. thesis, Fort Hays State University, Hays, Kansas, 68 pp.

Schumacher, B. A. 2011. A 'woollgari-zone mosasaur' (Squamata; Mosasauridae) from the Carlile Shale (Lower Middle Turonian) of central Kansas and the stratigraphic overlap of early mosasaurs and pliosaurid plesiosaurs. Transactions of the Kansas Academy of Science 114(1–2):1–14.

Schumacher, B. A., and D. W. Varner. 1996. Mosasaur Caudal Anatomy. Journal of Vertebrate Paleontology 16(3, Supplement):63A.

Schumacher, B. A., and D. W. Varner. 2007. Morphology and function of tailbends in mosasaurs. Abstracts, Second Mosasaur Meeting, Sternberg Museum of Natural History, Hays, Kansas, p. 24.

Schwimmer, D. R. 2002. King of the Crocodilians: The Paleobiology of *Deinosuchus*. Indiana University Press, Bloomington, Indiana, 220 pp.

Schwimmer, D. R., J. D. Stewart, and G. D. Williams. 1997. Scavenging by sharks of the genus *Squalicorax* in the Late Cretaceous of North America. Palaios 12:71–83.

Shaler, N. S. 1869. Report on the collection of fossil remains in general; pp. 41–45 in The Annual Report of the Trustees of the Museum of Comparative Zoology, together with the Report of the Director. Boston, Massachusetts.

Sheldon, M. A. 1996. Stratigraphic distribution of mosasaurs in the Niobrara Formation of Kansas. Paludicola 1:21–31.

Shimada, K. 1996. Ichthyosaur (Reptilia: Ichthyosauria) vertebra from the Kiowa Shale (Lower Cretaceous: Upper Albian), Clark County, Kansas. Transactions of the Kansas Academy of Science 99(1–2):39–44.

Shimada, K. 1997. Paleoecological relationships of the Late Cretaceous lamniform shark, *Cretoxyrhina mantelli* (Agassiz). Journal of Paleontology 71(5):926–933.

Shimada, K., and D. C. Parris. 2007. A long-snouted Late Cretaceous crocodyliform,

Terminonaris cf. *T. browni*, from the Carlile Shale (Turonian) of Kansas. Transactions of the Kansas Academy of Science 110(1–2):107–115.

Shor, E. N. 1971. Fossils and Flies: The Life of a Compleat Scientist—Samuel Wendell Williston, 1851–1918. University of Oklahoma Press, Norman, Oklahoma, 285 pp.

Simpson, G. G. 1942. The beginnings of vertebrate paleontology in North America. Proceedings of the American Philosophical Society 86(11):130–188.

Snow, F. H. 1878. On the dermal covering of a mosasauroid reptile. Transactions of the Kansas Academy of Science 6:54–58.

Sternberg, C. H. 1922. Field work in Kansas and Texas. Transactions of the Kansas Academy of Science 30(2):339–348.

Stewart, J. D. 1988. The stratigraphic distribution of Late Cretaceous *Protosphyraena* in Kansas and Alabama; pp. 80–94 in M. E. Nelson (ed.), Geology, Paleontology and Biostratigraphy of Western Kansas: Articles in Honor of Myrl V. Walker. Fort Hays Studies, 3rd ser., 10, Science, Fort Hays State University, Hays, Kansas.

Stewart, J. D. 1990. Niobrara Formation vertebrate stratigraphy; pp. 19–30 in S. C. Bennett (ed.), Niobrara Chalk Excursion Guidebook. University of Kansas Museum of Natural History and Kansas Geological Survey, Lawrence, Kansas.

Stewart, J. D. 1993. A skeleton of *Platecarpus* sp. (Lacertilia: Mosasauridae) with stomach contents and extensive integument. Journal of Vertebrate Paleontology 13(Supplement to 3):58A.

Stewart, J. D., and G. L. Bell Jr. 1994. North America's oldest mosasaurs are teleosts. Contributions to Science (Natural History Museum of Los Angeles County) 441:1–9.

Street, H. P., and M. W. Caldwell. 2016. Rediagnosis and redescription of *Mosasaurus hoffmannii* (Squamata: Mosasauridae) and an assessment of species assigned to the genus *Mosasaurus*. Geological Magazine 1:1–37.

Taft, R. 1953. Artists and Illustrators of the Old West, 1850–1900. Scribner, New York, New York, 492 pp.

Thurmond, J. T. 1969. Notes on mosasaurs from Texas. Texas Journal of Science 21(1):69–79.

Varricchio, D. J. 2001. Gut contents from a Cretaceous tyrannosaurid; implications for theropod dinosaur digestive tracts. Journal of Paleontology 75(2):401–406.

Webb, W. E. 1872a. Buffalo Land—An Authentic Account of the Discoveries,

Adventures, and Mishaps of a Scientific and Sporting Party in the Wild West. Hubbard Brothers, Philadelphia, Pennsylvania, 503 pp.

Webb, W. E. 1872b. Neb, the devil's own. Kansas Magazine 2:128–133.

Wetzel, C. R. 1960. Monument Station, Gove County. The Kansas Historical Quarterly 26:250–254.

Williston, S. W. 1891. The skull and hind extremity of *Pteranodon*. American Naturalist 25(300):1124–1126.

Williston, S. W. 1893. Mosasaurs, pt. 2: restoration of *Clidastes*. Kansas University Quarterly 2(2):83–84.

Williston, S. W. 1897. The Kansas Niobrara Cretaceous. University Geological Survey of Kansas 2:235–246.

Williston, S. W. 1898a. Mosasaurs. University Geological Survey of Kansas 4(5):81–347.

Williston, S. W. 1898b. Editorial notes. Kansas University Quarterly 7(4):235.

Williston, S. W. 1899. Some additional characters of the mosasaurs. Kansas University Quarterly 8(1):39–41.

Williston, S. W. 1902. Notes on some new or little-known extinct reptiles. Kansas University Science Bulletin 1(9):247–254.

Williston, S. W. 1914. Water Reptiles of the Past and Present. University of Chicago Press, Chicago, Illinois, 251 pp.

Williston, S. W. and E. C. Case. 1892. Kansas mosasaurs: Part I: *Clidastes*. Kansas University Quarterly 1:15-32, Pl. II-VI.

Wiman, C. J. 1920. Some reptiles from the Niobrara Group in Kansas. Bulletin of the Geological Institute of Uppsala 18:9–18.

Wright, K. R., and S. W. Shannon. 1988. *Selmasaurus russelli*, a new plioplatecarpine mosasaur (Squamata, Mosasauridae) from Alabama. Journal of Vertebrate Paleontology 8(1):102–107.

Yamashita, M., T. Konishi, and T. Sato. 2015. Sclerotic rings in mosasaurs (Squamata: Mosasauridae): structures and taxonomic diversity. PLoS ONE 10(2):16.

Yamashita, M., T. Konishi, and M. J. Everhart. 2015. Utility of sclerotic rings in mosasaur phylogeny and beyond. New insights from the subfamily Mosasaurinae. Abstracts, Annual Meeting of the Society of Vertebrate Paleontology, Dallas, Texas, p. 242.

Anonymous. 1872. On two new ornithosaurians from Kansas. American Journal of Science, ser. 3, 3(17):374–375.

Bennett, S. C. 1987. New evidence on the tail of the pterosaur *Pteranodon* (Archosauria: Pterosauria); pp. 18–23 in R. J. Currie and E. H. Koster (eds.), Fourth Symposium on Mesozoic Terrestrial Ecosystems. Occasional Papers of the Tyrrell Museum of Palaeontology No. 3, Drumheller, Alberta.

Bennett, S. C. 1992. Sexual dimorphism of *Pteranodon* and other pterosaurs, with comments on cranial crests. Journal of Vertebrate Paleontology 12(4):422–434.

Bennett, S. C. 1994a. Taxonomy and Systematics of the Late Cretaceous Pterosaur *Pteranodon* (Pterosauria, Pterodactyloida). Occasional Papers of the Natural History Museum, University of Kansas, no. 169. Natural History Museum, University of Kansas, Lawrence, Kansas, 70 pp.

Bennett, S. C. 1994b. The pterosaurs of the Smoky Hill Chalk. Earth Scientist 11(1):22–25.

Bennett, S.C. 2000a. Inferring stratigraphic position of fossil vertebrates from the Niobrara Chalk of western Kansas. Kansas Geological Survey, Current Research in Earth Sciences, Bulletin 244, part 1.

Bennett, S. C. 2000b. New information on the skeletons of *Nyctosaurus*. Journal of Vertebrate Paleontology 20(3, Supplement):29A.

Bennett, S. C. 2001. The osteology and functional morphology of the Late Cretaceous pterosaur *Pteranodon*, pts. 1 and 2. Palaeontographica Abteilung A 260:1–112, 113–153.

Bennett, S. C. 2003a. New crested specimens of the Late Cretaceous pterosaur *Nyctosaurus*. Paläontologische Zeitschrift 77:61–75.

Bennett, S. C. 2003b. A survey of pathologies of large pterodactyloid pterosaurs. Palaeontology 46: 185–198.

Bennett, S. C. 2007. Articulation and function of the pteroid bone of pterosaurs. Journal of Vertebrate Paleontology 27:881–891.

Betts, C. W. 1871. The Yale College Expedition of 1870. Harper's New Monthly Magazine 43(257):663–671.

Bonner, O. W. 1964. An osteological study of *Nyctosaurus* and *Trinacromerum* with a description of a new species of *Nyctosaurus*. M.S. thesis, Fort Hays State University, Hays, Kansas, 63 pp.

Bramwell, C. D., and G. R. Whitfield. 1970. Flying speed of the largest aerial vertebrate. Nature 225:660–661.

Brower, J. C. 1983. The aerodynamics of *Pteranodon* and *Nyctosaurus*, two large pterosaurs from the Upper Cretaceous of Kansas. Journal of Vertebrate Paleontology 3(2):84–124.

Brown, B. 1943. Flying reptiles. Natural History 52:104–111.

10. Pteranodons

Brown, G. W. 1986a. Reassessment of *Nyctosaurus*: new wings for an old pterosaur. Abstracts, Proceedings of the Nebraska Academy of Science, p. 47.

Brown, G. W. 1986b. Evidence of phalangeal reduction in the wing of the pterosaur *Nyctosaurus*. Abstracts, Annual Meeting, Society of Vertebrate Paleontology.

Carpenter, K. 1996. Sharon Springs Member, Pierre Shale (Lower Campanian) depositional environment and origin of its vertebrate fauna, with a review of North American plesiosaurs. Ph.D. dissertation, University of Colorado, Boulder, Colorado, 251 pp.

Carpenter, K. 2006. Comparative vertebrate taphonomy of the Pembina and Sharon Springs members (Middle Campanian) of the Pierre Shale, Western Interior. Paludicola 5:125–149.

Cope, E. D. 1872a. On the geology and paleontology of the Cretaceous strata of Kansas; pp. 318–349 in F. V. Hayden (ed.), [Fifth Annual] Preliminary Report of the United States Geological Survey of Montana and Portions of the Adjacent Territories, Being a Fifth Annual Report of Progress, pt. 3: Paleontology. Government Printing Office, Washington, D.C.

Cope, E. D. 1872b. On two new ornithosaurians from Kansas. Proceedings of the American Philosophical Society 12(88):420–422.

Cope, E. D. 1874. Review of the Vertebrata of the Cretaceous Period found west of the Mississippi River. Bulletin of the U.S. Geological and Geographical Survey of the Territories 1(2):3–48.

Cope, E. D. 1875. The Vertebrata of the Cretaceous Formations of the West; in F. V. Hayden (ed.), Report of the U.S. Geological Survey of the Territories, Volume 2. Government Printing Office, Washington, D.C., 302 pp.

Davidson, J. P., and M. J. Everhart. 2014. Fictionalized facts; "The young fossil hunters" by Charles H. Sternberg. Transactions of the Kansas Academy of Science 117(1–2):41–54.

Eaton, G. F. 1903. The characters of *Pteranodon*. American Journal of Science, ser. 4, 16(91):82–86.

Eaton, G. F. 1904. The characters of *Pteranodon* (second paper). American Journal of Science, ser. 4, 17(100):318–320.

Eaton, G. F. 1910. Osteology of *Pteranodon*. Memoirs of the Connecticut Academy of Arts and Sciences 2:1–38.

Elgin, R. A., D. W. E. Hone, and E. Frey, E. 2011. The extent of the pterosaur flight membrane. Acta Palaeontologica Polonica 56(1):99–111.

Frey, E., M.-C. Buchy, W. Stinnesbeck, A. González-González, and A. Stefano. 2006. *Muzquizopteryx coahuilensis* n.g., n.sp., a nyctosaurid pterosaur with soft tissue preservation from the Coniacian (Late Cretaceous) of northeast Mexico (Coahuila). Oryctos 6:19–39.

Habib, M. 2008. Comparative evidence for quadrupedal launch in pterosaurs; pp. 159–166 in E. Buffetaut and D. W. E. Hone (eds.), Wellnhofer Pterosaur Meeting: Zitteliana B28.

Hankin, E. H., and D. M. S. Watson. 1914. On the flight of pterodactyls. The Aeronautical Journal 18:324–335.

Hargrave, J. E. 2007. *Pteranodon* (Reptilia: Pterosauria): stratigraphic distribution and taphonomy in the lower Pierre Shale Group (Campanian), western South Dakota and eastern Wyoming; pp. 215–225 in J. E. Martin and D. C. Parris (eds.), The Geology and Paleontology of the Late Cretaceous Marine Deposits of the Dakotas. Geological Society of America, Special Paper 427. Boulder, Colorado, Geological Society of America.

Harksen, J. C. 1966. *Pteranodon sternbergi*, a new fossil pterodactyl from the Niobrara Cretaceous of Kansas. Proceedings South Dakota Academy of Science 45:74–77.

Hattin, D. E. 1982. Stratigraphy and Depositional Environment of the Smoky Hill Chalk Member, Niobrara Chalk (Upper Cretaceous) of the Type Area, Western Kansas. Kansas Geological Survey Bulletin 225. University of Kansas, Lawrence, Kansas, 108 pp.

Henderson, D. M. 2010. Pterosaur body mass estimates from three-dimensional mathematical slicing. Journal of Vertebrate Paleontology 30(3):768–785.

Humphries, S., Bonser, R.H.C., Witton, M.P., and Martill, D.M. 2007. Did pterosaurs feed by skimming? Physical modeling and anatomical evaluation of an unusual feeding method. PLoS Biology 5(8):1648–1655.

Kellner, A. W. A. 2010. Comments on the Pteranodontidae (Pterosauria, Pterodactyloidea) with description of two new species. Anais da Academa Brasileira de Ciências 82(4):1063–1084.

Konuki, R. 2008. Biostratigraphy of sea turtles and possible bite marks on a *Toxochelys* (Testudine, Chelonioidea) from the Niobrara Formation (Late Santonian), Logan County, Kansas and paleoecological implications for predator-prey relationships among large marine vertebrates. M.S. thesis, Fort Hays

State University, Hays, Kansas, 141 pp.

Lane, H. H. 1946. A survey of the fossil vertebrates of Kansas, pt. 3: the reptiles. Transactions of the Kansas Academy of Science 49(3):289–332.

Liggett, G. A., S. C. Bennett, K. Shimada, and J. Huenergarde. 1997. A Late Cretaceous (Cenomanian) fauna in Russell County, Kansas. Transactions of the Kansas Academy of Science 16:26.

Liggett, G. A., K. Shimada, S. C. Bennett, and B. Schumacher. 2005. Cenomanian (Late Cretaceous) reptiles from northwestern Russell County, Kansas. PaleoBios 25(2):9–17.

Lü, J., D. M. Unwin, D. C. Deeming, X. Jin, Y. Liu, and Q. Ji. 2011. An egg-adult association, gender, and reproduction in pterosaurs. Science 331(6015):321–324.

Lucas, F. A. 1901. The greatest flying creature, the great pterodactyl *Ornithostoma*; pp. 654–659 in Annual Report of the Board of Regents of the Smithsonian Institution, Washington, D.C.

Marsh, O. C. 1871a. Scientific expedition to the Rocky Mountains. American Journal of Science, ser. 3, 1(6):142–143.

Marsh, O. C. 1871b. Notice of some new fossil reptiles from the Cretaceous and Tertiary formations. American Journal of Science, ser. 3, 1(6):447–459.

Marsh, O. C. 1871c. Note on a new and gigantic species of pterodactyle. American Journal of Science, ser. 3, 1(6):472.

Marsh, O. C. 1872. Discovery of additional remains of Pterosauria, with descriptions of two new species. American Journal of Science, ser. 3, 3(16):241–248.

Marsh, O. C. 1876a. Notice of a new sub-order of Pterosauria. American Journal of Science, ser. 3, 3 11(65):507–509.

Marsh, O. C. 1876b. Principal characters of American pterodactyls. American Journal of Science, ser. 3, 12(72):479–480.

Marsh, O. C. 1881. Note on American pterodactyls. American Journal of Science, ser. 3, 21(124):342–343.

Marsh, O. C. 1884. Principal characters of American Cretaceous pterodactyls, pt. 1: the skull of *Pteranodon*. American Journal of Science, ser. 3, 27(161):422–426.

Mateer, N. J. 1975. A study of *Pteranodon*. Bulletin of the Geological Institutions of the University of Uppsala 6:23–33.

Miller, H. W. 1971a. The taxonomy of the *Pteranodon* species from Kansas. Transactions of the Kansas Academy of Science 74(1):1–19.

Miller, H. W. 1971b. A skull of *Pteranodon* (*Longicepia*) *longiceps* Marsh associated with wing and body parts. Transactions of the Kansas Academy of Science 74(10):20–33.

Miller, H. W. 1978. *Geosternbergia*, a new name for *Sternbergia* Miller, 1972: non Paulo Couto 1970; non Jordon, 1925. Journal of Paleontology 52(1):194.

Myers, T. S. 2010. Earliest occurrence of the Pteranodontidae (Archosauria: Pterosauria) in North America: new material from the Austin Group of Texas. Journal of Paleontology 84(6):1071–1081.

Owen, R. 1851. Monograph on the fossil Reptilia of the Cretaceous formations. Palaeontographical Society, pp. 1–118.

Padian, K. 1983. A functional analysis of flying and walking in pterosaurs. Paleobiology 9(3):218–239.

Russell, D. A. 1988. A Check List of North American Marine Cretaceous Vertebrates Including Fresh Water Fishes. Occasional Paper of the Tyrrell Museum of Palaeontology No. 4, Drumheller, Alberta, Canada, 57 pp.

Schultze, H.-P., L. Hunt, J. Chorn, and A. M. Neuner. 1985. Type and Figured Specimens of Fossil Vertebrates in the Collection of the University of Kansas Museum of Natural History, pt. 2: Fossil Amphibians and Reptiles. Miscellaneous Publications of the University of Kansas Museum of Natural History 77, Lawrence, Kansas, 66 pp.

Seeley, H. G. 1871. Additional evidence of the structure of the head in ornithosaurs from the Cambridge Upper Greensand; being a dupplement to "The Ornithosauria." The Annals and Magazine of Natural History, ser. 4, 7:20–36.

Seeley, H. G. 1901. Dragons of the Air: An Account Extinct Flying Reptiles. London, 239 pp.

Shor, E. N. 1971. Fossils and Flies: The Life of a Compleat Scientist—Samuel Wendell Williston, 1851–1918. University of Oklahoma Press, Norman, Oklahoma, 285 pp.

Stein, R. S. 1975. Dynamic analysis of *Pteranodon ingens*: a reptilian adaptation to flight. Journal of Paleontology 49(3):534–548.

Sternberg, C. H. 1881. The Niobrara Group. Kansas City Review of Science and Industry 5(1):1–4.

Sternberg, C. H. 1889. The young fossil-hunters: a true story of western exploration and adventure. The Swiss Cross 5(1–6):1–3, 37–39, 72–75, 98–100, 136–139, 157–160.

Sternberg, C. H. 1909. The Life of a Fossil Hunter. Henry Holt and Company, New York, New York, 286 pp.

Sternberg, G. F., and M. V. Walker. 1958. Observation of articulated limb bones of a recently discovered *Pteranodon* in the Niobrara Cretaceous of Kansas. Transactions of the Kansas Academy of Science 61(1):81–85.

Stewart, J. D. 1990. Niobrara Formation vertebrate stratigraphy; pp. 19–30 in S. C. Bennett (ed.), Niobrara Chalk Excursion Guidebook. University of Kansas Museum of Natural History and Kansas Geological Survey, Lawrence, Kansas.

Wang, X., and Z. Zhou. 2004. Pterosaur embryo from the Early Cretaceous. Nature 429:621.

Webb, W. E. 1872. Buffalo Land: An Authentic Account of the Discoveries, Adventures, and Mishaps of a Scientific and Sporting Party in the Wild West. Hubbard Brothers, Philadelphia, Pennsylvania, 503 pp.

Wellnhofer, P. 1991. The Illustrated Encyclopedia of Pterosaurs. Crescent Books, New York, New York, 192 pp.

Williston, S. W. 1891. The skull and hind extremity of pteranodon. American Naturalist 25(300):1124–1126.

Williston, S. W. 1892. Kansas pterodactyls, pt. 1. Kansas University Quarterly 1:1–13.

Williston, S. W. 1893. Kansas pterodactyls, pt. 2. Kansas University Quarterly 2:79–81.

Williston, S. W. 1894. On various vertebrate remains from the lowermost Cretaceous of Kansas. Kansas University Quarterly 3(1):1–4.

Williston, S. W. 1895. Note on the mandible of *Ornithostoma*. Kansas University Quarterly 4:61.

Williston, S. W. 1896. On the Skull of *Ornithostoma*. Kansas University Quarterly 4(4):95–197.

Williston, S. W. 1897. Restoration of *Ornithostoma* (*Pteranodon*). Kansas University Quarterly 6:35–51.

Williston, S. W. 1898. Crocodiles. The University Geological Survey of Kansas 4(4):75–78.

Williston, S. W. 1902a. On the skeleton of *Nyctodactylus*, with restoration. American Journal of Anatomy 1:297–305.

Williston, S. W. 1902b. On the skull of *Nyctodactylus*, an Upper Cretaceous pterodactyl. Journal of Geology 10:520–531.

Williston, S. W. 1902c. Winged reptiles. Popular Science Monthly 60: 314–322.

Williston, S. W. 1903. On the osteology of *Nyctosaurus* (*Nyctodactylus*), with notes on American pterosaurs. Field Columbian Museum Geological Series 2(3):125–163.

Williston, S. W. 1904. The fingers of pterodactyls. Geology Magazine, ser. 5, 1(2):59–60.

Williston, S. W. 1910. A mounted skeleton of *Platecarpus*. Journal of Geology 18(6):537–541.

Williston, S. W. 1911. The wing-finger of pterodactyls, with restoration of *Nyctosaurus*. Journal of Geology 19:696–705.

Williston, S. W. 1912. A review of G. B. Eaton's "Osteology of *Pteranodon*." Journal of Geology 20:288.

Williston, S. W. 1925. The Osteology of the Reptiles. Edited by W. K. Gregory. Harvard University Press, 300pp.

Wilson, L.E. 2015. Osteohistological insight into *Pteranodon* ontogeny. Program and Abstracts, Society of Vertebrate Paleontology Annual Meeting, p. 239.

Wiman, C. J. 1920. Some reptiles from the Niobrara Group in Kansas. Bulletin of the Geological Institute of Uppsala 18:9–18.

11. Feathers and Teeth

Aotsuka, K., and T. Sato. 2016. Hesperornithiformes (Aves: Ornithurae) from the Upper Cretaceous Pierre Shale, Southern Manitoba, Canada. Cretaceous Research 63:154–169.

Bell, A., and L. M. Chiappe. 2015. Identification of a new Hesperornithiform from the Cretaceous Niobrara Chalk and implications for ecologic diversity among early diving birds. PLoS ONE 10(11):34 pp. e0141690.

Bell, A., and M. J. Everhart. 2009. A new specimen of *Parahesperornis* (Aves: Hesperornithiformes) from the Smoky Hill Chalk (Early Campanian) of western Kansas. Transactions of the Kansas Academy of Science 112(1–2):7–14.

Bell, A., and M. J. Everhart. 2011. Remains of small ornithurine birds from a Late Cretaceous (Cenomanian) microsite in Russell County, north central Kansas. Transactions of the Kansas Academy of Science 114(1–2):115–123.

Bell, A., K. J. Irwin, and L. C. Davis. 2015. Hesperornithiform birds from the Late Cretaceous (Campanian) of Arkansas, USA. Transactions of the Kansas Academy of Science 118(3–4): 219–229.

Burnham, D. 2005. Transfer preparation of an *Ichthyornis* specimen from the Niobrara Formation. Journal of Vertebrate Paleontology 25(Supplement to 3):41A.

Chinsamy, A., L. D. Martin, and P. Dodson. 1998. Bone microstructure of the diving *Hesperornis* and the volant *Ichthyornis* from the Niobrara Chalk of western Kansas. Cretaceous Research 19:225–235.

Clarke, J. A. 2004. Morphology, phylogenetic taxonomy, and systematics of *Ichthyornis* and *Apatornis* (Avialae: Ornithurae). Bulletin of the American Museum of Natural History 286:1–179.

Cope, E. D. 1872. Note of some cretaceous Vertebrata in the State Agricultural College of Kansas. Proceedings of the American Philosophical Society 12(87):168–170.

Cumbaa, S. L., C. Schröder-Adams, R. G. Day, and A. Phillips, A. 2006. Cenomanian bonebed faunas from the Northeastern margin, Western Interior Seaway, Canada; pp. 139–155 in S. G. Lucas, and R. M. Sullivan (eds.), Late Cretaceous vertebrates from the Western Interior. New Mexico Museum of Natural History and Science Bulletin 35.

Davis, L. C., and K. Harris. 1997. Discovery of fossil Cretaceous bird in southwest Arkansas. Journal of the Arkansas Academy of Science 51:197–198.

Everhart, M. J. 2002. New data on cranial measurements and body length of the mosasaur, *Tylosaurus nepaeolicus* (Squamata; Mosasauridae), from the Niobrara Formation of western Kansas. Transactions of the Kansas Academy of Science 105(1–2):33–43.

Everhart, M. J. 2011. Rediscovery of the *Hesperornis regalis* Marsh 1871 holotype locality indicates an earlier stratigraphic occurrence. Transactions of the Kansas Academy of Science 114(1–2): 59–68.

Everhart, M. J. 2015. Elias Putnam West (1820–1892)—lawyer, attorney general, militia commander, judge, postmaster, archaeologist, and paleontologist. Transactions of the Kansas Academy of Science 118(3–4):285–294.

Everhart, M. J., and A. Bell. 2009. A hesperornithiform limb bone from the basal Greenhorn Formation (Late Cretaceous; Middle Cenomanian) of north central Kansas. Journal of Vertebrate Paleontology 28(3):952–956.

Fox, R. C. 1984. *Ichthyornis* (Aves) from the Early Turonian (Late Cretaceous) of Alberta. Canadian Journal of Earth Sciences 21:258–260.

Hanks, D. H., and K. Shimada. 2002. Vertebrate fossils, including non-avian dinosaur remains and the first shark-bitten bird bone from a Late Cretaceous (Turonian) marine deposit of northeastern South Dakota. Journal of Vertebrate Paleontology 22(3, Supplement):62A.

Heilmann, G. 1926. Origin of Birds. H. F. and B. Witherby, London, England, 208 pp

Kear, B. P., W. E. Boles, and E. T. Smith. 2003. Unusual gut contents of a Cretaceous ichthyosaur. Proceedings of the Royal Society of London B (Supplement) 270:S206–S208.

Lane, H. H. 1946. A survey of the fossil vertebrates of Kansas, pt. 4: birds. Transactions of the Kansas Academy of Science 49(4):390–400.

Marsh, O. C. 1872a. Discovery of a remarkable fossil bird. American Journal of Science, ser. 3, 3(13):56–57.

Marsh, O. C. 1872b. Notice of a new and remarkable fossil bird. American Journal of Science, ser. 3, 3 4(22):344.

Marsh, O. C. 1872c. Notice of a new reptile from the Cretaceous. American Journal of Science, ser. 3, 4(23):406.

Marsh, O. C. 1873a. On a new sub-class of fossil birds (Odontornithes). American Journal of Science, ser. 3, 5(25):161–162.

Marsh, O. C. 1873b. Fossil birds from the Cretaceous of North America. American Journal of Science, ser. 3, 5(27):229–231.

Marsh, O. C. 1875. On the Odontornithes, or birds with teeth. American Journal of Science 10(59):403–408.

Marsh, O. C. 1877. Notice of some new vertebrate fossils. American Journal of Science 14(81):249–256.

Marsh, O. C. 1880. Synopsis of American Cretaceous birds; pp. 191–199 in Odontornithes: A Monograph on the Extinct Toothed Birds of North America. U.S. Geological Exploration of the Fortieth Parallel, Clarence King, Geologist-in-charge, vol. 7. xv + 201 pp.

Marsh, O. C. 1883. Birds with Teeth. United States Geological Survey, 3rd Annual Report of the Secretary of the Interior, 3:43–88. Government Printing Office, Washington, D.C.

Marsh, O. C. 1897. The affinities of "*Hesperornis*." Nature 55 (1432):534.

Martin, J. E., and P. R. Bjork. 1987. Gastric residues associated with a mosasaur from the Late Cretaceous (Campanian) Pierre Shale in South Dakota. Dakoterra 3:68–72.

Martin, J. E., and A. Cordes-Person. 2007. A new species of the diving bird, *Baptornis* (Ornithurae: Hesperornithiformes), from the lower Pierre Shale Group (Upper Cretaceous) of southwestern South Dakota; pp. 227–237 in J. E. Martin and D. C. Parris (eds.), The Geology and Paleontology of the Late Cretaceous Marine Deposits of the Dakotas. Geological Society of America Special Paper 427. Boulder, Colorado, Geological Society of America.

Martin, J. E., and D. W. Varner. 1992. The highest stratigraphic occurrence of the fossil bird *Baptornis*. Abstracts,

Proceedings of the South Dakota Academy of Science, 71:167.

Martin, L. D. 1983. The origin of birds and of avian flight; pp. 105–129 in R. F. Johnston (ed.), Current Ornithology, vol. 1, Plenum Press, New York, New York.

Martin, L. D. 1984. A new hesperornithid and the relationships of the Mesozoic birds. Transactions of the Kansas Academy of Science 87:141–150.

Martin, L. D. 1987. The beginning of the modern avian radiation; pp. 9–19 in C. Mourer-Chauviré (ed.), L'évolution des oiseaux d'après le témoignage des ossils. Documents des Laboratoires de Géologie de Lyon, 99, Lyon, France.

Martin, L. D., and O. Bonner. 1977. An immature specimen of *Baptornis advenus* from the Cretaceous of Kansas. Auk 94(4):787–789.

Martin, L. D., and V. L. Naples. 2008. Mandibular kinesis in *Hesperornis*. Oryctos 7:61–65.

Martin, L. D., and J. D. Stewart. 1977. Teeth in *Ichthyornis* (Class: Aves). Science 185(4284):1331–1332.

Martin, L. D., and J. Tate Jr. 1966. A bird with teeth. Museum Notes, University of Nebraska State Museum, 29:1–2.

Martin, L. D., and J. Tate Jr. 1976. The skeleton of *Baptornis advenus* (Aves: Hesperornithiformes). Smithsonian Contributions to Paleobiology 27:35–66.

Martin, L. D, E. N. Kurochkin, and T. T. Tokaryk. 2012. A new evolutionary lineage of diving birds from the Late Cretaceous of North America and Asia. Palaeoworld 21:59–63.

Martin, L. D., B. M. Rothschild, and D. A. Burnham. 2016. *Hesperornis* escapes plesiosaur attack. Cretaceous Research 63. doi:10.1016/j.cretres.2016.02.005.

Mudge, B. F. 1866a. Discovery of fossil footmarks in the Liassic(?) Formation in Kansas. American Journal of Science, ser. 2, 41(122):174–176.

Mudge, B. F. 1866b. First Annual Report on the Geology of Kansas. State Printer, Lawrence, Kansas, 57 pp.

Mudge, B. F. 1876. Notes on the Tertiary and Cretaceous periods of Kansas. US Geological and Geographical Survey of the Territories Bulletin 2(3):211–221.

Mudge, B. F. 1877. Annual Report of the Committee on Geology, for the Year Ending November 1, 1876; pp. 4–5 in Transactions of the Kansas Academy of Science, Ninth Annual Meeting.

Nicholls, E. L. 1988. Marine vertebrates of the Pembina Member of the Pierre Shale (Campanian, Upper Cretaceous) of Manitoba and their significance to the biogeography. Ph.D. dissertation, University of Calgary, Calgary, Alberta, Canada, 317 pp.

Olson, S. L. 1975. *Ichthyornis* in the Cretaceous of Alabama. Wilson Bulletin 87(1):103–105.

Parris, D. C., and J. Echols. 1992. The fossil bird *Ichthyornis* in the Cretaceous of Texas. Texas Journal of Science 44:201–212.

Peterson, J. M. 1987. Science in Kansas: the early years, 1804–1875. Kansas History Magazine 10(3):201–240.

Porras-Múzquiz, H. G., S. Chatterjee, and T. M. Lehman. 2014. The carinate bird *Ichthyornis* from the Upper Cretaceous of Mexico. Cretaceous Research 51:148–152.

Reynaud, F. N. 2006. Hind limb and pelvis proportions of *Hesperornis regalis*: a comparison with extant diving birds. M.S. thesis, Fort Hays State University, Hays, Kansas, 45 pp.

Russell, D. A. 1967. Cretaceous vertebrates from the Anderson River N.W.T. Canadian Journal of Earth Sciences 4:21–38.

Russell, D. A. 1993. Vertebrates in the Western Interior Sea; pp. 665–680 in W. G. E. Caldwell and E. G. Kauffman (eds.), Evolution of the Western Interior Basin. Geological Association of Canada, Special Paper 39, St. John's, Newfoundland.

Schufeldt, R. W. 1897. On the feathers of "*Hesperornis*" (Letters to Editor). Nature 56:30.

Shimada, K., and M. V. Fernandes. 2006. *Ichthyornis* sp. (Aves: Ichthyornithiformes) from the lower Turonian (Upper Cretaceous) of western Kansas. Transactions of the Kansas Academy of Science 109(1–2):21–26.

Shimada, T. R., and L. E. Wilson. 2016. A new specimen of the Late Cretaceous bird, cf. *Ichthyornis* sp., from the Cenomanian of central Kansas, with comments on the size distribution of *Ichthyornis* in North America. Transactions of the Kansas Academy of Science 119(2):213–237.

Snow, F. H. 1887. On the discovery of a fossil bird track in the Dakota Sandstone. Transactions of the Kansas Academy of Science 10:3–6.

Sternberg, C. H. 1909. The Life of a Fossil Hunter. New York: Henry Holt and Company, 286 pp.

Tokaryk, T. T., S. L. Cumbaa, and J. E. Storer. 1997. Early Late Cretaceous birds from Saskatchewan, Canada: the oldest diverse avifauna known from North America. Journal of Vertebrate Paleontology 17:172–176.

Walker, M. V. 1967. Revival of interest in the toothed birds of Kansas. Transactions of the Kansas Academy of Science 70(1):60–66.

Williston, S.W. 1896. On the dermal covering of *Hesperornis*. Kansas University Quarterly 5(1):53–54.

Williston, S. W. 1898a. Addenda to Part I. University Geological Survey of Kansas 4:28–32.

Williston, S. W. 1898b. Birds. University Geological Survey of Kansas 4:43–49.

Williston, S. W. 1898c. Bird tracks from the Dakota Cretaceous. University Geological Survey of Kansas 4:50–53.

Wilson, L. E., and K. Chin. 2014. Comparative osteohistology of *Hesperornis* with reference to pygoscelid penguins: the effects of climate and behaviour on avian bone microstructure. Royal Society Open Science 1:140245, 16 pp.

Wilson, L. E., K. Chin, and S. L. Cumbaa. 2016. A new hesperornithiform (Aves) specimen from the Late Cretaceous Canadian High Arctic with comments on high latitude hesperornithiform diet. Canadian Journal of Earth Sciences 53(11):8 pp.

Wilson, L. E., K. Chin, S. Cumbaa, and G. Dyke. 2011. A high latitude hesperornithiform (Aves) from Devon Island: palaeobiogeography and size distribution of North American hesperornithiforms. Journal of Systematic Palaeontology 9(1):9–23.

Zhou, Z., and L. D. Martin. 2010. Distribution of the predentary bone in Mesozoic ornithurine birds. Journal of Systematic Paleontology 9:25–31.

Buskuskie, T. R. 2015. New dinosaur material from the Smoky Hill Chalk from the Smoky Hill Member (Niobrara Formation: Upper Cretaceous) referred to *Niobrarasaurus coleii*. Transactions of the Kansas Academy of Science 118(1–2):136.

Carpenter, K., and M. J. Everhart. 2007. Skull of the ankylosaur *Niobrarasaurus coleii* (Ankylosauria: Nodosauridae) from the Smoky Hill Chalk (Coniacian) of western Kansas. Transactions of the Kansas Academy of Science 110(1–2):1–9.

Carpenter, K., D. Dilkes, and D. B. Weishampel. 1995. The dinosaurs of the Niobrara Chalk Formation (Upper Cretaceous, Kansas). Journal of Vertebrate Paleontology 15(2):275–297.

Cole, V. B. 2007. Field notes regarding the 1930 discovery of the type specimen of *Niobrarasaurus coleii*, Gove County,

Kansas. Edited by M. J. Everhart. Transactions of the Kansas Academy of Science 110(1–2):132–134.

Eaton, T. H., Jr., 1960. A new armored dinosaur from the Cretaceous of Kansas. University Kansas Paleontology Contributions 8:1–14.

Everhart, M. J. 2004. Notice of the transfer of the holotype specimen of *Niobrarasaurus coleii* (Ankylosauria; Nodosauridae) to the Sternberg Museum of Natural History. Transactions of the Kansas Academy of Science 107(3–4):173–174.

Everhart, M. J. 2015. Elias Putnam West (1820–1892)—lawyer, attorney general, militia commander, judge, postmaster, archaeologist, and paleontologist. Transactions of the Kansas Academy of Science 118(3–4):285–294.

Everhart, M. J., and K. Ewell. 2006. Shark-bitten dinosaur (Hadrosauridae) vertebrae from the Niobrara Chalk (Upper Coniacian) of western Kansas. Transactions of the Kansas Academy of Science 109(1–2):27–35.

Everhart, M. J., and S. A. Hamm. 2005. A new nodosaur specimen (Dinosauria: Nodosauridae) from the Smoky Hill Chalk (Upper Cretaceous) of western Kansas. Transactions of the Kansas Academy of Science 108(1–2):15–21.

Hamm, S. A., and M. J. Everhart. 2001. Notes on the occurrence of Nodosaurs (Ankylosauridae) in the Smoky Hill Chalk (Upper Cretaceous) of western Kansas. Journal of Vertebrate Paleontology 21(3, Supplement):58A.

Hattin, D. E. 1982. Stratigraphy and depositional environment of the Smoky Hill Chalk Member, Niobrara Chalk (Upper Cretaceous) of the type area, western Kansas. Kansas Geological Survey Bulletin No. 225. University of Kansas, Lawrence, Kansas, 108 pp.

Lane, H. H. 1946. A survey of the fossil vertebrates of Kansas, pt. 3: the reptiles. Transactions of the Kansas Academy of Science 49(3):289–332.

Lesquereux, L. 1891. Flora of the Dakota Group. Edited by F. H. Knowlton. Monograph of the United States Geological Survey, 17, Government Printing Office, Washington, D.C., 400 pp.

Liggett, G. A. 2005. A review of the dinosaurs from Kansas. Transactions of the Kansas Academy of Science 108(1–2):1–14.

Marsh, O. C. 1872. Notice of a new species of *Hadrosaurus*. American Journal of Science, ser. 3, 3(16):301.

Marsh, O. C. 1890. Additional characters of the Ceratopsidae, with notes on new

12. Dinosaurs?

Cretaceous dinosaurs. American Journal of Science 3(39):418–425.

Mehl, M. G. 1931. Aquatic dinosaur from the Niobrara of western Kansas. Bulletin of the Geologic Society of America 42:326–327.

Mehl, M. G. 1936. *Hierosaurus coleii*: a new aquatic dinosaur from the Niobrara Cretaceous of Kansas. Denison University Bulletin, Journal of the Scientific Laboratory 31:1–20.

Parmenter, C. S. 1899. Fossil turtle cast from the Dakota Epoch. Kansas Academy of Science, Transactions 16:67.

Reisdorf, A. G., R. Bux, D. Wyler, M. Benecke, C. Klug, M. W. Maisch, P. Fornaro, and A. Wetzel. 2012. Float, explode or sink: postmortem fate of lung-breathing marine vertebrates. Palaeobiodiversity and Palaeoenvironments 92:67–81.

Sternberg, C. H. 1909. An armored dinosaur from the Kansas Chalk. Transactions of the Kansas Academy of Science 22:257–258.

Walters, R. F. 1986. Memorial: Virgil Bedford Cole (1897–1984). American Association of Petroleum Geologists Bulletin 70(2):208–209.

Wieland, G. R. 1909. An armored saurian. American Journal of Science 27: 250–252.

Wieland, G. R. 1911. Notes on the armored Dinosauria. American Journal of Science 31:112–124.

13. The Big Picture

Beeson, E., and K. Shimada. 2004. Vertebrates from a unique bonebed of the Cretaceous Niobrara Chalk, western Kansas. Journal of Vertebrate Paleontology 24(Supplement to 3):37A.

Bell, A., and L. M. Chiappe. 2015. Identification of a new hesperornithiform from the Cretaceous Niobrara Chalk and implications for ecologic diversity among early diving birds. PLoS ONE 10(11):34 pp. e0141690.

Bennett, S. C. 1992. Sexual dimorphism of *Pteranodon* and other pterosaurs, with comments on cranial crests. Journal of Vertebrate Paleontology 12(4):422–434.

Bennett, S. C. 2000. Inferring stratigraphic position of fossil vertebrates from the Niobrara Chalk of western Kansas. Current Research in Earth Sciences, Kansas Geological Survey Bulletin No. 244. University of Kansas, Lawrence, Kansas, 26 pp.

Bennett, S. C. 2003. New crested specimens of the Late Cretaceous pterosaur *Nyctosaurus*. Paläontologische Zeitschrift 77:61–75.

Bourdon, J., and M. J. Everhart. 2011. Analysis of an associated *Cretoxyrhina mantelli* dentition from the Late Cretaceous (Smoky Hill Chalk, Late Coniacian) of western Kansas. Transactions of the Kansas Academy of Science 114(1–2):15–32.

Carpenter, K. 1990. Upward continuity of the Niobrara fauna with the Pierre Shale fauna; pp. 73–81 in S. C. Bennett (ed.), Niobrara Chalk Excursion Guidebook. University of Kansas Museum of Natural History and Kansas Geological Survey, Lawrence, Kansas.

Carpenter, K. 2003. Vertebrate biostratigraphy of the Smoky Hill Chalk (Niobrara Formation) and the Sharon Springs Member (Pierre Shale); pp. 421–437 in P. J. Harries (ed.), Approaches in High-Resolution Stratigraphic Paleontology. Kluwer Academic Publishers, Dordrecht, the Netherlands.

Cicimurri, D. J., D. C. Parris, and M. J. Everhart. 2008. Partial dentition of a chimaeroid fish (Chondrichthyes, Holocephali) from the Upper Cretaceous Niobrara Chalk of Kansas, USA. Journal of Vertebrate Paleontology 28(1):34–40.

Clarke, J. A. 2004. Morphology, phylogenetic taxonomy, and systematics of *Ichthyornis* and *Apatornis* (Avialae: Ornithurae). Bulletin of the American Museum of Natural History 286:1–179.

Davidson, J. P. 2003. Edward Drinker Cope, Professor Paleozoic and *Buffalo Land*. Transactions of the Kansas Academy of Science 106(3–4):177–191.

Dutel, H., J. G. Maisey, D. R. Schwimmer, P. Janvier, M. Herbin, and G. Clément. 2012. The giant coelacanth (Actinistia, Sarcopterygii) *Megalocoelacanthus dobiei* Schwimmer, Stewart and Williams 1994, and its bearing on Latimerioidei interrelationships. PLoS ONE 7(11):27 pp.

Elias, M. K. 1931. The Geology of Wallace County, Kansas. State Geological Survey Bulletin 18, 254 pp.

Everhart, M. J. 2000. Gastroliths associated with plesiosaur remains in the Sharon Springs Member of the Pierre Shale (Late Cretaceous), western Kansas. Transactions of the Kansas Academy of Science 103(1–2):58–69.

Everhart, M. J. 2001. Revisions to the biostratigraphy of the Mosasauridae (Squamata) in the Smoky Hill Chalk Member of the Niobrara Chalk (Late Cretaceous) of Kansas. Transactions of the Kansas Academy of Science 104(1–2):56–75.

Everhart, M. J. 2002. Remains of immature mosasaurs (Squamata; Mosasauridae) from the Niobrara Chalk (Late

Cretaceous) argue against nearshore nurseries. Journal of Vertebrate Paleontology 22(3, Supplement):52A.

Everhart, M. J. 2003. First records of plesiosaur remains in the Lower Smoky Hill Chalk Member (Upper Coniacian) of the Niobrara Formation in western Kansas. Transactions of the Kansas Academy of Science 106(3–4):139–148.

Everhart, M. J. 2004a. Late Cretaceous interaction between predators and prey: evidence of feeding by two species of shark on a mosasaur. PalArch Journal of Vertebrate Paleontology 1(1):1–7.

Everhart, M. J. 2004b. Plesiosaurs as the food of mosasaurs: new data on the stomach contents of a *Tylosaurus proriger* (Squamata; Mosasauridae) from the Niobrara Formation of western Kansas. Mosasaur 7:41–46.

Everhart, M. J. 2005a. Earliest record of the genus *Tylosaurus* (Squamata; Mosasauridae) from the Fort Hays Limestone (Lower Coniacian) of western Kansas. Transactions of the Kansas Academy of Science 108(3–4):149–155.

Everhart, M. J. 2005b. *Tylosaurus kansasensis*, a new species of tylosaurine (Squamata: Mosasauridae) from the Niobrara Chalk of western Kansas, U.S.A. Netherlands Journal of Geosciences / Geologie en Mijnbouw 84(3):231–240.

Everhart, M. J. 2006. The occurrence of elasmosaurids (Reptilia: Plesiosauria) in the Niobrara Chalk of western Kansas. Paludicola 5(4):170–183.

Everhart, M. J. 2007a. Remains of a pycnodont fish (Actinopterygii: Pycnodontiformes) in a coprolite: an upper record of *Micropycnodon kansasensis* in the Smoky Hill Chalk, western Kansas. Transactions of the Kansas Academy of Science 110(1–2):35–43.

Everhart, M. J. 2007b. New stratigraphic records (Albian–Campanian) of the guitarfish, *Rhinobatos* sp. (Chondrichthyes; Rajiformes), from the Cretaceous of Kansas. Transactions of the Kansas Academy of Science 110(3–4):225–235.

Everhart, M. J. 2008. Rare occurrence of a *Globidens* sp. (Reptilia; Mosasauridae) dentary in the Sharon Springs Member of the Pierre Shale (Middle Campanian) of western Kansas; pp. 23–29 in G. H. Farley and J. R. Choate (eds.), Unlocking the Unknown; Papers Honoring Dr. Richard Zakrzewski. Fort Hays Studies, Special Issue No. 2, Fort Hays State University, Hays, Kansas, 153 pp.

Everhart, M. J. 2013. A new specimen of the marine turtle, *Protostega gigas* Cope (Cryptodira; Protostegidae), from the Late

Cretaceous Smoky Hill Chalk of western Kansas. Transactions of the Kansas Academy of Science 116(1–2):73.

Everhart, M. J. 2016. William E. Webb— Civil War correspondent, railroad land baron, town founder, Kansas legislator, adventurer, fossil collector, author. Transactions of the Kansas Academy of Science 119(2):179–192.

Everhart, M. J., and M. M. Everhart. 2014. A new specimen of *Saurodon leanus* and a brief historical review of the discovery of saurocephalid fishes. Transactions of the Kansas Academy of Science 117(1–2):116.

Everhart, M. J., and Pearson, G. 2014. An isolated squamate dorsal vertebra from the Late Cretaceous Greenhorn Formation of Mitchell County, Kansas. Transactions of the Kansas Academy of Science 117(3–4):261–269.

Everhart, M. J., S. A. Hageman, and B. L. Hoffman. 2010. Another Sternberg 'fish-within-a-fish' discovery: first report of *Ichthyodectes ctenodon* (Teleostei; Ichthyodectiformes) with stomach contents. Transactions of the Kansas Academy of Science 113(3–4): 197–205.

Everhart, M. J., M. Triebold, A. Maltese, J. Jett, and B. Small. 2015. First occurrence of *Mosasaurus* (Squamata; Mosasauridae) from the Late Cretaceous (upper Campanian) of western Kansas. Transactions of the Kansas Academy of Science 118(1–2):139.

Hattin, D. E. 1982. Stratigraphy and depositional environment of the Smoky Hill Chalk Member, Niobrara Chalk (Upper Cretaceous) of the type area, western Kansas. Kansas Geological Survey Bulletin No. 225. University of Kansas, Lawrence, Kansas, 108 pp.

Hoganson, J. W. 2009. First report and biogeographic significance of an extremely large gladius of *Tusoteuthis longa* Logan (Coleoidea, Teuthida) from the Pembina Member of the Pierre Formation (Campanian) in North Dakota. Abstracts with programs, Geological Society of America, p. 106

Konishi, T., and M. W. Caldwell. 2007. New specimens of *Platecarpus planifrons* (Cope, 1874) (Squamata: Mosasauridae) and a revised taxonomy of the genus. Journal of Vertebrate Paleontology 27(1):59–72.

Konishi, T., M. W. Caldwell, and G. L. Bell Jr. 2010. Redescription of the holotype of *Platecarpus tympaniticus* Cope 1869 (Mosasauridae: Plioplatecarpinae), and its implications for the alpha taxonomy of the genus. Journal of Vertebrate Paleontology 30(5):1410–1421.

Liggett, G. A., S. C. Bennett, K. Shimada, and J. Huenergarde. 1997. A Late Cretaceous (Cenomanian) fauna in Russell County, Kansas. Transactions of the Kansas Academy of Science 16:26.

Logan, W. N. 1897. The Upper Cretaceous of Kansas: With an introduction by Erasmus Haworth. University Geological Survey of Kansas 2:194–234.

Martin, L. D. 1984. A new hesperornithid and the relationships of the Mesozoic birds. Transactions of the Kansas Academy of Science 87:141–150.

Martin, L. D., and J. D. Stewart. 1977. The oldest (Turonian) mosasaurs from Kansas. Journal of Paleontology 51(5):973–975.

Moodie, R. L. 1912. The stomach stones of reptiles. Science, n.s., 35(897):377–378.

O'Keefe, F. R., and L. M. Chiappe. 2011. Viviparity and K-selected life history in a Mesozoic marine plesiosaur (Reptilia, Sauropterygia). Science 333(6044):870–873.

Polcyn, M. J., and M. J. Everhart. 2008. Description and phylogenetic analysis of a new species of *Selmasaurus* (Mosasauridae: Plioplatecarpinae) from the Niobrara Chalk of western Kansas. Proceedings of the Second Mosasaur Meeting, Fort Hays Studies Special Issue 3, Fort Hays State University, Hays, Kansas, 13–28.

Polcyn, M. J., G. L. Bell Jr., K. Shimada, and M. J. Everhart. 2008. The oldest North American mosasaurs (Squamata: Mosasauridae) from the Turonian (Upper Cretaceous) of Kansas and Texas with comments on the radiation of major mosasaur clades; pp. 137–155 in Proceedings of the Second Mosasaur Meeting, Fort Hays Studies Special Issue 3, Fort Hays State University, Hays, Kansas.

Russell, D. A. 1988. A Check List of North American Marine Cretaceous Vertebrates Including Fresh Water Fishes. Occasional Paper of the Tyrrell Museum of Palaeontology, no. 4, Drumheller, Alberta, Canada, 57 pp.

Schumacher, B. A. 2011. A '*woollgari*-zone mosasaur' (Squamata; Mosasauridae) from the Carlile Shale (Lower Middle Turonian) of central Kansas and the stratigraphic overlap of early mosasaurs and pliosaurid plesiosaurs. Transactions of the Kansas Academy of Science 114(1–2):1–14.

Shimada, K. 1996. Selachians from the Fort Hays Limestone Member of the Niobrara Chalk (Upper Cretaceous), Ellis County, Kansas. Transactions of the Kansas Academy of Science 99(1–2):1–15.

Shimada, K. 1997. Paleoecological relationships of the Late Cretaceous lamniform shark *Cretoxyrhina mantelli* (Agassiz). Journal of Paleontology 71(5):926–933.

Shimada, K. 2007. Skeletal and dental anatomy of lamniform shark, *Cretalamna appendiculata*, from upper Cretaceous Niobrara Chalk of Kansas. Journal of Vertebrate Paleontology 27(3): 584–602.

Shimada, K., and M. J. Everhart. 2003. *Ptychodus mammillaris* (Elasmobranchii) and *Enchodus* cf. *E. schumardi* (Teleostei) from the Fort Hays Limestone Member of the Niobrara Chalk (Upper Cretaceous) in Ellis County, Kansas. Transactions of the Kansas Academy of Science 106(3–4):171–176.

Shimada, K., and M. J. Everhart. 2009. First record of *Anomoeodus* (Osteichthyes: Pycnodontiformes) from the Upper Cretaceous Niobrara Chalk of western Kansas. Transactions of the Kansas Academy of Science 112(1–2):98–102.

Shimada, K., and C. Fielitz. 2006. Annotated checklist of fossil fishes from the Smoky Hill Chalk of the Niobrara Chalk (Upper Cretaceous) in Kansas; pp. 193–213 in S. G. Lucas and R. M. Sullivan (eds.), Late Cretaceous Vertebrates from the Western Interior. New Mexico Museum of Natural History and Science Bulletin 35, Albuquerque, New Mexico.

Shimada, K., M. J. Everhart, R. Decker, and P. D. Decker. 2009. A new skeletal remain of the durophagous shark, *Ptychodus mortoni*, from the Upper Cretaceous of North America: an indication of gigantic body size. Cretaceous Research 31(2):249–254.

Siverson, M., J. Lindgren, M. G. Newbrey, P. Cederstrom, and T. D. Cook. 2015. Cenomanian-Campanian (Late Cretaceous) mid-plaeolatitude sharks of *Cretalamna appendiculata* type. Acta Palaeontologica Polonica 60(2):339–384.

Stahl, B. J. 1999. Chondrichthyes III: Holocephali; in H.-P. Schultze (ed.), Handbook of Paleoichthyology, Volume 4. Verlag Dr. Friedrich Pfeil, München, Germany, 164 pp.

Sternberg, C. H. 1917. Hunting Dinosaurs in the Badlands of the Red Deer River, Alberta, Canada. World Company Press, Lawrence, Kansas, 261 pp.

Stewart, J. D. 1990a. Niobrara Formation vertebrate stratigraphy; pp. 19–30 in S. C. Bennett (ed.), Niobrara Chalk Excursion Guidebook. University of Kansas Museum of Natural History and Kansas Geological Survey, Lawrence, Kansas.

Stewart, J. D. 1990b. Niobrara Formation symbiotic fish in inoceramid bivalves; pp. 31–41 in S. C. Bennett (ed.), Niobrara Chalk Excursion Guidebook. University of Kansas Museum of Natural History and Kansas Geological Survey, Lawrence, Kansas.

Stewart, J. D. 1990c. Preliminary account of holecostome-inoceramid commensalism in the Upper Cretaceous of Kansas; pp. 51–58 in A. J. Boucot, Evolutionary Paleobiology of Behavior and Coevolution. Elsevier, Amsterdam, the Netherlands.

Stewart, J. D. 1999. A new genus of Saurodontidae (Teleostei: Ichthyodectiformes) from Upper Cretaceous rocks of the western interior of North America; pp. 335–360, in G. Arrantia and H-P Schultze (eds.), Mesozoic Fishes 2—Systematics and Fossil Record. Verlage Dr. Friedrich Pfeil, München, Germany.

Webb, W. E. 1872. Buffalo Land: An Authentic Account of the Discoveries, Adventures, and Mishaps of a Scientific and Sporting Party in the Wild West. Hubbard Brothers, Philadelphia, Pennsylvania, 503 pp.

Williston, S. W. 1890a. Structure of the plesiosaurian skull. Science 16(405):262.

Williston, S. W. 1890b. A new plesiosaur from the Niobrara Cretaceous of Kansas. Transactions of the Kansas Academy of Science 12:174–178. 2 figs.

Williston, S. W. 1897. The Kansas Niobrara Cretaceous. University Geological Survey of Kansas 2:235–246.

Williston, S. W. 1898a. Birds. University Geological Survey of Kansas 4:43–49.

Williston, S. W. 1898b. Mosasaurs. University Geological Survey of Kansas 4(5):81–221.

Index